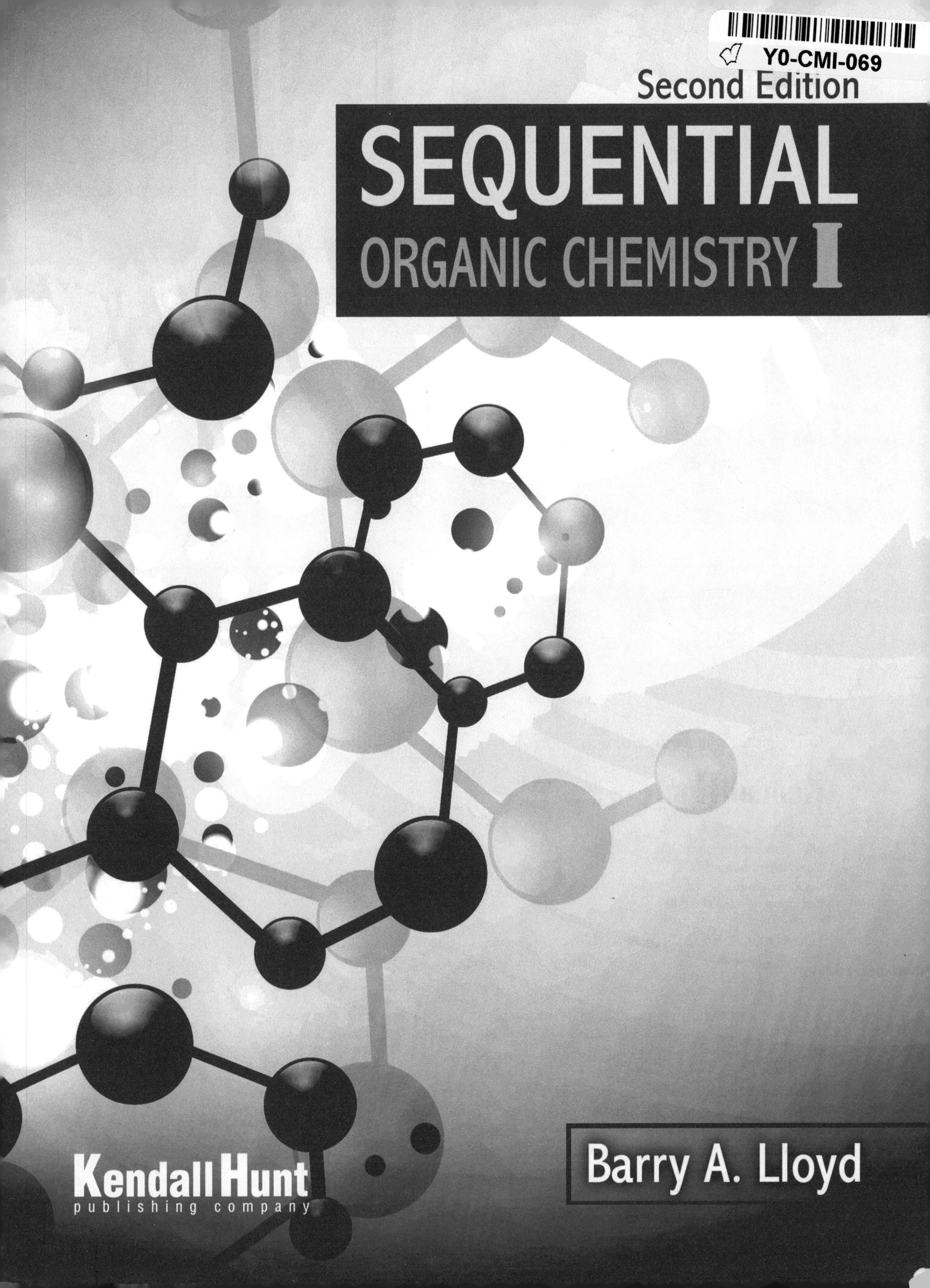

SEQUENTIAL
ORGANIC CHEMISTRY I

Second Edition

Barry A. Lloyd

Kendall Hunt
publishing company

Cover image © Shutterstock.com

Kendall Hunt
publishing company

www.kendallhunt.com
Send all inquiries to:
4050 Westmark Drive
Dubuque, IA 52004-1840

Copyright © 2016, 2020 by Barry Lloyd

ISBN: 978-1-9724-0044-5

Kendall Hunt Publishing Company has the exclusive rights to reproduce this work,
to prepare derivative works from this work, to publicly distribute this work,
to publicly perform this work and to publicly display this work.

All rights reserved. No part of this publication may be reproduced,
stored in a retrieval system, or transmitted, in any form or by any
means, electronic, mechanical, photocopying, recording, or otherwise,
without the prior written permission of the copyright owner.

Published in the United States of America

CONTENTS

Chapter 1 1

Chapter 2 27

Chapter 3 65

Chapter 4 107

Chapter 5 141

Chapter 6 173

Chapter 7 225

Chapter 8 249

Chapter 9 291

Chapter 10 311

Chapter 11 335

Chapter 12 359

Chapter 13 399

CHAPTER 1

1.1.1. Organic chemistry definition – study of covalent carbon compounds.

1.1.2. J. J. Berzelius (1807) – Vitalism: must organic chemicals come from living organisms?

1.1.3. F. Wöhler (1828) –

$$NH_4OCN \xrightarrow[\text{exclude } O_2]{\text{heat}} H_2N-\underset{\underset{O}{\|}}{C}-NH_2$$

ammonium cyanate (inorganic) → urea (organic) previously only from urine

1.1.4. Sources of organic compounds: biomass, petroleum, coal, and from inorganics (NH_4OCN, CO_2, Na_2CO_3, KCN).

1.1.5. C – 13th most abundant element in Earth's crust (0.09 mass %). Si ranks second (25.8 mass %).

1.1.6. C allotropes
  a. soot (amorphous)
  b. diamond (tetrahedral)
  c. graphite (planar)
  d. fullerenes (such as C_{60}) soccer ball structure

1.1.7. Strong C-C bonds ~371 kJ/mol (~88.6 kcal/mol) vs. Si-Si ~226 kJ/mol (~54.0 kcal/mol). Catenation – chain formation.

a branched chain   a ring

1.1.8. C most commonly bonds to: C, H, F, Cl, Br, I, O, N, S, and P.

Section numbering is not in numerical order because it is correlated with, sections in Wade, L. G. Jr., Simek, J. W. *"Organic Chemistry,"* 9th edition.

1.1.9. A. Kekulé's London "bus" ride dream (1854):

"One fine summer evening, I was returning by the last omnibus, 'outside' as usual, through the deserted streets of the metropolis, which are at other times so full of life. I fell into a reverie and lo! the atoms were gambolling before my eyes....I saw how, frequently, two smaller atoms united to form a pair, how a larger one embraced two smaller ones; how still larger ones kept hold of three or even four of the smaller; whilst the whole kept whirling in a giddy dance. I saw how the larger ones formed a chain....I spent part of the night putting on paper at least sketches of these dream forms."

– August Kekulé.*

1.1.10. Age of carbon – >20 million organic compounds known. New one catalogued by the Chemical Abstracts Service ~ every nine sec.

1.1.11. Wöhler's unexplored jungle (1835).

"Organic chemistry nowadays almost drives me mad. To me it appears like a primeval tropical forest full of the most remarkable things, a dreadful endless jungle into which one does not dare enter for there seems to be no way out."

– Friedrich Wöhler (1835).*

*Cited in "Organic Chemistry," Morrison, R. T. and Boyd, R. N., 6th ed., p 3, Prentice Hall, N.J., (1992).

1.1.12. How to organize it?
- **a.** Freshman chemistry: formulas (H_2O, CH_4, C_2H_6O).
- **b.** Organic chemistry: formulas and structures.
- **c.** Structural theory – assumes physical and chemical properties are related to molecular structure.

1.2.1. Review: quantum mechanics results (1920's).

1.2.2. Atomic electron orbitals fill in order of increasing potential energy (*aufbau* diagram).

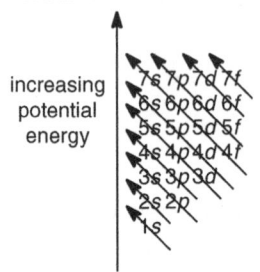

1.2.3. Hund's rule – electrons will not pair until all orbitals in a given subshell are half-full (to minimize electron repulsions).
- **a.** Problem 1. Write electron configurations for Na, Al, and S.
  1. How are electron configurations similar for two atoms in the same periodic group (column)?

  2. Why do an element's physical and chemical properties depend on its electron configuration?

1.2.4. Octet rule – in chemical bond formation, atoms lose, gain or share electrons until each valence (outermost) shell has eight electrons.

1.2.5. Completely full shells (and subshells) are more stable than if they are only partly full.

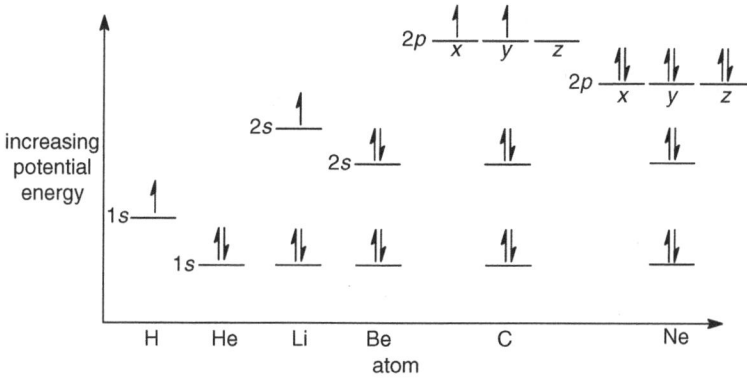

1.3.1. Two broad bonding and compound types – ionic and covalent (1915).
1.3.2. Lewis dot structures – valence electrons are shown as dots.
1.4.1. Electron pushing arrows – show direction electrons move as bonds form.

    Li· + ·F: ⟶ Li⊕ + :F:⊖ ionic bond

    H· + ·H ⟶ H:H covalent bond (similar to He)

    :F· + ·F: ⟶ :F:F:

    H· + ·O· + ·H ⟶ H:O:H

    H· + ·C· + ·H ⟶ H:C:H (with H above and below)

    a. Half arrowheads ⇒ one electron moves.

1.4.2. Covalent Lewis dot structure drawing rules:
    a. Correct bonding geometry must be assumed from experience or found experimentally. Examples:
        1. The lower numbered atom will be central in a formula that has only two elements. Examples: SO_3, H_2O, CH_4, PF_3. Central atom: S, O, C, and P, respectively.
        2. N, S, and P, respectively, are central in HNO_3, H_2SO_4, H_3PO_4 (ternary compounds). O is bonded to the central atom, and H is bonded to O.
        3. H and F are never central atoms.
    b. Draw valence electron dots on each atom.
        1. Include electrons lost or gained in polyatomic ions as valence electrons.
    c. Move two dots (one from each atom) between each bonded atom pair.
    d. Move lone electron pairs to noncentral elements (except H) until they have eight electrons.

octet rule violation

e. Move any remaining valence electrons to central atom.
f. *Verify that structure has same number of dots as valence electrons on atoms at beginning.*
g. Problem 2. Which of the following would you expect to be ionic, and which nonionic? Draw a Lewis dot structure showing only valence electrons.
   a. KBr

   b. NF_3

   c. H_2S

   d. $CHCl_3$

   e. $CaSO_4$

   f. PH_3

   g. NH_4Cl

   h. CH_3OH

h. Problem 3. Draw a Lewis dot structure for each of the following. Assume that every atom has a complete octet (except H), and that some atoms may share more than one electron pair.
   a. N_2H_4

   b. $COCl_2$

   c. HONO

   d. $NaNO_2$

   e. K_2CO_3

   f. C_2H_4

g. C_2H_2

h. CH_2O

i. CH_2O_2

j. C_3H_8

1.5.1. If a central atom doesn't have eight electrons, shift lone pair electrons and make double bonds. Examples: C_2H_4, CH_2O, CH_3N and CO_2.

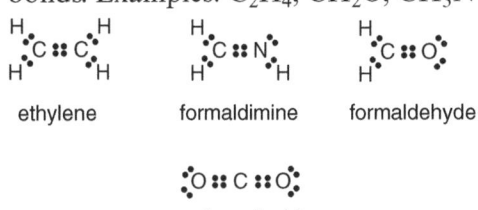

```
    H      H        H              H
     ••C::C••        ••C::N••       ••C::O••
    H      H        H              H
   ethylene      formaldimine   formaldehyde

        ••O::C::O••
        carbon dioxide
```

a. Sometimes three electron pairs must be shared in triple bonds to satisfy octet rule. Examples: C_2H_2, HCN and N_2.

```
   H:C:::C:H      H:C:::N:      :N:::N:
   acetylene    hydrocyanic acid   nitrogen
```

1.5.2. Coordinate covalent bond – both bonding electrons come from one atom.

a. Ammonia N-H bonds – one electron is from N (·), and one from H (x). Fourth N-H bond – both electrons originally from N.

```
                                              coordinate covalent bond
                                           H
   H:N:H   +   H—Cl:    ⇌    H:N:H  +  :Cl:⁻
     x                       ⊕ x
     H                         H
```

b. Once formed, all four ammonium ion N-H bonds are indistinguishable.

c. Examples: chlorous acid, nitrous oxide, and carbon monoxide.

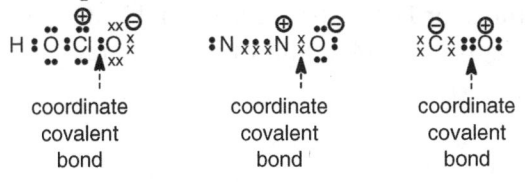

```
   H:O:Cl:O:⁻     :N:::N:O:      :C:::O:
         ⊕           ⊕ ⁻         ⁻ ⊕
   coordinate    coordinate    coordinate
   covalent      covalent      covalent
   bond          bond          bond
```

1.7.1. Formal charge – theoretical electrical charge on an atom that is bonded within a compound or polyatomic ion (see a. and c. above).

a. Covalent compound – electrical charge an atom *would have* if its bonding electrons were all *equally* shared with other bonded atoms. (Equal electron sharing actually only occurs between identical atoms.)

b. Ionic compound – electrical charge an atom *would have* if all its bonds to other atoms were 100% ionic (with no covalent character).

c. All real bonds actually have some covalent and some ionic character. Although formal charges often deviate from reality, they are useful.

d. Note: organic chemists often do not show all lone pair electrons. *You should be able to figure out how many electrons are present even if they are not shown.*

1.7.2. Formal charge determination:
  a. Remember how many valence electrons each atom has (group number).
  b. Assume that an atom "owns" half of all electrons in covalent bonds to it.
  c. Assume that an atom "owns" all lone pair (nonbonding) electrons.
  d. Subtract the electrons in steps b and c from the number in step a.

  $$\text{formal charge} = \text{group number} - \text{nonbonding electrons} - \frac{1}{2}\text{shared electrons}$$

  e. Atomic formal charges must all add to zero for a neutral uncharged molecule, or to the ionic charge for a polyatomic ion.

1.7.3. Verify formal charges on example species below:

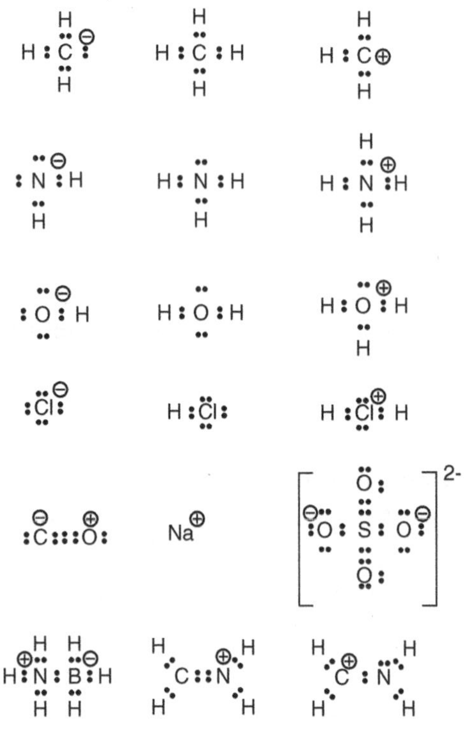

[Lewis dot structures shown for: CH₃⁻, CH₄, CH₃⁺, CH₃• ; NH₂⁻, NH₃, NH₄⁺ ; OH⁻, H₂O, H₃O⁺ ; Cl⁻, HCl, H₂Cl⁺ ; :C⁻≡O⁺:, Na⁺, SO₄²⁻ ; NH₄BH₄ (H₃N⁺–B⁻H₃), H₂C=N⁺H₂ (iminium), H₂C–NH₂ (aminomethyl)]

1.7.4. More examples: draw Lewis dot structures for H₂SO₄ (sulfuric acid), HClO₃ (chloric acid), SO₃²⁻ (sulfite anion), O₃ (ozone), SO₃ (sulfur trioxide). Show formal charges on atoms where needed. Answers:

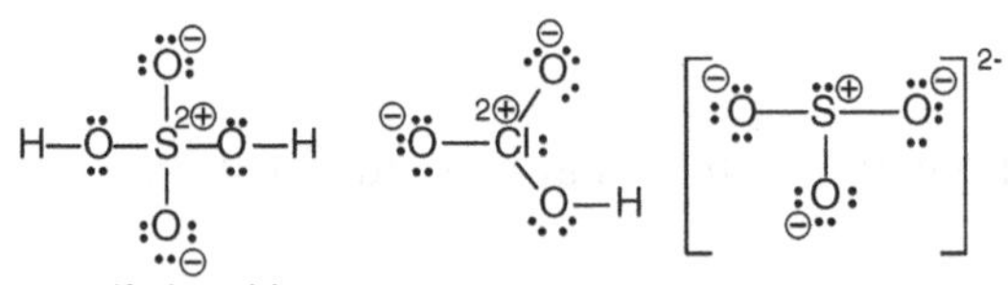

[Structures shown:]

sulfuric acid — H–O–S(2+)(=O⁻)(=O⁻)–O–H with central S bearing 2+ formal charge and two terminal O⁻

chloric acid — Cl(2+) with two O⁻ and one O–H

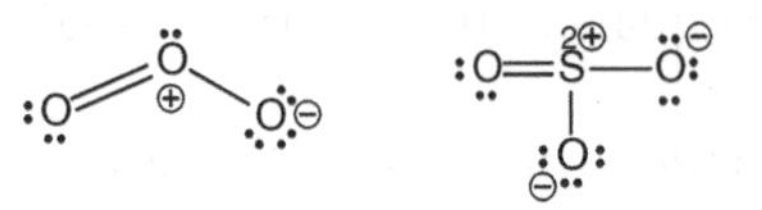

sulfite anion — [O⁻–S⁺–O⁻ with O⁻ below]²⁻

ozone — :O=O⁺–O⁻:

sulfur trioxide — O=S(2+)(–O⁻)(–O⁻)
(one possible structure)

1.7.5. Problem 4. Draw likely electron dot structures for the following formulas. Assume that every atom (except hydrogen) has a complete octet, and that two atoms may share more than one electron pair.

    a. H_2O_2

    b. N_2

    c. $HONO_2$

    d. NO_3^-

    e. HCN

    f. CO_2

    g. H_2CO_3

    h. C_2H_6

4.6.1. Bond dissociation enthalpy (energy) (*BDE* or *D*) – the energy (kJ or kcal) needed to break one mol of bonds.

    a. Bond *homolysis* – one electron goes with each group:

$$A:B \longrightarrow A\cdot + \cdot B \qquad \Delta H° = \text{bond dissociation energy}$$

    b. Bond *heterolysis* (ionic cleavage) – both electrons go with one group.

$$A:B \longrightarrow A^{\oplus} + :B^{\ominus} \qquad \Delta H° = \text{bond dissociation energy}$$

        1. Bond heterolysis (gas phase) usually has larger bond dissociation energy than homolysis. Bond heterolysis energy depends on solvent.

        2. Separating a + charged ion from a – charged ion is endothermic.

3. – charge normally goes with more electronegative atom or group. See electronegativity section below.

c. Examples:

homolysis		BDE (kJ/mol or kcal/mol)
H—H →energy→ H• + H•		+436 +104
H—Cl: →energy→ H• + :Cl:•		+432 +103
H₃C—H →energy→ H₃C• + H•		+439 +105
:Cl—Cl: →energy→ 2 :Cl:•		+240 +57

heterolysis

| | |
---|---|---
H—H →energy→ H⁺ + H:⁻ | | +1680 +401
H—Cl: →energy→ H⁺ + :Cl:⁻ | | +1400 +334
H—Cl: →energy→ H:⁻ + :Cl:⁺ | | higher higher

1. *BDE*'s above are from "Organic Chemistry," Wade, L. G. Jr., 9th ed., p 167, Pearson, (2017), and "Organic Chemistry," Morrison, R. T. and Boyd, R. N., 6th ed., inside front cover, Prentice Hall, N.J., (1992).

d. Half arrowhead ⇒ one electron moves. Full arrowheads ⇒ two electrons move.

1.6.1. Electronegativity – measure of an atom's ability to attract bonding electrons (in a chemical bond to another atom) to itself.

a. Calculated from *D*'s.

1.6.2. Table of Pauling electronegativities:

H 2.2						
Li 1.0	Be 1.6	B 2.0	C 2.5	N 3.0	O 3.4	F 4.0
Na 0.9	Mg 1.3	Al 1.6	Si 1.9	P 2.2	S 2.6	Cl 3.2
K 0.8						Br 3.0
						I 2.7

1.6.3. Electronegativity causes:
   a. More nearly full outermost electron shell ⇒ more electronegative.
   b. More inner shell electrons "screen" outer shell electrons from being attracted to nucleus ⇒ less electronegative.
   c. Result – top right corner elements (except H and noble gases) are most electronegative.

1.6.4. Unequal electron pair sharing occurs unless two chemically bonded atoms are identical. δ is a variable fraction between 0 and +1 or –1. Electrostatic potential maps show δ+ and δ– regions as colors (blue positive, red negative).

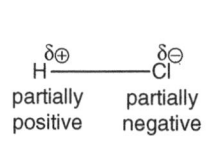

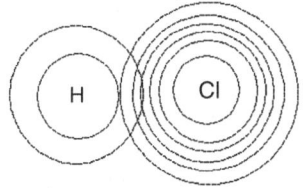

$$\underset{\substack{\text{partially}\\\text{positive}}}{\overset{\delta\oplus}{H}}\text{———}\underset{\substack{\text{partially}\\\text{negative}}}{\overset{\delta\ominus}{Cl}}$$

electron cloud is more dense on chlorine

1.6.5. Bond polarity continuum:

$$H-H \;<\; \overset{\delta\ominus}{H_3C}-\overset{\delta\oplus}{H} \;<\; \overset{\delta\oplus}{H}-\overset{\delta\ominus}{Cl}: \;<\; \overset{\delta\ominus}{HO}-\overset{\delta\oplus}{H} \;<\; \overset{\oplus}{Na}\;:\overset{\cdot\cdot\ominus}{Cl}:$$

nonpolar   polar (ionic character)

   a. Problem 5. Use δ+ and δ– symbols to show the following bond polarizations. *Hint*: consider formal charges in CO.

   :Ï—Br: H—Br: Li—Cl: :C≡O:

1.9.1. Resonance – a nonmathematical bonding theory describing molecules that have more than one reasonable Lewis dot structure.
   a. The true structure is pictured as a composite (average, or hybrid) of two (or more) Lewis dot structures (*resonance contributors*) in such cases.

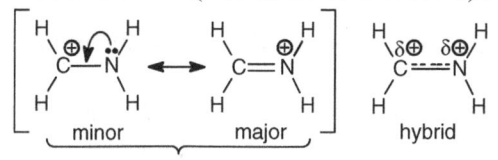

   [minor ↔ major]   hybrid
   contributing structures

   1. "Iminium" cation is an unstable intermediate product during imine formation (Chapter 18).
   2. + charge is *delocalized* (shared between the C and N atoms).

1.9.2. Resonance contributing structures – two (or more) Lewis dot structures that differ only in the electron dot positions. Each structure must have the same number of odd electrons (if any).

1.9.3. A Summary of Rules for Writing Resonance Contributing Structures
   1. All atomic (nuclear) positions are assumed to be fixed (immovable) in space. Nuclei can then be distinguished from each other (and numbered). *This assumption is not true for real molecules. Atoms of the same element are actually indistinguishable from each other. Atom numbering is only a bookkeeping tool useful for drawing theoretical resonance contributors.*
      a. Example: allyl free radical (below) is formed as a reactive intermediate during allyl halogenation (section 6.6.1). A free radical is an atom or molecule that has an unpaired electron. Free radicals are usually highly reactive because the atom with the unpaired electron violates the octet rule.

   $$\left[\overset{1}{H_2\dot{C}}-\overset{2}{CH}=\overset{3}{CH_2} \;\longleftrightarrow\; \overset{1}{H_2C}=\overset{2}{CH}-\overset{3}{\dot{C}H_2} \right] \quad \overset{1/2\;1}{H_2C}\text{====}\overset{2}{CH}\text{====}\overset{1/2\;3}{\dot{C}H_2}$$

   hybrid (real)

b. Doubly bonded or lone pair electrons are moved (shifted) to generate new structures. Some common electron shifts are shown below (X represents any element). Shifts that yield equal potential energy structures are more important. Electrons can be shifted only to an adjacent atom or bond position, and no further.

   c. Using electron pushing arrows is strictly inconsistent with the idea of resonance since it suggests equilibria. It is ok to use the arrows to generate imaginary resonance contributors as long as we realize that the electrons do not actually move – the true picture of the molecule is the average of the most stable contributors (the hybrid).
   d. Only lone pair, double and triple bonding electrons are shifted when writing conventional resonance contributing structures as in a–c.
2. Numerous resonance contributing structures can be written for almost any molecule. Which ones make significant contributions to the hybrid structure? Criteria for judging importance (stability):
   a. Forms with fewer formal charge separations are generally more important.
   b. Forms with a + formal charge on an electropositive atom, and a − formal charge on an electronegative atom are more important that the reverse.
   c. Forms with fewer octet rule violations are more important than those with more violations.
   d. Forms with the maximum possible number of bonds are more important than those with fewer bonds.
      1. Forms with stronger bonds are more important than those having the same number of weaker bonds.
   e. Structures are more stable when intermediate electronegativity (C and H) atoms have no formal electrical charges.
   f. Examples:

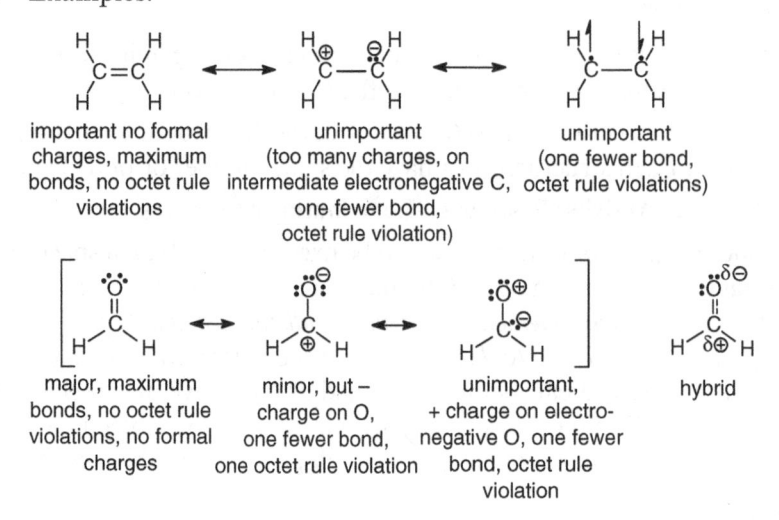

3. If two or more important (low potential energy) resonance contributors exist, the hybrid (real molecule) potential energy is lower than any one of the (theoretical) resonance forms.
   a. Only the most stable Lewis dot structures make significant contributions to the hybrid structure.
   b. Electrical charge delocalization results in more stabilization than electron delocalization.
4. Imaginary contributing structures (inside square brackets) have full bonds, full electrons and full (integer) formal charges (for example, ·, +, –, or 2+). Hybrid structures usually contain partial (fractional) bonds, fractional electrons, and fractional electrical charges (for example, δ·, or δ+).

1.9.4. Example: ethanoate (acetate) anion. Equivalent (or equal energy) structures. Large stabilization.

a. Example: nitromethane. Equivalent structures. Large stabilization.

1.9.5. Equivalent contributing structures of ozone, sulfur dioxide, and sulfur trioxide. Large stabilization.

a. Each hybrid bond is identical – consisting of a single bond and a fractional double bond.
b. Some electrical charges are *delocalized* (over more than one atom). π-Electrons are also delocalized (over more than two atoms).

1.9.6. Contributing forms are <u>not</u> in equilibrium with each other (changing back and forth into each other).
  a. No one Lewis dot contributing structure represents reality. They are imaginary and do not really exist. The best real structure picture is the hybrid.

12 SEQUENTIAL ORGANIC CHEMISTRY I

1.9.7. Analogy (J. D. Roberts) – a rhinoceros (real animal) could be described as a cross between a unicorn and a dragon (imaginary).

    a. Problem 6. Nitromethane is used as a fuel in high performance racing cars. A Lewis dot structure is shown.

        1. Fill in lone pair electrons and formal electrical charges. Draw another resonance contributing structure.

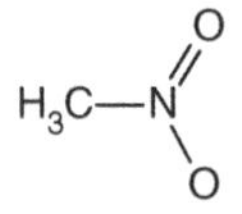

        2. Actual measurement shows that the two nitrogen-oxygen bonds are exactly the same length (1.21 Å), compared to 1.36 Å for an N–O single bond and 1.18 Å for an N=O double bond. Draw a more accurate nitromethane (resonance hybrid) structure.

    b. Problem 7. A carbonate anion Lewis dot structure is drawn below.

        1. Fill in lone pair electrons, formal electrical charges, and draw the other important resonance contributing structures.

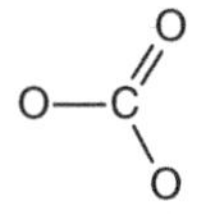

        2. Actual measurement shows that all the carbon-oxygen bonds in $CaCO_3$ are the same length, 1.31 Å, as compared with 1.36 Å for a C–O single bond and 1.23 Å for a C=O double bond. Draw a more accurate (resonance hybrid) picture for carbonate anion.

    c. Problem 8. Methanoic acid, HCOOH, contains one carbon-oxygen bond (1.36 Å) and one carbon-oxygen double bond (1.23 Å). Sodium methanoate (HCOONa) contains two equal length carbon-oxygen bonds (1.27 Å). Draw resonance contributing structures that account for these facts.

1.8.1. Ionic bonds are not shown. If they were they would sometimes violate the octet rule.

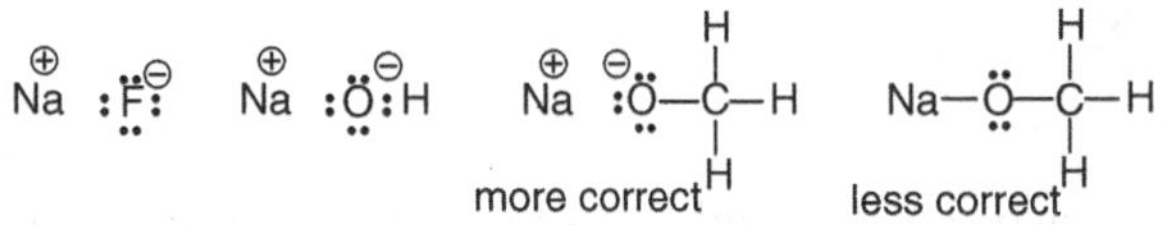

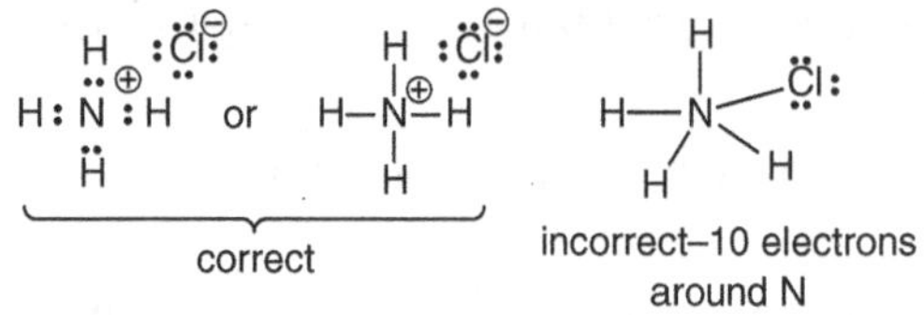

2.4.1. S. Arrhenius (Nobel Prize, 1903) – ions in aqueous solution are completely dissociated and are solvated by solvent molecules.

    a. Arrhenius acid – Latin *acidus* (sour) – any substance that yields H_3O^+ (hydronium ions) when dissolved in water.

- b. Brønsted/Lowry acid (1923) – a proton donor.
- c. Lewis acid – an electron pair acceptor.
- d. Arrhenius base – neutralizes acids, Arabic *al kalai* (plant ashes, bitter) – any substance that yields ⁻OH (hydroxide ions) when dissolved in water.
- e. Brønsted/Lowry base – a proton acceptor.
- f. Lewis base – an electron pair donor.

2.5.1. Brønsted-Lowry definitions are broader than Arrhenius definitions, allowing substances other than H_3O^+ to be classified as acids, and substances other than ⁻OH to be classified as bases.

  a. Lewis acid/base definitions are emphasized with electron pushing arrows – arrows point direction that electrons move.

2.13.1. When an acid/base reaction involves bond formation to an element other than hydrogen, use Lewis definitions. An electron donor is called a *nucleophile* (nucleus lover), and an electron pair acceptor an *electrophile* (electron lover).

2.13.2. All Arrhenius and Brønsted-Lowry acid/base reactions are also Lewis acid/base reactions.

2.13.3. Many organic reactions are Lewis acid/base reactions. Example:

a. A Lewis acid/base reaction, (where Brønsted/Lowry definitions do not apply).

Lewis base   Lewis acid
(stronger base) (stronger acid)

weaker acid
internal salt

weaker base

b. B in BF_3 violates the octet rule (which is why it easily accepts an electron pair). Other Lewis acids: $ZnCl_2$, $FeCl_3$, $AlCl_3$, and $SnCl_4$.

c. An atom must have a lone electron pair to be a Lewis base.

2.6.1. Example – water self-ionization at 25°C.

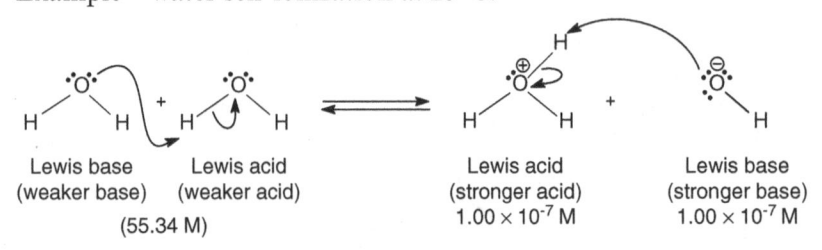

Lewis base   Lewis acid   Lewis acid   Lewis base
(weaker base) (weaker acid)   (stronger acid) (stronger base)
(55.34 M) 1.00×10^{-7} M 1.00×10^{-7} M

a. Formal charges – trends for oxygen.

Formal charge	Number of bonds to oxygen	Number of lone electron pairs
1+	3	1
0	2	2
1–	1	3

b. Review example: calculate concentration (molarity) of H_2O in 1.00 L of pure H_2O. Water density is 0.9970 kg/L at 25°C.

$$\frac{0.9970 \text{ kg } H_2O}{1 \text{ L } H_2O} \times \frac{1000 \text{ g } H_2O}{1 \text{ kg } H_2O} \times \frac{1 \text{ mol } H_2O}{18.02 \text{ g } H_2O} = \frac{55.34 \text{ mols } H_2O}{1 \text{ L } H_2O}$$

c. Equilibrium constant for any reaction (section 4.4) is defined as:

$$aA + bB \rightleftharpoons cC + dD$$

$$K_{eq} = \frac{[C]^c[D]^d}{[A]^a[B]^b}$$

d. Molarity (M) – mols of solute per liter of solution. Square brackets indicate molar concentration.

e. Equilibrium constant for water self-ionization is defined as:

$$K_{eq} = \frac{[H_3O^+][^-OH]}{[H_2O]^2} = \frac{[1.0 \times 10^{-7}]}{[55.34]^2} = 3.3 \times 10^{-18}$$

f. $[H_2O]$ = 55.34 M is much larger than $[H_3O^+]$ and $[^-OH]$ = 1.0×10^{-7} M, respectively. For dilute solutions, $[H_2O]$ is nearly a constant (55.34 M), and chemists agreed long ago to omit $[H_2O]$ from equilibrium constant expression. K_w (water ion-product constant) is defined as below at 25°C.

$$K_w = [H_3O^+][^-OH] = 1.00 \times 10^{-14} \text{ M}^2$$

g. K_a (acidity constant) is a measure of acid strength for any acid (HA).

$$HA + H_2O \rightleftharpoons H_3O^+ + A:^-$$

$$K_a = \frac{[H_3O^+][A:^-]}{[HA]}$$

h. Since K_a's vary over an enormous range, a logarithmic scale is often used.

$$pK_a = -\log_{10} K_a$$

i. [H_2O] is left out of K_a expression by convention. K_a values are larger than true K_{eq} by a factor of 55.34, but chemists usually only need relative acidities.

2.6.2. K_b (basicity constant) is a measure of base strength.

$$K_b = \frac{[HA][^-OH]}{[A^-]}$$

a. [H_2O] is left out by convention.

2.6.3. K_b is related to K_a through the water ion-product expression in aqueous solution:

$$K_w = K_a \times K_b = \frac{[H_3O^+][A^-]}{[HA]} \times \frac{[HA][^-OH]}{[A^-]} = [H_3O^+][^-OH] = 1.0 \times 10^{-14} \text{ M}^2$$

or taking $-\log_{10}$ of both sides:

$$pK_a + pK_b = -\log_{10} 10^{-14} = 14$$

a. K_a expression for water self-ionization (section 1.12.6) is (for water acting as an acid):

$$K_a = \frac{[H_3O^+][^-OH]}{[H_2O]} = \frac{[1.0 \times 10^{-7}]^2}{[55.34]} = 1.8 \times 10^{-16} \text{ M}$$

b. The self-ionization reaction (section 1.12.6) can also be interpreted in terms of K_b (for water acting as a base).

$$K_b = \frac{[H_3O^+][^-OH]}{[H_2O]} = \frac{[1.0 \times 10^{-7}]^2}{[55.34]} = 1.8 \times 10^{-16} \text{ M}$$

c. Water can act as a weak acid in some reactions, and a weak base in other reactions.

2.6.4. For strong acids, $K_a > 10$. For moderate acids, $1 < K_a < 10$. For weak acids, $K_a < 1$. The pK_a of a strong acid is small (or even negative), while it is large for a weak acid.

2.7.1. Acidity is a chemical property of a particular acid. Strong acids ionize almost completely, whereas weak acids do not ionize very much when dissolved in water.

a. Example: HCl is a strong acid ($K_a = 1.6 \times 10^2$).

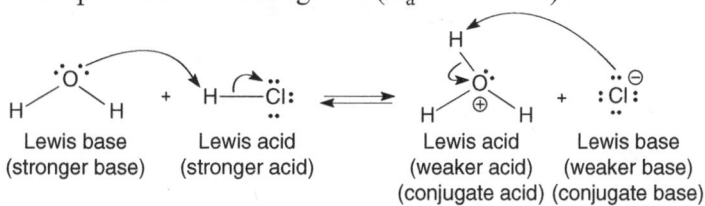

1. Equilibrium lies to right, i.e., [H₃O⁺] and [Cl⁻] are high, while [HCl] is low at equilibrium.
  b. Example: acetic acid is a weak acid ($K_a = 1.75 \times 10^{-5}$).

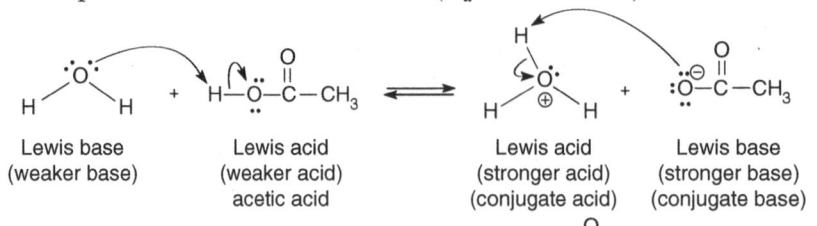

  [Reaction scheme: H₂O (Lewis base, weaker base) + acetic acid CH₃COOH (Lewis acid, weaker acid) ⇌ H₃O⁺ (Lewis acid, stronger acid, conjugate acid) + CH₃COO⁻ (Lewis base, stronger base, conjugate base)]

  1. Equilibrium lies to left, i.e., [H₃O⁺] and [:Ö–C(=O)–CH₃]⁻ are low and [H–Ö–C(=O)–CH₃] is high at equilibrium.
  2. Acetic acid is water soluble, but since it is a weak acid, it does not ionize very much.
  c. Example: H₃O⁺ is a strong acid ($K_a = 5.534 \times 10^{1}$).

  [Reaction scheme: H₃O⁺ (Lewis acid) + H₂O (Lewis base) + :Cl:⁻ (spectator) ⇌ H₂O (conjugate base) + H₃O⁺ (conjugate acid) + :Cl:⁻ (spectator)]

  $$K_a = \frac{[H_3O^+][H_2O]}{[H_3O^+]} = [H_2O] = 55.34 \text{ M}$$

2.8.1. Neutralization reactions – acids react with bases. Equilibrium lies far to right.
  a. *Equilibrium always lies towards the weaker acid/base pair.*
  b. Example:

  [Reaction scheme: NH₃ (Lewis base, stronger base) + HCl (Lewis acid, stronger acid) ⇌ NH₄⁺ (Lewis acid, weaker acid, conjugate acid) + :Cl:⁻ (Lewis base, weaker base, conjugate base)]

  c. Example:

  [Reaction scheme: Na⁺ (spectator ion) + H–Ö:⁻ (Lewis base, stronger base) + NH₄⁺ (Lewis acid, stronger acid) + :Cl:⁻ (spectator ion) ⇌ H₂O (Lewis acid, weaker acid, conjugate acid) + NH₃ (Lewis base, weaker base, conjugate base) + Na⁺ + :Cl:⁻ (spectator ions)]

2.9.1. Any acid HA will be stronger if its conjugate base A⁻ is more stable.

$$H\text{-}A + H_2O \rightleftharpoons H_3O^+ + A{:}^-$$

Structural factors that stabilize bases and affect acid strength:

a. *A electronegativity*:

 less electronegative   more electronegative
    C < N < O < F

 stronger base   weaker base
 (less stable)   (more stable)
   :CH$_3^-$ > :NH$_2^-$ > :ÖH$^-$ > :F̈:$^-$

 weaker acid   stronger acid
  H—CH$_3$ < H—NH$_2$ < H—ÖH < H—F̈:

    H—SH < H—Cl

More polar H-A bond ⇒ more acidic.

b. *A:$^-$ size* – larger A:$^-$ ⇒ the more – charge is spread out (delocalized) over larger surface area ⇒ the more stable A:$^-$. If A:$^-$ is more stable, its equilibrium concentration is higher, and H-A is a stronger acid.

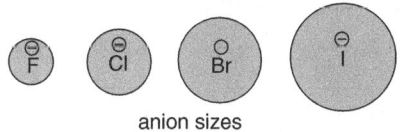

  F$^-$ Cl$^-$ Br$^-$ I$^-$
    anion sizes

weaker acid   stronger acid
H—F̈: < H—C̈l: < H—B̈r: < H—Ï:

  H—OH < H—SH < H—SeH

1. Size overwhelms electronegativity.

2.11.1. Hybridization effect (see section 9.6.2.).

2.12.1. Resonance is a size factor special case. Examples: ethanol (1 O atom), ethanoic acid (2 O atoms, larger), methanesulfonic acid (3 O atoms, largest).

                 pK_a

H$_3$CCH$_2$–Ö–H + H$_2$Ö ⇌ H$_3$CCH$_2$–Ö:$^-$ + H$_3$O$^+$ 15.9
ethanol

H$_3$C–C(=O)–Ö–H + H$_2$Ö ⇌ H$_3$C–C(=O)–Ö:$^-$ + H$_3$O$^+$ 4.74
ethanoic acid   ↕
       H$_3$C–C(–O:$^-$)=O

H$_3$C–S(=O)$_2$–Ö–H + H$_2$Ö ⇌ H$_3$C–S(=O)$_2$–Ö:$^-$ + H$_3$O$^+$ -1.2
methanesulfonic acid   ↕
     H$_3$C–S(=O)(–O:$^-$)=O ↔ H$_3$C–S(=O)(=O)–O:$^-$

2.10.1. Induction is another size factor special case. A:⁻ charge is stabilized (delocalized) through σ-bonds by nearby electronegative atoms. Examples (see section 20.4.21):

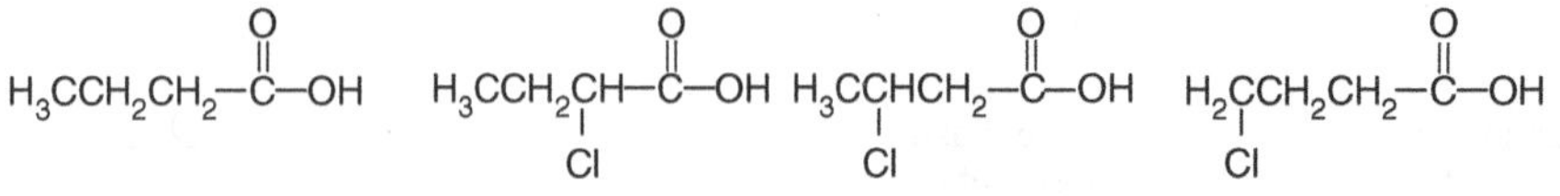

H₃CCH₂CH₂—C(=O)—OH   H₃CCH₂CH(Cl)—C(=O)—OH   H₃CCHCH₂(Cl)—C(=O)—OH   H₂C(Cl)CH₂CH₂—C(=O)—OH

$K_a = 1.52 \times 10^{-5}$ 1.39×10^{-3} 8.9×10^{-5} 2.96×10^{-5}

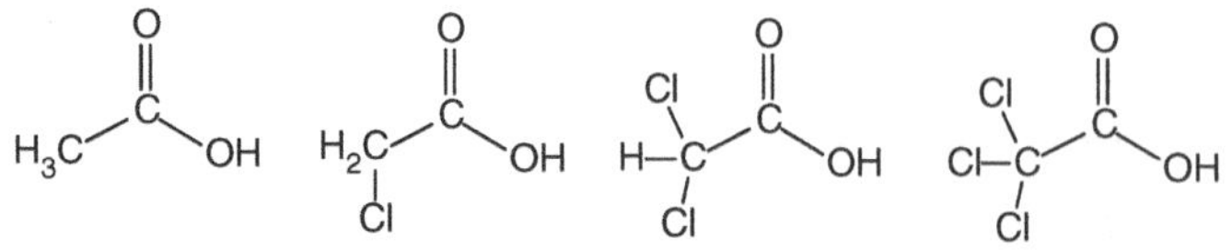

H₃C—C(=O)—OH   H₂C(Cl)—C(=O)—OH   HC(Cl)₂—C(=O)—OH   C(Cl)₃—C(=O)—OH

$K_a = 1.75 \times 10^{-5}$ 1.36×10^{-3} 5.53×10^{-2} 2.32×10^{-1}

2.7.2. Strong acid ⇒ weak conjugate base

Weak acid ⇒ strong (or moderately strong) conjugate base.

a. Review examples: if 1.00 mol of H-A were added to enough H₂O to make 1.00 L of solution at 25°C, what would [H₃O⁺] be?

acid HA	[H₃O⁺] at equilibrium	[H₂O] at equilibrium
HCl	1.00 M	55.34 M
acetic acid	4.2×10^{-3} M	55.34 M

b. The K_a value for HCl is large (1.6×10^2). It dissociates almost completely into ions. For every 1.00 mol of HCl added, 1.00 mol of H₃O⁺ forms, and [H₃O⁺] = 1.00 M.

    1. [H₃O⁺] can vary over an enormous range. A logarithmic scale (pH) is often used. For a 1.00 M HCl solution:

$$pH = -\log_{10}[H_3O^+] = -\log_{10}[1.00] = 0.0$$

c. K_a for acetic acid is 1.75×10^{-5}. Use the K_a expression for acetic acid to calculate the [H₃O⁺] value in the table above.

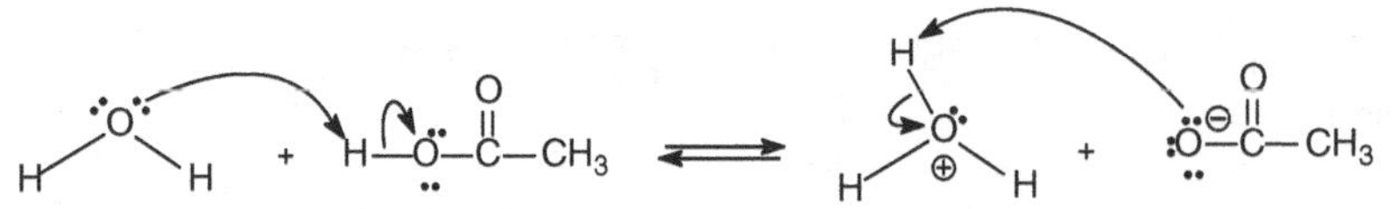

H₂O (Lewis base, weaker base) + H—O—C(=O)—CH₃ (Lewis acid, weaker acid) acetic acid ⇌ H₃O⁺ (Lewis acid, stronger acid) (conjugate acid) + ⁻O—C(=O)—CH₃ (Lewis base, stronger base) (conjugate base)

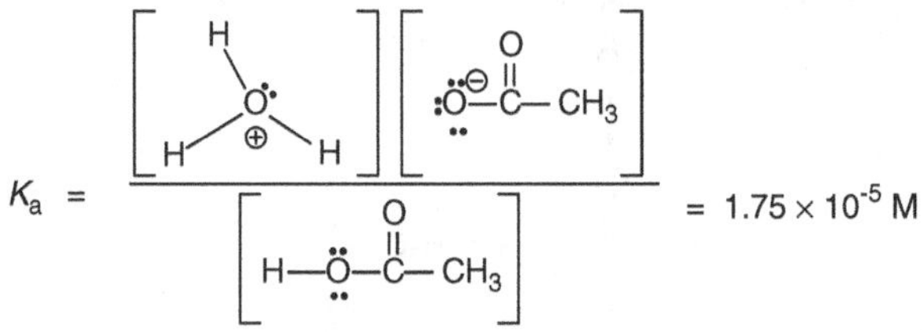

$$K_a = \frac{[H_3O^+][^{-}O-C(=O)-CH_3]}{[H-O-C(=O)-CH_3]} = 1.75 \times 10^{-5} \text{ M}$$

Let $x = [H_3O^+] = [^{-}O-C(=O)-CH_3]$ at equilibrium.

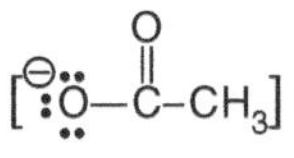

$$1.75 \times 10^{-5} = \frac{[x][x]}{[1.00 - x]} = \frac{x^2}{1.00 - x}$$

solve for x. (Use the quadratic formula. Or since K_a is small, x can be neglected in the denominator.)

$$x = [H_3O^+] = 4.2 \times 10^{-3} \text{ M}.$$

1. Find the pH of the solution:

$$pH = -\log_{10}[H_3O^+] =$$

$$-\log_{10}[4.2 \times 10^{-3}] = 2.38$$

2.7.3. Some reactions involving bases:
    a. Example (equilibrium lies to right).

    b. Example (equilibrium lies to left).

    NH_3 dissolves in H_2O, but does not ionize very much.

2.6.5. Acidity and basicity must be determined experimentally for each acid or base (by measuring K_a or K_b values). We will not memorize numerical values, but the following order of acidity/basicity should be memorized:

stronger acid
$H_2SO_4, HCl, HNO_3 > H_3O^+ > H_3C-C(=O)-OH > NH_4^+ > H_2CO_3 > HCN$

$\gg H_2O: > H-C \equiv C-H > NH_3 > H_2 > CH_4$
                                                            weaker acid

weaker base
$HSO_4^-, Cl^-, NO_3^- < H_2O: < H_3C-C(=O)-O:^- < NH_3 < HCO_3^- < :CN:^-$

$\ll :OH^- < :C \equiv C:^- < :NH_2^- < H:^- < :CH_3^-$
                                                        stronger base

2.13.4. Problem 9. Predict which acid would be stronger:
    a. methanol (CH_3OH) or methanamine (CH_3NH_2)

    b. methanol (CH_3OH) or methanethiol (CH_3SH)

    c. Hydronium (H_3O^+) or ammonium ($^+NH_4$) ions.

    d. Hydronium (H_3O^+) ion or water H_2O.

e. Ammonium ($^+NH_4$) ion or ammonia NH_3.

f. H_2S or HS^- ion.

g. H_2O or ^-OH ion.

h. What is the relation between electrical charge and acidity?

2.13.5. Problem 10. Arrange the following species in order of increasing basicity.
   a. $^-NH_2$, F^-, $^-CH_3$, ^-OH

   b. NH_3, HF, H_2O

   c. ^-SH, Cl^-

   d. I^-, F^-, Cl^-, Br^-

   e. ^-OH, ^-SeH, ^-SH

   f. CH_3F, CH_3NH_2, CH_3OH

   g. H_2O, H_3O^+, ^-OH

   h. $^-NH_2$, $^+NH_4$, NH_3

   i. H_2S, ^-SH, S^{2-}, H_3S^+

   j. What is the relation between electrical charge and basicity?

2.13.6. Problem 11. Use electron pushing arrows to illustrate Lewis acid/base definitions in the following reactions. Label stronger acids and bases, and weaker acids and bases. Do equilibria lie to the left or right?

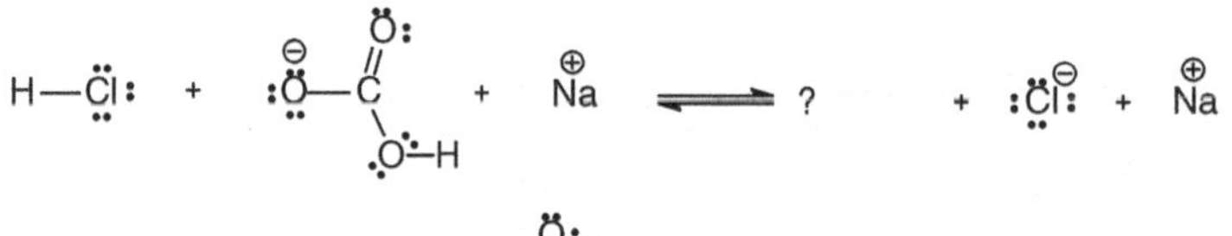

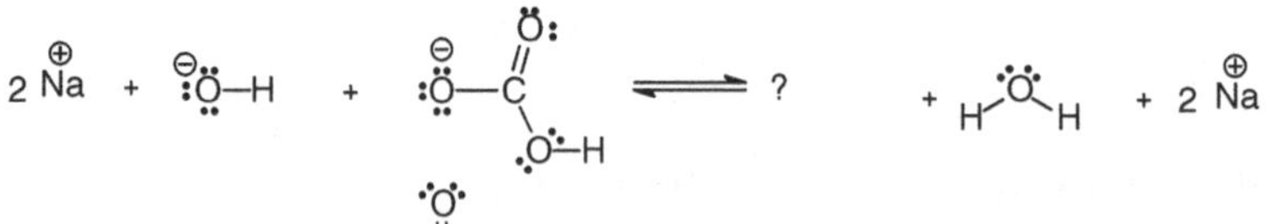

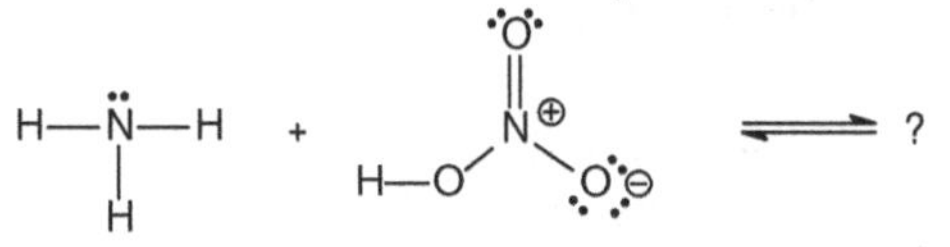

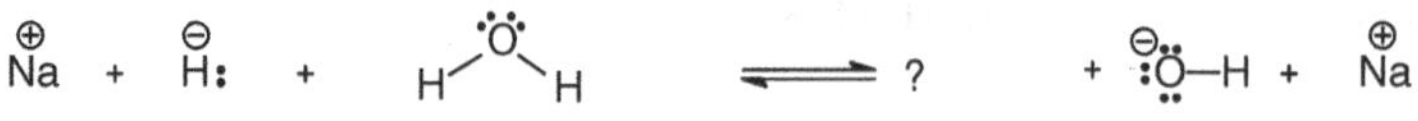

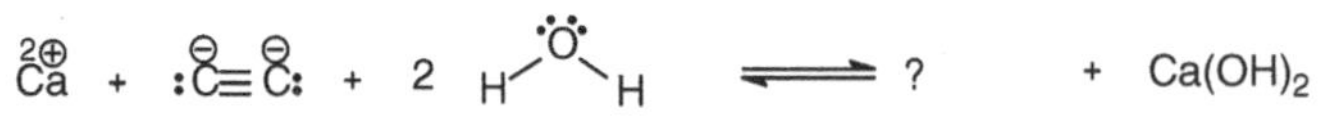

2.13.7. Problem 12. Write the HA reaction for each acid below. Account for the following acidity order by drawing the important resonance contributing structures (those having only one - formal electrical charge) for A⁻:

$$HA + H_2O \rightleftharpoons H_3O^+ + A^-$$

stronger HClO₄ > HClO₃ > HClO weaker

$$H_2SO_4 > H_2SO_3$$

1.11.1. The first steps in unknown structure determination are to find its molecular mass and formula. Methods:
  a. Quantitative analysis — weigh reactants and products, convert masses into mols, use stoichiometry to find empirical formula.
  b. Combustion (burning) – C, H analysis.

  $$C_xH_y + CuO \xrightarrow{600 - 800°C} Cu + xCO_2 + yH_2O$$

  1. H₂O is absorbed by Mg(ClO₄)₂; CO₂ is absorbed by NaOH.
  2. Empirical formula is calculated from the mol ratios.
  c. X (halogen), N, and S analyses:
  1. Sodium fusion: add a known compound mass to excess sodium metal and heat with a flame. As the compound reductively decomposes, X, N, and S are converted into X⁻, CN⁻, and S²⁻ ions which are then dissolved in water and analyzed. The mass percent X, N, and/or S in the original compound is calculated stoichiometrically.

  $$C_aH_bX_cN_dS_e + Na \xrightarrow{flame} \xrightarrow{H_2O(l)} NaX(aq) + NaCN(aq) + Na_2S(aq)$$
  compound (excess)
  containing X, N, or S

  2. Schöniger oxidation: analyze for X⁻, NO₂⁻, and SO₃²⁻ ions.

  $$C_aH_bX_cN_dS_e + O_2 + NaOH \xrightarrow{heat} \xrightarrow{H_2O(l)} NaX(aq) + NaNO_2(aq) + Na_2SO_3(aq)$$
  compound (excess)
  containing X, N, or S

  d. Problem 13. The first step in quantitative elemental analysis is combustion (with excess oxygen). Ethanol analysis: 52.1% carbon and 13.1% hydrogen by mass. How do you know it contains oxygen? What is the mass % O?

  e. Problem 14. When 7.36 mg of chloromethane are heated in a bomb with sodium peroxide (Na₂O₂), C-Cl bonds are broken and Cl is converted into Cl⁻ ions. Adding excess AgNO₃(aq) yielded 20.7 mg of AgCl(s) after filtration. What was the mass % Cl in chloromethane?

1.11.2. Example: empirical formula of ethanol – combustion of ethanol yields 52.1% C and 13.1 % H (by mass). The remaining 34.8 % is assumed to be oxygen. What is the empirical formula of ethanol?

    a. If a sample has a mass of 100. g, there are 52.1 g of C, 13.1 g of H, and 34.8 g O. Convert to mols:

$$52.1 \text{ g} \times \frac{1 \text{ mol C}}{12.011 \text{ g C}} = 4.34 \text{ mol C}$$

$$13.1 \text{ g H} \times \frac{1 \text{ mol H}}{1.008 \text{ g H}} = 13.0 \text{ mol H}$$

$$34.8 \text{ g O} \times \frac{1 \text{ mol O}}{16.00 \text{ g O}} = 2.18 \text{ mol O}$$

    b. An empirical formula is: $C_{4.34}H_{13.0}O_{2.18}$
    c. Simplifying the combining ratios:

$$\frac{C_{4.34}}{2.18}\frac{H_{13.0}}{2.18}\frac{O_{2.18}}{2.18} = C_2H_6O$$

    d. The empirical formula yields the simplest integer combining *ratios*. The true formula could be $C_4H_{12}O_2$, or $C_6H_{18}O_3$, etc. (Mass spectrometry shows it is actually C_2H_6O.)
    e. Problem 15. Complete combustion of a 3.02 mg compound sample yielded 8.86 mg of CO_2 and 5.43 mg of H_2O. What are C, H, and O mass percentages and an empirical formula?

    f. Problem 16. Complete combustion of a 5.17 mg sample of a compound gives 10.32 mg of CO_2 and 4.23 mg of H_2O. The molecular mass (determined by mass spectrometry) is 88 g/mol. What is the molecular formula?

    g. Problem 17. What is the 1-chloropropane mass percentage composition (molecular formula C_3H_7Cl)?

1.11.3. To know the true formula the molecular mass (weight) must be determined. Methods:
    a. Mass spectrometry.
        1. High resolution mass spectrometry can directly determine formula.
    b. Osmotic pressure.
    c. Freezing point depression.
    d. Boiling point elevation.
    e. Ideal gas law ($PV = nRT$).

1.10.1. Structural formula preview. Assume that C has linear, trigonal, or tetrahedral bonding geometry (Chapter 2).

1.16.1. Methane structures.

<u>3 dimensional</u>   <u>planar</u>   <u>condensed</u>

in plane (dashes) into plane / out of plane, sp³ C

H–C–H (with H above and below)   CH₄

a. Larger molecules.

full planar structure	partially condensed	condensed
H–C(H)(H)–C(H)(H)–H (ethane)	H₃C—CH₃	CH₃CH₃
H–C(H)(H)–C(H)(H)–C(H)(H)–H	H₃C—CH₂—CH₃	CH₃CH₂CH₃
H–C(H)(H)–C(H)(H)–C(H)(H)–C(H)(H)–H	H₃C—CH₂—CH₂—CH₃	CH₃CH₂CH₂CH₃ or CH₃(CH₂)₂CH₃
isobutane structure	H₃C—CH—CH₃ with CH₃	CH(CH₃)₃
isopropanol structure with :Ö: H	H₃C—CH—CH₃ with OH	H₃CCHOHCH₃ or (H₃C)₂CHOH

b. More examples.

full planar structure	partially condensed	condensed
2-butene full structure	H₃C\C=C/H with H/CH₃	CH₃CHCHCH₃
H–C(H)(H)–C≡N:	H₃C–C≡N	CH₃CN
H–C(H)(H)–C(=Ö:)–H	H₃C—C(=O)—H or CH₃CH(=O)	H₃CCHO
H–C(H)(H)–C(=Ö:)–C(H)(H)–H	H₃C—C(=O)—CH₃ or H₃CCCH₃	H₃CCOCH₃
H–C(H)(H)–C(=Ö:)–Ö:–H	H₃C—C(=O)—OH or H₃CCOH	H₃CCOOH or H₃CCO₂H
H–C(H)(H)–N(H)–C(H)(H)–H	H₃C—NH—CH or H₃CNHCH₃	(CH₃)₂NH

CHAPTER 1   23

c. Line structures – a C atom is located at each vertex. H atoms are not usually shown.

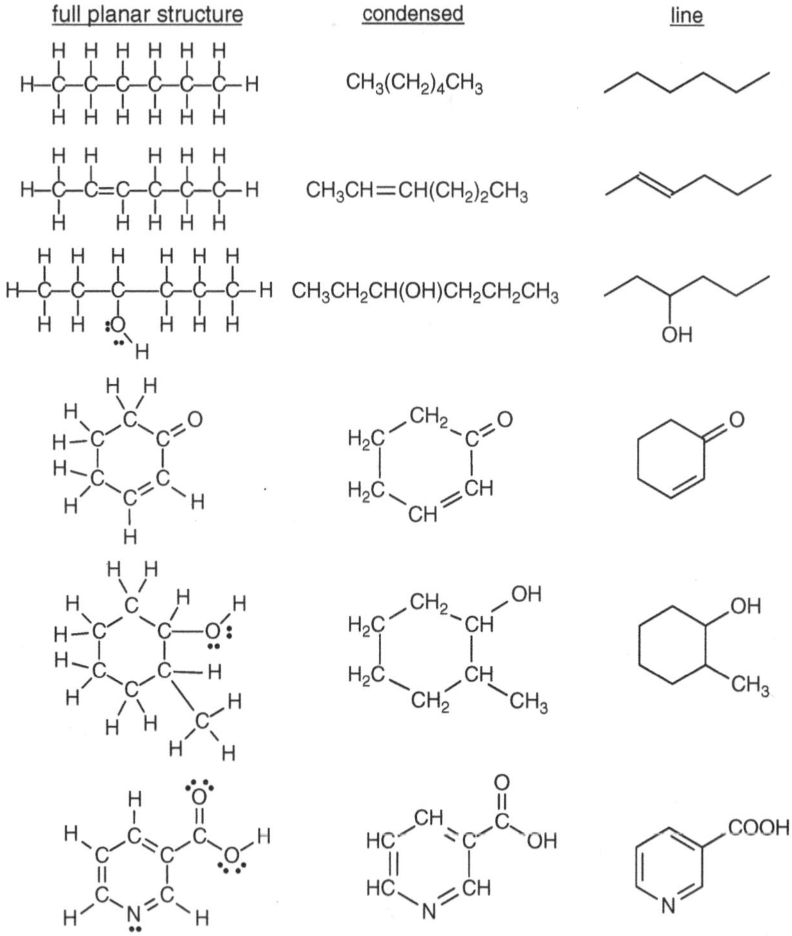
full planar structure
condensed
line
CH3(CH2)4CH3
CH3CH=CH(CH2)2CH3
CH3CH2CH(OH)CH2CH2CH3

1.19.1. Substance identity depends not only on composition (chemical formula) but also on structure (three dimensional bonding arrangement). Example:

a. Ethyl alcohol formula C_2H_6O

1. Colorless liquid, bp 78°C, reacts with Na to yield H_2 and C_2H_5ONa, and with HI to yield H_2O and C_2H_5I.

b. Dimethyl ether formula C_2H_6O.

1. Colorless gas, bp –24°C, no reaction with Na, reacts with HI, but no C_2H_5I product. H_3CI is one product.

c. Explanation – ethyl alcohol and dimethyl ether are two different compounds having the same formula.

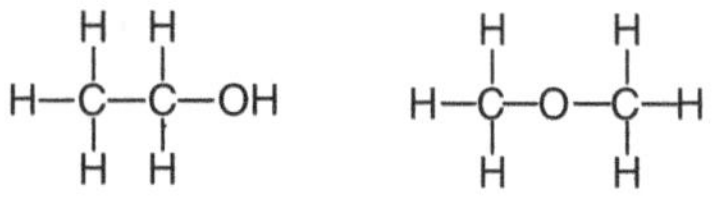

ethyl alcohol dimethyl ether

1.19.2. Structural (constitutional) isomers – distinct compounds with the same chemical formula (for example, C_2H_6O) but different structures and different physical and chemical properties (see also section 3.8.3).

    a. Problem 18. Draw all possible structural isomers for the chemical formulas below.
        1. C_2H_7N

        2. C_3H_7Cl

        3. C_3H_8O

        4. C_2H_4O

1.19.3. Organic chemists seek to understand how different properties arise from different structures.

CHAPTER 2

1.12.1. Lewis dot structures (including resonance hybrids) do not fully predict three dimensional molecular shapes.

1.12.2. Bonding electrons in molecules occupy molecular orbitals (MO's). MO's are derived mathematically by merging atomic orbitals (linear combination of atomic orbitals (AO's), LCAO theory).

1.12.3. Molecular models help visualize bonding and structure.

1.12.4. A half-full AO from one atom overlaps with a half-full AO of another atom in a molecular orbital. Electron spins must pair.

1.12.5. Wave property review. Graph the function $y = \sin x$.

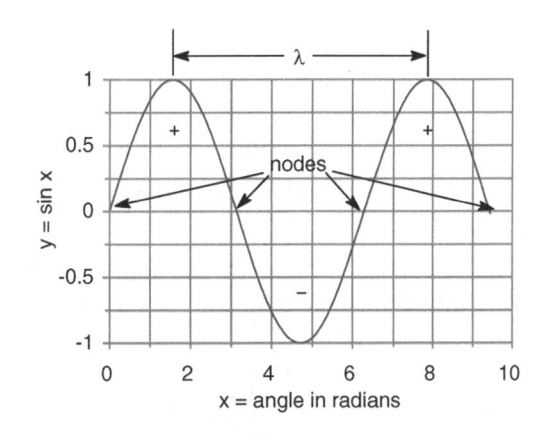

a. A + sign indicates positive phase (where y is positive). A – sign indicates negative phase (where y is negative). "Nodes" are points where $y = 0$.

1.12.6. $y = \sin x$ is an example of a simple wave function, i.e., it is a mathematical description of a wave.

1.12.7. Two (or more) wave functions can be added point by point to produce a new sum function.

a. When waves are "in phase," constructive interference results.

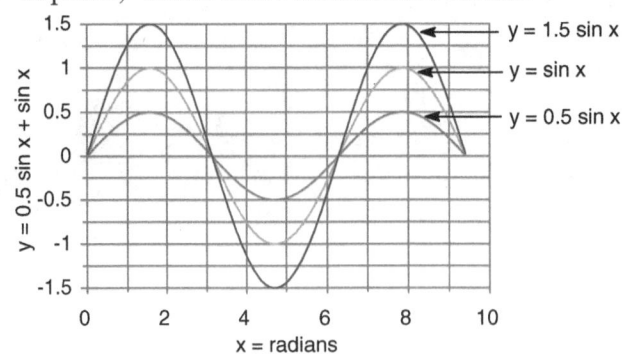

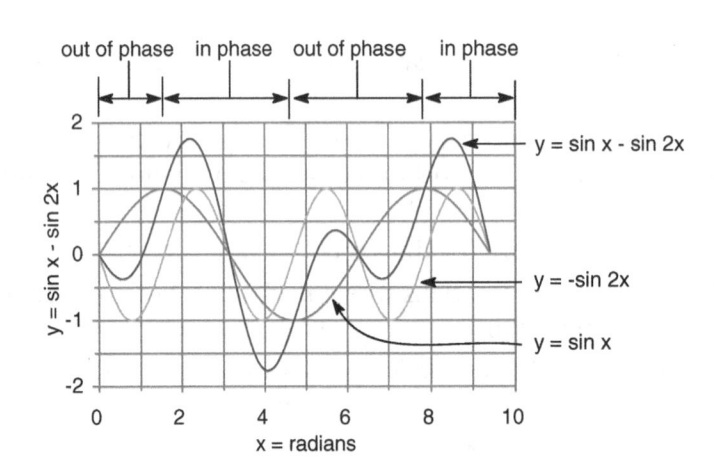

b. Destructive interference results for *x* values where waves are out of phase.

1.12.8. Atomic electron wave functions (atomic orbitals) are similar to wave functions above except they are three dimensional, and more complicated.

a. A hydrogen 1s orbital wave function ψ_{1s} (in polar coordinates) is a function of *r*, where *r* is the distance from the origin (nucleus),

$$\psi_{1s} = \sqrt{\left(\frac{1}{\pi a_0^3}\right)} e^{-r/a_0}$$

and a_0, *e*, and π are constants. Do not memorize this equation! In a ground state hydrogen atom, this 1s-orbital contains one electron.

b. The square of this wavefunction describes the probability of finding the electron at any point that is a distance *r* from the nucleus.
1. This erroneously suggests that the electron is most likely to be found *at the nucleus*.
2. When the probability of finding an electron inside a thin spherical shell (thickness Δr) is plotted versus the radius *r* (out from the nucleus), a more complete and realistic picture emerges. Using this model, the most probable radius (at which the electron will be found) is 0.529 Å (the Bohr radius) out from the nucleus. Electron density $\rightarrow 0$ as $r \rightarrow 0$ because the volume of the spherical shell ($4\pi r^2 \times \Delta r$) $\rightarrow 0$ as $r \rightarrow 0$. Remember, (1 Å = 1×10^{-8} cm).

probability of finding $\psi^2(r)$ a 1s electron at a point (distance r) from the nucleus (point probability density)

distance (r) from nucleus

$4\pi r^2 \psi^2(r)$ probability of finding a 1s electron inside a spherical shell (Δr thickness) at distance r from the nucleus (radial probability density)

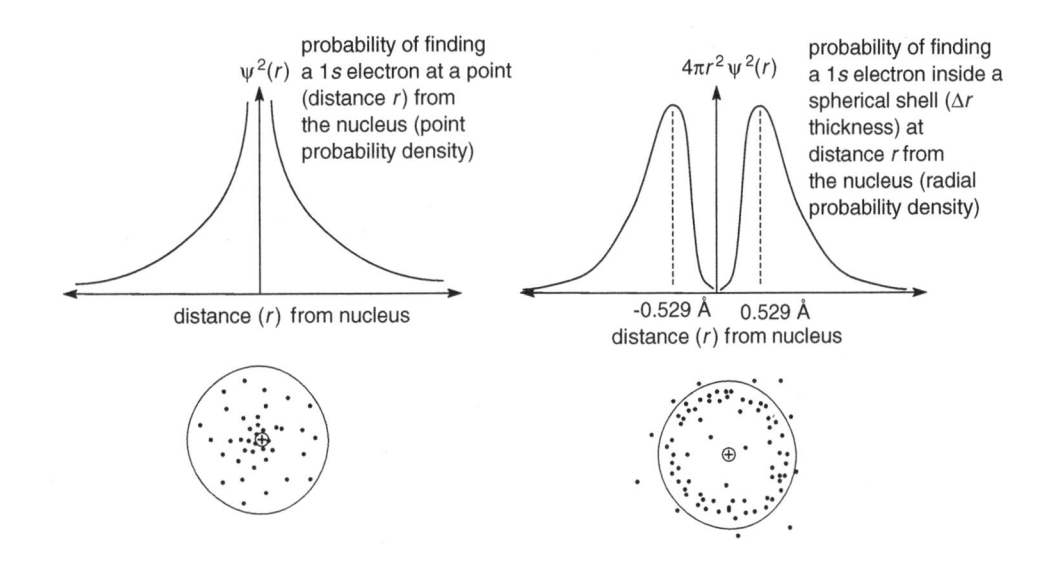

-0.529 Å   0.529 Å
distance (r) from nucleus

c. Example: a hydrogen 2p orbital wave function ψ_{2p} (in polar coordinates) is a function of r, θ, and φ, where r is the distance from the origin (nucleus), and θ and φ are angles:

$$\psi_{2p}(r,\theta,\varphi) = \frac{1}{4\sqrt{2\pi}} \cdot \left(\frac{Z}{a_0}\right)^{\frac{3}{2}} \cdot \frac{Z}{a_0} \cdot (r\sin\theta) \cdot e^{-\left(\frac{Z}{2a_0}\right)r} e^{ir}$$

and a_0, Z, e, and π are constants, and $i = \sqrt{-1}$. Do not memorize this equation! In a ground state hydrogen atom, this 2p-orbital is empty. The electron occupies the lower energy 1s orbital.

1. The graph of this wave function squared is the familiar dumbbell shaped 2p orbital, where the gray region indicates positive phase and the white region is negative phase.

1.12.9. Electron wave functions add and subtract similarly as in section 2.1.7ab above to describe hybrid and molecular orbitals (chemical bonds).

1.13.1. Two atomic hydrogen 1s orbitals combine in-phase to yield a σ-bonding MO. Out-of-phase combination produces a σ*-antibonding orbital.

1.13.2. Combining orbitals rule – when n AO's combine, n (same number) new MO's form.

a. H$_2$ molecule – two 1s AO's yield two new MO's (σ and σ^*).

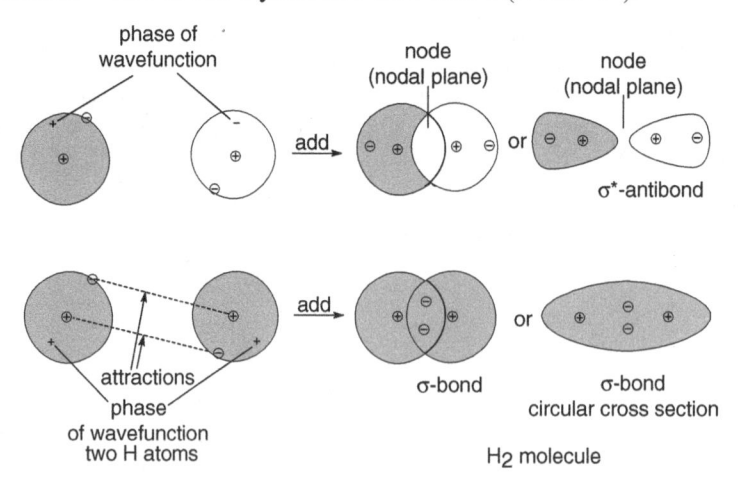

b. σ-Bonds have a circular cross section.

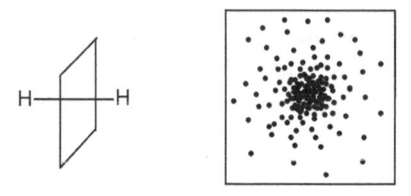

1.13.3. Correlation diagram – potential energy plot:

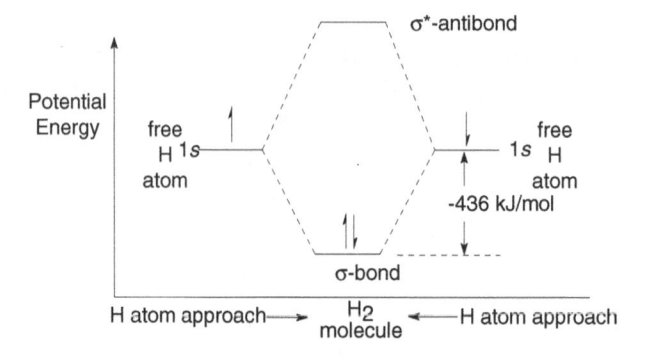

1.13.4. A σ-bonding orbital has minimal potential energy when the distance between two H nuclei is 0.74 Å (*bond distance*).

  a. Electron attraction to both nuclei is maximized and internuclear repulsion is minimized (by electron shielding) at this distance.

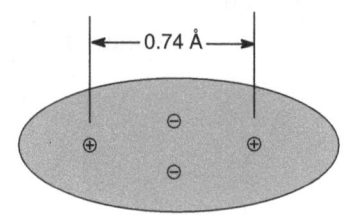

H$_2$ molecule bond distance

1.13.5. Only the σ-orbital is occupied, while σ* is empty. Although anti-bonding orbitals are empty in stable molecules, they often participate in reactions.

  a. Problem 1. Draw a correlation diagram similar to that in section 2.2.3 for a possible He_2^+ ion, which could form from an He^+ ion with an He atom in an electrical discharge through He gas. Should He_2^+ ion formation be exothermic or not?

1.13.6. Another σ-bond type is formed in the F_2 molecule due to $2p$–$2p$ overlap (σ* is not shown).

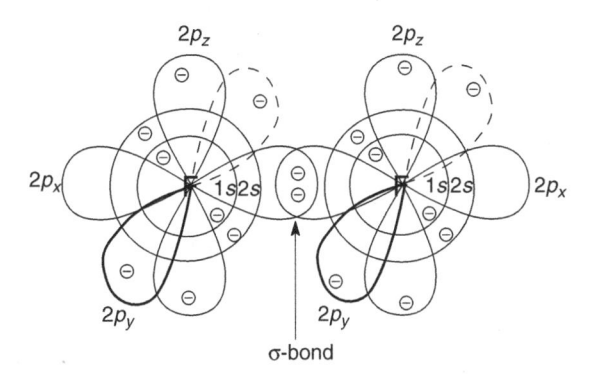

  a. Problem 2. Draw orbital geometries that would result from in phase and out of phase end-on overlap of a $2p$ orbital with a $2s$ orbital.

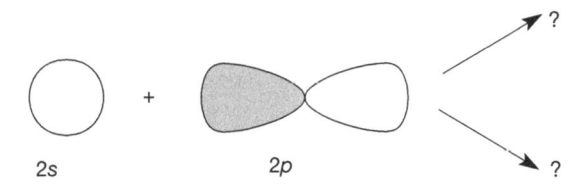

1.15.1. $BeCl_2$ forms when beryllium reacts with chlorine. The experimental Cl-Be-Cl bond angles is 180°.

  a. Valence shell electron pair repulsion (VSEPR) theory predicts 180° bond angles since two groups (of electrons) are as far apart as possible, minimizing electron repulsions.
  b. MO picture? Start with atomic electron configurations:
  Be – $1s^2 2s^2$
  Cl – $1s^2 2s^2 2p^6 3s^2 3p^5$

1.15.2. *sp*-Hybridized Be atom:

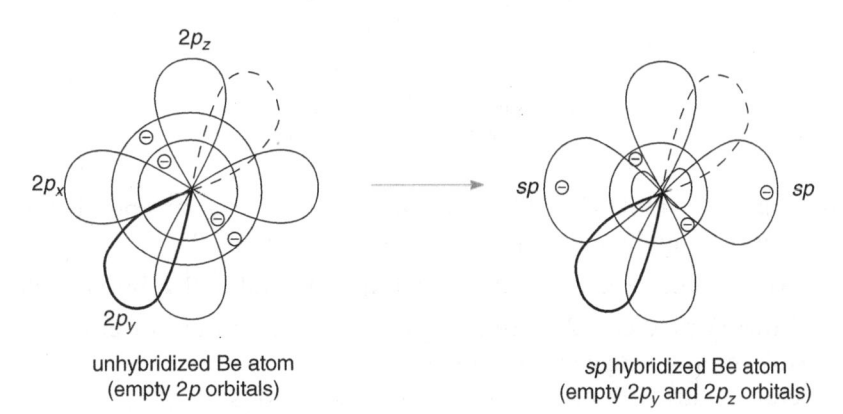

unhybridized Be atom
(empty $2p$ orbitals)

sp hybridized Be atom
(empty $2p_y$ and $2p_z$ orbitals)

1.15.3. Heavy lines (or wedges) are out of the paper plane in "perspective" drawings; dashed lines go into the paper plane (section 1.10.1a).

1.15.4. Two hybrid *sp* orbitals are created (mathematically) by adding or subtracting a $2s$ and a $2p$ wave function. We will not do the mathematics, but we will examine results graphically.

a. A hybrid orbital electron cloud is oriented favorably for bond formation, (for overlap with an atomic orbital on another atom.)

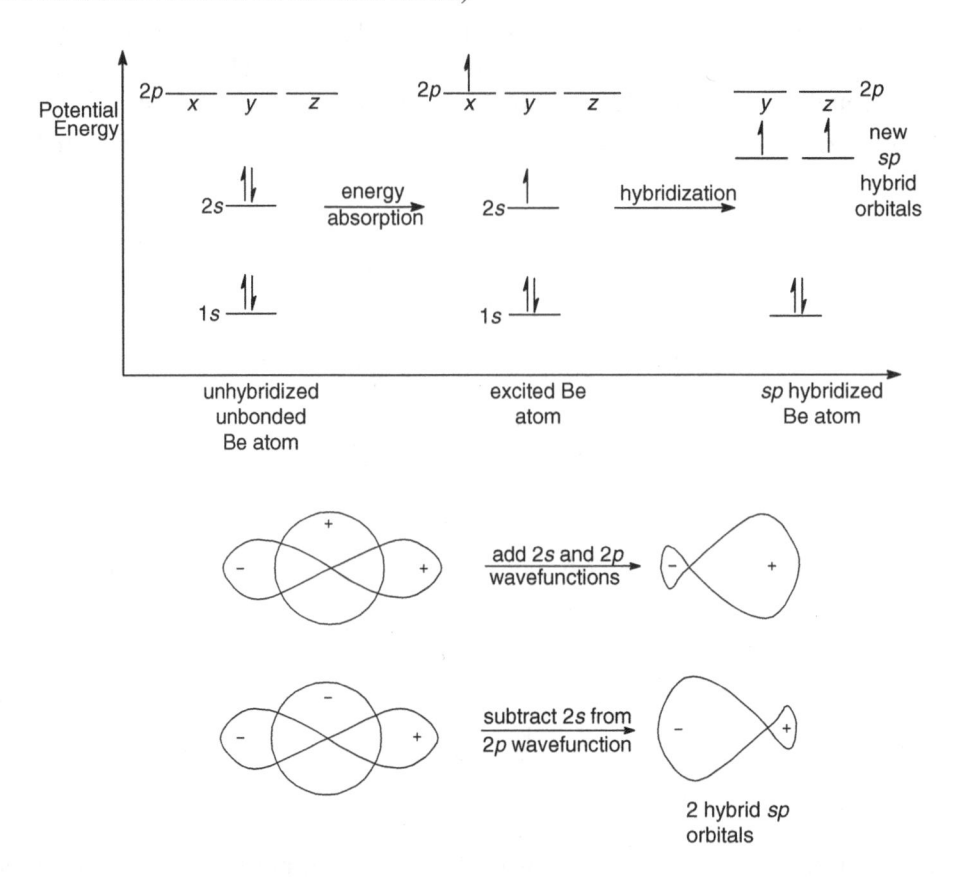

1.15.5. Be sp hybrid orbitals overlap with Cl $3p$ orbitals to form σ-bonds. (σ*-antibonds are not shown.). An LCAO picture:

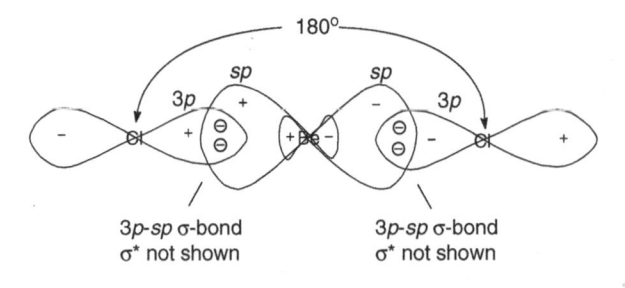

1.15.6. Acetylene (ethyne, formula C_2H_2) has two sp-hybridized C-atoms bonded to each other. Its structure is easier to understand after ethene's. See section 2.3.

1.15.7. BF_3 forms when boron reacts with fluorine. Experimental F-B-F bond angles are all 120°.

a. VSEPR theory predicts 120° bond angles since three groups of electrons are as far apart as possible, minimizing electron repulsions.

b. MO picture? Start with atomic electron configurations:

B – $1s^2 2s^2 2p_x^1$

F – $1s^2 2s^2 2p_x^2 2p_y^2 2p_z^1$

1.15.8. sp^2-Hybridized B atom:

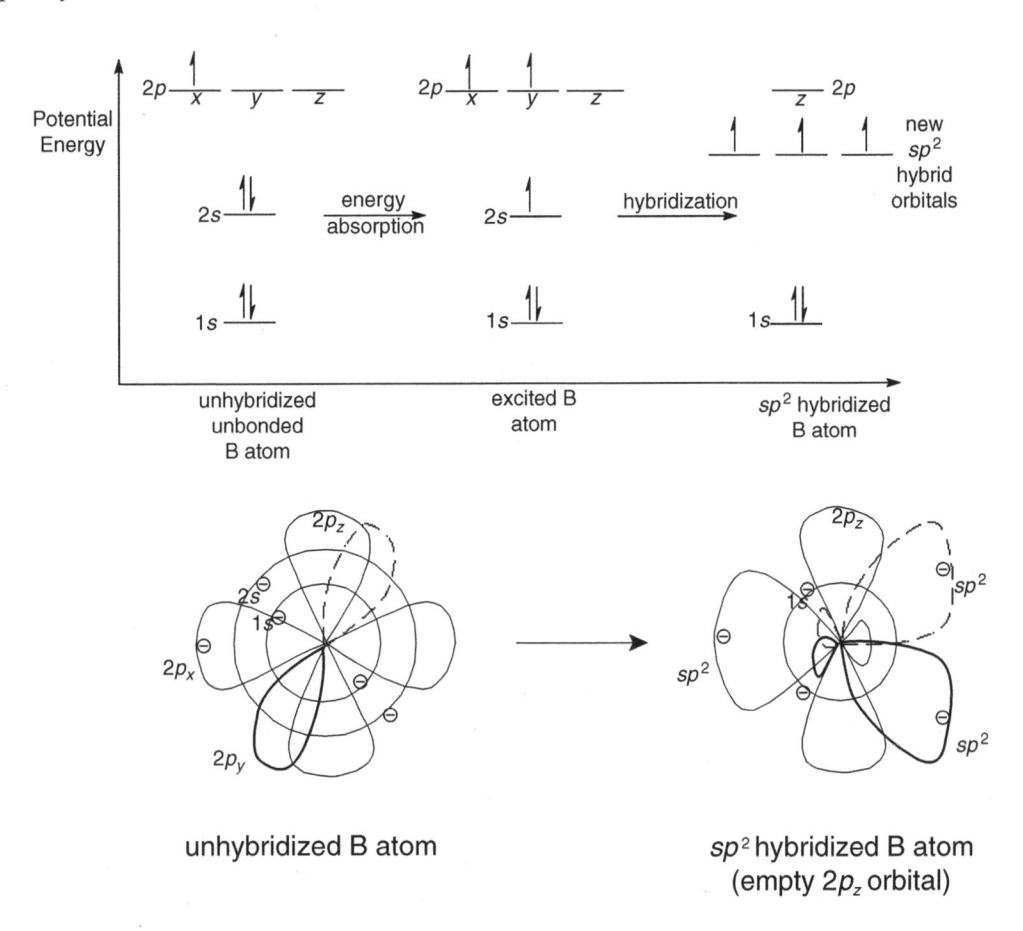

unhybridized B atom

sp^2 hybridized B atom
(empty $2p_z$ orbital)

1.15.9. B sp^2 hybrid orbitals overlap with F $2p$ orbitals to form σ-bonds. (σ*-antibonds are not shown.) BF$_3$ LCAO picture:

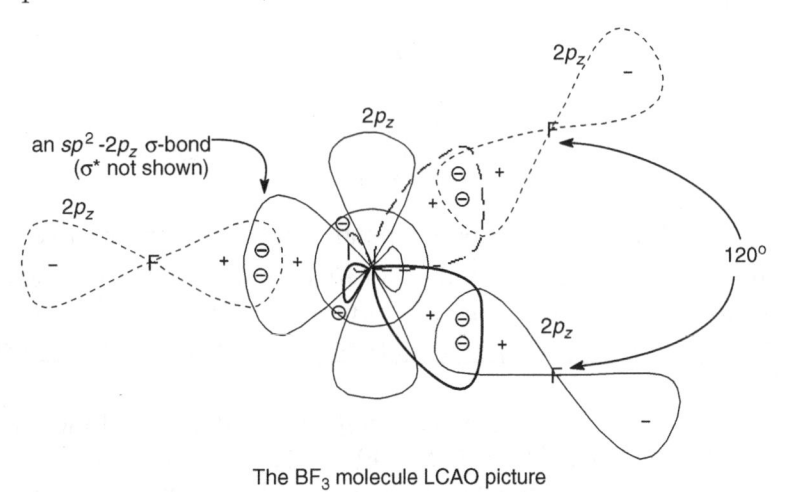

The BF$_3$ molecule LCAO picture

1.17.1. Ethylene (ethene) Lewis dot structure was drawn in section 1.5.1.

7.2.1. Ethene MO picture. Both C atoms are *sp²* hybridized (trigonal).

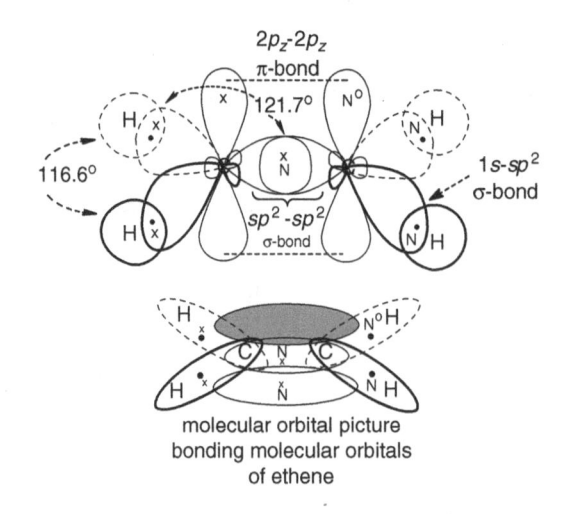

molecular orbital picture bonding molecular orbitals of ethene

7.2.2. Ethene characteristics:
  a. Two 1s-*sp²* H-C σ-bonds to each C atom. (Each σ-bond has an associated, unoccupied σ*-antibond.)
  b. One *sp²*-*sp²* C-C σ-bond to each C (and associated σ*-bond).
  c. H-C-H bond angle = 116.6°, H-C=C bond angle = 121.7° (from electron diffraction, spectroscopic studies), near the ideal 120° (predicted by VSEPR theory) for *sp²* atoms.
  d. All C and H atoms lie in the same plane.
  e. Two half-filled unhybridized $2p_z$-hybrid orbitals overlap (in phase) sideways (parallel) to form one π-bond (and out of phase to form an unoccupied π*-antibond).
    1. Two maximum π-electron density regions – half above and half below the internuclear line – a cross section is not cylindrically symmetrical.

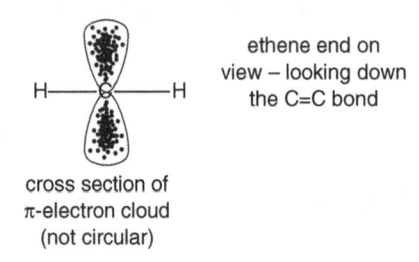

ethene end on view – looking down the C=C bond

cross section of π-electron cloud (not circular)

    2. π-Electrons are exposed, polarizable, and more reactive than σ-electrons.

7.1.1. Ethene's C=C "double" bond (728 kJ/mol or 174 kcal/mol) is less than twice as strong as two C-C ethane σ-bonds (377 kJ/mol × 2 = 754 kJ/mol or 90 kcal/mol × 2 = 180 kcal/mol). Sideways $2p_z$–$2p_z$ π-overlap is not as efficient as end on overlap, as in BF_3 *sp²*–2p B-F σ-bonds above.
  a. The C=C σ-bond energy has been estimated at 407 kJ/mol or 97 kcal/mol. π-Bond energy is then estimated to be (728 − 407 = 321 kJ/mol or 174 − 97 = 77 kcal/mol).
  b. The C=C ethene bond (1.33 Å long) is shorter than the ethane C-C bond ethane (1.54 Å). The shorter bond is consistent with its greater strength.
    1. The *sp²*-*sp²* σ-bond should be shorter even without the π-bond, because it has ⅓ *s*-character vs. ¼ *s*-character for the *sp³*-*sp³* σ-bond of ethane (see below). The more *s*-character, the more strongly electrons are attracted and the closer they are to the nuclei.
    2. Ethene's *sp²*–1s C-H σ-bonds (1.08 Å) are shorter than ethane's *sp³*–1s C-H σ-bonds (1.10 Å) due to their greater *s*-character.

CHAPTER 2   35

c. Correlation diagram:

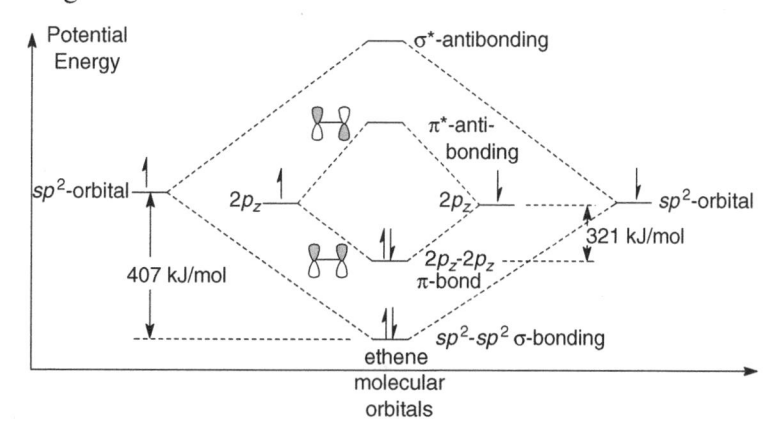

1.19.4. Ethene does not have enough thermal energy (ca. 285 kJ/mol at 25°C) for rotation about the C=C bond. (π-Bond would have to break.) See also section 7.2.

π-bond must break

1.19.5. Restricted rotation about π-bonds causes *cis/trans* isomerism. (E_a is activation energy.)

cis-but-2-ene → trans-but-2-ene
E_a 293 kJ/mol, very slow at 25°C

a. *Cis* and *trans*-isomers can be isolated and have different properties, i.e., they are different compounds.
 1. Two different groups must be bonded to each *sp²*-C atom in order for *cis* and *trans* isomers to exist.

   two identical groups — no *cis/trans* isomers
   no *cis/trans* isomers

 2. *Cis* has like groups on the same side of the double bond – *trans* has like groups on opposite sides.
 3. *Cis* and *trans* isomers are one type of *geometrical* isomers – compounds that differ only in the geometry of groups on the double bond.

1.19.6. Stereoisomers – isomers that are identical except they have different three dimensional atom orientations (Chapter 5).

a. Stereoisomers have identical chemical bonding (connectivity). Geometrical isomers are one class of stereoisomers.

1.19.7. Ethene and higher alkenes will be studied further in Chapter 7.

1.17.2. Acetylene (ethyne) Lewis dot structure was drawn in section 1.5.1a.

1.17.3. Ethyne LCAO and MO pictures. Both C atoms are *sp*-hybridized (linear).

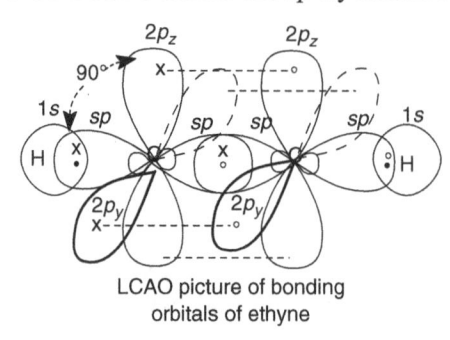

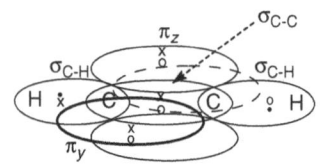

LCAO picture of bonding orbitals of ethyne

molecular orbital picture bonding orbitals of ethyne

1.17.4. Ethyne characteristics:
  a. One 1s-sp H-C σ-bond to each C atom. (Each σ-bond has an associated, unoccupied σ*-antibond.)
  b. One *sp-sp* C-C σ-bond to each C (and associated σ*-bond).
  c. H-C≡C bond angle = 180° (from electron and X-ray diffraction, and spectroscopic studies) ideal (predicted by VSEPR theory) for *sp*-hybridized atoms.
  d. All C and H atoms lie on a line (linear geometry).
  e. Two half-filled unhybridized $2p_y$-hybrid orbitals overlap (in phase) sideways (parallel) to form the $π_y$-bond (and out of phase to form an unoccupied $π_y$*-antibond). At 90° to $π_y$, two half-filled unhybridized $2p_z$ orbitals overlap to form a second $π_z$-bond (and associated $π_z$*-antibond).
  f. The ethyne "triple" bond (965 kJ/mol or 231 kcal/mol) is less than three times as strong as the σ-bond of ethane (377 × 3 = 1130 kJ/mol or 90 × 3 = 270 kcal/mol).
    1. The σ-bond should be stronger than in ethene, about 428 kJ/mol or 102 kcal/mol. The two π-bonds account for additional 965 – 428 = 537 kJ/mol or 128 kcal/mol (about 268 kJ/mol or 64 kcal/mol each).
  g. The C≡C bond distance is 1.20 Å, shorter than ethene C=C bond (1.33 Å).
    1. The *sp-sp* σ-bond of ethyne has ½ s-character and should be shorter than the sp^2-sp^2 σ-bond in ethene even if it didn't have the π-bonds.
    2. The ethyne ≡C-H bond energy (558 kJ/mol) is stronger than the ethene =C-H bond (463 kJ/mol).
    3. The ≡C-H bond length is 1.06 Å, shorter than in ethene (1.08 Å).
  h. Ethyne will be studied further in section 9.1.

1.15.10. CH_4 is one product that can form when C and H react. The experimental H-C-H bond angles are all 109.5°. Methane MO picture? Start with atomic electron configurations:

C – $1s^2 2s^2 2p_x^1 2p_y^1$

H – $1s^1$

1.15.11. C has two unpaired electrons. Does CH$_2$ exist? Yes, but it is highly reactive since C violates octet rule.

$$H:\overset{..}{\underset{..}{C}}:H$$

1.15.12. CH$_4$ is a more common, stable molecule. C uses hybrid AO's to form 4 bonds. The sp^3-hybridized C atom:

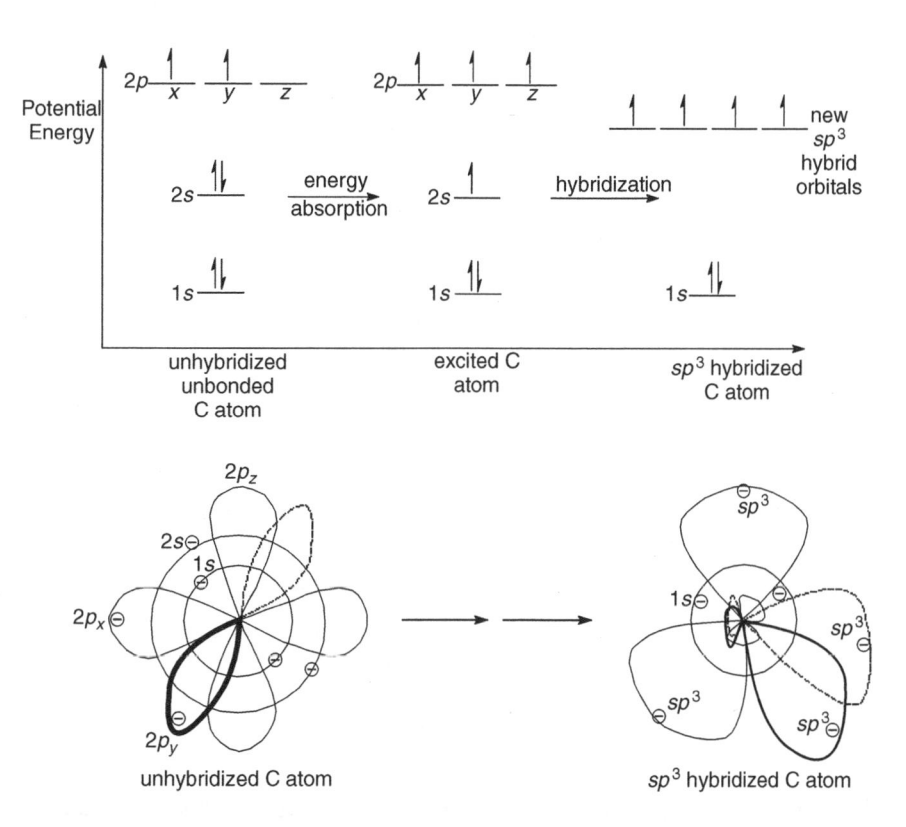

1.15.13. 2p–1s σ-bonds would be longer than 2s–1s σ-bonds if C used unhybridized AO's to form bonds to H. Bond angle between $2p_x$–1s and $2p_y$–1s σ-bonds would be 90°.

1.15.14. Experimentally determined C-H bond lengths are 1.10 Å; all bond angles are 109.5°.

   a. Methane LCAO picture.

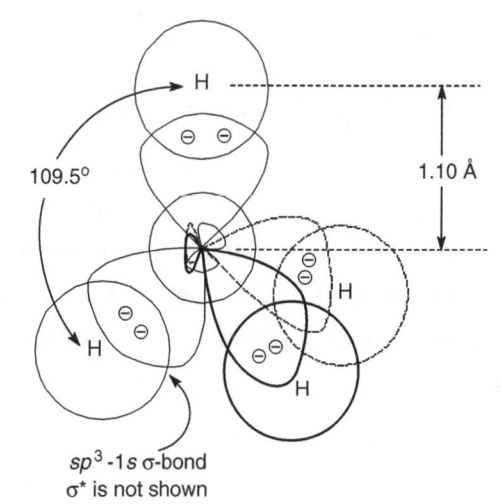

1.15.15. VSEPR theory predicts that four electron pairs will be as far apart as possible in a "tetrahedral" geometry.

1.15.16. If C formed only two bonds to H, only $2 \times -439 = -878$ kJ/mol would be released.

2.4.17. Although hybridization is endothermic (+402 kJ/mol), formation of four bonds is more exothermic overall:

$$+402 + 4 \times (-439) = -1354 \text{ kJ/mol}$$

1.15.18. Analogy – four ugly people and the hair salon parable.
   a. Suppose that they can't get dates.
   b. A parent suggests a new hair style.
   c. They go to a salon offering three hair styles, called sp, sp^2, and sp^3. They each cost $402.
   d. One says it's too expensive and goes home. The other three each choose one hair style (waves).
   e. The sp person then gets four dates. Two were worth about $268, one about $428, and one about $558. The net profit is about:
   $$-402 + 2 \times 268 + 428 + 558 = \$1120$$
      1. The $ amounts actually represent bond enthalpies (kJ/mol).

   $$H-C\equiv C-H$$
   ethyne

   f. The sp^2 person also gets four dates. Two are worth $463, one about $407, and one $321. The net profit is about:
   $$-402 + 2 \times 463 + 407 + 321 = +\$1252$$
      1. Analogous to a C atom bonded in an ethene molecule.

   g. The sp^3 person also gets four dates. Each one is worth $439. The net profit is
   $$-402 + 4 \times 439 = \$1354$$
      1. This is analogous to the C atom in a methane molecule.

   h. The person who did not get a new hair style could only get two dates, each worth about $439.
   $$2 \times 439 = \$878$$
      1. This is analogous to the reactive CH_2 molecule.

1.15.19. Conclusion – it is better to get a new hair style (new electron waves).

1.15.20. It is almost always more exothermic for atoms to use hybrid atomic orbitals to form bonds.

1.16.2. Ethane Lewis dot structure was drawn in section 1.4.2d.

1.16.3. Ethane LCAO and MO perspective pictures. Both C atoms are *sp*³-hybridized (linear).

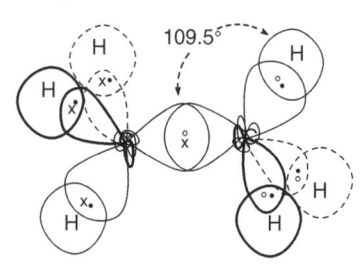

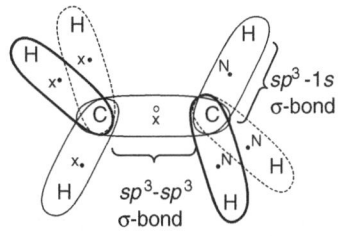

molecular orbital picture
bonding orbitals of
ethane

1.16.4. Ethane characteristics:
   a. Three 1*s*-*sp*³ H-C σ-bonds to each C atom (Each σ-bond has an associated, unoccupied σ*-antibond.)
   b. One *sp*³-*sp*³ C-C σ-bond to each C (and associated σ*-bond).
   c. H-C-H and H-C-C bond angles = 109.5° (from electron and X-ray diffraction, and spectroscopic studies) ideal (predicted by VSEPR theory) for *sp*³-hybridized atoms.
   d. All C atoms are tetrahedral.
   e. Bond dissociation energies are 377 and 423 kJ/mol for the C-C and C-H bonds, respectively.
   g. C-C and C-H bond distances are 1.54 Å and 1.10 Å, respectively, typical also of higher alkanes (with longer chains).
   h. A perpendicular cut (cross section) through the electron cloud is circular with maximum density on the internuclear line.

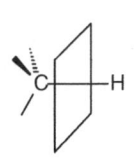

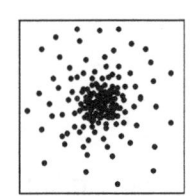

1.18.1. Conformations – different three dimensional atom arrangements in compounds due to rotations about single bonds. (See also section 3.7.)
   a. An infinite number of ethane conformations exist. Staggered is the most stable and eclipsed is the least stable.
      1. Three different conformation representations: "perspective" or 3 dimensional (left), Andiron or sawhorse (middle), and Newman projection (right).
         a. Newman projections are created by looking down a bond end on. Bonds to front C are drawn to center – bonds to back C are partially obscured.

Staggered conformation

Eclipsed conformation

3 dimensional drawing

Andiron projection drawing

Newman projection drawing

angle of rotation = 0°

1.17.5. Hybridized heteroatoms. Example: ammonia (NH_3) molecule. Some possible N electron configurations:

$$1s^2 2s^2 2p_x^1 2p_y^1 2p_z^1 \ ?$$

$$1s^2(sp^3)^2(sp^3)^1(sp^3)^1(sp^3)^1 \ ?$$

$$1s^2(sp^2)^2(sp^2)^1(sp^2)^1 2p_z^1 \ ?$$

$$1s^2(sp^2)^1(sp^2)^1(sp^2)^1 2p_z^2 \ ?$$

1.17.6. Experimental evidence – consistent with sp^3 hybridized N atom.

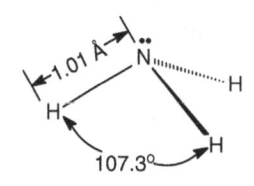

    a. N-H bond energy is 450 kJ/mol.

1.17.7. Three bonds would be formed for either sp^2 or sp^3-hybridized N. Why is N sp^3-hybridized?

    a. VSEPR theory – four electron pairs are farther apart and electron repulsions are minimized if N is tetrahedral.

1.17.8. Why are bond angles not 109.5°? Lone pair is attracted only to N nucleus. It takes up more space than bonding electrons.

1.17.9. N atom rapidly inverts at 27°C: $sp^3 \rightarrow sp^2 \rightarrow sp^3$.

a. $^+NH_4$ ion is similar to methane. N no longer inverts.

1.17.10. Example: water (H_2O) molecule. Some possible O electron configurations:

$$1s^2 2s^2 2p_x^2 2p_y^1 2p_z^1 \; ?$$

$$1s^2 (sp^3)^2 (sp^3)^2 (sp^3)^1 (sp^3)^1 \; ?$$

$$1s^2 (sp^2)^2 (sp^2)^2 (sp^2)^1 2p_z^1 \; ?$$

$$1s^2 (sp^2)^2 (sp^2)^1 (sp^2)^1 2p_z^2 \; ?$$

1.17.11. Experimental evidence – consistent with sp^3 hybridized O atom. However, only two bonds forming to O is not exothermic enough to pay for the hybridization process. The energy gap between $2s$ and $2p$ is also larger for O than for C and N. O and X are now believed to use unhybridized orbitals to form bonds to H.

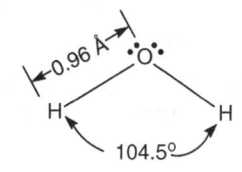

a. O-H bond energy is 497 kJ/mol.

1.17.12. VSEPR theory – four electron pairs are farther apart if O is near tetrahedral.

1.17.13. Why are bond angles not 109.5°? The lone pairs take up more space than bonding electrons.

1.17.14. Both electrons in a *coordinate covalent bond* (new O-H bond below) come from oxygen.

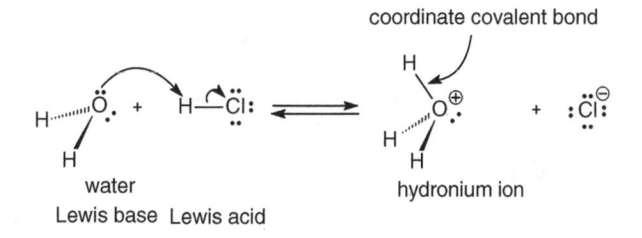

1.17.15. Will C and N always be sp^3-hybridized in compounds? No. Although somewhat less stable (more reactive), many compounds having sp and sp^2-hybridized atoms exist.

1.17.16. Optimal atomic bonding geometry in a given molecule maximizes attractions and minimizes repulsions between nuclei and electrons (minimizes potential energy).

   a. Atoms often remain in a less than optimal geometry (sp or sp^2 hybridized rather than sp^3) because rehybridization energy barrier is high.

1.15.1. Molecular properties (potential energies, bond lengths and angles) are mathematically modeled using hybrid orbitals.

1.17.17. C and N, almost always use hybrid (sp, sp^2, or sp^3) orbitals to form bonds to other atoms.

1.17.18. Recognizing a given atom's hybridization in stable molecules, where atoms obey octet rule:

Atom description	Hybridization	Angle °
no π-bonds to the atom and four single bonds or three single bonds and one lone pair or two single bonds and two lone pairs or one single bond and three lone pairs	sp^3	109.5
one π-bond to the atom and three single bonds two single bonds and one lone pair or one single bond and two lone pairs	sp^2	120.0
two π-bonds to the atom and two single bonds or one single bond and one lone pair	sp	180.0

   a. Some exceptions exist.
   b. Where resonance occurs, examine the form with the most π-bonds.

1.17.19. Problem 3. Use perspective drawings (wedged and dashed bonds to the central atom) to illustrate the three dimensional shape (geometry) expected for each of the following species. Indicate the central atom's hybridization (as unhybridized, sp, sp^2, or sp^3). Show lone electron pair orbital electron clouds

(for example, ⊂⊃ :).

   a. $(CH_3)_3B$

   b. methyl anion, $^-$:CH_3

   c. methyl cation, $^+CH_3$

   d. H_2S

   e. amide anion, $^-$:NH_2

   f. $(H_3C)_2O$

   g. fluoroborate anion, $^-BF_4$

   h. $(CH_3)_3N$

2.1.1. Bond classification (Pauling).

  a. A-A. 100% covalent (nonpolar). Identical atoms. Zero electronegativity difference (section 1.6).
  b. A-B. If electronegativity difference (ΔEN) between two bonded atoms < 2.0, bond is more covalent than ionic (polar covalent bond). If ΔEN is less than 0.5, bond is mostly covalent.
  c. A-B. If ΔEN > 2.0, bond is more ionic than covalent (ionic bond).

Example	ΔEN	Classification
C-C	2.5 – 2.5 = 0.0	100% covalent
C-O	3.4 – 2.5 = 0.9	polar covalent
C-H	2.5 – 2.2 = 0.3	covalent
C-Cl	3.2 – 2.5 = 0.7	polar covalent
Na-Cl	3.2 – 0.9 = 2.3	ionic

2.1.2. C has intermediate electronegativity (2.5) ⇒ rarely forms ionic bonds.

2.1.3. Dipole moment (μ) – a measure of turning force experienced by a polar bond in an electric field:

$$\mu = e \times d$$

where μ units are debyes = electrostatic units (esu) × distance (cm). Alternatively, 1 debye = 3.34×10^{-30} coulomb · meters, where e is the absolute value of one of the partial charges ($\delta+$ or $\delta-$) and d is the distance between two bonded nuclei. Dipole moment arrow points towards $\delta-$ end.

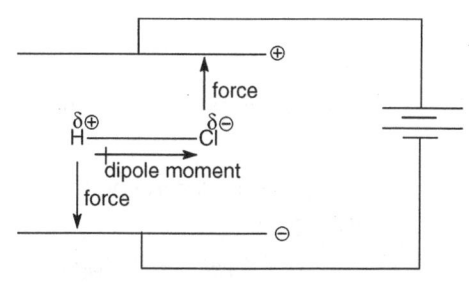

  a. Example: a hypothetical electron and proton separated by 1 Å.

  $$\mu = (1.60 \times 10^{-19} \text{ coulomb}) \times 10^{-10} \text{ meter} = 1.60 \times 10^{-29} \text{ coulomb} \cdot \text{meter}$$

  in debyes:

  $$\mu = 1.60 \times 10^{-29} \text{ C} \cdot \text{m} \times \frac{D}{3.34 \times 10^{-30} \text{ C} \cdot \text{m}} = 4.8 \text{ D}$$

2.1.4. Dipole moment = 4.8 D for a –1 and +1 charge separated by 1. Any dipole moment can be calculated in terms of charge on an electron (or proton) using the equation:

$$\mu \text{ (in debyes)} = \frac{4.8 \text{ debyes}}{\text{angstrom}} \times \delta \text{ (electron charge)} \times d \text{ (in angstroms)}$$

a. Example: the C-O bond experimental dipole moment is 0.86 D, and bond length is 1.43 Å. What is the fractional charge (on C and O)?

$$0.86 \text{ D} = \frac{4.8 \text{ D}}{\text{angstrom}} \times \delta \times 1.43 \text{ angstrom}$$

Solving for $\delta = 0.125$, or about ⅛ of a charge on C and O.

2.1.5. Larger $\mu \Rightarrow$ a more polar bond.

Bond	Dipole moment μ	Bond	Dipole moment μ
C-N	0.22	H-C	0.3
C-O	0.86	H-N	1.31
C-F	1.51	H-O	1.53
C-Cl	1.56	C=O	2.4
C-Br	1.48	C≡N	3.6
C-I	1.29		

a. Problem 4. Although H-Cl (1.27 Å) is a longer molecule than H-F (0.92 Å), it has a smaller dipole moment (1.03 D vs. 1.75 D). What two factors contribute to dipole moment? Explain.

2.1.6. An entire molecule can be polar if the bond dipole moment (vector) sum is nonzero.

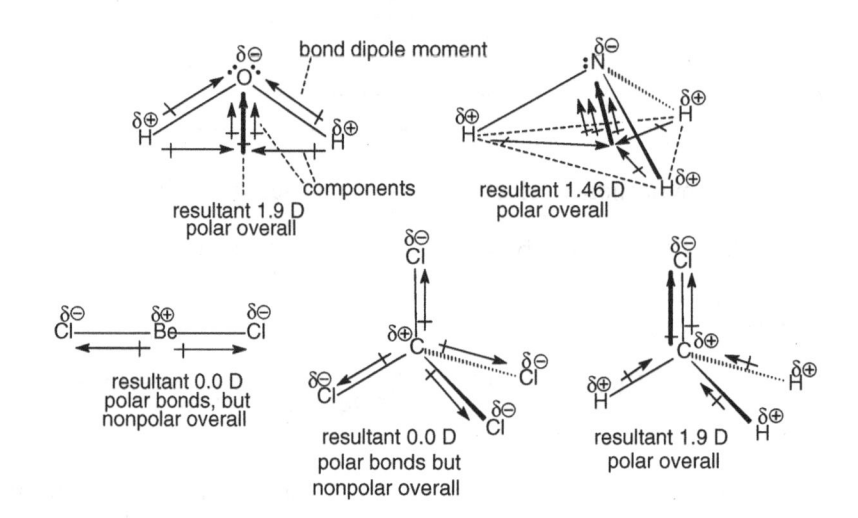

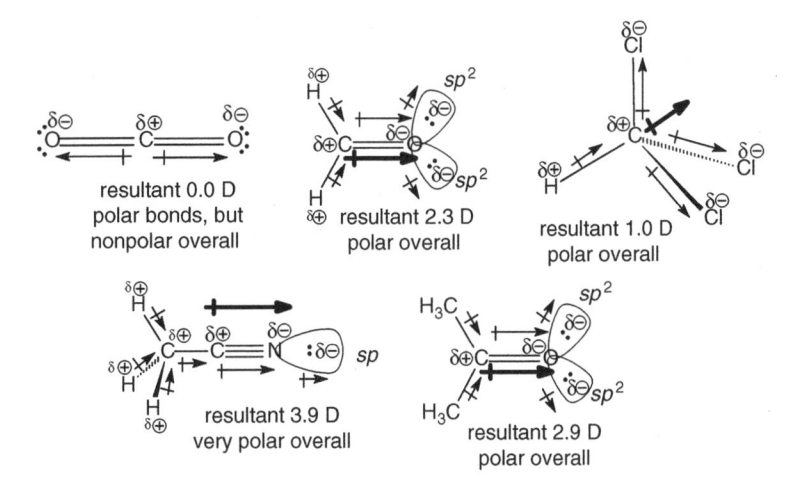

resultant 0.0 D
polar bonds, but
nonpolar overall

resultant 2.3 D
polar overall

resultant 1.0 D
polar overall

resultant 3.9 D
very polar overall

resultant 2.9 D
polar overall

a. Computer generated electrostatic potential maps can be calculated from quantum mechanics.

2.1.7. Problem 5. The structures for NF$_3$ and NH$_3$ are shown below.

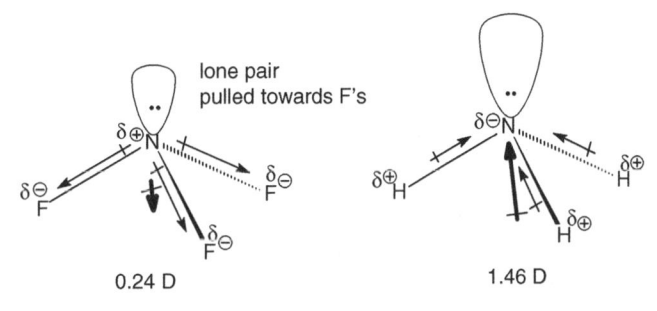

lone pair pulled towards F's

0.24 D

1.46 D

a. What dipole moment would be expected if the N atom in NH$_3$ were sp^2-hybridized?

b. If N used $2p$ orbitals to form N-H and N-F bonds (instead of sp^3-orbitals), would molecular dipole moments be larger or smaller than in structures above? What would bond angles be?

2.1.8 a. Problem 6. Which of the following conceivable CCl$_4$ structures would have zero molecular dipole moment?

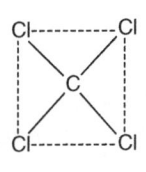

square planar

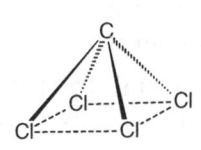

square pyramidal

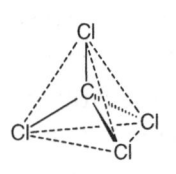

tetrahedral

b. Problem 7. Draw a geometry for the CO_2 molecule that accounts for its zero molecular dipole moment.

c. Problem 8. Draw three dimensional (perspective) structures (using wedged and dashed bonds) for the compounds below. Indicate the molecular dipole moment direction (if any), expected for each compound.
   1. CH_2Cl_2

   2. $CHCl_3$

   3. CH_3OH

   4. $(CH_3)_2O$

   5. $(CH_3)_3N$

   6. CF_2Cl_2

2.1.9. Assumption – a compound's physical and chemical properties are related to its molecular structure (bonding geometries).
    a. Some aspects of this relationship are not fully understood, and are under development.
2.1.10. Goal – to predict physical and chemical properties from structure.
    a. Example – what properties must be predicted in order to separate a mixture by
       1. distillation? (boiling point)
       2. solvent extraction? (solubility)
       3. filtration? (solubility)
       4. crystallization? (melting point, solubility)
       5. chromatography? (solubility)
2.1.11. Molecular polarity is a fundamental structural characteristic that helps predict properties such as boiling points, solubilities, etc. We want to focus on such ideas.
2.2.1. What forces hold particles (atoms, ions, or molecules) together, or cause repulsions?
    a. Example: two noble gas atoms are weakly attracted when farther apart than sum of their radii. If closer than this they strongly repel. Similar interactions occur in organic molecules.

2.6.14. Attractive and repulsive forces:
   a. Electrical (Coulombic) attractions or repulsions (opposite charges attract, like charges repel).
   b. Magnetic – an electrically charged particle that is spinning has an associated magnetic field. Opposite poles attract (north attracts south) but like poles repel. Opposite spin particles attract, while like spin particles repel.
   c. Electrical attractions/repulsions are much larger than magnetic attractions/repulsions or gravity, and the latter are rarely considered.
2.2.2. Interparticle forces are most important for condensed phases (solids and liquids).
2.2.3. Secondary (weak) forces (weakest to strongest):
   a. gravity – all objects (particles) are attracted to each other. For one mol of He atoms, the sum of all gravitational attractions ≈ 10^{-33} kJ, negligibly small compared to other forces below.
   b. London (dispersion) force – attraction of electrons in one atom to the nucleus of another atom. Instantaneously induced dipoles cause one He atom to weakly stick to another.

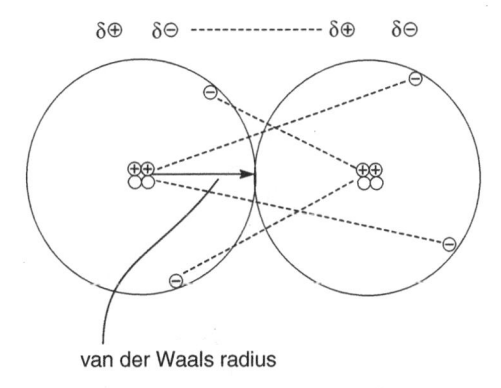

   1. van der Waals radius – ½ the distance between two identical nuclei, where attractions are maximized and repulsions are minimized.
   2. For methane (CH_4) molecules separated by 3.00 Å, London attractions amount to about 0.4 kJ/mol.
   c. Dipole-dipole force – attraction of the δ+ end of a polar molecule for the δ– end of another molecule. Molecules rotate to maximize attractions and minimize repulsions. Attractions amount to about 0.8 to 1.2 kJ/mol.

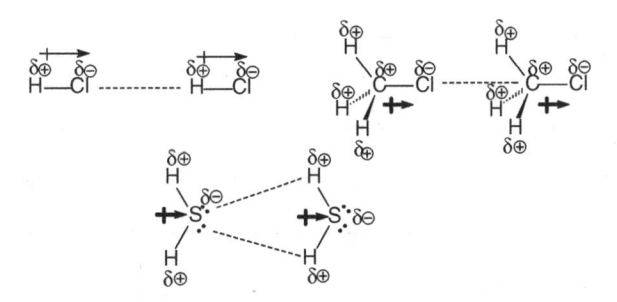

   1. van der Waals forces – combined secondary forces (London plus dipole-dipole).
   2. Example. CCl_4 (bp 77°C) has a higher boiling point than $CHCl_3$ (bp 62°C), despite being nonpolar (molecular dipole moment $\mu = 0$), while $CHCl_3$ is somewhat polar ($\mu = 1.0$).
      a. London attractions in CCl_4 must be greater than London plus dipole-dipole attractions in $CHCl_3$. With one more Cl atom, CCl_4 has larger surface area and more electrons involved in London attractions than in $CHCl_3$.

3. *Greater molecular mass and surface area ⇒ greater London attractions, and higher boiling point.*
d. Hydrogen bond – a special type of dipole-dipole attraction occurring only when H is covalently bonded to N, O, or F.
   1. Hydrogen bonds are still weak (23 kJ/mol for ice), only 2 to 10% as strong as a covalent bond. *Note: N, O and F all have δ- lone electron pairs to which the δ+ H atoms are attracted.* H-bonds are about twice as long (about 2 Å) as an O-H covalent bond. Examples: some hydrogen bonds (shown as dashed bonds).

   1. Mixed hydrogen bonds (shown as bold dashes).

   2. Some H-bond types are stronger than others.

Type of hydrogen bond	Approximate bond dissociation energy (kJ/mol)
—Ö—H------:N—	29
—Ö—H------:O—	21
\N—H------:N—	13
\N—H------:O—	8

3. Why are hydrogen bonds different from dipole-dipole forces? Factors a-c below strengthen hydrogen bond dipole-dipole attractions above usual.
   a. H is small and can approach O, N or F very closely.
   b. A partial coordinate covalent bond (section 1.5.2) forms between N, O, or F lone pair and H atom on an adjacent molecule.
   c. Electronegative N, O, and F atoms highly polarize H-N, H-O, or H-F covalent bonds.
e. Problem 9. Which liquids form hydrogen bonds. Draw hydrogen bonds. You may draw methyl groups as condensed structures.
   a. CH_3OCH_3

   b. CH_3F

   c. CH_3Cl

   d. CH_3NH_2

   e. $(CH_3)_2NH$

   f. $(CH_3)_3N$

f. Problem 10. Suggest a reason for representing hydrogen bonding in aqueous ammonia as:

$$H-\overset{\overset{H}{|}}{\underset{\underset{H}{|}}{N}}{:}^{\delta\ominus}\text{------}H^{\delta\oplus}-\overset{\delta\ominus}{O}{:}^{H} \quad \text{rather than} \quad H-\overset{\overset{H}{|}}{N}-\overset{\delta\oplus}{H}\text{------}{:}\overset{\overset{H}{\diagdown}{\delta\ominus}}{O}-H$$

2.2.4. Primary (strong) forces.
   e. Metallic bond – strong force (about 300–500 times van der Waals force) holding atoms together in metals. Similar to covalent bonds.
   f. Covalent bond – shared electron pair between two nuclei (about 300–500 times van der Waals forces).
   g. Ionic bond – (an) electron(s) lost by one atom (or group of atoms) is (are) gained by another atom (or group of atoms). Resulting ions strongly attract (Coulombically) each other. Some ionic bonds are very strong (up to three times covalent bonds).

2.2.5. Solids – consist of attracting particles in a repeating "crystal lattice."

2.2.6. Melting point – temperature at which particle kinetic energy overcomes lattice attractive forces.

2.2.7. Ionic compounds usually have melting points (mp's) > 300°C. Example: NaCl mp = 801°C, boiling point (bp) = 1413°C. No discrete molecules exist in the solid state. Each Na^+ ion is surrounded by six Cl^- ions, and vice versa.

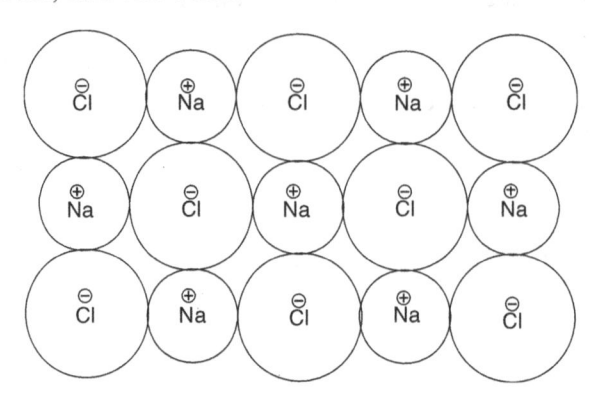

2.2.8. Methane (CH_4) is a nonpolar covalent compound, mp = –183°C, and bp = –161.5°C. Only weak (London) forces hold molecules together. It is a gas at 25°C (1 atm).

2.2.9. Liquids – particles have enough kinetic energy to translate and rotate. Interparticle attractions must be completely overcome to vaporize into the gas phase. Stronger interparticle attractions ⇒ higher bp.

  a. Review. Bp – temperature at which liquid vapor pressure equals external (atmospheric) pressure.

2.2.10. Example: $H^{\delta+}$—$Cl^{\delta-}$ bp = –84.9°C, much higher than CH_4 due to dipole-dipole attractions.

2.2.11. Examples:

Compound	Mol. mass	Bp (°C)
HCl	36.46	–84.9
HBr	80.92	–67.0
HI	127.91	–35.4
HF	20.01	19.5

2.2.12. Higher molecular mass ⇒ higher bp due to greater London attractions. Exception:

2.2.13. Why is HF bp so high? (Hydrogen bonding).

a. Problem 11. *n*-Butanol (bp 118°C) and ethoxyethane (bp 35°C) are isomers (formula $C_4H_{10}O$), but both compounds have nearly the same water solubility (8 g/100 g H_2O at 25°C). Explain.

2.2.14. Other H-bonding examples:

Compound	Mol. mass	Bp (°C)
H_2O	18.02	100
H_2S	34.08	−80
H_3COH	32.04	63
H_3CSH	48.10	−123

All have much higher bp's than CH_4.

a. H-Bonding effect on bp is larger for O than N because O is more electronegative.

Compound	H-bonding?	Bp (°C)
H_3CCH_2OH	yes	78
H_3C-O-CH_3	no	−25
$H_3CCH_2CH_2$-NH_2	yes (two)	49
H_3CCH_2-NH-CH_3	yes	37
$(H_3C)_3N$	no	3.5

2.2.15. Most organic compounds have bp's below 350°C. Decomposition begins above this.

a. How might decomposition be minimized when distilling a liquid with a high bp? (Lower pressure.)

2.3.1. Dissolution – similar to melting since solid lattice is destroyed. Solute (X) is dispersed in solvent (S) *down to the molecular level*.

2.3.2. For dissolution to occur, the following must be true:

solute-solute attractions and
solvent-solvent attractions and
solute entropy and
solvent entropy

$<$

solute-solvent attraction and
solution entropy

2.3.3. "Like polarity dissolves like polarity."

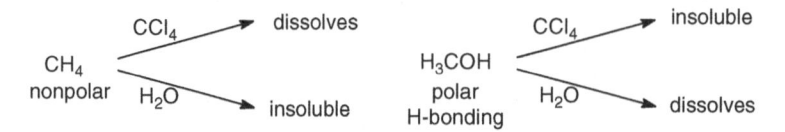

CH_4 nonpolar: $CCl_4 \rightarrow$ dissolves; $H_2O \rightarrow$ insoluble

H_3COH polar H-bonding: $CCl_4 \rightarrow$ insoluble; $H_2O \rightarrow$ dissolves

2.3.4. Polar (or ionic) solute in polar solvent (dissolves).
Collective solvent-ion attractions (and entropy) must be stronger than ion-ion attractions (and entropy) in the solid, plus solvent-solvent attractions (and entropy).

  a. Example: water (polar solvent) and NaCl.

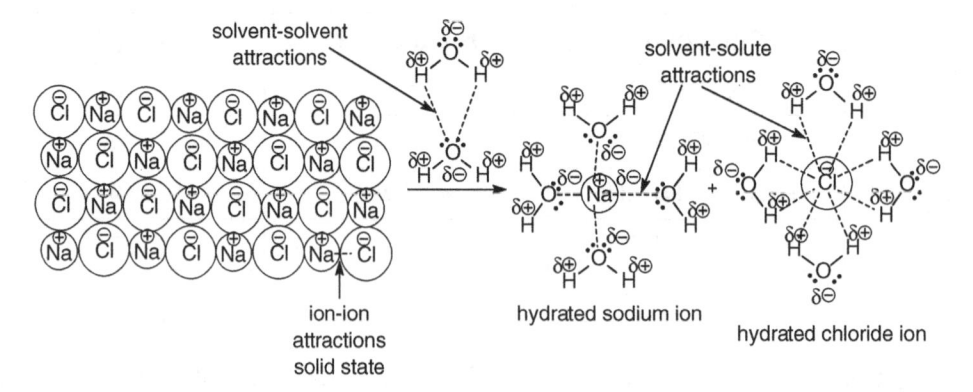

  b. Example: ethanol in water.

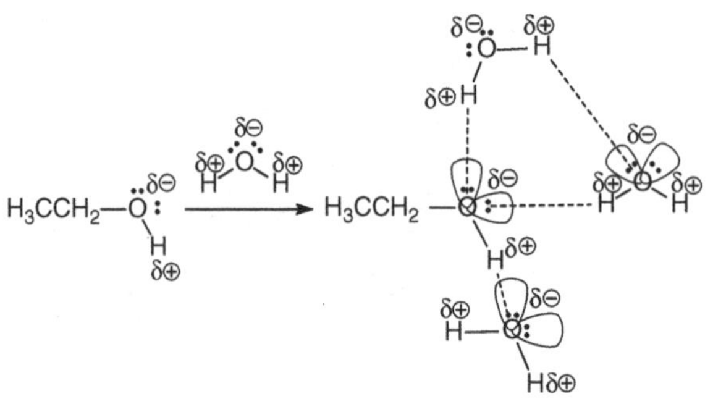

  c. Many weak solute-solvent attractions can overpower fewer strong solute-solute attractions.
  d. Entropy change (ΔS) for dissolution is positive, because particles in solution are more randomly distributed.

e. Solvents with high *dielectric constant* (insulating ability) are needed to dissolve ionic compounds.
   1. Solvent polarity can be measured by its dielectric constant. Larger dielectric constant ⇒ more polar. Examples:

Substance	Dielectric constant
vacuum	1.000
pentane	1.844
carbon tetrachloride	2.238
diethyl ether	4.335
acetone	20.7
ethyl alcohol	24.3
water	78.54

   2. Cations and anions experience some mutual attraction in solution, depending on solvent insulating ability.
      a. Some attraction exists even in a strongly insulating solvent. Ions are separated as a *loose ion pair*.

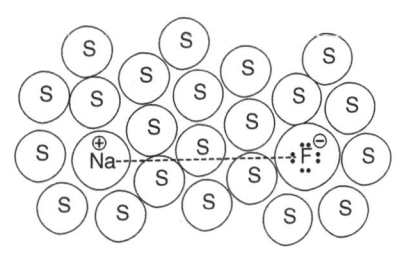

      solvent separated (loose) ion pair

f. Ions can be in direct contact in a weakly insulating solvent, with solvent molecules surrounding the *tight ion pair*.

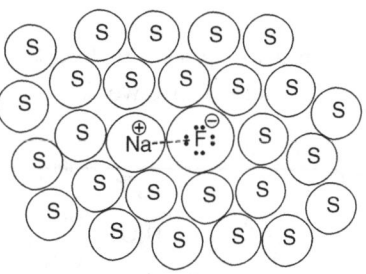

   tight ion pair

g. H_2O has a high dielectric constant. Oxygen can use an sp^3 lone electron pair to form a stronger than usual cation-dipole bond. Oxygen is uncrowded, bonded only to small H atoms, which helps it approach cations closely. Because hydrogen is small, water can also approach anions closely, and forms stronger than usual hydrogen-ion bonds. Water dissolves many ionic compounds.

relatively strong
ion-dipole
bond

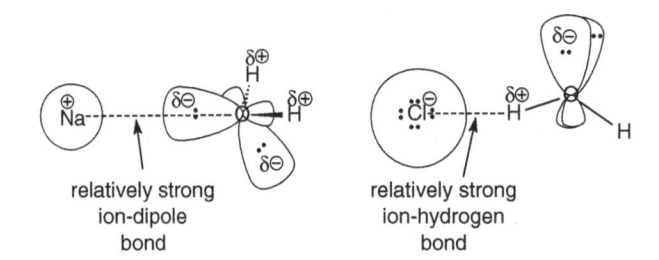

relatively strong
ion-hydrogen
bond

    h. Methanol, H₃COH, solvates ions, but not as well as water. The larger H₃C– group (vs. H– in water) results in more crowding, and it can H-bond with only one H atom (vs. two for H₂O).

2.3.5. Polar solute in nonpolar solvent (does not dissolve). Solute-solvent attractions (and entropy) are not strong enough to break up lattice.

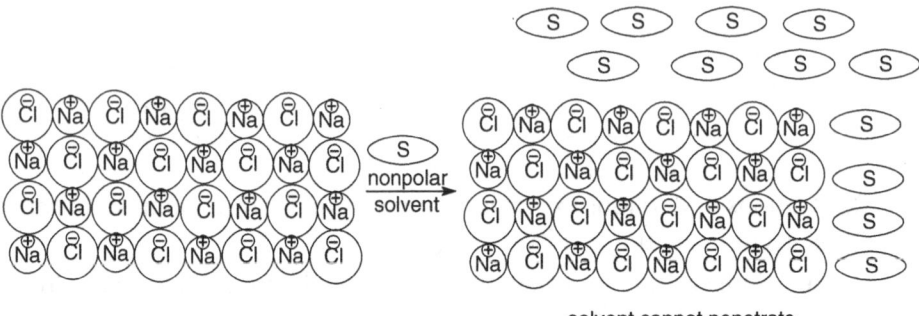

solvent cannot penetrate

2.3.6. Nonpolar solute in nonpolar solvent (dissolves). Solute-solute, solvent-solvent, and solute-solvent attractions are weak (van der Waals). Dissolution is mainly driven by entropy increase.

nonpolar solid

2.3.7. Nonpolar solute in polar solvent. Does not dissolve. Solvent-solvent attractions are too strong and solute-solvent attractions are weak.

nonpolar solid   nonpolar solid

2.3.8. Most organic reactions are performed in solution (liquid phase).

2.15.1. Organic compounds having similar structures and properties can be organized into families.

2.15.2. Hydrocarbons – compounds containing only C and H. Subgroups:

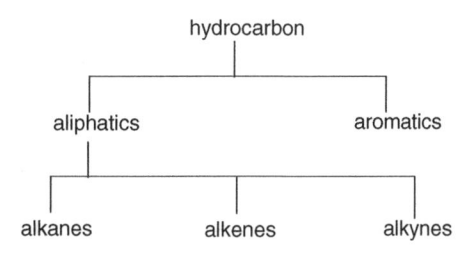

```
                    hydrocarbon
                   /           \
             aliphatics       aromatics
            /    |    \
       alkanes alkenes alkynes
```

    a. Aliphatics – may or may not contain rings.
    b. Aromatics – must contain at least one ring.

2.15.3. Alkanes (family) will be studied first – hydrocarbons with all sp^3 hybridized C atoms and only single bonds.

    a. In 1892, the International Union of Pure and Applied Chemistry (IUPAC) group first met to develop rules for naming compounds systematically based on their structures. We will learn IUPAC names (listed first) for our class, but not common names (below IUPAC name in examples below).

    b. Straight (normal) chain IUPAC parent (base) alkane names:

CH_4	H_3CCH_3	$H_3CCH_2CH_3$	$H_3CCH_2CH_2CH_3$	$H_3CCH_2CH_2CH_2CH_3$
methane	ethane	propane	butane	pentane

$H_3C(CH_2)_4CH_3$	$H_3C(CH_2)_5CH_3$	$H_3C(CH_2)_6CH_3$	$H_3C(CH_2)_7CH_3$	$H_3C(CH_2)_8CH_3$
hexane	heptane	octane	nonane	decane

    c. Alkyl groups – branches attached to a parent chain. They have formulas C_nH_{2n+1}, and are not stable molecules.

    d. Alkyl group names (branch or substituent names). Drop "ane" ending and add "yl."

- methyl: H_3C-
- ethyl: H_3CCH_2-
- n-propyl: $H_3CCH_2CH_2-$
- isopropyl: $(H_3C)_2CH-$
- n-butyl: $H_3C(CH_2)_2CH_2-$
- n-pentyl: $H_3C(CH_2)_3CH_2-$
- secondary-butyl: $H_3CCH_2CHCH_3$
- isobutyl: $(H_3C)_2CHCH_2-$
- tertiary-butyl: $(H_3C)_3C-$
- isopentyl: $(H_3C)_2CHCH_2CH_2-$
- neopentyl: $(H_3C)_3CCH_2-$

2.15.4. Cycloalkanes are ring hydrocarbons with all sp^3 C atoms.

cyclopropane   cyclobutane   cyclopentane   cyclohexane

a. Branched cycloalkane examples:

ethylcyclohexane   line structure   methylcyclopentane   line structure

b. R- is a general symbol representing any alkyl branch (methyl, ethyl, *n*-propyl, isopropyl, etc.).

⬠–R might represent ⬠–CH₃ or ⬠–CH(CH₃)₂

2.15.5. Functional group: an atom or group of atoms common to all compounds in a given family, which imparts similar properties to all family members.
  a. Functional group properties are largely independent of the R- structure to which it is bonded.
  b. Compounds with multiple functional groups manifest properties of each functional group that they contain.

2.15.6. Alkene functional group: C=C double bond.
  a. Parent alkene names – start with alkane name, drop "ane" and add "ene."
    1. Use a hyphenated number before the -ene parent suffix to show double bond location.
    2. IUPAC changed the naming rule in 1997 so the number comes directly before parent suffix.
    3. Number 1 is optional, and is often left out.

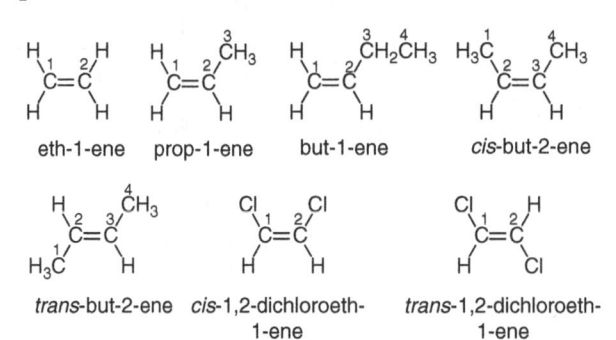

eth-1-ene   prop-1-ene   but-1-ene   *cis*-but-2-ene

trans-but-2-ene   *cis*-1,2-dichloroeth-1-ene   *trans*-1,2-dichloroeth-1-ene

b. Cycloalkenes.

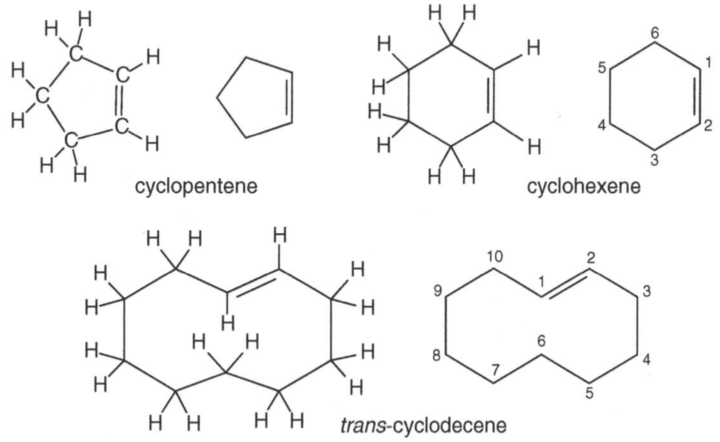

cyclopentene   cyclohexene

trans-cyclodecene

2.15.7. Alkyne functional group: C≡C triple bond.

a. Parent alkyne names – start with alkane name, drop "ane" and add "yne."

1. Use a hyphenated number before -yne parent suffix to show triple bond location.

H—C≡C—H   H—C≡C—CH₃   H—C≡C—CH₂CH₃   H₃C—C≡C—CH₃

eth-1-yne   prop-1-yne   but-1-yne   but-2-yne
acetylene

2. Alkynes are usually only found in rings with eight or more C atoms, i.e., cyclooctyne. Otherwise, the linear alkyne structure would have to be severely bent (with large "angle strain").

2.15.8. Most common aromatic (arene): benzene ring, with three alternating double bonds.

a. Aromatics do not behave like alkenes, but have their own unique properties.

b. Parent arene name: often "benzene."

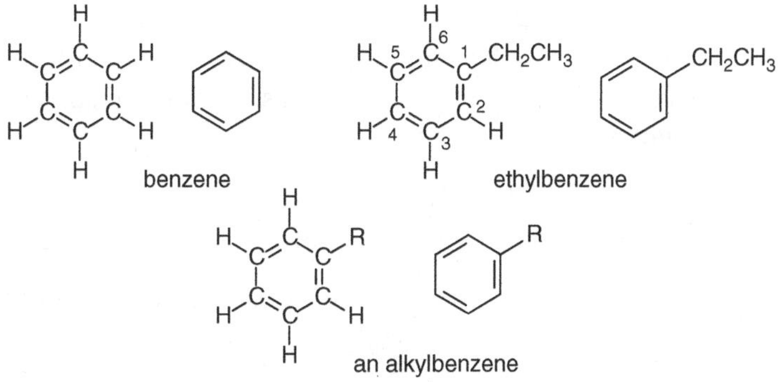

benzene   ethylbenzene

an alkylbenzene

c. Ar- is a general symbol to represent any aromatic (aryl) branch. Ph- is a shorthand for phenyl, the benzene branch name.

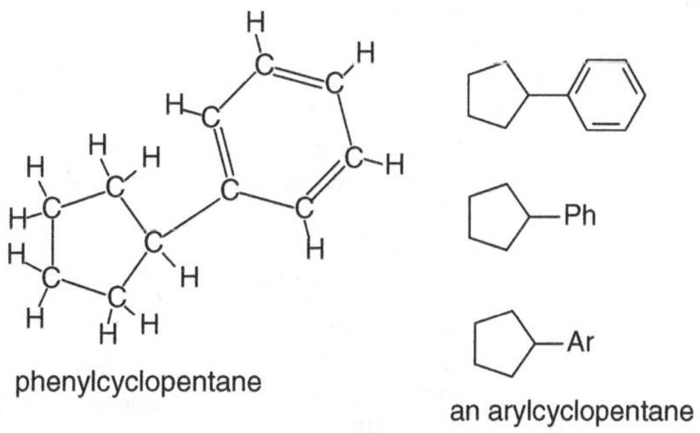

phenylcyclopentane

an arylcyclopentane

58 SEQUENTIAL ORGANIC CHEMISTRY I

1. Simplest aryl group: phenyl. Aryl group examples:

phenyl   p-methylphenyl

p-nitrophenyl   4-pyridyl

2.16.1. R-OH is the alcohol functional group. Alcohols are common, polar, and capable of hydrogen bonds. Methanol and ethanol are miscible with water. They are common antiseptics, H-bonding solvents, fuels, and synthetic reactants.

   a. Parent alcohol names – start with alkane name, drop last "e" and add "ol." Branch name: "hydroxy."

      1. Use a hyphenated number prefix to show alcohol location.

general alcohol   methanol / methyl alcohol / wood alcohol   ethanol / ethyl alcohol / grain alcohol

propan-1-ol   propan-2-ol / isopropanol / rubbing alcohol

2.16.2. R-O-R' is the ether functional group – cannot hydrogen bond to itself, but can to water, alcohols, amines etc. – more polar than alkanes, – common solvents, anesthetics and fast start fuels for engines.

crossed hydrogen bond

   a. No ether parent name. Named as alkanes with alkoxy (R-O-) branches in IUPAC system. Common names are often used.

general ether   1-methoxymethane / dimethyl ether   methoxyethane / ethyl methyl ether

ethoxyethane / diethyl ether (anesthetic and fast-start)   1-methoxycyclohexane / cyclohexyl methyl ether

b. Problem 12. Write IUPAC names for the following compounds:

a. CH₃—CH(CH₂CH₃)—CH₂—CH(CH₃)—CH₃   b. CH₃—CH(CH₃)—CH(CH₃)—Br   c. (CH₃)(H)C=C(H)(CH₂CH₃)

d. CH₃—C(CH₃)(H)—CH(OH)—CH₃   e. H₃C—CH₂—O—CH₂CH₂CH₃

2.16.3. Aldehydes and ketones contain a carbonyl group (C=O) – cannot hydrogen bond to themselves, but can to water, alcohols, amines, etc. More polar than alkanes and ethers. Used as solvents, in drug syntheses, and are metabolites.

(diagram showing crossed hydrogen bond between two water molecules and carbonyl oxygen lone pairs on R–C(=O)–R')

a. Parent aldehyde names – start with alkane name, drop last "e" and add "al."

- general aldehyde: R–CHO
- methan-1-al / formaldehyde: H–CHO
- ethanal / acetaldehyde: H₃C–CHO
- propanal / propionaldehyde: H₃CCH₂–CHO
- butanal / butyraldehyde: H₃CCH₂CH₂–CHO

b. Parent ketone names – start with alkane name, drop last "e" and add "one."

- general ketone: R–CO–R'
- propan-2-one / acetone: H₃C–CO–CH₃
- butan-2-one / methyl ethyl ketone (MEK): H₃C–CO–CH₂CH₃
- cyclohexan-1-one

2.16.4. -COOH is the carboxylic acid functional group. Hydrogen bonding. More polar than alkanes, alcohols, ethers, aldehydes and ketones. With four or fewer C atoms, usually miscible with water. H-bonding solvents, sour flavoring agents, metabolites and used in synthesis.

a. Ionize somewhat in water solution (pK_a = 5). Conjugate base carboxylate anion is resonance stabilized, which explains why a carboxylic acid is acidic.

b. Parent carboxylic acid names – start with alkane name, drop last "e" and add "oic acid." Branch name: carboxyl. Conjugate base name: drop the "ic acid" suffix and add "ate."

general carboxylic acid

methan-1-oic acid
formic acid
(from ants)

ethanoic acid
acetic acid
(in vinegar)
Latin *acetum*
(sour)

propanoic acid
propionic acid
(in cheeses)

butanoic acid
butyric acid
(in rancid butter)

benzoic acid

sodium methanoate
sodium formate

potassium ethanoate
potassium acetate

2.16.5. Several more carbonyl functional groups are carboxylic acid "derivatives" (converted into carboxylate anions when heated with a strong acid or base catalyst in boiling water (hydrolysis)).

a. *Hydrolysis* (cleavage by water) – a process where a substance reacts with (usually hot) water.
b. Three common families are:

general acid chloride

general ester

general primary amide

ethanoyl chloride
acetyl chloride

ethyl ethanoate
ethyl acetate

ethanamide
acetamide

c. Acid chloride names: start with carboxylic acid name, drop "ic acid" suffix and add "oyl chloride."
d. Ester names: start with alkyl group name bonded to sp^3 oxygen. Name carboxylic acid as a conjugate base. Drop "ic acid" suffix and add "ate."
e. Amide names: start with carboxylic acid name. Drop "oic acid" suffix and add "amide."

2.17.1. Amine functional group also contains nitrogen. Amines are one class of alkaloids (basic organic compounds), similar to ammonia ($pK_b = 4$). They are often toxic but are important in biochemistry.

a primary amine ⇌ an ammonium cation

- **a.** Three amine classes, primary (1°), secondary (2°), and tertiary (3°).
    1. A 1° amine has one R-group bonded to N.
    2. A 2° amine has two R-groups bonded to N.
    3. A 3° amine has three R-groups bonded to N.
- **b.** Parent amine names – start with alkane name, drop last "e" and add "amine." Branch name: amino. Conjugate acid name: drop "amine" suffix and add "ammonium ion."

H_3C-NH_2 methan-1-amine (1°) methylamine

$H_3CCH_2-NH_2$ ethanamine (1°) ethylamine

$H_3CCH_2CH_2-NH_2$ propanamine (1°) propylamine

$H_3C-NH-CH_3$ N-methylmethanamine (2°) dimethylamine

$H_3C-NH-CH_2CH_3$ N-methylethanamine (2°) ethylmethylamine

$H_3CCH_2-NH-CH_2CH_3$ N-ethylethanamine (2°) diethylamine

N,N-dimethylmethanamine (3°) trimethylamine

N,N-diethylethanamine triethylamine

methanammonium chloride

N,N-dimethylethanammonium bisulfate

2.17.2. Similar to amines, three amide classes exist. Amides are important in biochemistry.

ethanamide (1°)   N-methylethanamide (2°)   N,N-dimethylbenzamide (3°)

- **a.** Amides are not as basic as amines due to resonance stabilization.

- **b.** Protein molecules are polyamides (containing many amide bonds).

2.17.3. -C≡N is the nitrile functional group. It is important in synthesis, solvents, and in nitrile plastics.
   a. Parent nitrile name – start with alkane name and add "nitrile." Branch name: cyano.

R—C≡N: H₃C—C≡N: H₃CCH₂—C≡N: C₆H₅—C≡N:
general nitrile   ethanenitrile   propanenitrile   benzonitrile

2.17.4. Common functional group name summary table. Families are listed in order of decreasing IUPAC nomenclature priority (highest priority on top, lowest on the bottom).

Full name	Structure	Parent name (suffix)	Branch name (usually prefix)
carboxylic acid	R–C(=O)–OH	-oic acid	carboxy-
ester	R–C(=O)–O–R'	-oate	alkoxycarbonyl-
amide	R–C(=O)–NH₂	-amide	amido-
nitrile	R–C≡N:	-nitrile	cyano-
aldehyde	R–C(=O)–H	-al	oxo-
ketone	R–C(=O)–R'	-one	oxo- or keto-
alcohol	R–OH	-ol	hydroxy-
amine	R–NH₂	-amine	amino-
alkene*	R₂C=CR₂	-ene	-en
alkyne*	R–C≡C–R'	-yne	-yn
ether	R–O–R	named as alkane	alkoxy-
alkane§	R₂C–CR₂ (with R₁–R₆)	-ane	-yl
halogen§	R–X:	named as alkane	halo-
nitroalkane§	R–N⁺(=O)(O⁻)	named as alkane	nitro-
benzene	R–C₆H₅	benzene	phenyl- (C₆H₅-)

*Alkenes and alkynes have the same nomenclature priority.
§These groups have the same nomenclature priority.

2.17.5. Naming priorities are needed for compounds with more than one functional group.
  a. Problem 13. Write IUPAC names for the following compounds.

a. 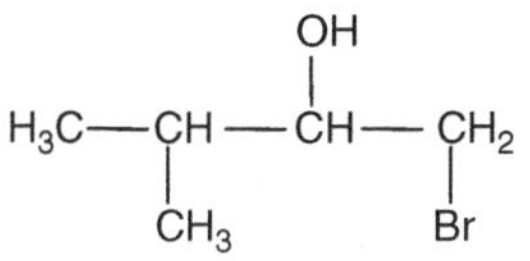 H₃C—CH(CH₃)—CH(OH)—CH₂Br

b. 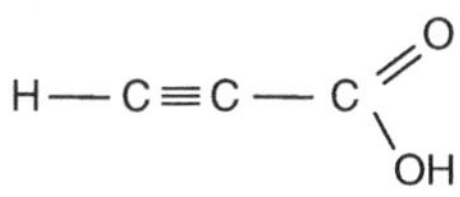 H—C≡C—C(=O)OH

c. H₃C—C(OH)(H)—C(=O)OH

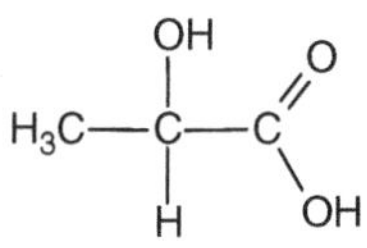

d. 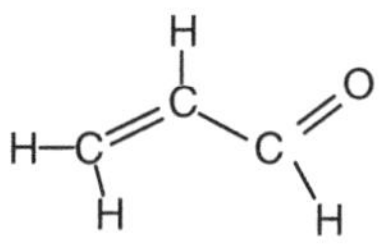 H(H)C=C(H)—C(=O)H

e. 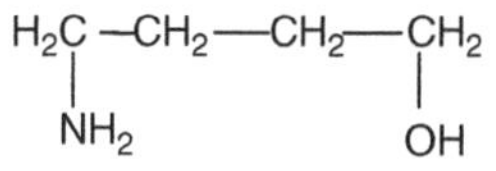 H₂C(NH₂)—CH₂—CH₂—CH₂(OH)

CHAPTER 3

3.7.1. Methane characteristics (simplest alkane):
   a. C is *sp³* hybridized.
   b. C shares two electrons with each H atom.
   c. C obeys octet rule.
   d. C is tetrahedral. Bond angles are 109.5° (from electron diffraction data.)
   e. All four C-H bond lengths are 1.10 Å.

Lewis dot pictures   3 dimensional picture

3.7.2. If two bonds are oriented in paper plane, the other two will be out of plane. Other orientations do *not* change methane's identity.

three different methane molecule orientations

   a. A methane molecule is nonpolar. Small C-H bond dipole moments are equal and cancel due to symmetry.

mp = -183°C
bp = -164°C

van der Waals attractions for one molecule to the next are weak

65

b. Little kinetic energy is needed to break molecules apart. Methane is a gas at 25°C (1 atm).
   c. Compare to NaCl.

$$\overset{\oplus}{Na} \text{---------} \overset{\ominus}{Cl} \quad \begin{array}{l} mp = -801°C \\ bp = -1413°C \end{array}$$

ionic bond
strong attraction

3.7.3. Other methane physical properties:
   a. Colorless.
   b. Low density (0.4 g/mL for liquid, below −164°C).
      1. Water density is 1.00 g/mL at 4°C.
   c. Insoluble in polar solvents (H_2O). Soluble in less polar solvents (ether, gasoline, alcohols).
   d. Extremely weak acid and base.

3.5.1. Methane source: (1) plant material anaerobic decay.
   a. Natural gas – a CH_4, H_3CCH_3, $H_3CCH_2CH_3$ mixture in ratios about 12 : 1 : 2 (by mass), also 3% higher alkanes, sometimes He, sulfur and nitrogen compounds. Composition depends on source.
      1. Once a waste product, burned off. Now a valuable fuel.
      2. Found in swamp gas, mines, and petroleum deposits.
   b. Methane (and other alkanes) purification. Fractional distillation:

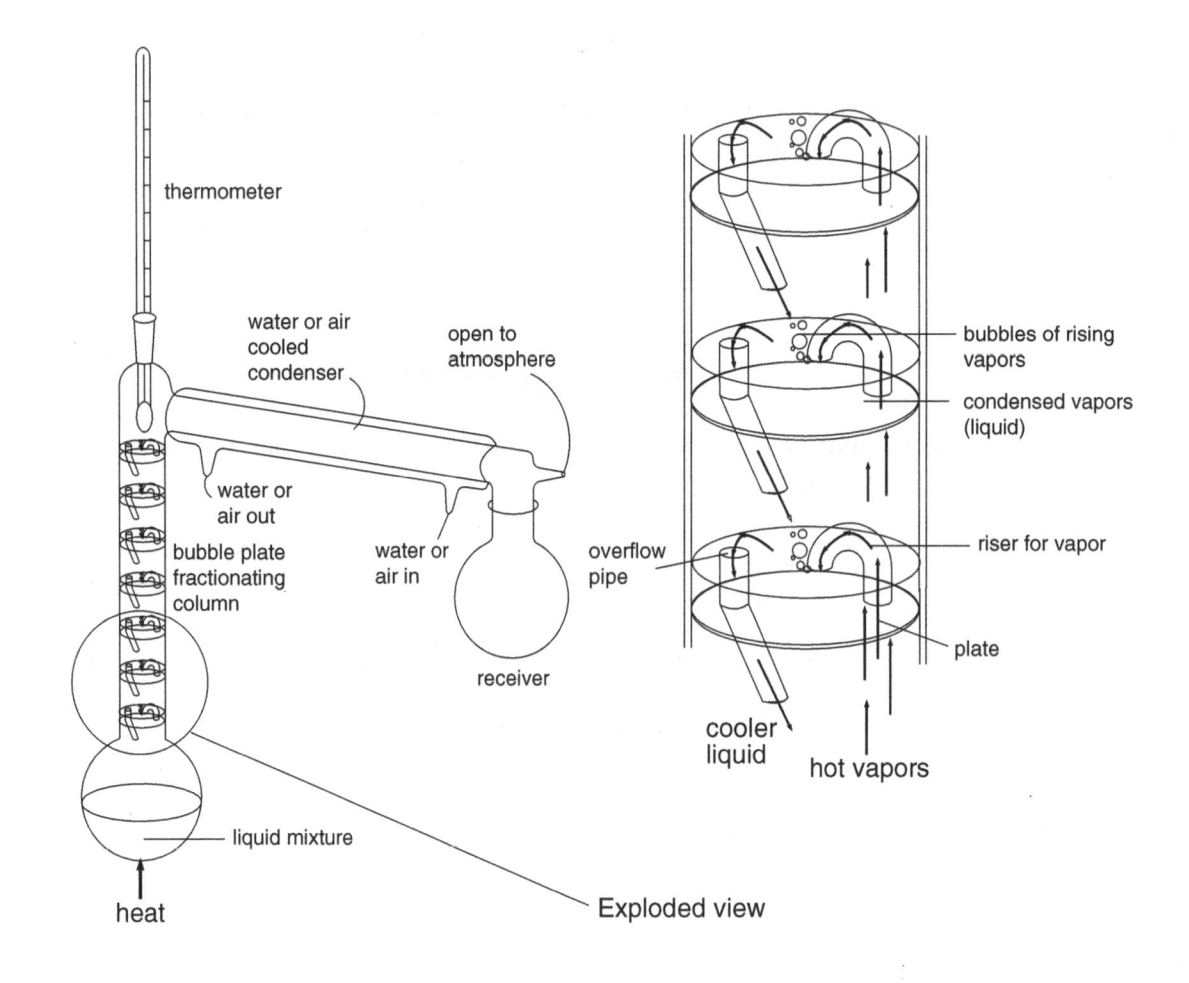

c. If fractionating column is long enough, lowest boiling liquid (methane) rises to top and is collected before any higher boiling compounds. When all methane has been collected out of mixture, higher boiling liquids begin to distill (indicated by rising head temperature) as heat is added. Receiver containing pure CH_4 is removed and a new, empty collection flask is installed to collect the next higher boiling compound.

d. Harold C. Urey, Stanley Miller (U. of Chicago, 1953). Methane: "mother of all organic compounds."

$$CH_4 + H_2O + NH_3 + H_2 \longrightarrow \text{amino acids + other products}$$

3.7.4. Chemical properties:
   a. Methane (and other alkanes) is (are) inert compared to other organic compounds.

3.6.1. Only a few methane reactions are useful.
   a. Combustion (excess O_2).

$$CH_4(g) + 2\,O_2(g) \xrightarrow[\text{burn}]{\text{flame}} CO_2(g) + 2\,H_2O(g) + 891 \text{ kJ/mol}$$

   b. Less O_2, 1500°C:

$$6\,CH_4(g) + O_2(g) \xrightarrow{1500°C} 2\,H-C\equiv C-H(g) + 10\,H_2(g) + 2\,CO(g)$$
$$\text{acetylene (ethyne)} \qquad\qquad \text{toxic!}$$

   c. No O_2, H_2O, Ni catalyst, 850°C.

$$CH_4(g) + H_2O(g) \xrightarrow{850°C} \underbrace{CO(g) + 3\,H_2(g)}_{\text{syngas}}$$

$$CO(g) + 2\,H_2(g) \xrightarrow{\text{catalyst}} H_3COH$$
$$\text{wood alcohol (methanol)}$$
$$\text{solvent, liquid fuel}$$

3.6.2. H atom substitution by halogen (halogenation):

$$CH_4(g) \xrightarrow{X_2} H_3C-X \xrightarrow{X_2} H_2C(X)-X \xrightarrow{X_2} HC(X)_2-X \xrightarrow{X_2} X-C(X)_2-X$$
$$\qquad\qquad +H-X \qquad\quad +H-X \qquad\quad +H-X \qquad\quad +H-X$$

3.7.6. Combustion (burning) – practical uses:
   a. Empirical formula determination.
   b. Heat – once started, reaction self-perpetuates.
      1. Heat of combustion – heat released when one mol of a compound is burned completely with O_2 to CO_2 and H_2O.
      2. Methane heat of combustion (891 kJ/mol) is fairly high.
   c. Products from partial oxidation (uses).
      1. Acetylene – welding, synthesis.
      2. Hydrogen:
         a. Methanol production:
            $CO + 2\,H_2 \rightarrow H_3COH$
         b. Ammonia product (Haber process):
            $N_2 + 3\,H_2 \rightarrow 2\,NH_3$ (fertilizers)

3.7.7. Ethane structure and some conformations were shown in section 2.7.1.

    a. Most stable conformation (staggered) minimizes C-H σ-bond electron repulsions (VSEPR theory).

    b. Rotating about σ-bond axis from staggered towards eclipsed increases σ-electron repulsions, or *torsional strain energy* – potential energy stored in molecule.

    c. Potential Energy diagram for ethane c—c rotation.

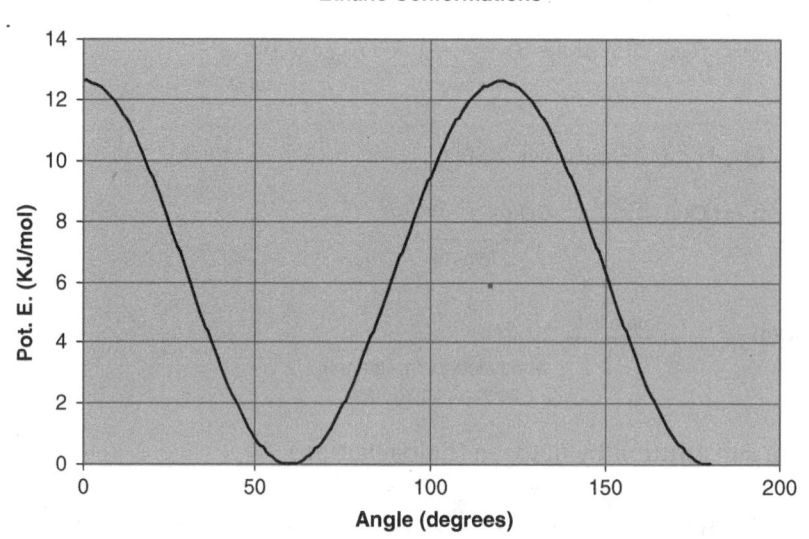

    d. A 12.6 kJ/mol energy barrier is small. Molecules have ~84 kJ/mol thermal (kinetic) energy at 298 K – more than enough to overcome rotation barrier. Millions of rotations / sec occur.

    e. *Dihedral* angle (between H atoms on adjacent C atoms) is 0° in an eclipsed conformation, 60° for staggered. It can be any other angle in *skew* conformations.

3.7.8. Propane structure:

$$H_3C\!-\!C\!-\!C\!-\!H \quad H_3CCH_2CH_3$$

Lewis dot   expanded   condensed

  a. C—C rotation barrier = 13.8 kJ/mol (caused by σ-electron repulsions, not by crowding). Potential energy diagram otherwise similar to ethane's (section 3.7.7c).
    1. Two eclipsed C-H bonds contribute 4.2 kJ/mol torsional energy (from ethane, 12.6 / 3 C-H interactions = 4.2 kJ/mol per C-H interaction). A C-H bond eclipsed with a C-CH$_3$ bond contributes 13.8 – (4.2 × 2) = 5.4 kJ/mol.

3.8.1. *normal*-Butane structure (straight chain):

Lewis dot   expanded   condensed

H$_3$CCH$_2$CH$_2$CH$_3$ or H$_3$C(CH$_2$)$_2$CH$_3$

3.8.2. 2-Methylpropane (isobutane) structure:

Lewis dot   expanded   condensed

H$_3$C — CH — CH$_3$
        |
       CH$_3$

or

HC(CH$_3$)$_3$

3.8.3. Review: structural isomers (section 2.8.2) – compounds with identical formulas but different structures (3 dimensional bonding arrangements), and different physical and chemical properties.

Isomers	Bp °C	Mp °C	Density g/mL (−20°C)	Solubility in 100 mL alcohol
n-butane	0	−138	0.622	1813 mL
isobutane	−12	−159	0.604	1320 mL

3.8.4. Which butane is which?
  a. Problem 1. Monochlorination (Chapter 4) yields two chloroalkane isomers from *n*-butane and isobutane, respectively. Draw these structures.

3.8.5a. Problem 2. How many dichloro isomers would be expected from *n*-butane? Draw them.

b. Isobutane?

3.8.6. *Anti (staggered)* is the most stable *n*-butane conformation. Highest energy eclipsed conformation is 20.9 kJ/mol less stable than *anti*. Two different but equal energy *gauche* (crooked, or left) *(staggered)* conformations exist, 3.8 kJ/mol less stable than *anti*. The other two eclipsed conformations have equal energy, 15.1 kJ/mol less stable than *anti*.

eclipsed forms

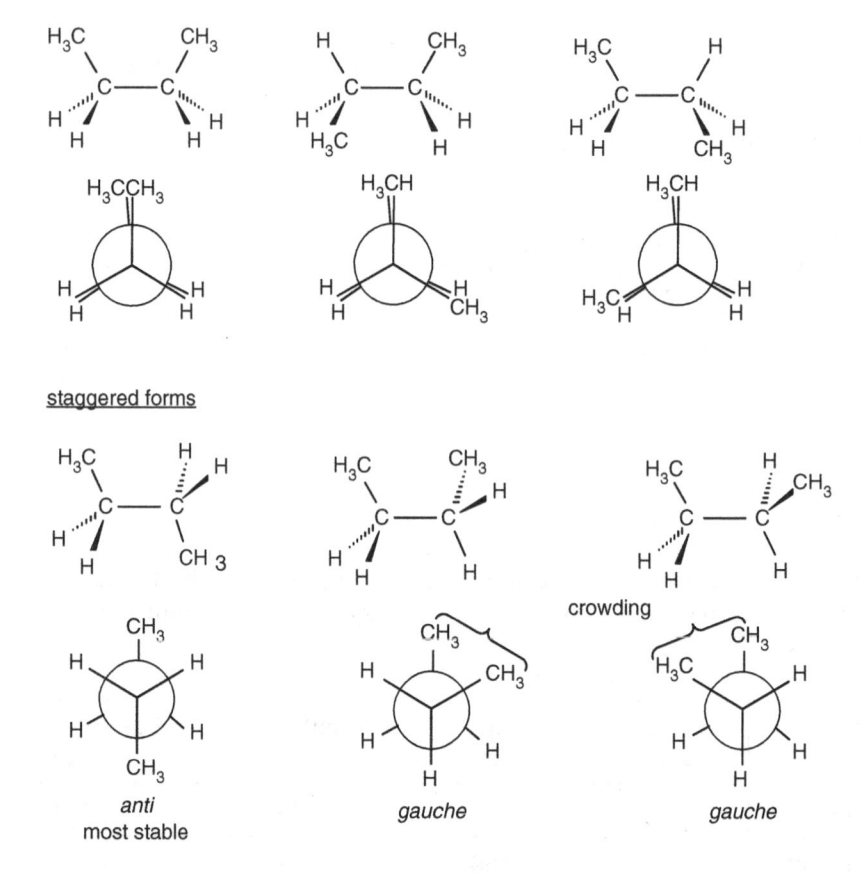

staggered forms

anti
most stable

gauche

gauche

3.8.7. No torsional strain exists in any staggered conformations, but *steric crowding* raises *gauche* form potential energy above *anti* form.

gauche conformation

a. Potential energy diagram.*

Butane Conformations

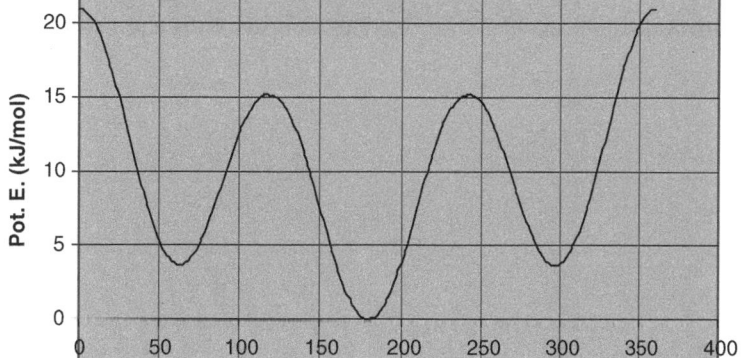

[Graph: Pot. E. (kJ/mol) vs. Angle (degrees), showing butane conformation energy curve with minima near 60°, 180°, 300° and maxima near 0°/360°, 120°, 240°.]

[Newman projections shown at 0°, 60°, 120°, 180°, 240°, 300°, and 360°.]

b. If each 2,3- H-------H eclipsing interaction is 4.2 kJ/mol, then 1,4- H$_3$C-------CH$_3$ eclipsing interaction must be 12.5 kJ/mol. H-------CH$_3$ is 5.4 kJ/mol (section 3.8a1).
c. Problem 3. Draw a potential energy vs. rotation curve for:
   1. (H$_3$C)$_2$CH-CH(CH$_3$)$_2$

---

* This graph was generated using quatro pro. Data points were plotted in increments of 1° (in a column). In the b column, the following formula was used:
    1. 2.95 + 2.05*(@cos(@radians(+a1*3))) from 0 to 60°.
       and
    2. 2.25 + 1.35*(@cos(@radians(+a62*3))) from 61 to 120°.
       and
    3. 1.8 + 1.8*(@cos(@radians(+a122*3))) from 121 to 240°.

    Function 2 repeats from 241 to 300°, and function 1 repeats for 301 to 420°, then function 2 repeats from 421 to 480°.

2. (H₃C)₂CH-CH₂CH₃

d. Compare the various energy barrier heights with each other and with those for butane.

3. (H₃C)₃C-C(CH₃)₃

3.8.8. Torsional strain – an intramolecular (within the same molecule) force tending to cause a rotation about a σ-bond. Four possible causes are:
a. Bonding electron intramolecular repulsions (section 3.7.7b).
b. Steric (van der Waals) interaction – force acting through space between two atoms (or groups of atoms) within a molecule. Two groups held closer than the sum of respective van der Waals radii strongly repel. Otherwise, they weakly attract. See section 3.8.7.
c. Dipole-dipole interaction – attractive or repulsive (Coulombic) force acting through space between two atoms (or groups of atoms) within a molecule due to polar bonds (partial atomic electrical charges).

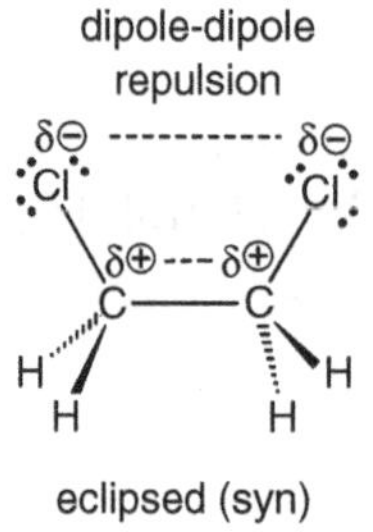

dipole-dipole repulsion

eclipsed (syn)

d. Hydrogen bonding interaction – attractive force (fairly strong) between two H-bonding functional groups within a molecule.

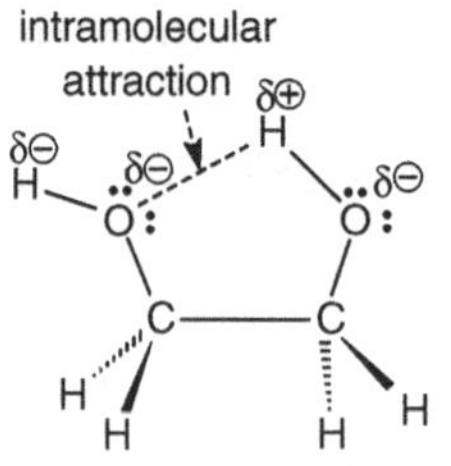

intramolecular attraction

eclipsed

3.9.1. Higher alkanes contain additional -CH$_2$- (methylene) groups. Structures are most stable when all C atoms are *anti* to each other, yielding a zigzag structure for *normal* (straight) chain alkanes. Rotation about σ-bonds is easy at 298 K, interconverting conformations.

normal (straight) chain octane
n-octane
most stable all *anti*-conformations

3.2.1. Alkane formulas: always C$_n$H$_{2n+2}$ where *n* is any integer. Hydrocarbons with all *sp*^{3} C atoms.
3.3.1. Structural isomers:

pentanes	C$_5$H$_{12}$	3 isomers
hexanes	C$_6$H$_{14}$	5 isomers
heptanes	C$_7$H$_{16}$	9 isomers
octanes	C$_8$H$_{18}$	18 isomers
decanes	C$_{10}$H$_{22}$	75 isomers
isocanes	C$_{20}$H$_{42}$	366,319 isomers

a. Problem 4. **1.** Draw all isomers having the formula C$_5$H$_{11}$Cl (eight).

2. C$_7$H$_{16}$ (nine)

b. Alkyl group names (branch or substituent names) were shown in section 2.15.3c-d.
3.3.3. Common names – did not develop logically.
3.3.4. Systematic IUPAC names – should be able to construct structure from the name, and vice versa.

Structure	Common name	IUPAC name
H$_3$C–Cl	methyl chloride	chloromethane
H$_3$C–OH	methyl alcohol	methanol
H$_3$CCH$_2$–Cl	ethyl chloride	chloroethane
H$_3$CCH$_2$–OH	ethyl alcohol	ethanol
(H$_3$C)(H$_3$CCH$_2$)CH–Cl	*sec*-butyl chloride	2-chlorobutane
H$_3$C–C(CH$_3$)(CH$_3$)–Cl	*tert*-butyl chloride	2-chloro-2-methylpropane

3.3.5. IUPAC alkane naming rules.
  a. Find longest carbon chain and assign corresponding parent alkane name (section 2.15.3b).
    1. If two or more chains are the same length, the one with most branches (and smallest branches) is parent (base).

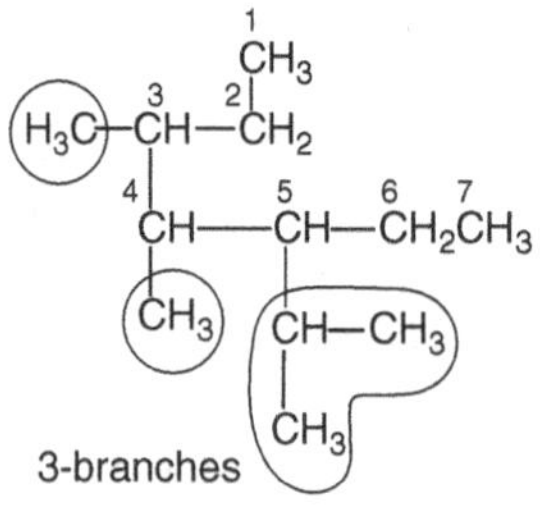

3-branches

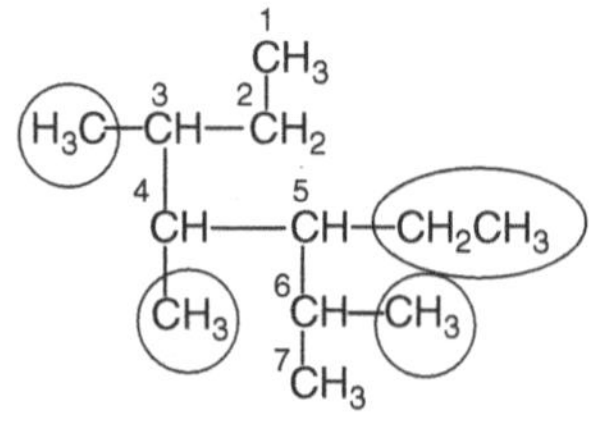

4-branches
(correct parent chain)

  b. Number parent chain so that first branch has smallest number possible.
    1. If first branch has the same number starting from either end, then start from the end that gives the second branch the smaller number, etc.
  c. If the same branch occurs more than once, use di, tri, tetra, etc. as prefixes.
  d. Branch prefixes are listed in alphabetical order.
    1. *Di, tri, tetra,* etc. prefixes do *not* count when alphabetizing.
    2. *Iso* and *neo* prefixes are alphabetized.
    3. *Sec* and *tert* prefixes are not alphabetized.
  e. Examples:

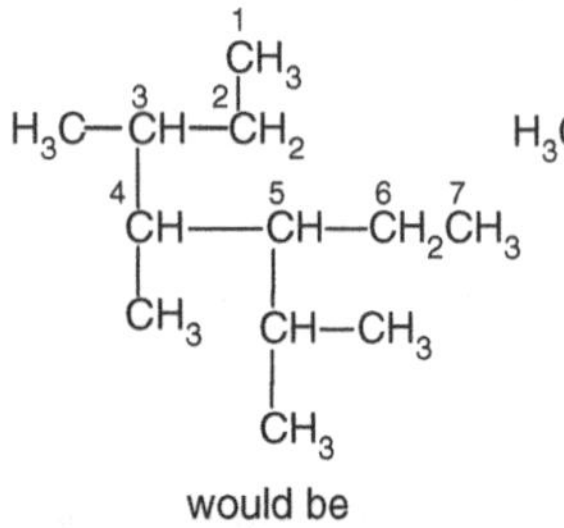

would be
5-isopropyl-3,4-dimethyl-heptane
(incorrect)

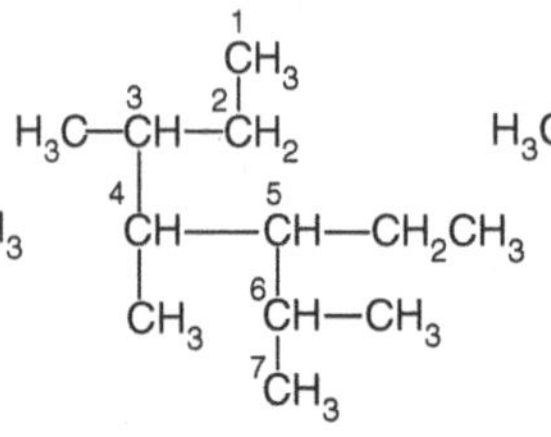

would be
5-ethyl-3,4,6-trimethyl-heptane
(incorrect)

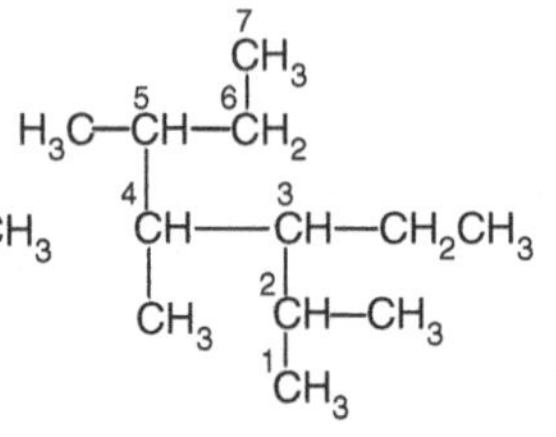

3-ethyl-2,4,5-trimethyl-heptane
(correct)
(smaller numbers)

  f. Example:

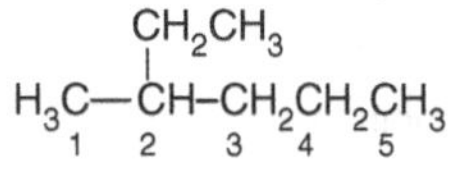

2-ethylpentane?
(incorrect – wrong parent chain)

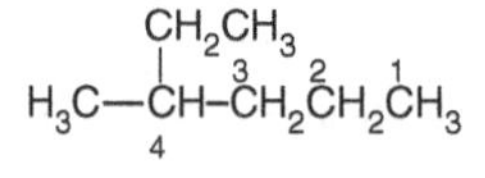

4-methylhexane?
(incorrect – methyl branch number too large)

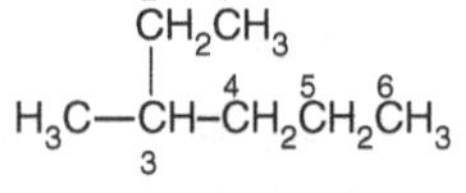

3-methylhexane
(correct)

g. Example:

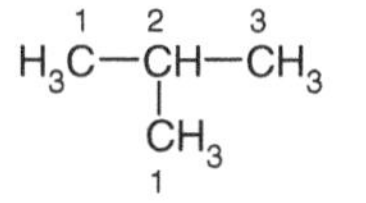

$$H_3C-\underset{\underset{CH_3}{|}}{CH}-CH_3$$
2-methylpropane
isobutane

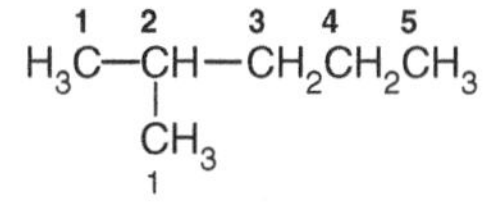

$$H_3C-\underset{\underset{CH_3}{|}}{CH}-CH_2CH_2CH_3$$
2-methylpentane

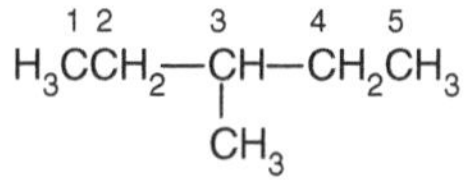

$$H_3CCH_2-\underset{\underset{CH_3}{|}}{CH}-CH_2CH_3$$
3-methylpentane

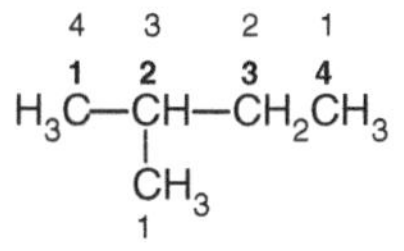

$$H_3C-\underset{\underset{CH_3}{|}}{CH}-CH_2CH_3$$
2-methylbutane
isopentane

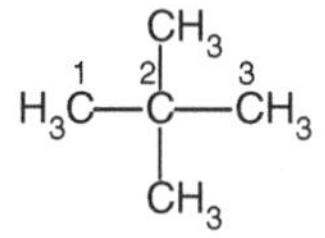

$$H_3C-\underset{\underset{CH_3}{|}}{\overset{\overset{CH_3}{|}}{C}}-CH_3$$
2,2-dimethylpropane
neopentane

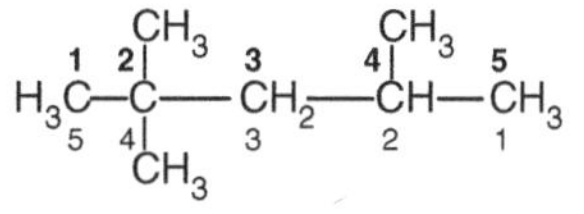

$$H_3C-\underset{\underset{CH_3}{|}}{\overset{\overset{CH_3}{|}}{C}}-CH_2-\underset{}{\overset{\overset{CH_3}{|}}{CH}}-CH_3$$
2,2,4-trimethylpentane
isooctane

h. More examples:

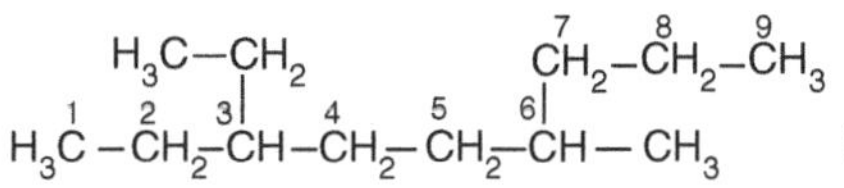

3-ethyl-6-methylnonane

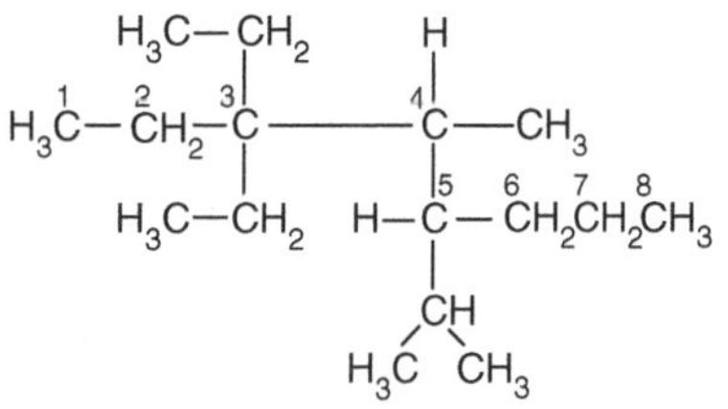

3,3-diethyl-5-isopropyl-4-methyl-octane

i. Problem 5. Draw structures for all isomers with the formula C_6H_{14}. Give them IUPAC names.

j. Problem 6. Name the nine heptane isomers drawn in section 3.3.1.

k. Problem 7. Draw structural formulas for the following names:
   1. 2,2,3,3-tetramethylpentane

   2. 2,3-dimethylbutane

3. 3,4,4,5-tetramethylheptane

4. 4-ethyl-3,4-dimethylheptane

5. 4-ethyl-2,4-dimethylheptane

6. 2,5-dimethylhexane

7. 3-ethyl-2-methylhexane

8. 2,2,4-trimethylpentane

9. 3-chloro-2-methylpentane

10. 1,2-dibromo-2-methylpropane

l. Problem 8. Write IUPAC names for the following compounds. Expand structures as needed.
   1. (H$_3$C)$_2$CHCH$_2$CH$_2$CH$_3$

   2. H$_3$CCBr$_2$CH$_3$

   3. H$_3$CCH$_2$C(CH$_3$)$_2$CH$_2$CH$_3$

   4. (C$_2$H$_5$)$_2$C(CH$_3$)CH$_2$CH$_3$

   5. H$_3$CCH$_2$CH(CH$_3$)CH(CH$_3$)CH(CH$_3$)$_2$

   6. H$_3$CCH$_2$CHCH$_2$CHCH$_2$CH$_3$
               | |
             CH$_3$ CH$_2$CH$_2$CH$_3$

   7. (H$_3$C)$_3$CCH$_2$C(CH$_3$)$_3$

8. (H₃C)₂CClCH(CH₃)₂

9. (H₃C)₂CHCH₂CH₂CH(C₂H₅)₂

10. (H₃C)₃CCH₂C(C₂H₅)₂CH₃

11. (H₃C)₂CHC(C₂H₅)₂CH₂CH₂CH₃

m. Problem 9. Which alkanes above (problem 8), if any, contain:
1. methyl group(s).
2. ethyl group(s).
3. *n*-propyl group(s).
4. isopropyl group(s).
5. *n*-butyl group(s).
6. isobutyl group(s).
7. *sec*-butyl group(s).
8. *tert*-butyl group(s).

3.3.6. Naming branches more complex than the butyl groups. When the branch has a branch:
  a. Find the longest chain bonded to the "head" carbon (the one bonded to the parent chain). The head carbon is numbered 1 in the branch.
  b. Name the parent branch as an alkyl group, methyl, ethyl, propyl, butyl, etc.
  c. Name branches on the branch also as alkyl groups.
  d. Enclose the complex branch in parentheses.
  e. Examples:

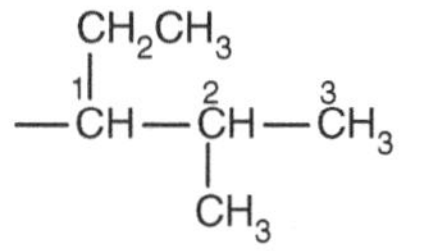

CHAPTER 3   79

(1-ethyl-2-methylpropyl) group

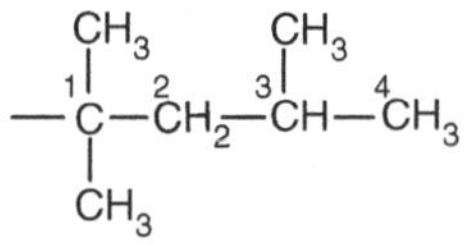

(1,1,3-trimethylbutyl) group

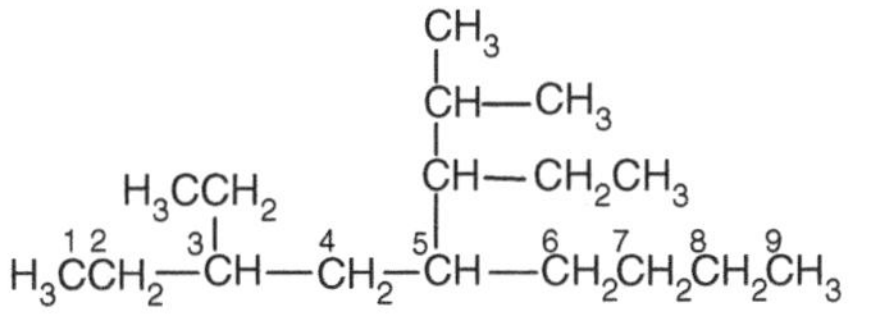

3-ethyl-5-(1-ethyl-2-methylpropyl)nonane

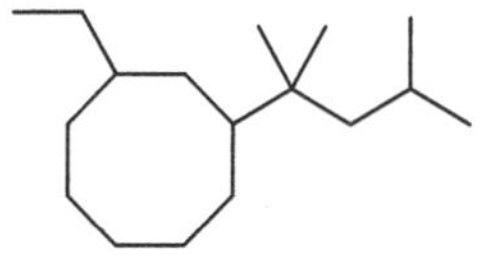

1-ethyl-3-(1,1,3-trimethylbutyl)-cyclooctane

3.3.7. Alkyl halide names (IUPAC) – use alkane parent name with "halo" branch. No halogen parent names.

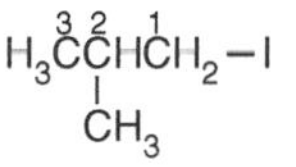

1-iodo-2-methylpropane

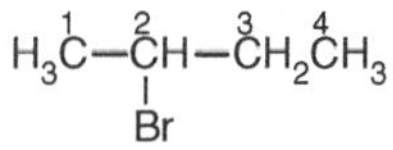

2-bromobutane

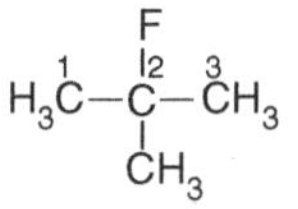

2-fluoro-2-methylpropane

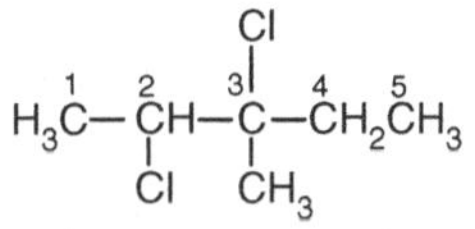

2,3-dichloro-3-methylpentane

a. Problem 10. Write IUPAC names for the eight $C_5H_{11}Cl$ isomers drawn in section 3.3.1.

3.4.1. Alkane properties. "Higher" alkanes have properties similar to methane.

Name	Structure	Bp (°C)	Mp (°C)	Density (20°C)
methane	CH_4	−164	−183	0.55 (at bp)
ethane	H_3CCH_3	−89	−183	0.51 (at bp)
propane	$H_3CCH_2CH_3$	−42	−189	0.50
butane	$H_3C(CH_2)_2CH_3$	0	−138	0.58
pentane	$H_3C(CH_2)_3CH_3$	36	−130	0.63
hexane	$H_3C(CH_2)_4CH_3$	69	−95	0.66
heptane	$H_3C(CH_2)_5CH_3$	98	−91	0.68
octane	$H_3C(CH_2)_6CH_3$	126	−57	0.70
nonane	$H_3C(CH_2)_7CH_3$	151	−51	0.72
decane	$H_3C(CH_2)_8CH_3$	174	−30	0.73

3.4.2. Do not memorize numbers, but be familiar with trends.
   a. Relatively low boiling points–only weak van der Waals attractions between molecules.
      1. C-C and C-H bonds are covalent (largely nonpolar).
   b. Bp's increase about 20 – 30°C for each additional CH_2 group.
      1. Larger surface area implies stronger intermolecular attractions, which implies a higher boiling point.
      2. More branching lowers the bp (section 2.2.3c3) for isomers.

Common name	IUPAC name	Bp °C
isobutane	2-methylpropane	−12
	n-butane	0
neopentane	2,2-dimethylpropane	9.5
isopentane	2-methylbutane	28
	n-pentane	36

   c. n-Alkanes melting points are low and increase irregularly with molecular mass. Alkanes with an even number of C atoms have slightly higher mp's than those with odd numbers of C atoms.
   d. Branched alkanes have higher mp's than normal chain alkane because molecules can pack more tightly.

3.4.3. Low densities that increase with molecular mass. They are usually colorless liquids at 25°C. Smell – similar to gasoline.
   a. Pentane (C_5H_{12}): a component of gasoline, colorless, low bp (36°C), low density (0.63 g/mL (20°C)), insoluble in H_2O (forms a distinct layer that floats on top), flammable, and reacts with X_2/heat.
   b. Physical state:

Number of C atoms	Physical state at 25°C (1 atm)
0–4	gases
5–17	liquids
≥18	waxy solids

   c. Almost completely insoluble (hydrophobic, or water hating) in H_2O (polar solvent). Soluble in relatively nonpolar solvents such as benzene (C_6H_6), ether ($C_4H_{10}O$), or chloroform ($CHCl_3$).
      1. Alkanes are often used as solvents. They will dissolve grease and oil (nonpolar).
   d. Problem 11. Without referring to tables, rank the following hydrocarbons in order of decreasing boiling point.
      1. 3,3-dimethylpentane
      2. n-heptane
      3. 2-methylbutane
      4. n-pentane
      5. 2-methylhexane

3.1.1. Alkanes are "saturated" hydrocarbons – have a maximum number of bonded hydrogen atoms, and cannot absorb any more.

   a. Compounds with multiple bonds are said to be unsaturated.

3.5.2. Petroleum – mainly an alkane mixture, 5 to 40 C atoms, and some cycloalkanes.

   a. Crude oil refinement – straight run (fractional) distillation gives partial separation based on bp differences.

Boiling range (°C)	Number of C atoms	Name	Use
under 20	1–4	gas fraction	heating fuel
20–60	5–6	petroleum ether	solvent
60–100	6–7	ligroin (light naphtha)	solvent
40–205	5–10	natural gasoline	automobile fuel
160–230	8–16	kerosene	heating oil, jet fuel (further refined)
200–320	10–18	diesel	motor fuel
>275	>12	gas oil	diesel fuel, heating fuel
305–405	18–25	heavy gas oil	heating fuel
nonvolatile liquids	>25	mineral oils	lubricating oils, paraffin wax, petroleum jelly
nonvolatile "solids"	high	asphalt, petroleum coke	roofing tar, paving roads

   b. Pure C_1 - C_5 alkanes are available from fractional distillation. Methane and ethane are usually handled as compressed gases. Liquefied petroleum gas (LPG) is mainly propane and butane, used as fuel.

3.5.3. Cyclics sources – some (California) petroleum deposits are rich in cycloalkanes (separated by fractional distillation).

   a. Catalytic reforming – dehydrogenation converts cycloalkanes into aromatics.

1. Hydrogenation (reverse reaction) is also useful (different catalyst and temperature).

benzene + 3 H₂ →(Ni, 150 - 200°C, 25 atm)→ cyclohexane

phenol →(Ni, 150 - 200°C, 150 atm)→ cyclohexanol

3.5.4. Alkanes can be made from coal (Fischer-Tropsch "gasoline" synthesis):

C (coal) + H₂O (steam) →(heat)→ CO + H₂ ("syngas") →(H₂, Fe catalyst, heat/pressure)→ alkanes + H₂O

3.5.5. Two broad categories of organic syntheses: (1) industrial (large scale) and (2) laboratory (small scale). We will normally worry only about laboratory methods for this class.
  a. Factors involved in industrial organic chemical production – costs, volume, time, availability, separations, repetition, environmental pollution.
  b. Factors involved in laboratory (small scale) organic chemical production – labor costs, product yields, purity, and general applicability.

3.5.6. Laboratory synthetic methods for making alkanes:
  a. Alkene hydrogenation (section 8.10).

$$R_1R_2C=CR_3R_4 \xrightarrow{H_2 \text{ (1 atm)}, \text{ Pt (catalyst)}} R_1R_2CH-CHR_3R_4$$

an alkene (unsaturated) → an alkane (saturated)

  1. Excellent yields.
  2. Simple alkenes are available from crude oil catalytic reforming. Higher alkenes are not always readily available.

  b. Alkyl halide reduction. Examples:

H₃CCH₂CHCH₃ (Br) —Mg/dry ether→ H₃CCH₂CHCH₃ (MgBr) —H₂O→ H₃CCH₂CHCH₃ (H)
2-bromobutane → n-butylmagnesium bromide (a Grignard reagent) → n-butane

or

H₃CCH₂CHCH₃ (Br) —Zn / HCl (dil) / H₂O→ H₃CCH₂CHCH₃ (H)
2-bromobutane → n-butane

  1. Alkyl halides are usually made from alcohols (section 11.7). They are not always available.

3.3.8 and 3.3.9. Carbon classes.
  a. Primary (1°) carbon atom – bonded to one other C atom and three H atoms.
  b. Secondary (2°) carbon atom – bonded to two other C atoms and two H atoms.
  c. Tertiary (3°) carbon atom – bonded to three other C atoms and one H atom.

  d. Problem 12. For the structures in problems 7 and 8, identify and label the:
     1. Primary (1°) C atoms.

     2. Secondary (2°) C atoms.

     3. Tertiary (3°) C atoms.

     4. Quaternary (4°) C atoms (with four other C atoms bonded and no H atoms).

3.5.7. Organometallic chemistry (coupling reactions) – a huge field.
  a. See sections 10.8 and 10.9.
  b. Victor Grignard (1905) first made organometallics (organomagnesiums).

  c. Other useful metals: Li, Na, K, Zn, Cu, Hg, Pb, Al, and Tl. – lower electronegativity than C.

Element	Pauling Electronegativity
K	0.8
Na	0.9
Li	1.0
Mg	1.3
Cu	1.5

(*Continued*)

Element	Pauling Electronegativity
Zn	1.6
Al	1.6
Hg	1.7
Tl	1.8
Pb	1.9
C	2.5

3.5.8. A Grignard reagent is a very strong base (δ- charge on C) – reaction with H$_2$O replaces alkyl halide X with H. Example:

$$H_3C-Mg-Br + H-O-H \rightleftharpoons H_3C-H + {}^-O-H + Mg^{2+} + Br^-$$

stronger base + stronger acid ⇌ weaker acid + weaker base

a. Not a very useful reaction except to replace H with D (heavy hydrogen) – reaction with water is sometimes a nuisance that occurs while trying make alcohols (section 10.10.1).

b. Any alcohol R-OH or amine NRR'H reacts similarly to H$_2$O above.

$$H_3C-Mg-Br + H-O-R \rightleftharpoons H_3C-H + {}^-O-R + Mg^{2+} + Br^-$$

stronger base + stronger acid ⇌ weaker acid + weaker base

$$H_3C-Mg-Br + H-NRR' \rightleftharpoons H_3C-H + {}^-NRR' + Mg^{2+} + Br^-$$

stronger base + stronger acid ⇌ weaker acid + weaker base

c. Problem 13. What would the product alkane be when *n*-propylmagnesium chloride reacts with water? Isopropylmagnesium chloride? *tert*-Butyl magnesium chloride? What would the products be with heavy water (D$_2$O)?

d. Problem 14. Which alkyl halide reactant isomers would yield the alkane from the Grignard reagent and water?
   1. *n*-pentane?

   2. 2-methylbutane?

   3. 2,3-dimethylbutane?

   4. 2,2-dimethylpropane?

e. Problem 15. A solution containing an unknown amount of methanol (CH$_3$OH) dissolved in *n*-octane is added to H$_3$CMgI (excess) in *n*-butyl ether (high boiling point solvent, H$_3$C(CH$_2$)$_3$O(CH$_2$)$_3$CH$_3$). A gas is evolved and collected. Its mass and volume are 0.743 g and 1.04 mL, respectively, at STP. What is the gas? Write a reaction to show how it was formed. What was the unknown methanol mass in the original sample?

f. Grignard syntheses are very useful, for example to make alcohols from aldehydes and ketones (section 10.9).

g. Examples:

[Mechanism scheme showing:]

H₃CCH₂CH₂CH₂Br →(Mg, dry ether (solvent))→ H₃CCH₂CH₂CH₂—MgBr (δ⁻/δ⁺) →(acetone, dry ether (solvent))→ intermediate magnesium alkoxide →(H₂O (solvent), NH₄⁺Cl⁻)→

Cl⁻ + Br⁻ + Mg²⁺ + H₃C—C(OH)(CH₃)—CH₂CH₂CH₂CH₃

2-methylhexan-2-ol (3°)

3.5.9. If Grignard reagents stand in dry ether solution for a prolonged period (without adding an aldehyde or ketone), they slowly couple.

2 H₃C—Mg—Br ⟶ H₃C—CH₃ (new C–C bond) + other products

a. An unwanted side reaction when making alcohols, but can be done purposely to make alkanes.

3.5.10. Wurtz reaction – old coupling reaction using sodium metal.

2 H₃CCH₂CH₂CH₂Br —(2 Na / heat)→ H₃CCH₂CH₂CH₂CH₂CH₂CH₂CH₃
n-octane 48%
+ 2 NaBr

a. Simplified (partial) mechanism (reaction steps). Some aspects are not fully understood.

H₃CCH₂CH₂CH₂—Br + Na ⟶ H₃CCH₂CH₂CH₂• + Na⁺ + :Br:⁻

H₃CCH₂CH₂CH₂—Br ↓

H₃CCH₂CH₂CH₂CH₂CH₂CH₂CH₃ + •Br:

:Br• + Na ⟶ Na⁺ + :Br:⁻

b. Limitations:
   1. Organosodiums are so reactive that they couple with R-X before they can be isolated.
   2. The two R-groups must be identical.
   3. Yields are often poor, and many side reactions occur.

3.5.11. E. J. Corey, H. O. House (1960) – A general alkane synthesis (see Wade, 9th ed., problem 10–21). Examples:

2 H₃C–Br —(4 Li)→ 2 H₃C—Li (δ⁻/δ⁺) + 2 LiBr

↓ CuI
+ LiI

H₃C(CH₂)₆CH₂—I + [H₃C—Cu—CH₃]⁻ Li⁺ ⟶ H₃C(CH₂)₆CH₂—CH₃
n-nonane
+ H₃C—Cu + LiI

CHAPTER 3   87

$$2\ H_3CCH_2CHCH_3 \xrightarrow{4\ Li} 2\ H_3CCH_2\overset{\delta\ominus}{C}HCH_3 + 2\ LiCl$$
$$\quad\quad |\quad\quad\quad\quad\quad\quad\quad\quad\quad |$$
$$\quad\ Cl\quad\quad\quad\quad\quad\quad\quad\quad Li\delta\oplus$$

↓ CuI

$$\left[\begin{array}{c} H_3C\quad\quad\quad CH_3 \\ \overset{\delta\ominus}{H}C-Cu-\overset{\delta\ominus}{C}H \\ |\quad\quad\quad\quad\quad | \\ CH_2\quad\quad\quad CH_2 \\ |\quad\quad\quad\quad\quad | \\ CH_3\quad\quad\quad CH_3 \end{array}\right]^{\ominus} Li^{\oplus} + LiI$$

$$H_3C(CH_2)_3\overset{\delta\oplus}{C}H_2 - \ddot{\underset{..}{Br}}: \longrightarrow H_3C(CH_2)_3CH_2-\underset{|}{\overset{CH_3}{\underset{|}{CH}}}$$
$$\quad\quad\quad\quad\quad\quad\quad\quad\quad\quad\quad\quad\quad\quad CH_2$$
$$\quad\quad\quad\quad\quad\quad\quad 3\text{-methyl-}\quad\quad\quad |$$
$$\quad\quad\quad\quad\quad\quad\quad\quad\text{octane}\quad\quad\quad CH_3$$

$$\quad\quad\quad\quad\quad\quad\quad\quad\quad\quad H_3C$$
$$\quad\quad\quad\quad\quad\quad + \quad HC-Cu + LiBr$$
$$\quad\quad\quad\quad\quad\quad\quad\quad\quad\quad |$$
$$\quad\quad\quad\quad\quad\quad\quad\quad\quad\quad CH_2$$
$$\quad\quad\quad\quad\quad\quad\quad\quad\quad\quad |$$
$$\quad\quad\quad\quad\quad\quad\quad\quad\quad\quad CH_3$$

a. Features/limitations:
   1. Reaction steps are incompletely understood.
   2. Key step – a carbanion-like alkyl group substitution for X. A carbanion has a negative electrical charge on C, and is strongly basic (nucleophilic).
   3. Yields are good only if both alkyl halides are methyl or 1°. Yields are lower for 2°, and 3° R-X's do not work.

b. Problem 16. Outline two conceivable 2-methylpentane Corey-House syntheses from halopropane isomers.

c. Problem 17. Write reactions showing how you would make *n*-butane from:
   1. 1-bromobutane

   2. 2-bromobutane

   3. chloroethane

   4. but-1-ene

   5. but-2-ene (*cis* or *trans*)

d. An unknown compound whose formula (from mass spectrometry) is $C_{10}H_{22}$ is believed to be 2,7-dimethyloctane. How could you use a Corey-House reaction to confirm this guess?

3.6.3. Chemical properties. Alkanes (paraffins - Latin *parum* "too little" and *affinis* "affinity or reactivity") do not react with acids, bases, or most other reagents.

    a. Alkanes sometimes react under energetic or high temperature conditions. Reactions a-c and e (section 3.6.8) involve free radicals (section 1.9.3.1a).

        1. Alkane reactions are not usually synthetically useful – hard to control and often give multiple product mixtures.

3.6.4. Burning, or free radical (section 1.11.1b and Chapter 4) combustion with O_2 is not fully understood. Simplified steps:

$$H_3C-CH_2-CH_3 + 5\,O_2 \xrightarrow{heat} 3\,CO_2 + 4\,H_2O + heat$$

$$C_nH_{(2n+2)} + \text{excess } O_2 \xrightarrow{heat} n\,CO_2 + (n+1)\,H_2O + heat$$
general alkane

3.6.5. Alkanes do not oxidize at 25°C with common oxidizing agents, $KMnO_4$ or $K_2Cr_2O_7$.

3.6.6. Enzymatic oxidation (with O_2). Example:

3.6.7. Gasoline terminology –
   a. Knocking – rapid (explosive) combustion, worse in higher compression engines.
   b. Octane number (rating) – *n*-heptane and isooctane (2,2,4-trimethylpentane) are used as standards.

Compound	Octane rating
n-heptane	0
unleaded regular	85
unleaded premium	92
isooctane	100
ethanol	106

   c. Some octane boosting additives:

   $Pb(CH_2CH_3)_4$
   leaded gas

   $H_3COC(CH_3)_3$
   methyl *tert*-butyl ether (MTBE)
   (in unleaded, was made by Huntsman)
   an oxygenated additive

   H_3CCH_2OH
   ethanol
   an oxygenated additive

   d. Some pollutants from gasoline combustion:

   hydrocarbons + O_2 ⟶ O_3 + CO + N_xO_y
   ozone (corrosive) + $C_xH_yO_z$
   partially burned hydrocarbons

   e. Catalytic converters:

   stage 1   hydrocarbons + O_2 $\xrightarrow{catalyst}$ CO_2 + H_2O + more N_xO_y

   stage 2 N_xO_y $\xrightarrow{Pt\ (catalyst)}$ N_2 + O_2

   1. Some greenhouse gases: CO_2, CH_4, and CF_2Cl_2 (a freon or chlorofluorocarbon (CFC)). Allow visible light to pass in, but trap infrared light (heat) on the way out.
   2. Sources:

CO_2	burning
CH_4	natural gas, volcanos, cows
CFC's	synthetic (man made)

3.6.8. Petroleum refining after distillation.
   a. Cracking (pyrolysis) – heat alkanes (400–600°C) without O_2. Products – smaller alkanes, alkenes (ethene), and H_2. Example (section 7.11):

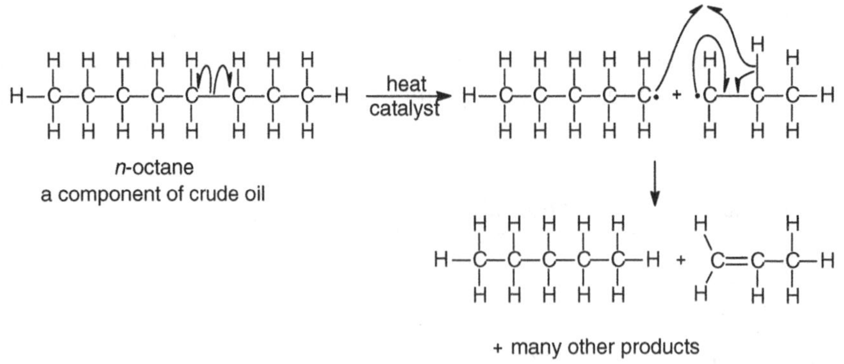

[Diagram: n-octane (a component of crude oil) undergoing cracking with heat/catalyst to give radical fragments, then forming a shorter alkane and an alkene + many other products]

b. Steam cracking – heat alkanes (700–900°C), add steam, and cool rapidly. Products – ethene, propene, butadiene, 2-methylbuta–1,3-diene (isoprene), and cyclopentadiene (valuable synthetic precursors).

c. Hydrocracking – heat alkanes (250–450°C). Products – small sulfur and nitrogen free hydrocarbons. Example:

$$H_3C-(CH_2)_{12}-CH_3 \xrightarrow[SiO_2 \text{ or } Al_2O_3 \text{(catalyst)}]{H_2, \text{ heat}} 2\ H_3C-(CH_2)_5-CH_3$$

$$H_3C-(CH_2)_{10}-CH_3 \xrightarrow[SiO_2 \text{ or } Al_2O_3 \text{(catalyst)}]{H_2, \text{ heat}} H_3C-(CH_2)_5-CH_3 + H_3C-(CH_2)_3-CH_3$$

d. Catalytic cracking – high boiling petroleum fractions pass over powdered silica-alumina clay (450–550°C), to increase gasoline yield. Products – smaller, more branched, lower boiling alkanes and alkenes. Example:

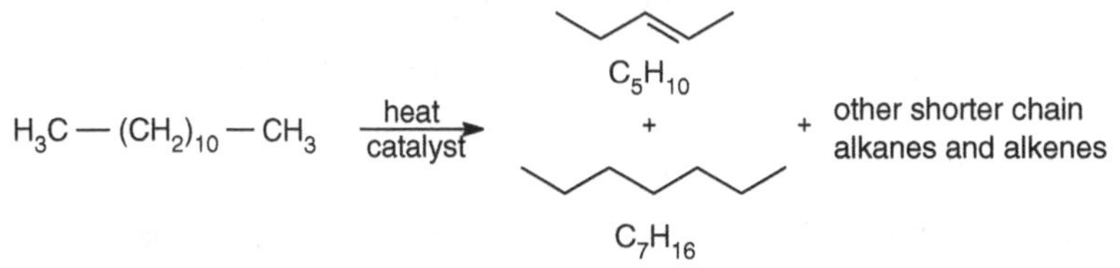

$$H_3C-(CH_2)_{10}-CH_3 \xrightarrow[\text{catalyst}]{\text{heat}} C_5H_{10} + C_7H_{16} + \text{other shorter chain alkanes and alkenes}$$

e. Alkylation:

small alkanes and alkenes $\xrightarrow{\text{catalysts}}$ larger branched alkanes

f. Catalytic reforming:

alkane $\xrightarrow{\text{catalysts}}$ aromatics, high octane fuels, synthetic precursors

3.6.9. Halogenation (high temp. or UV light). See examples in sections 3.6.2, 3.8.4 and Chapter 4.

3.6.10. Carbocation (+ ion) formation – reaction with super acids: $HF-SbF_5$, FSO_3H-SbF_5. Example:

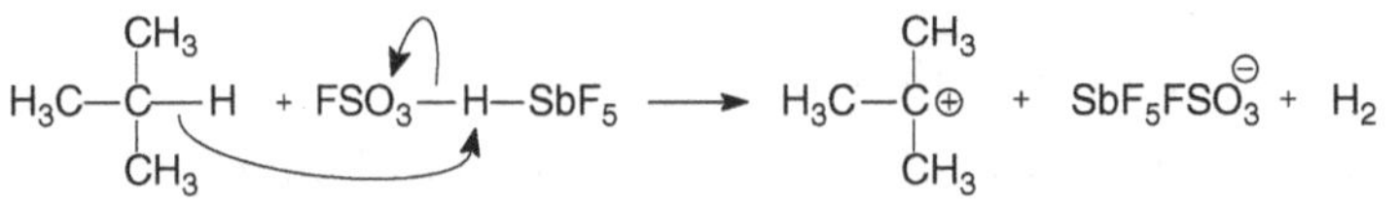

[Diagram: $(CH_3)_3C-H + FSO_3-H-SbF_5 \longrightarrow (CH_3)_3C^\oplus + SbF_5FSO_3^\ominus + H_2$]

3.10.1. Open (normal, or straight) chain compounds – contain no rings.

3.10.2. Cyclic compounds were introduced in sections 2.15.4 and 3.5.3. Alicylic (from aliphatic, meaning "fat like," and cyclic) compound – a cyclic compound containing all sp^3 hybridized atoms.

  a. Homocyclic compound – a ring structure that contains only C and H. Examples:

  cyclopentane   cyclohexene

  b. Heterocyclic compound – a ring structure that contains other atoms (O, N, or S) besides C and H. Examples:

  furan   tetrahydrofuran (a solvent)   pyridine   pyrrole   thiophene

3.10.3. Single ring homocyclic hydrocarbons have formula C_nH_{2n}, and properties similar to alkanes. Ring imparts some unusual properties.

  a. Cyclopropane was used as a general anesthetic, but is flammable.

3.10.4. Naming – count ring C atoms. Name as parent alkane but insert *cyclo* prefix.

  a. Remember functional group nomenclature priorities (section 2.17.5).
  b. Keep branch numbers as small as possible. Examples:

  1-chlorocyclopropane   1,1-dimethylcyclopentane   1-ethyl-3-methyl-cyclohexane

  3-ethylcyclopentene   cycloheptane   1,1,3-trimethylcyclopentane

  1,1-diethyl-4-isopropylcyclohexane

c. If an attached chain has more C atoms than a ring, or if the attached chain has a functional group, name ring as a branch.
   1. Cycloalkyl groups (branch or substituent names). Examples:

   cyclopropyl   cyclobutyl   cyclopentyl   cyclohexyl

   2-cyclopentylpropan-1-ol

   3-cyclohexyl-2,3-dimethylpentane?
   or
   (1-ethyl-1,2-dimethylpropyl)cyclohexane?

   cyclopentylcyclohexane

   4-cyclopropyl-3-methyloctane

   5-cyclobutylpent-1-yne

3.10.5. Line structures were introduced in section 1.16.1c. H atoms are not all shown. It is assumed that you know where they are. More examples:

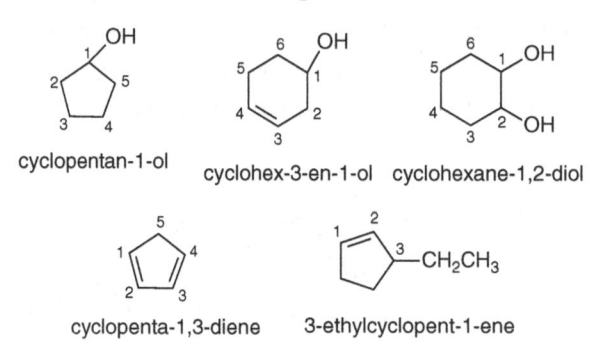

cyclopentan-1-ol   cyclohex-3-en-1-ol   cyclohexane-1,2-diol

cyclopenta-1,3-diene   3-ethylcyclopent-1-ene

3.10.6. Rings can be made from open chain compounds.

a. One method is to cyclize (intramolecularly – within the same molecule) a dihalide by making an organometallic.

b. Other methods (such as cycloadditions) will be discussed later.

3.12.1. Adolf von Baeyer (Nobel Prize 1905) – ring strain theory. Angle + torsional strain = ring strain.

  a. Angle strain – if sp^3 hybridized C atoms are forced (by a ring structure) to deviate from ideal (109.5°) bond angles, σ-electron repulsions destabilize structure (potential energy increases).
   1. Cyclopropane – C-C-C bond angles are 60° (49.5° from ideal). Cyclobutane – C-C-C bond angles are 90° (19.5° from ideal).

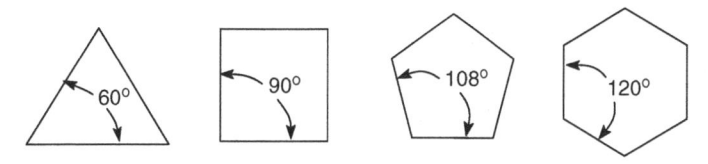

   2. Bond angles would be 108 and 120°, respectively, in planar cyclopentane and cyclohexane.
   3. Baeyer thought that cyclopentane would be very stable since its angles are near 109.5°, but (wrongly) that cyclohexane would be less stable since its angles are farther from 109.5°. Larger rings are more difficult to synthesize.
  b. Torsional strain – due to eclipsing interactions (section 3.7.7).

3.12.2. Ring strain energy can be measured from heat of combustion data – heat given off into water surrounding a *bomb calorimeter* (sealed container) when one mol of a compound is burned (with excess O_2) to CO_2 and H_2O.

$$(H_2C)_n \begin{pmatrix} CH_2 \\ | \\ CH_2 \end{pmatrix} + \frac{3}{2}nO_2 \longrightarrow nCO_2 + nH_2O + n \text{ (energy per CH}_2\text{)}$$

cycloalkane, heat of combustion

3.12.3. More -CH_2- units ⇒ more CO_2, H_2O, and heat evolved from a given molecule.
  a. Total heat evolved must be divided by the number of -CH_2- units per molecule to make a fair comparison.

3.12.4. Ring hydrocarbon heats of combustion.

Ring size	Molar heat of combustion (kJ/mol)	Heat of combustion per CH₂ unit (kJ/mol)	Ring strain per CH₂ group (kJ/mol)	Total ring strain (kJ/mol)
3	2091	697.1	38.5	115
4	2744	686.1	27.5	110.
5	3320	664.0	5.4	27
6	3951	658.6	0.0	0.0
7	4637	662.4	3.8	27
8	5309	663.6	5.1	41
10	6636	663.6	5.0	50.
12	7913	659.4	0.8	9.8
15	9885	659.0	0.4	5.7
Open chain		658.6	0.0	0.0

3.12.5. Products are always the same (CO_2 and H_2O). Any evolved heat differences (per -CH_2- unit) reflect reactant potential energy differences (see reaction diagram below).

    a. Cyclopropane contains ca. (697.1 – 658.6) = 38.5 kJ/mol per CH_2 unit, or 38.5 × 3 = 115 kJ/mol of "ring strain" energy.

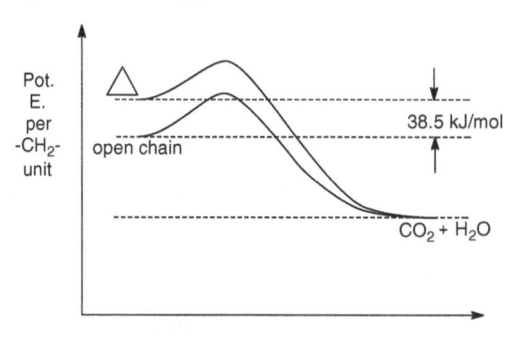

    b. Cyclobutane ring strain energy is ca. (686.1 – 658.6) = 27.5 kJ/mol per CH_2 unit, or 27.5 × 4 = 110 kJ/mol.

    c. Cyclopentane ring strain: (664.0 – 658.6) = 5.4 kJ/mol per CH_2; 5.4 × 5 = 27 kJ/mol.

    d. Cyclohexane ring strain: (658.6 – 658.6) = 0 kJ/mol.

    e. Ring sizes 7 – 11: ring strain goes back up a little.

    f. Ring sizes > 12: very little ring strain.

3.12.6. Baeyer thought that all rings are planar (which is <u>not</u> true).

    a. All rings larger than 3 C atoms are puckered to some extent. For rings with more than 5 C atoms puckering allows bond angles to more nearly approach 109.5°.

3.12.7. Why are large rings hard to synthesize from open chain reactants?

    a. Larger ring ⇒ harder for two ends to find and react with each other – large negative entropy factor.

        1. Trick (for making large rings) – use a dilute reactant solution to improve intramolecular vs. intermolecular reaction probability.

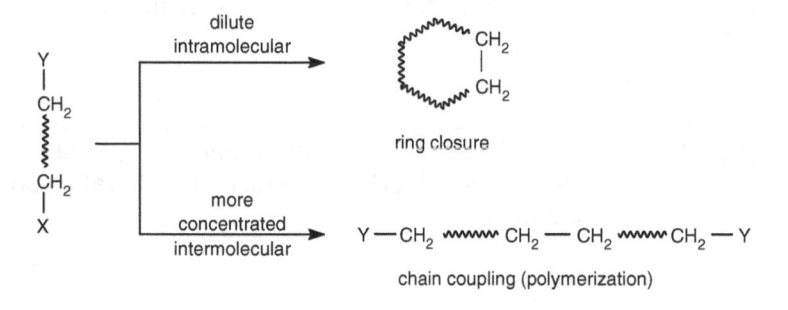

3.12.8. C. A. Coulson and W. A. Moffitt – "banana" cyclopropane bonds. Less sp^3-sp^3 overlap occurs when C-C-C bond angles are not 109.5°, and C-C bonds are weaker. (Supported by X-ray studies.)

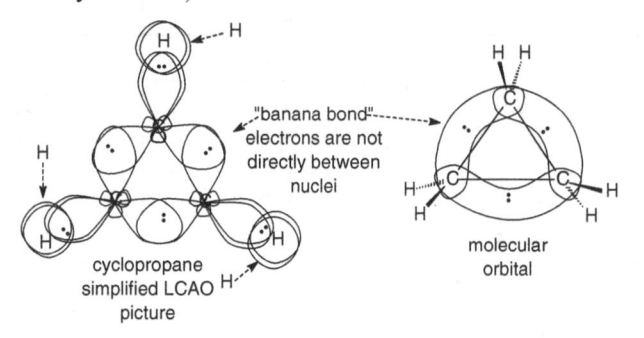

3.12.9. H-C bonds are all eclipsed:

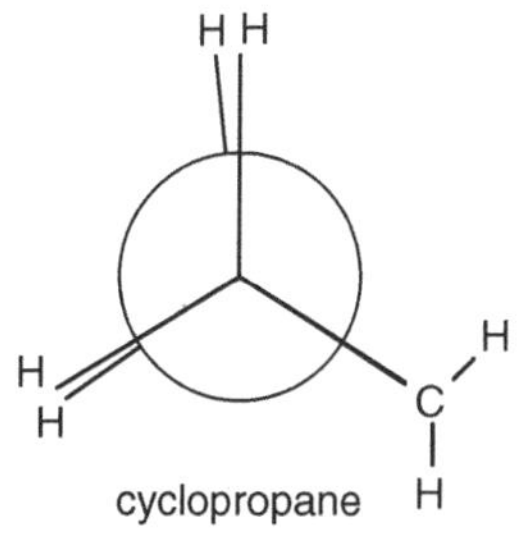

cyclopropane

3.12.10. Most stable conformation is usually a compromise between torsional and angle strain factors. Intramolecular forces can also result in bond length compression or elongation.

    a. Problem 18: Calculations and experimental evidence show that the dihedral angle in the most stable *gauche n*-butane conformer > 60°. How do you account for this fact? What two opposing factors must balance?

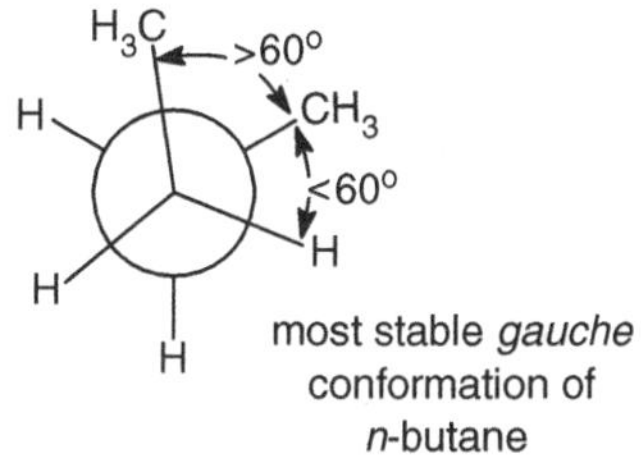

most stable *gauche* conformation of *n*-butane

    b. Molecular mechanics (or molecular force field) computer programs incorporate all known ways (above) molecules can be distorted to find (approximately) the lowest potential energy molecular geometry, including conformation (molecular modeling).

3.12.11. Cyclobutane – puckers (butterfly, or folded conformation). Ring apparently accepts angle strain to relieve eclipsed H-atom torsional strain.

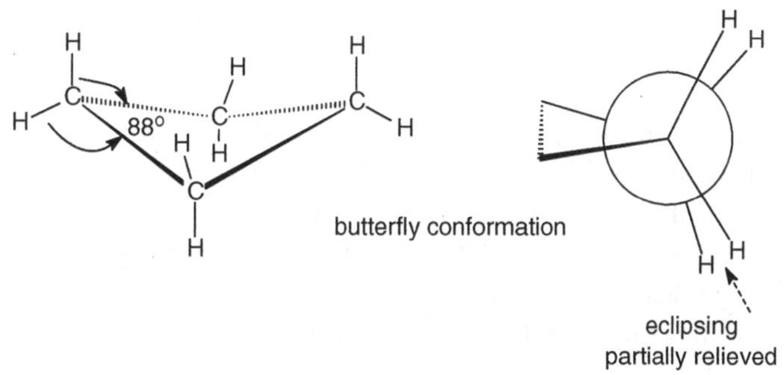

butterfly conformation

eclipsing partially relieved

3.12.12. Cyclopentane – puckers (envelope conformation) ⇒ larger angle strain but less eclipsing. Molecule constantly flexes.

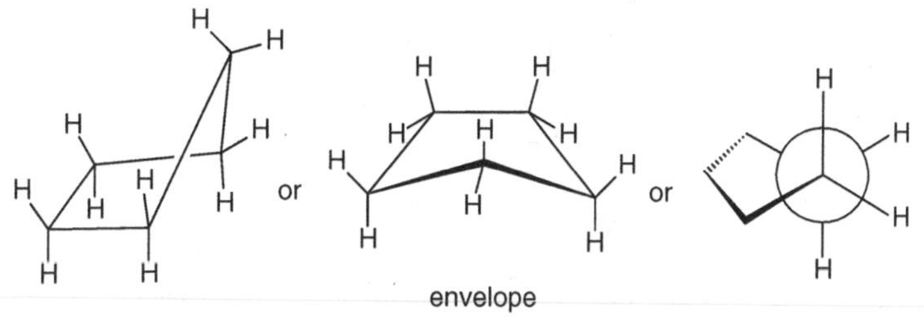

envelope

3.13.1. Angle strain free cyclohexane conformations (be able to recognize):

chair ⇌ other chair (flip)

Newman

flagpole hydrogens steric repulsion

boat conformation ⇌ twist-boat

or other twist-boat

- **a.** Chair – most stable conformer – torsional and steric strain free (alternate CH_2 units are staggered, H atoms 2.3 Å apart). Only one in 10,000 molecules is not in chair conformation at any instant at 25°C.
- **b.** Boat – some eclipsing and van der Waals interactions (flagpole H atoms are 1.83 Å apart); 29 kJ/mol less stable than chair.
- **c.** Twist boat – some eclipsing, but less than boat. Two equal energy (mirror image) forms, 23 kJ/mol less stable than chair.

3.13.2. Potential energy diagram.

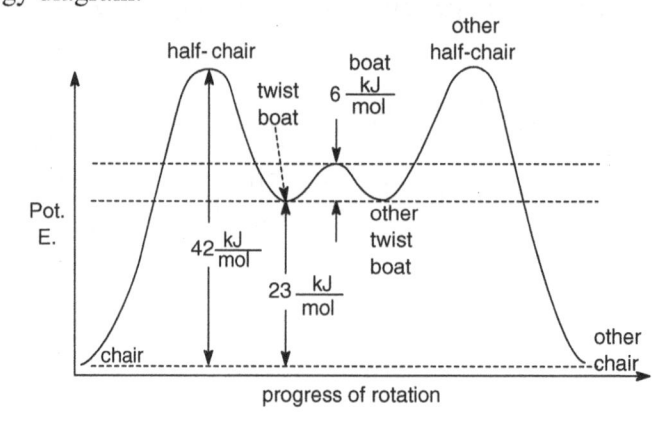

a. Half chair has angle and torsional strain (half way between two chair forms), 42 kJ/mol above chair.

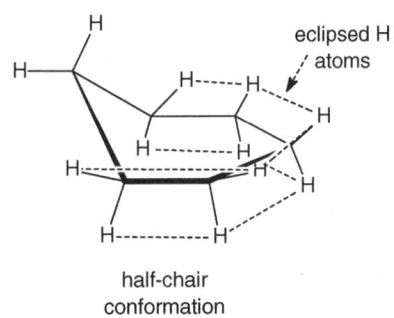

half-chair conformation

3.13.3. Cyclohexane chair has six <u>equitorial</u> H atoms (labeled "e" in section 3.13.1) – near approximate C atom plane, around outside. It also has six <u>axial</u> H atoms (labeled "a") – above and below C atom plane. When a ring flips (via twist-boat and boat) all equitorial positions become axial, and all axials become equitorials.

a. Problem 19. Decalin ($C_{10}H_{18}$) consists of two fused cyclohexane rings.

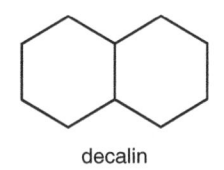

decalin

a. Use models to make *cis* and *trans* decalins. *b.* Draw all angle strain free conformations (andiron drawings, chair and twist boat) for *cis* and *trans* decalins. *c.* What is the most stable *cis*- and *trans*-decalin conformation? *d.* Why is *trans* decalin more stable (~8 kJ/mol) than *cis*? *Hint*: consider each ring in turn. What are the largest substituents on each ring? *e.* Explain why *cis-trans*-decalin interconversion occurs only under very vigorous conditions, while cyclohexane chair and twist-boat forms (energy difference ~25 kJ/mol) readily interconvert at 25°C. Illustrate with approximate potential energy diagrams.

3.13.4. 7 - 12 Carbon rings – subject to eclipsing interactions and H-atom crowding inside rings (raises energy slightly).

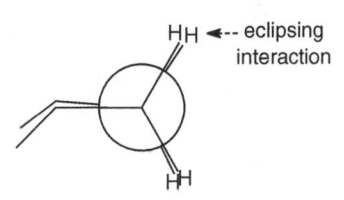

3.14.1. Methylcyclohexane – is 7.1 kJ/mol more stable with CH₃ in equitorial position (experimental value) due to two 1,3 diaxial steric interactions. Only 5 out of 100 molecules have CH₃ axial at 25°C.

a. Alternative view – in axial position, CH₃ group is *gauche* with respect to ring C atoms. Two "butane *gauche* interactions" should raise energy by 2 × 3.8 kJ/mol = 7.6 kJ/mol (≈ 7.1 kJ/mol, experimental value). See section 3.8.7. In equitorial position, -CH₃ group is *anti* (more stable).

3.14.2. For *t*-butylcyclohexane, only one in 10,000 molecules has *t*-butyl group axial at 25°C. Equitorial is 23 kJ/mol more stable than axial.

  a. S. Winstein – a *t*-butyl group can be used to effectively "lock" another, (smaller) substituent in axial position.

  c. For *cis*-1,4-di-*tert*-butylcyclohexane, twist boat is the most stable conformation:

b. Groups larger than H are more stable in equitorial positions:

Group	$E_{ax} - E_{eq}$ (kJ/mol)
-F	0.8
-C≡N	0.8
-Cl	2.1
-Br	2.5
-OH	4.1
-COOH	5.9
-CH$_3$	7.6
-CH$_2$CH$_3$	7.9
-CH(CH$_3$)$_2$	8.8
-C(CH$_3$)$_3$	23

c. Problem 20. Use models to explain why the energy difference for *t*-butyl is so large compared to ethyl and isopropyl.

3.11.1. Rotation about σ-bonds is restricted in ring compounds in a similar manner to alkenes. Ring compounds can exhibit *cis/trans* isomerism.

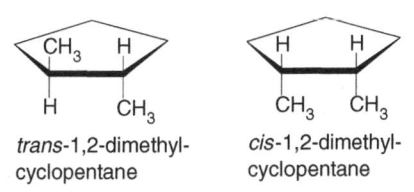

trans-1,2-dimethyl-cyclopentane   *cis*-1,2-dimethyl-cyclopentane

Haworth projection drawings

3.15.1. *Cis/trans* isomerism can be harder to recognize when a ring is puckered (as in cyclohexane).
  a. Two possible chair conformers exist for *trans*-1,2-dimethylcyclohexane; diequitorial is more stable.
     1. No 1,3-diaxial interactions occur in the diequitorial conformer, and one butane *gauche* interaction (3.8 kJ/mol) between the two methyl groups. Total torsional strain = 3.8 kJ/mol.

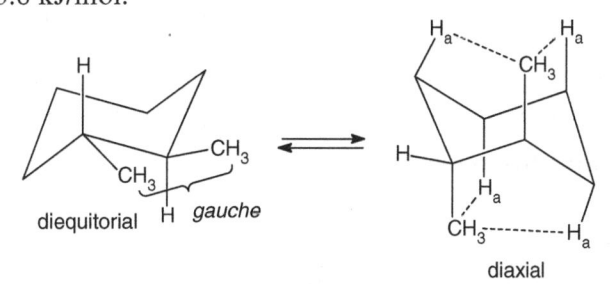

     diequitorial   diaxial

     2. No butane *gauche* interaction between methyl groups, but four 1,3-diaxial interactions occur in the diaxial conformer; torsional strain = (4 × 3.8 = 15.2 kJ/mol).

b. The two chair forms have equal energy for *cis*-1,2-dimethylcyclohexane. Two 1,3-diaxial interactions occur (2 × 3.8 = 7.6 kJ/mol) and one butane-*gauche* interaction (between the two methyl groups, 3.8 kJ/mol). Total torsional strain = 11.4 kJ/mol.

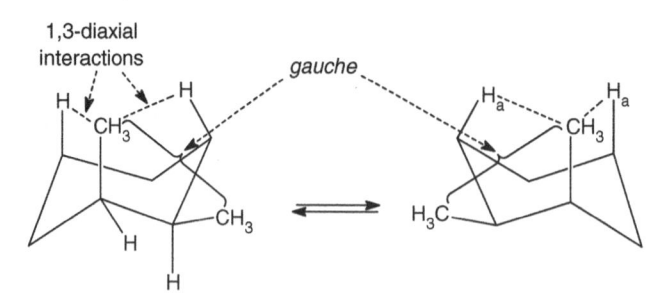

1. Dieq. *trans* is more stable than *cis* by ~7.6 kJ/mol (measured 7.82 kJ/mol).

3.15.2. General rules for disubstituted cyclohexanes:

Substitution pattern	*Geometrical isomer*	One chair conformer	Other chair conformer
1,2 or 1,4	cis	a,e	e,a
1,2 or 1,4	trans	a,a	e,e
1,3	cis	a,a	e,e
1,3	trans	a,e	e,a

a. Problem 21. Assuming that each butane *gauche* interaction or 1,3-methyl – hydrogen diaxial interaction increases potential energy by 3.8 kJ/mol, estimate the energy difference (if any) between the two chair conformations for:
1. *cis*-1,3-dimethylcyclohexane

2. *trans*-1,3-dimethylcyclohexane

3. *cis*-1,4-dimethylcyclohexane

4. *trans*-1,4-dimethylcyclohexane

b. Problem 22. *a.* K. S. Pitzer calculated that the 1,3-methyl – methyl diaxial interaction in *cis*-1,3-dimethyl cyclohexane (neglected in previous problem) must be taken into account and should raise the energy difference much higher (to 22.6 kJ/mol). How large is a 1,3-methyl – methyl diaxial interaction? (Subtract the number calculated in previous problem part 1 from 22.6 kJ/mol.) *b.* N. L. Allinger experimentally measured a 15.5 kJ/mol energy difference between *cis*- and *trans*-1,1,3,5-tetramethylcyclohexane isomers. Does Allinger's measurement support Pitzer's calculation? Explain.

c. Problem 23. Counting 3.8 kJ/mol for each butane *gauche* or 1,3-methyl – hydrogen diaxial interaction, calculate the potential energy difference (if any) between *a. cis*-1,3-dimethylcyclohexane (dieq) and *trans*-1,3-dimethylcyclohexane (ax-eq), and *b. cis*-1,4-dimethylcyclohexane (ax-eq) and *trans*-1,4-dimethylcyclohexane (dieq).

d. **Problem 24.** Use the energy differences from the section 3.14.2b table to calculate 1,3-alkyl – hydrogen interaction values for the following dialkylcyclohexanes. From these values, calculate the energy difference between the two chair conformations.
   1. *cis*-4-*tert*-butylmethylcyclohexane

   2. *trans*-4-*tert*-butylmethylcyclohexane

   3. [structure] ⇌ other chair conformation?

3.16.1. Three bicyclic alkane classes:
   a. Fused rings – share two adjacent C atoms and the bond between them. (See below for naming rules.)

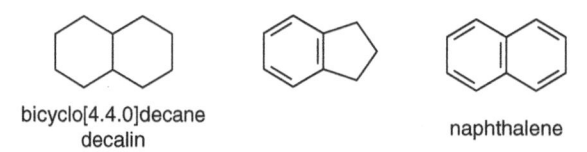

   bicyclo[4.4.0]decane
   decalin

   naphthalene

   b. Bridged rings – share two nonadjacent C atoms and one or more C atoms (bridge) between them.

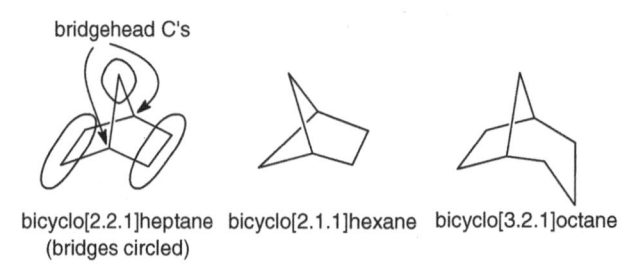

   bridgehead C's

   bicyclo[2.2.1]heptane   bicyclo[2.1.1]hexane   bicyclo[3.2.1]octane
   (bridges circled)

   c. Spirocyclics (rare) – two rings share one C atom.

   spiro[4.3]octane

3.16.2. Bicyclic naming rules:
  a. Count C atoms and name as parent alkane.
  b. Find C atoms common to both rings (bridgehead C atom(s)).
  c. Count C atoms between (linking) bridgehead carbons for each bridge.
     1. Place numbers in square brackets (decending order), separated by periods.
     2. Use "bicyclo[*x.y.z*] as a prefix to the parent alkane name for fused and bridged rings; use "spiro[*x.y*] for spirocyclics. Examples:

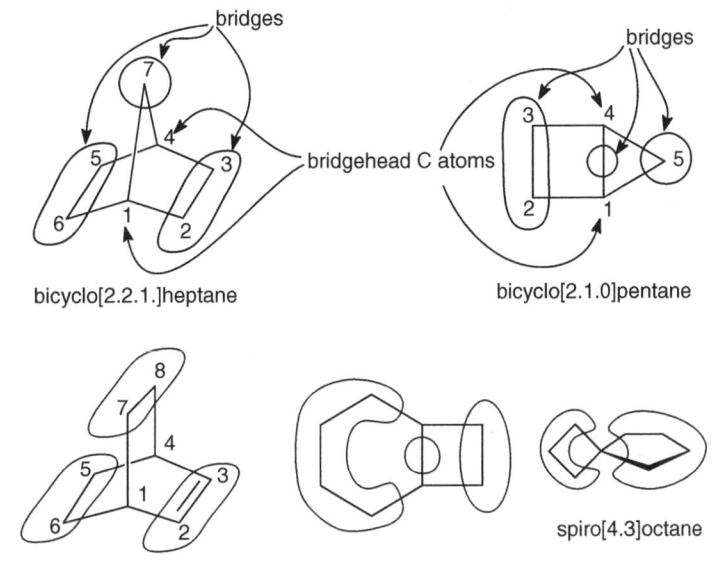

  bicyclo[2.2.1.]heptane   bicyclo[2.1.0]pentane

  bicyclo[2.2.2]oct-2-ene   bicyclo[4.2.0]octane   spiro[4.3]octane

  d. Problem 25. Draw structural formulas for:
     1. methylcyclopentane

     2. 1-methylcyclopentene

     3. 3-methylcyclopentene

     4. *trans*-1,3-dichlorocyclobutane

5. *cis*-1-bromo-2-methylcyclopentane

6. cyclohexylcyclohexane

7. 1-cyclopentylethyne

8. 4-chloro-1,1-dimethylcycloheptane

9. bicyclo[2.2.1]hepta-2,5-diene

10. 1-chlorobicyclo[2.2.2]octane

3.16.3. Two isomeric decalin forms (*cis* and *trans*) exist. *Cis* isomer chair conformations can easily flip to the other chair, but *trans* isomer chairs cannot easily flip. (Verify using models.)

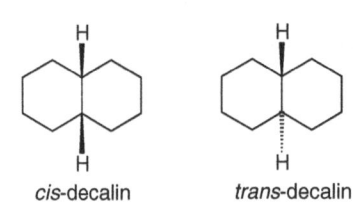

cis-decalin   *trans*-decalin

3.16.4. A few polycyclic examples (common names shown):

"cubane" "basketane" "adamantane"

    a. Diamonds (adamantane like structure) continue to find new uses.

CHAPTER 4

4.1.1. Organic reactants are transformed into products (in reactions) by specific steps (mechanisms).
   a. Reactions can be organized by mechanism (pathway). Relatively few reaction pathways exist.
   b. Mechanism – a complete step-by-step description of all the collisions during which reactant bonds break and product bonds form.
      1. A mechanism includes *thermodynamic* (enthalpy, entropy, and equilibrium position) and *kinetic* (reaction rate) analyses.
   c. Molecules are too small to observe collisions directly. Mechanisms cannot (yet) be proven.

4.1.2. Tools for gathering indirect evidence:
   a. Kinetic rate law – reaction *order* elucidates some of the collisions.
   b. Reaction product structures – provide hints about their formation.
      1. Three dimensional product structure (stereochemistry) is described in Chapter 5.
   c. Heavy isotope containing molecules – can provide information that is not otherwise obvious.
   d. Other observations – absorption of light, *initiator* (substances that help get a reaction started) effects, *catalysts* (substances that speed up a reaction), and *inhibitors* (substances that slow down or stop a reaction), can help construct a pathway.

4.1.3. Understanding mechanisms allows us to:
   a. predict origins of rate differences.
   b. predict untried reaction outcomes.
   c. Alter reaction conditions to improve outcome.

4.4.1. Thermodynamic laws describe any equilibrium reaction enthalpy and entropy.

4.4.2. During any chemical reaction:

$$\text{Total Energy} = \text{Potential Energy} + \text{Kinetic Energy}$$

   a. Total energy must be conserved.
      1. If Pot. E. decreases, K. E. must increase, i.e., heat is released into surroundings, (an exothermic reaction, enthalpy, $\Delta H^\circ_{rxn} < 0$).

$$\Delta H^\circ_{rxn} = H^\circ_{products} - H^\circ_{reactants}$$

a. The ° indicates reactants and products in their standard states (pure substances at 25°C, 1.00 M concentrations, or 1 atm pressures).
2. If Pot. E. increases, K. E. must decrease and heat is absorbed from the surroundings (an endothermic reaction, enthalpy, $\Delta H°_{rxn} > 0$).
b. If reaction products have higher entropy (freedom of motion, or randomness) than reactants, that reaction has a greater tendency to be spontaneous than it would based on $\Delta H°_{rxn}$ alone.

4.5.1. Enthalpy and entropy both affect spontaneity (at a given absolute temperature T) according to the Gibbs free energy expression:

$$\Delta G°_{rxn} = \Delta H°_{rxn} - T\Delta S°_{rxn}$$

where

$$\Delta G°_{rxn} = G°_{products} - G°_{reactants}$$

and

$$\Delta S°_{rxn} = S°_{products} - S°_{reactants}$$

a. If $\Delta G°_{rxn} < 0$, the equilibrium will lie to the right, i.e., conversion of reactants to products is favored (*exergonic* reaction).
b. If $\Delta G°_{rxn} > 0$, the equilibrium will lie to the left, i.e., reversion of products to reactants is favored (*endergonic* reaction).
c. If $\Delta G°_{rxn} = 0$, the system is at equilibrium (no net reaction).
d. Near 298 K, the $T\Delta S°_{rxn}$ term is often small compared to $\Delta H°_{rxn}$. It is sometimes neglected.

4.4.3. Gibbs free energy is related to the equilibrium constant (section 2.6.1c).

$$K_{eq} = e^{\frac{-\Delta G°_{rxn}}{RT}} \quad \text{or}$$

$$\Delta G°_{rxn} = -RT(\ln K_{eq}) = -2.303 RT(\log_{10} K_{eq})$$

where $R = 8.314$ J · kelvin^{-1} · mol^{-1} (gas constant)

T = absolute temperature

e = 2.718, the base of naturallogarithms

a. If $\Delta G°_{rxn} < -12.0$ kJ · mol^{-1}, the equilibrium will lie > 99% towards products.

$\Delta G°_{rxn}$ (kJ/mol)	$K_{eq} = e^{\frac{-\Delta G°_{rxn}}{RT}}$	Conversion to product
+4.0	0.20	17%
+2.0	0.45	31%
0.0	1.0	50%
−2.0	2.2	69%
−4.0	5.0	83%

(*Continued*)

ΔG^o_{rxn} (kJ/mol)	$K_{eq} = e^{\frac{-\Delta G^o_{rxn}}{RT}}$	Conversion to product
−8.0	25	96%
−12.0	127	99.2%
−16.0	638	99.8%
−20.0	3200	99.96%

4.4.4. Equilibrium constant for overall methane chlorination reaction:

$$CH_4 + Cl_2 \rightleftharpoons H_3C-Cl + H-Cl$$

$$K_{eq} = \frac{[H_3CCl][HCl]}{[CH_4][Cl_2]} = 1.1 \times 10^{19}$$

a. Large K_{eq} implies that $[H_3CCl]$ and $[HCl]$ are large and $[CH_4]$ and $[Cl_2]$ are small at equilibrium.

b. ΔG^o_{rxn} can be calculated at 298 K from K_{eq}:

$$\Delta G^o_{rxn} = -(298\ K)(8.314 \frac{J}{mol \cdot K})(\ln 1.1 \times 10^{19}) \cdot \frac{kJ}{1000\ J} = -108.7 \frac{kJ}{mol}$$

4.5.2. K_{eq} is calculated from ΔG^o_{rxn}. ΔG^o_{rxn} can be calculated from ΔH^o_{rxn} and ΔS^o_{rxn}.

a. ΔH^o_{rxn} is measured experimentally by placing known methane and chlorine amounts in a bomb calorimeter and measuring temperature increase after reaction.

b. $\Delta S^o_{rxn} = +12.1\ J \cdot kelvin^{-1} \cdot mol^{-1}$. A given substance's absolute entropy value S^o is calculated from its heat capacity data.

$$\Delta G^o_{rxn} = -105.1 \frac{kJ}{mol} - (298\ K)(12.1 \frac{J}{mol \cdot K})(\frac{kJ}{1000\ J})$$

$$= (-105.1 - 3.61) \frac{kJ}{mol} = -108.7 \frac{kJ}{mol}$$

c. Note that the $T\Delta S^o_{rxn}$ term is rather small.

4.8.1. A reaction does not necessarily occur at an appreciable rate just because ΔG^o_{rxn} is negative (i.e., just because its equilibrium is favorable).

4.8.2. Some factors that influence methane halogenation rate.

a. Bond dissociation energies – broken and formed during reaction.
b. Reactant concentrations.
c. Absolute temperature.

4.8.3. Any reaction rate is linked to collision types, energies, and frequencies between reactant particles.

a. Speed (rate) unit analysis:

$$\frac{distance}{time}$$

1. Example: $\frac{\text{miles}}{\text{hour}}$

b. A chemical reaction speed (rate) is concentration change (increase or decrease) over time. Unit analysis:

$$\frac{\text{concentration}}{\text{time}}$$

c. Example: a "first order" reaction rate (see e below):

$$\frac{\text{molarity (M)}}{\text{sec}} = \frac{\frac{\text{mols of reactant}}{\text{liter of solution}}}{\text{sec}}$$

d. Non-first order rate laws have more complex units involving concentrations and time.
e. A rate law for the general reaction below has the form:

$$aA + bB \rightarrow cC + dD$$

$$\text{rate} = k[A]^x[B]^y$$

1. The rate is xth order in [A], and yth order in [B], and $x + y$ order overall, and k is the *rate constant*. (Square brackets mean molarity.)
2. Order must be determined experimentally by varying reactant concentrations. For example, doubling a given reactant concentration might have no effect, double, or quadruple the rate, etc.
3. Example: an experimentally determined rate law is shown for the reaction below.

$$H_3C\text{-}Br + {}^{\ominus}\!\ddot{\underset{..}{O}}\text{-}H + \overset{\oplus}{Na} \xrightarrow[\text{acetone (solvent)}]{H_2O} H_3C\text{-}\underset{..}{\overset{H}{\overset{|}{O}}}{}^{\oplus} + \overset{\oplus}{Na} + :\!\ddot{\underset{..}{Br}}\!:^{\ominus}$$

$$\text{rate} = k[H_3C\text{-}Br][^-OH]$$

a. Reaction follows a rate law that is first order in [H$_3$C-Br] and first order in [$^-$OH], second order overall.

4. Example: an experimentally determined rate law is shown for the reaction below.

$$H_3C\text{-}\underset{\underset{CH_3}{|}}{\overset{\overset{CH_3}{|}}{C}}\text{-}\ddot{\underset{..}{Br}}: + {}^{\ominus}\!\ddot{\underset{..}{O}}\text{-}H + \overset{\oplus}{Na} \xrightarrow[\text{acetone (solvent)}]{H_2O} H_3C\text{-}\underset{\underset{CH_3}{|}}{\overset{\overset{CH_3}{|}}{C}}\text{-}\underset{..}{\overset{H}{\overset{/}{O}}}{}^{\oplus} + \overset{\oplus}{Na} + :\!\ddot{\underset{..}{Br}}\!:^{\ominus}$$

$$\text{rate} = k[(H_3C)_3C\text{-}Br]$$

a. Reaction is first order in [(H$_3$C)$_3$C-Br], zeroth order in [$^-$OH], first order overall.

4.9.1. k is not really constant, but is a function of several variables, including temperature. Arrhenius was first to describe "rate constant" k mathematically:

$$k = Ae^{\frac{-E_{act}}{RT}} \quad \text{and for the reaction in e above}$$

$$rate = k[A]^x[B]^y = Ae^{\frac{-E_{act}}{RT}}[A]^x[B]^y \text{ where}$$

E_{act} or E_a = *Activation energy* – minimum kinetic energy that reactant molecules must possess to overcome electron cloud repulsions during an *effective collision* (see section 4.9.2 below).

A = a constant, *frequency factor*, i.e., how many effective collisions per sec would occur if E_a were zero.

R = gas constant (8.314 J · mol^{-1} · kelvin^{-1})

T = the absolute temperature

4.9.2. *Effective collision* – transfers sufficient energy (E_a) to reactant molecules that they reach the *activated complex* or *transition state* (section 4.9.3), an unstable transient structure that reorganizes (by breaking and forming bonds) into products.
   a. For ineffective collisions, reactant molecules simply bounce off each other and no reaction occurs.
   b. Strictly, E_a is independent of ΔH^o_{rxn}. However, for bond homolysis, $E_a = \Delta H^o_{rxn}$. For reactions involving *both* bond forming and breaking, $E_a > \Delta H^o_{rxn}$.
      1. Example: chlorine homolysis.

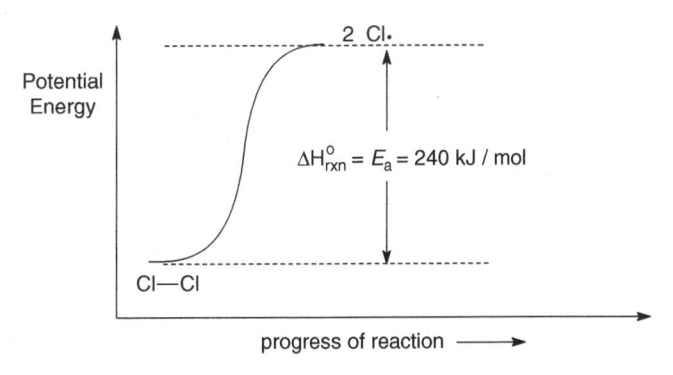

   a. *Reaction diagrams* – invented by Henry Eyring and co-workers.
   2. Example: acid/base neutralization.

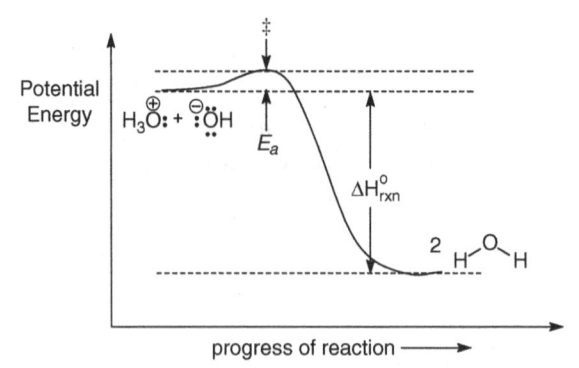

3. E_a is always positive.
4. Principle of microscopic reversibility – reverse reaction (right to left) follows the same pathway as forward reaction.
   a. Problem 1: what is E_a for reverse Cl_2 homolysis?

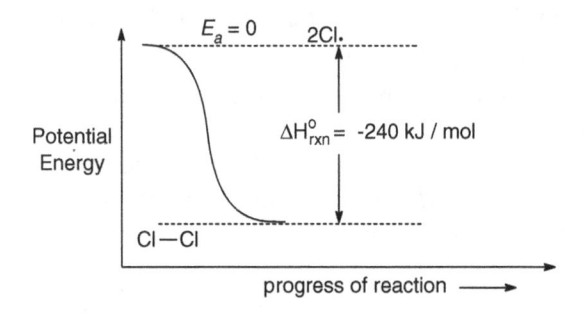

4.9.3. *Transition state* (represented by a ‡ symbol) – a fleeting structure that is part way between reactants and products during an effective collision. It has partially broken and partially formed bonds (and partial electrons and partial charges), and corresponds to a peak on a reaction diagram.

   a. Highly unstable – cannot normally be isolated or studied because it only exists for a tiny fraction (ca. 10^{-12}) of a sec.

4.9.4. The Arrhenius rate equation was later expanded by Henry Eyring and others into the general reaction rate equation:

$$\text{rate of any reaction} = \text{collision frequency} \times \text{energy factor} \times \text{probability (orientation factor)}$$

   a. Unit analysis:

Term	Description
rate	number of effective collisions per liter per sec.
collision frequency	total number of collisions per liter per sec.
energy factor	fraction of collisions that have sufficient energy (E_a) to break bonds.
probability factor	fraction of collisions that have the proper orientation to break bonds.

- **b.** Collision frequency – depends on temperature, concentration*, pressure*, particle size, and speed. *(more important)
  1. Higher concentration or pressure ⇒ more collisions.
  2. Particle size and mass (speed) effects tend to cancel out. Large particles move slower, small particles move faster.
- **c.** Energy factor – depends on temperature and E_a.
  1. Higher temperature ⇒ more effective collisions.
  2. Lower E_a ⇒ less energy is needed in a collision for reaction to occur, i.e., an effective collision.
  3. A Boltzman curve describes kinetic energies of a particle collection at a given temperature:

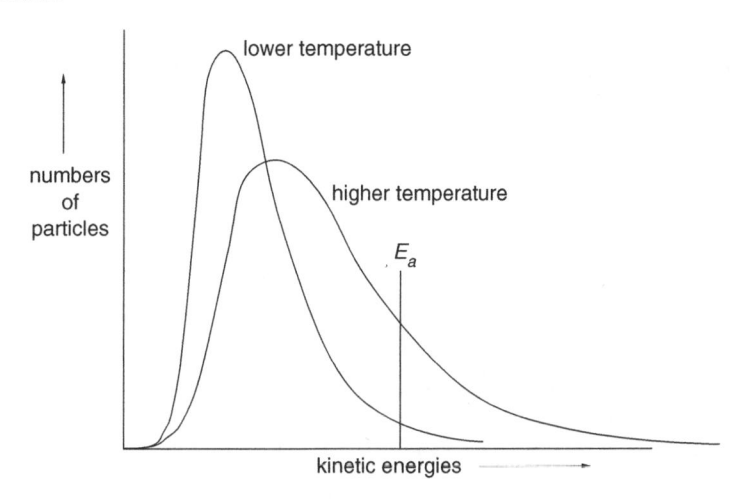

  4. Particle fraction with energy $< E_a$ is $e^{-\frac{E_a}{RT}}$.
- **d.** Probability factor – depends on collision geometry and mechanism type.
  1. It is nearly constant for closely related mechanisms such as CH_4 chlorination vs. bromination.

4.9.5. H. Eyring wrote a general reaction rate expression for the section 4.9.1e general reaction. Using math symbols:

$$rate = PZe^{-\frac{E_a}{RT}} = k[A]^x[B]^y$$

P = probability (orientation factor)

Z = collision frequency (concentration/pressure)

exponential term = energy/temperature factor

4.9.6. The exponential term is usually the most important of the three factors (b-d above) in determining a reaction rate.

- **a.** Example: E_a and T effects on reaction rate:

Total Collisions	T (°C)	E_a (kJ/mol)	Collisions having energy ≥ E_a	Relative reaction rate
1,000,000	275	20.9	10,000	10,000
1,000,000	275	41.8	100	100
1,000,000	275	62.8	1	1

b. A factor of 3 increase in E_a changes a rate by a factor of 10,000 at 275°C.
c. Example: temperature effect on three different reaction rates (with different E_a's).

T (K)	E_a (kJ/mol)	Relative rate
523	20.9	1.0
573	20.9	1.5
523	41.8	1.0
573	41.8	2.0
523	62.8	1.0
573	62.8	3.0

~10% change in temperature (from 523 to 573 K) changes the rate by a factor of 3.0 (last two lines).

4.10.1. H. Eyring – E_a is related to activation enthalpy, $\Delta H^{\ddagger}$,

$$E_a = \Delta H^{\ddagger} + RT$$

and probability (P) is related to activation entropy $\Delta S^{\ddagger}$.

$$P = e^{\frac{\Delta S^{\ddagger}}{R}}$$

a. High $\Delta H^{\ddagger}$ ⇒ high E_a at a given absolute temperature, T, and slow reaction.
b. Small or negative $\Delta S^{\ddagger}$ ⇒ small P and slow reaction.
c. $\Delta H^{\ddagger}$ and $\Delta S^{\ddagger}$ are related to $\Delta G^{\ddagger}$.

$$\Delta G^{\ddagger} = \Delta H^{\ddagger} - T\Delta S^{\ddagger}$$

Smaller $\Delta G^{\ddagger}$ ⇒ faster reaction.

d. Catalysts speed up a reaction by lowering $\Delta G^{\ddagger}$, but do not affect reactant or product potential energies.

4.2.1. Methane halogenation – first mechanism that we will write.

4.2.2. Reasons to study this reaction.

a. Methane is a simple reactant.
b. The mechanism has been well studied.
c. Reaction rates, activation energies, enthalpies (see below), and yields are known.
d. No solvent effects occur in the gas phase.
e. Other free radical reactions on larger molecules are similar.

4.2.3. Chlorination conditions: 250–400°C, gas phase. Overall double replacement (substitution) reaction:

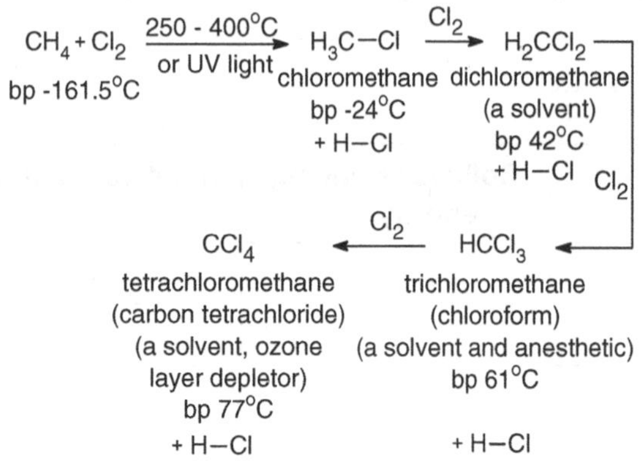

$CH_4 + Cl_2$ $\xrightarrow[\text{or UV light}]{250 - 400°C}$ H_3C-Cl $\xrightarrow{Cl_2}$ H_2CCl_2 $\xrightarrow{Cl_2}$ $HCCl_3$ $\xrightarrow{Cl_2}$ CCl_4

bp -161.5°C

chloromethane
bp -24°C
+ H–Cl

dichloromethane
(a solvent)
bp 42°C
+ H–Cl

trichloromethane
(chloroform)
(a solvent and anesthetic)
bp 61°C
+ H–Cl

tetrachloromethane
(carbon tetrachloride)
(a solvent, ozone layer depletor)
bp 77°C
+ H–Cl

4.2.4. Stoichiometry:
- **a.** A 1:1 CH_4 to Cl_2 mol ratio yields all four products and unreacted CH_4.
  1. H_3CCl, H_2CCl_2, and $HCCl_3$ products compete with CH_4 for Cl_2.
  2. Trick – use a large CH_4 molar excess (relative to Cl_2) for a good H_3CCl yield. Excess CH_4 can be distilled out and recycled.

4.2.5. Mechanism steps must be written so they are consistent with experimentally observed facts.

4.2.6. Methane chlorination facts:
- **a.** No reaction occurs in the dark at 25°C.
- **b.** Reaction will go in the dark at T > 250°C.
- **c.** Reaction will go at 25°C with blue/violet or ultraviolet (UV) light.
  1. Known fact: these photons have ≥ 250 kJ/einstein, according to Planck's equation ($E_{photon} = h\nu$). (An einstein is a mol of photons.) This is enough energy to cause Cl_2 molecules to break apart (homolyze) into Cl· atoms. Cl_2 bond dissociation enthalpy (section 4.6.1, Chapter 1 (after section 1.7.4.)) is 240 kJ/mol.
- **d.** Each UV photon absorbed causes several thousand product molecules to form (i.e., reaction has a high *quantum yield*).
  1. A high quantum yield suggests a *chain reaction* – a multistep reaction where each step yields a reactive intermediate substance that is needed for the next step.
  2. Example: a high energy neutron reacts with a $^{235}_{92}U$ atom. Each nuclear fission yields more neutrons that react with more $^{235}_{92}U$ nuclei.
     - **a.** Nuclear fission:
       $$^{235}_{92}U + ^{1}_{0}n \rightarrow ^{135}_{53}I + ^{97}_{39}Y + 4^{1}_{0}n$$
       and many other products.
- **e.** A trace of O_2 slows reaction for a period, then it proceeds normally.

4.3.1. Methane chlorination mechanism:

(1) $:\!\ddot{Cl}\!:\!\ddot{Cl}\!: \xrightarrow{\text{UV light } (h\nu)} 2 :\!\ddot{Cl}\cdot$ ΔH°_{rxn} (kJ/mol) +240

(2) $:\!\ddot{Cl}\cdot + H\!:\!CH_3 \longrightarrow :\!\ddot{Cl}\!:\!H + \cdot CH_3$ +7

(3) $H_3C\cdot + :\!\ddot{Cl}\!:\!\ddot{Cl}\!: \longrightarrow H_3C\!:\!\ddot{Cl}\!: + :\!\ddot{Cl}\cdot$ –110

recycles in step 2

- **a.** Step (1) has a high activation energy ($E_a = \Delta H^\circ_{rxn} = +240$ kJ/mol for bond homolysis). Step (1) is an *initiation* step – free radical *intermediates* are *initiated* (formed) by photon absorption or heat from non-free radical reactants.
  1. Free radical (section 1.9.3) intermediate – an atom or molecule with (at least) one unpaired electron. Characteristics:
     - **a.** Usually extremely reactive (contains energy absorbed from homolysis).
     - **b.** Violates octet rule.
     - **c.** Tries to pair lone electron and complete octet rule.
- **b.** What will Cl· collide with once formed? Most likely CH_4 or Cl_2. Collision with another Cl· atom is unlikely (Cl· concentration is relatively low).

1. Problem 2. When Cl· collides with Cl_2, chlorine exchange might occur.

$$:\overset{35}{\ddot{C}l}\cdot \; + \; :\overset{36}{\ddot{C}l}-\ddot{\ddot{C}l}: \longrightarrow :\ddot{\ddot{C}l}-\ddot{\ddot{C}l}: + \; :\ddot{\ddot{C}l}\cdot$$

Naturally occurring chlorine is a mixture of ^{35}Cl and ^{37}Cl isotopes, but ^{36}Cl is absent. If you had some $^{36}Cl_2$ available (from a nuclear reactor), draw the reaction that would occur when $^{35}Cl\cdot$ reacts with $^{36}Cl_2$. Include electron pushing arrows and electron dots as above.

$$^{35}\ddot{C}l - ^{36}\ddot{C}l: \; + \; ^{36}\ddot{C}l\cdot$$

If you had a mass spectrometer, what molecule (molecular mass) would you try to detect (that would prove that this reaction actually occurs)?

$$^{35}Cl - ^{36}Cl$$

2. Effective collision with CH_4 yields step (2):

(2) $:\ddot{C}l\cdot \; + \; H-CH_3 \longrightarrow :\ddot{C}l-H \; + \; \cdot CH_3$

c. What will $\cdot CH_3$ collide with once formed? CH_4 or Cl_2? Effective collision with CH_4 just exchanges one $\cdot CH_3$ for another.

$\cdot CH_3 \; + \; H-CH_3 \longrightarrow H-CH_3 \; + \; \cdot CH_3$

d. Effective collision with Cl_2 yields step (3):

(3) $\cdot CH_3 \; + \; :\ddot{C}l-\ddot{C}l: \longrightarrow Cl-CH_3 \; + \; \cdot \ddot{C}l:$

e. Steps (2) and (3) are *propagation* steps – start with a free radical and yield a new free radical.

4.3.2. How does reaction stop? Although unlikely, two free radicals occasionally collide.

termination (4) $:\ddot{C}l\cdot \; + \; \cdot \ddot{C}l: \longrightarrow :\ddot{C}l-\ddot{C}l:$

termination (5) $H_3C\cdot \; + \; \cdot CH_3 \longrightarrow H_3C-CH_3$
                                                                  ethane

termination (6) $H_3C\cdot \; + \; \cdot \ddot{C}l: \longrightarrow H_3C-\ddot{C}l:$
                                                                  chloromethane

a. Steps (4-6) occur only once for every ~10,000 cycles through steps (2) and (3). Only one in ~10,000 H₃CCl molecules is formed in step (6).
   1. All *termination* steps are exothermic. The Cl₂ molecule in step (4) cannot dissipate the energy to atomic vibrations (because Cl· has no bonds).
   2. Termination steps become more likely near the end of the reaction.
b. Oxygen effect:

$$H_3C\cdot \;+\; \cdot\ddot{O}-\ddot{O}\cdot \longrightarrow H_3C-\ddot{O}-\ddot{O}\cdot$$
$$\text{less reactive}$$

   1. Inhibition – one ·CH₃ consumption breaks the chain (steps (2) and (3)), and prevents thousands of product molecules from forming.

4.7.1. Methane chlorination $\Delta H°_{rxn}$ measurement was described in section 4.5.2a. $\Delta H°_{rxn}$ can also be approximately calculated from homolytic bond dissociation energies (see Chapter 1, section 4.6.1 (after section 1.7.4)).
   a. *Hess' law of constant heat summation*: $\Delta H°_{rxn} \approx$ *D*'s (bond dissociation enthalpies) of bonds broken minus *D*'s of bonds formed.
   b. Bond homolysis (dissociation) energies (enthalpies):

Bond	kJ/mol	Bond	kJ/mol	Bond	kJ/mol
H-H	436	H₃C-OH	385	(H₃C)₃C-H	403
D-D	440	H₃C-NH₂	356	(H₃C)₃C-F	464*
F-F	154	H₃CCH₂-H	423	(H₃C)₃C-Cl	355
Cl-Cl	240	H₃CCH₂-F	464*	(H₃C)₃C-Br	304
Br-Br	190	H₃CCH₂-Cl	355	(H₃C)₃C-I	233
I-I	149	H₃CCH₂-Br	303	(H₃C)₃C-OH	401
H-F	570	H₃CCH₂-I	238	PhCH₂-H	376
H-Cl	432	H₃CCH₂-OH	393	H₂C=CHCH₂-H	372
H-Br	366	H₃CCH₂CH₂-H	423	H₂C=CH-H	463
H-I	298	H₃CCH₂CH₂-F	464*	HC≡C-H	558
HO-H	497	H₃CCH₂CH₂-Cl	355	Ph-H	472
HS-H	382	H₃CCH₂CH₂-Br	303	H₃C-CH₃	377
HOO-H	367	H₃CCH₂CH₂-I	238	H₃CCH₂-CH₃	372
HO-OH	213	(H₃C)₂CH-H	413	H₃CCH₂-CH₂CH₃	368
H₂N-H	450	(H₃C)₂CH-F	463	(H₃C)₂CH-CH₃	371
H₃C-H	439	(H₃C)₂CH-Cl	356	(H₃C)₃C-CH₃	366
H₃C-F	481	(H₃C)₂CH-Br	309	H₂C=CH₂	728
H₃C-Cl	350	(H₃C)₂CH-I	238	H₂C=NH	736*

Bond	kJ/mol	Bond	kJ/mol	Bond	kJ/mol
H$_3$C-Br	302	(H$_3$C)$_2$CH-OH	399	H$_2$C=O	749
H$_3$C-I	241				

*approximate
From Wade, L. G. Jr., Simek, J. W., "Organic Chemistry," 9th edition, Pearson, U.S.A., (2017), p 167.

4.7.2. Methane chlorination $\Delta H°_{rxn}$: step 1 is not included since it occurs only once for every ≈10,000 cycles through steps (2) and (3), i.e., the contribution of step (1) is negligible.

$$\begin{array}{llll}
& & & \Delta H°_{rxn} \text{ (kJ/mol)} \\
(2) \; :\ddot{Cl}\cdot \; + \; H\overset{439}{—}CH_3 & \longrightarrow & H\overset{432}{—}Cl \; + \; \cdot CH_3 & +7 \\
(3) \; \cdot CH_3 \; + \; Cl\overset{240}{—}Cl & \longrightarrow & H_3C\overset{350}{—}Cl \; + \; \dot{Cl} & -110 \\
\hline
H—CH_3 \; + \; Cl-Cl & \longrightarrow & H_3C-Cl \; + \; H—Cl & -103
\end{array}$$

4.7.3. Could chloromethane possibly form by alternate steps (2a) and (3a) below?

$$\begin{array}{llll}
& & & \Delta H°_{rxn} \text{ (kJ/mol)} \\
(2a) \; :\ddot{Cl}\cdot \; + \; H\overset{439}{\vdots}CH_3 & \longrightarrow & :\ddot{Cl}:CH_3 \overset{350}{} \; + \; H\cdot & 89 \\
(3a) \; H\cdot \; + \; :\ddot{Cl}\overset{240}{\vdots}\ddot{Cl}: & \longrightarrow & H:\ddot{Cl}: \overset{432}{} + :\ddot{Cl}\cdot & -192 \\
\hline
H-CH_3 \; + \; Cl-Cl & \longrightarrow & H_3C-Cl \; + \; H-Cl & -103
\end{array}$$

a. E_a (step 2a) ≥ $\Delta H°_{rxn}$ = +89 kJ/mol, 82 kJ/mol more endothermic than step (2).
   1. Only one effective (2a) collision occurs for every 1.4×10^7 step (2) effective collisions at 275°C. *Step (2a) is much slower than step (2)*.

4.7.4. Competing reactions – two (or more) different reactions that are occurring simultaneously.

a. The pathway that has the low E_a is easier (faster), and predominates.

4.9.7. H-Cl bond formation process (exothermic) lags behind H-C bond breaking (endothermic) in step (2). Energy release is not synchronized with energy absorption ⇒ more than +7 kJ/mol is needed for reaction to occur (E_a = +17 kJ/mol).

a. Although steps (2) and (3) could occur easily at 25°C in the dark, they cannot occur without step (1).

4.9.8. For step (2) only one in 40 collisions has E_a at 275°C.

4.9.9. Of these, only 1 in 8 have correction orientation (trajectory) to break a C-H bond.

a. Reaction occurs for only 1 out of 40 × 8 = 320 collisions.

4.10.2. Transition state structure for step 2 (‡$_2$), methane chlorination:

$$(2) \; :\ddot{Cl}\cdot \; + \; H\overset{sp^3}{\vdots}CH_3 \longrightarrow \left[:\ddot{Cl}^{\delta\cdot}\text{-------}H\text{----}^{\delta\cdot}CH_3 \right]^{‡_2}$$

partly formed   partly broken

$$\downarrow$$

$$:\ddot{Cl}:H \; + \; \cdot CH_3 \quad (sp^2)$$

a. Cl· is losing its unpaired electron and C is acquiring an unpaired electron in ‡₂. The H-Cl bond is forming, and the C-H bond is breaking.

4.11.1. Intermediate (general definition) – a highly reactive product (such as R· free radical) that reacts further to yield a more stable product in the end. It corresponds to a high energy valley on a reaction diagram.

    b. Examples: reaction diagrams for methane chlorination and bromination.

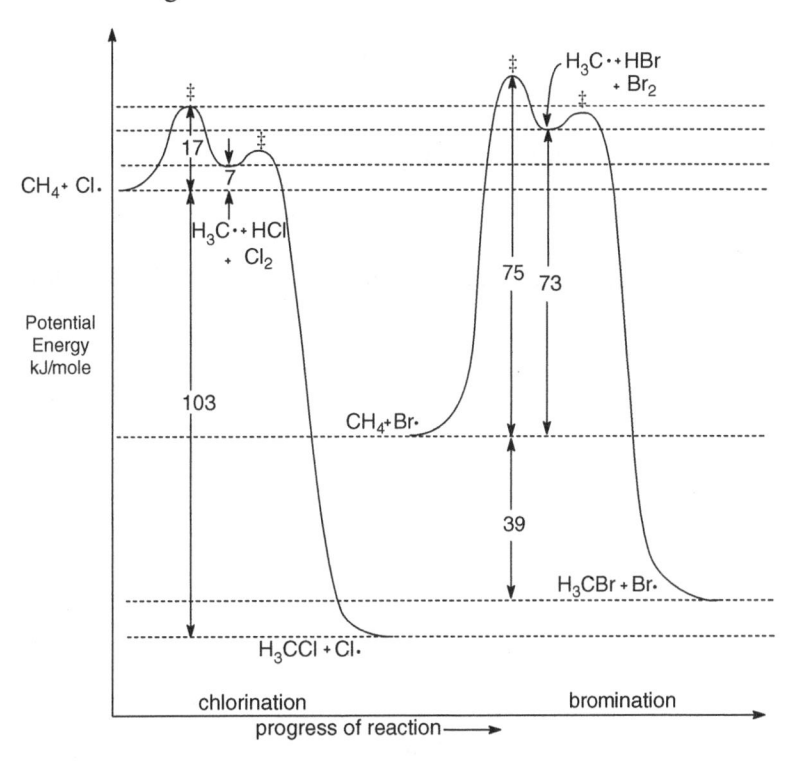

        1. Do not confuse a transition state (peak maximum) with an intermediate (high energy minimum).

        2. *Rate determining (limiting) step* transition state corresponds to the highest point on a graph (step (2)) – a single elementary step (collision) whose rate limits overall multistep reaction rate.

4.16.1. Methyl free radical C atom hybridization – sp^2 or sp^3?

    a. Calculations do not give a clear answer.

    b. Spectroscopic studies are consistent with sp^2-hybridized C.

c. Problem 3: What would E_a be for the reverse of step (2) for *a.* chlorination and *b.* bromination?

chlorination = 17-7 = 10 kJ/mol
bromination = 75-73 = 2 kJ/mol

4.12.1. Relative reactivity – rate ratio of two similar reactions *carried out under identical conditions*.
   a. F_2 – explosion (very fast and exothermic) unless at low pressure, diluted with Ar.
   b. Cl_2 – fast and exothermic, but controllable at 300 K.
   c. Br_2 – slower and less exothermic. Usually must be heated.
   d. I_2 – very slow (does not react) and endothermic.

4.12.2. Methane bromination – occurs by the same steps as chlorination. Step (2) is rate determining.

4.12.3. Bromination is much slower than chlorination – step (2) E_a = 75 kJ/mol.

4.12.4. Reaction diagram comparison for chlorination *vs.* bromination step (2).

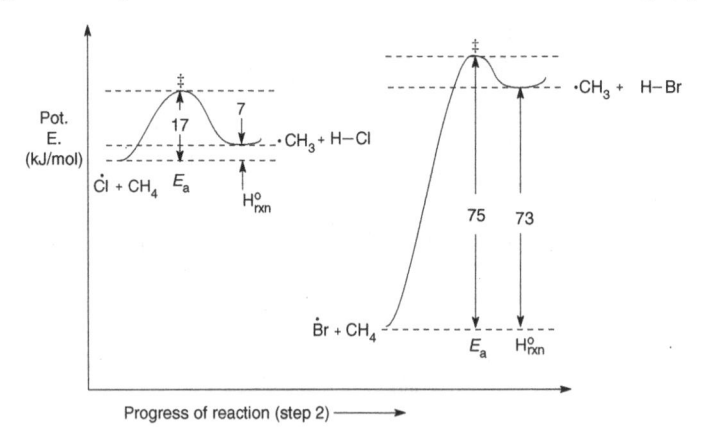

4.12.5. Rate theory applied to CH_4 chlorination *vs.* bromination:

4.12.6. Although Br· has a larger surface area for collisions than Cl·, it is also heavier and moves slower at a given T. Result – collision frequency is nearly size and mass independent. *Slightly different Z values (section 4.9.5) for chlorination vs. bromination cannot account for the large reaction rate difference.*

4.12.7. About one in eight collisions has proper orientation for *either* chlorination *or* bromination at 275°C. *The slightly different P values (section 4.9.5) for chlorination vs. bromination cannot account for the large reaction rate difference either.*

4.12.8. Compare the effect of activation energies:

Total Collisions	T (°C)	E_a (kJ/mol)	Collisions having energy ≥ E_a
chlorination 1.5×10^7	275	17	375,000
bromination 1.5×10^7	275	75	1

The large rate difference between chlorination and bromination is mainly due to the exponential terms (section 4.9.5), i.e. the respective step (2) E_a values.

4.12.9. Evidence that step (2) is rate determining:

4.12.10. E_a data have not been measured for all of the reactions below. Missing data are estimated from ΔH°_{rxn} data.

ΔH°_{rxn} values for X = halogen (kJ/mol)				
Reaction	F	Cl	Br	I
(1) X$_2$ $\xrightarrow{\text{UV light or 275°C}}$ 2 X·	+154	+240	+190	+149
(2) X· + CH$_4$ ⟶ HX + H$_3$C·	–131	+7	+73	+141
(3) H$_3$C· + X$_2$ ⟶ H$_3$CX + X·	–327	–110	–112	–92

a. $E_a = \Delta H^\circ_{rxn}$ for step (1), a bond homolysis. If step (1) were rate determining, relative rates would be:

(faster) I$_2$ > Br$_2$ > Cl$_2$ (slower) < F$_2$ (faster)

(contrary to facts). *Step (1) must not be rate determining (except possibly for F).*

b. E_a is probably small for the very exothermic step (3). *Step (3) is probably not the rate determining step.*

c. Observed relative rates are in the same order as ΔH°_{rxn}'s for step (2):

(faster) F$_2$ > Cl$_2$ > Br$_2$ > I$_2$ (slower).

Step (2) is likely the rate determining step.

		$e^{\frac{-E_a}{RT}} \times 10^6$	
Reaction	E_a (kJ/mol)	rel. rate 300 K	rel. rate 500 K
F$_2$ + CH$_4$ → HF + ·CH$_3$	5.0	140,000	300,000
Cl$_2$ + CH$_4$ → HCl + ·CH$_3$	17	1300	18,000
Br$_2$ + CH$_4$ → HBr + ·CH$_3$	75	9 × 10^{-8}	0.015
I$_2$ + CH$_4$ → HI + ·CH$_3$	140	2 × 10^{-19}	2 × 10^{-9}

Characteristics of step 2				
	F	Cl	Br	I
E_a (kJ/mol)	+5.0	+17	+75	≥+140
ratio of collisions having E_a at 275°C	1:10	1:40	1:1 × 10^6	1:1 × 10^{13}
chain length (cycles before termination)	> 10^4	10^3 - 10^4	10^2	no reaction

d. Problem 4. Methane bromination is retarded by HBr addition, owing to the reaction (reverse step 2):

What test could be done to see if this actually happens? Add excess D-Br. After awhile, analyze for CH₃D

4.12.1. Further evidence. A large deuterium isotope rate effect ($k^H / k^D = 12$) is observed for CH_4 versus CD_4 chlorination. (Deuterium = D = $_1^2H$.)

  a. Deuterium is "heavy hydrogen," a stable hydrogen isotope with both a proton and a neutron in the nucleus). A C-D bond is somewhat stronger than a C-H bond.
  b. If a C-H bond is broken in the rate determining step of a mechanism, then replacement of H by D should slow the reaction down.
  c. CH_4 reacts 12 times faster than CD_4. This observation strongly supports the idea that step (2) is the rate determining step.

4.13.1. Higher molecular mass alkane halogenation yields isomeric product mixtures.

Alkane	Number of monohalogenated isomers
CH_4	1
H_3CCH_3	1
$H_3CCH_2CH_3$	2
$H_3C(CH_2)_2CH_3$	2
2-methylpropane	2
$H_3C(CH_2)_3CH_3$	3
2-methylbutane	4

  a. H atom class (section 3.3.8, after 3.5.6) largely determines reactivity towards Cl· or Br· free radicals.

4.13.2. Examples: propane, n-butane, 2-methylpropane.

$H_3CCH_2CH_3$ $\xrightarrow{Cl_2}{hv / 25°C}$ $H_3CCH_2CH_2-Cl$ + H_3CCHCH_3
propane   1-chloropropane (bp 47°C) |
(excess)   40%   Cl
                                                  2-chloropropane (bp 36°C)
                                                            60%

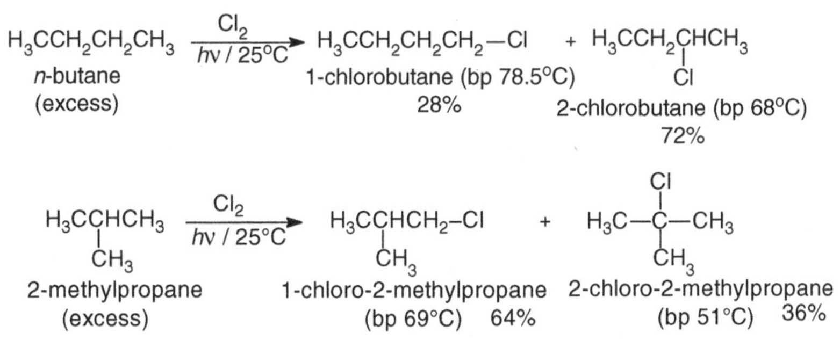

H₃CCH₂CH₂CH₃ →[Cl₂][hv / 25°C] H₃CCH₂CH₂CH₂—Cl + H₃CCH₂CHCH₃
                                                                           |
                                                                           Cl
n-butane   1-chlorobutane (bp 78.5°C)   2-chlorobutane (bp 68°C)
(excess)   28%   72%

H₃CCHCH₃ →[Cl₂][hv / 25°C] H₃CCHCH₂—Cl + H₃C—C(Cl)(CH₃)—CH₃
    | |
    CH₃   CH₃
2-methylpropane   1-chloro-2-methylpropane   2-chloro-2-methylpropane
(excess)   (bp 69°C) 64%   (bp 51°C) 36%

4.13.3. One isomer often predominates with Br₂.

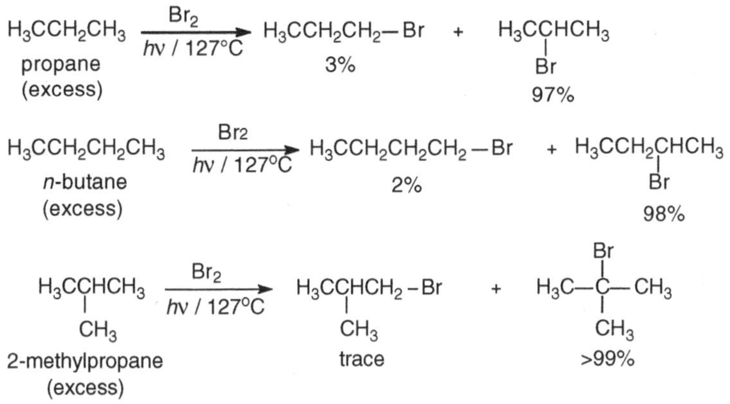

H₃CCH₂CH₃ →[Br₂][hv / 127°C] H₃CCH₂CH₂—Br + H₃CCHCH₃
                                                      |
                                                      Br
propane   3%   97%
(excess)

H₃CCH₂CH₂CH₃ →[Br₂][hv / 127°C] H₃CCH₂CH₂CH₂—Br + H₃CCH₂CHCH₃
                                                              |
                                                              Br
n-butane   2%   98%
(excess)

H₃CCHCH₃ →[Br₂][hv / 127°C] H₃CCHCH₂—Br + H₃C—C(Br)(CH₃)—CH₃
    | |
    CH₃   CH₃
2-methylpropane   trace   >99%
(excess)

4.13.4. Bromination is occasionally a useful synthetic reaction (as for 2-bromo-2-methylpropane above).

4.13.5. Propane chlorination mechanism:

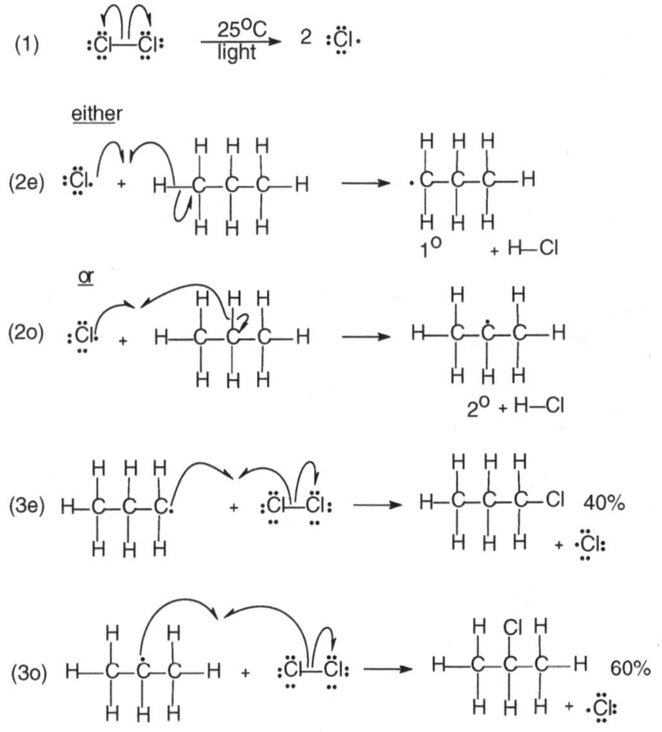

(1) :Cl—Cl: →[25°C][light] 2 :Cl·

either

(2e) :Cl· + H₃C—CH₂—CH₃ → ·CH₂—CH₂—CH₃ (1°) + H—Cl

or

(2o) :Cl· + H₃C—CH₂—CH₃ → H₃C—·CH—CH₃ (2°) + H—Cl

(3e) ·CH₂—CH₂—CH₃ + :Cl—Cl: → Cl—CH₂—CH₂—CH₃ 40% + ·Cl:

(3o) H₃C—·CH—CH₃ + :Cl—Cl: → H₃C—CHCl—CH₃ 60% + ·Cl:

a. If more 2° product forms than 1°, 2° product must form faster.
b. Step 2e or 2o is rate limiting. Step 3e or 3o is fast by comparison.

4.13.7. Orientation – site where reaction occurs (1° *vs.* 2° H abstraction).

4.13.8. Propane – has six 1° H atoms : two 2° H atoms. Cl· collision with a 1° H atom is 3 times more likely than with a 2° H atom.

    a. Yields would be 75% 1-chloropropane : 25% 2-chloropropane if all H atoms reacted at the same rate.

    b. 2° product actually forms in 60% yield. A 2° H atom must experience 4.5 times more effective Cl· collisions than a 1° H atom. E_a must be smaller for 2° H atoms than for 1° H atoms.

4.13.9. Isobutane – nine 1° H atoms vs. one 3° H atom. If all H atoms reacted at the same rate, products should be 90% 1°, and 10% 3°. Yields are actually 64% 1° and 36% 3°. 5.5 times more effective collisions occur for each 3° H atom than for each 1° H atom.

4.13.10. Relative H atom reactivities towards Cl· have been averaged for each class over a large number of alkanes:

Class	3° H	2° H	1° H	methane
Rel. rate	5.5	4.5	1.0	0.0037*

*from Morrison, R. T. and Boyd, R. N., "Organic Chemistry," 6th ed., Prentice-Hall, Inc., New Jersey, (1992), or March, J., Smith, M. B., "Advanced Organic Chemistry, Reactions, Mechanisms, and Structure", 6th ed., Hoboken, N. J., Wiley-InterScience (2007).

    a. An ethane H atom reacts 270 times faster than a methane H atom.

4.13.11. Example: estimate the relative % monochloro products formed in 2,3-dimethylbutane chlorination.

$$3° \text{ product ratio} = \frac{rate\ 3°\ H\ abstr.}{rate\ 3°\ H\ abstr.\ +\ rate\ 1°\ H\ abstr.}$$

$$= \frac{2\ 3°\ H\ atom \times \dfrac{5.5\ effective\ collisions}{3°\ H\ atom}}{2\ 3°\ H\ atom \times \dfrac{5.5\ effective\ collisions}{3°\ H\ atom} + 12\ 1°\ H\ atom\ \dfrac{1.0\ effective\ collisions}{1°\ H\ atom}}$$

$$= \frac{2 \times 5.5}{(12 \times 1.0) + (2 \times 5.5)} = 0.48$$

% yield 3° product = 0.48 × 100 = 48%

A similar calculation for the % 1° product yield = 52%.

a. Problem 5. Predict the isomeric product proportions for the following alkane monochlorinations (at 25°C with excess alkane vs. Cl$_2$). Use average reaction rates from section 4.13.10.

1. propane H–C–C–C–H (with H's)

1-chloropropane: $\dfrac{6 \times 1.0 \ (1°H)}{(6 \times 1.0)+(2 \times 4.5)} = 0.4$

2-chloropropane: $\dfrac{2 \times 4.5 \ (2°H)}{(6 \times 1.0)+(2 \times 4.5)} = 0.6$

40% 1-chloropropane
60% 2-chloropropane

2. isobutane

1-chloro-2-methylpropane: $\dfrac{9 \times 1.0 \ (1°H)}{(9 \times 1.0)+(1 \times 5.5)} = 0.62$

2-chloro-2-methylpropane: $\dfrac{1 \times 5.5 \ (3°H)}{(9 \times 1.0)+(1 \times 5.5)} = 0.38$

62% 1-chloro-2-methylpropane
38% 2-chloro-2-methylpropane

3. n-pentane

1-chloropentane: $\dfrac{6 \times 1.0 \ (1°H)}{(6 \times 1.0)+(4 \times 4.5)+(2 \times 4.5)} = 0.18$

2-chloropentane: $\dfrac{4 \times 4.5 \ (2°H)}{(6)+(18)+(9)} = 0.55$

3-chloropentane: $\dfrac{2 \times 4.5 \ (2°H)}{6+18+9} = 0.27$

1-chloropentane: 18%
2-chloropentane: 55%
3-chloropentane: 27%

4. 2,2,3-trimethylbutane

1-chloro-2,2,3-trimethylbutane: $\dfrac{9 \times 1.0 \ (1°H)}{9+6+5.5} = 0.44$

1-chloro-2,3,3-trimethylbutane: $\dfrac{6 \times 1.0 \ (1°H)}{20.5} = 0.29$

2-chloro-2,3,3-trimethylbutane: $\dfrac{5.5 \ (3°H)}{20.5} = 0.27$

44% 1-chloro-2,2,3-trimethylbutane
29% 1-chloro-2,3,3-trimethylbutane
27% 2-chloro-2,3,3-trimethylbutane

5. 2,2,4-trimethylpentane

1-chloro-2,2,4-trimethylpentane: $\dfrac{9 \times 1.0 \ (1°H)}{9+9+4.5+6} = 0.32$

3-chloro-2,2,4-trimethylpentane: $\dfrac{2 \times 4.5 \ (2°H)}{28.5} = 0.32$

1-chloro-2,4,4-trimethylpentane: $\dfrac{6 \times 1.0 \ (1°H)}{28.5} = 0.21$

2-chloro-2,4,4-trimethylpentane: $\dfrac{5.5 \ (3°H)}{28.5} = 0.19$

32% 1-chloro-2,2,4-trimethylpentane
32% 3-chloro-2,2,4-trimethylpentane
21% 1-chloro-2,4,4-trimethylpentane
19% 2-chloro-2,4,4-trimethylpentane

4.13.12. How was methane rate 0.0037 measured?

4.13.13. By a competition experiment – where two reactants compete for a limited amount of reagent.

a. An equimolar methane/ethane mixture is chlorinated, with methane/ethane mixture in large excess relative to Cl$_2$. Product amounts are determined.

$$CH_4 + H_3CCH_3 \xrightarrow[25°C]{\underset{UV \ light}{Cl_2}} H_3C\text{–}Cl + H_3CCH_2\text{–}Cl + H\text{–}Cl$$

equimolar, large excess   1 : 400

1. Chloroethane : chloromethane ratio =

$$\dfrac{400}{1.0} = \dfrac{\text{rate ethane H}}{\text{rate methane H}} = \dfrac{6 \ 1° \ H \ \text{atoms} \times \dfrac{1.0 \ \text{effective collisions}}{1° \ H \ \text{atom}}}{4 \ \text{methane H atoms} \times \dfrac{n \ \text{effective collisions}}{\text{methane H atom}}}$$

$$400 = \dfrac{6}{4n}$$

solve for n: $n = 0.0037$

b. An ethane H atom reacts $\frac{1}{n}$ = 270 times faster than a methane H atom.

c. Problem 6. Limited chlorine (Cl$_2$) reacts with an equimolar ethane / 2,2-dimethylpropane mixture yielding 2.3 times more mols of 1-chloro-2,2-dimethylpropane than 1-chloroethane. Do 2,2-dimethylpropane 1° H atoms react at the same rate as ethane 1° H atoms? What is the relative rate ratio per 1° H atom?

4.13.14. Chlorination rate differences between 1°, 2°, and 3° H atoms are not large (1.0 - 5.5).

a. Higher alkane free radical chlorination usually yields mixtures – not usually useful synthetically.

4.13.15. Average hydrogen atom relative reactivity toward bromine (125°C) for each class over a large number of alkanes.

3° H	2° H	1° H	methyl
1600*	97	1.0	0.0007*

*from Morrison, R. T. and Boyd, R. N., "Organic Chemistry," 6th ed., Prentice-Hall, Inc., New Jersey, (1992), or March, J., Smith, M. B., "Advanced Organic Chemistry, Reactions, Mechanisms, and Structure", 6th ed., Hoboken, N. J., Wiley-Interscience (2007).

a. Bromination rate differences are much larger than for chlorination.
b. Problem 7. Repeat problem 5 for bromination, using average relative rates in section 4.13.15 above.

4.13.16. First propagation step (step 2) activation energies for free radical halogenation:

(2) X• + H–R ⟶ R• + H–X

R	E_a (X = Cl) kJ / mol	E_a (X = Br) kJ / mol
CH$_3$	17	75
1°	6.1	54*
2°	2.1*	42*
3°	0.42*	31*

*(from Morrison, R. T. and Boyd, R. N., "Organic Chemistry," 6th ed., Prentice-Hall, Inc., New Jersey, (1992).)

    a. Propane chlorination, step (2), 1° vs. 2° reaction diagram.

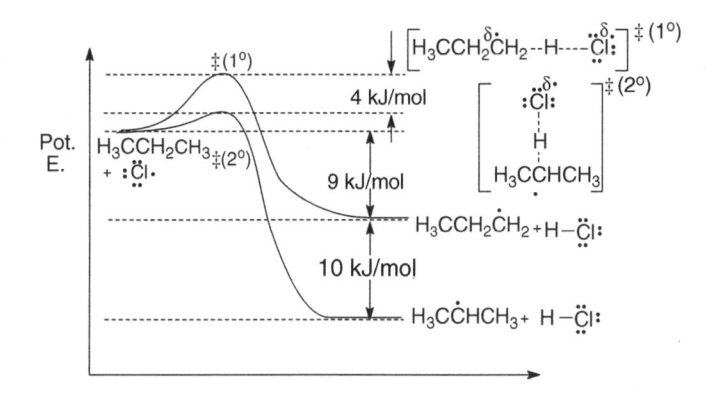

4.13.17. When $E_a \gg T$, reaction rate will change rapidly with temperature, according to the Eyring general rate equation, section 4.9.5.

$$rate = PZe^{-\frac{E_a}{RT}} = k[A]^x[B]^y$$

4.13.18. When $E_a \ll T$ (i.e., at high T), exponential term $\to 1$ as $(-E_a/RT) \to 0$, and all H (methane, 1°, 2°, and 3°) classes will react at nearly the same rate.

    a. Leveling effect – methane, 1°, 2°, and 3° H atom chlorination rates are "leveled" (nearly the same) at very high temperatures.

4.13.19. Rate differences between methane, 1°, 2°, and 3° H atoms will be larger when E_a is larger and when temperatures are lower.

    a. Selectivity – Cl·'s or Br·'s ability to distinguish a more reactive from a less reactive H atom during reaction.

    b. E_a is often higher for endothermic than for exothermic reactions.

    c. Propane bromination, step (2), 1° vs. 2° reaction diagram.

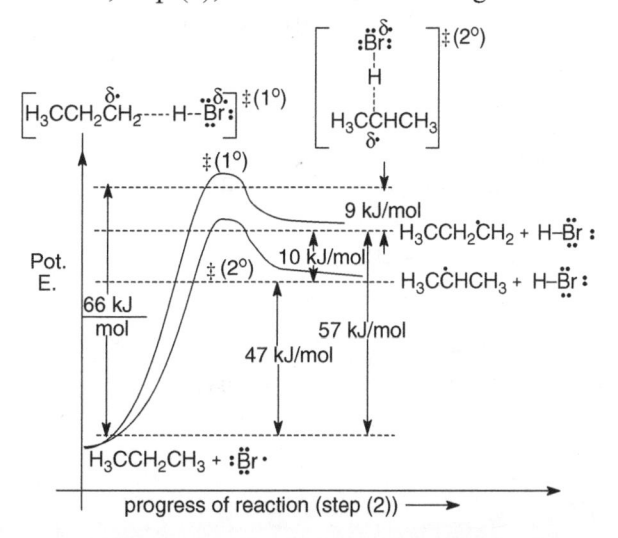

d. A transition state structure cannot normally be investigated experimentally, because it doesn't survive long enough to isolate and study.

e. The Hammond postulate. A transition state structure is assumed to be similar to other more stable structures that are near it (in potential energy) on a reaction curve.
   1. When comparing similar reactions (such as alkane chlorination *vs.* bromination), greater $E_a \Rightarrow \ddagger_2$ structure looks more like products (and less like reactants).

4.14.2. The two transition state structures for propane chlorination are near reactants on the reaction curve (section 4.13.16a), and the transition state structure is assumed to resemble reactants, with little free radical character (unpaired electron density) on C.

4.14.3. The two transition state structures for propane bromination are near the alkyl free radical on the reaction curve (section 4.13.19c), and the transition state structure is assumed to resemble the alkyl free radical intermediate, having much free radical character on C.

4.13.20. If an alkyl free radical intermediate is more stable (relative to the alkane from which it is formed), the transition state structure leading to its formation is also assumed to be more stable.

a. Transition state stability parallels intermediate free radical stability.

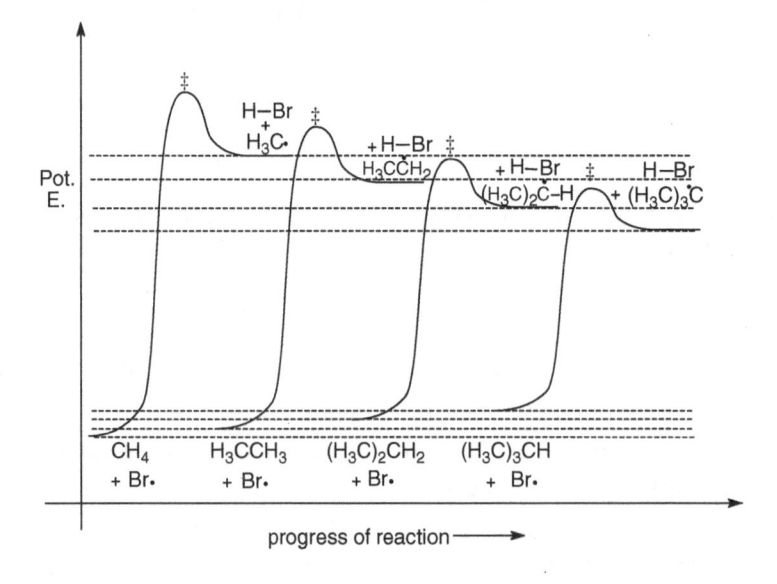

4.13.21. If a transition state structure is more stable, E_a will be lower, and reaction will be faster.

4.13.22. Halogenation rates parallel free radical stabilities – measured from bond homolysis energies ($\Delta H^°_{rxn} = E_a$).

	$\Delta H^°_{rxn}$ (kJ/mol)
$H_3C-H \longrightarrow H_3C\cdot + \cdot H$	439
$H_3CCH_2-H \longrightarrow H_3C\dot{C}H_2 + \cdot H$	423
$H_3CCH_2CH_2-H \longrightarrow H_3CCH_2\dot{C}H_2 + \cdot H$	423
$H_3C-CH(H)-CH_3 \longrightarrow H_3C-\dot{C}H-CH_3 + \cdot H$	413
$H_3C-C(CH_3)(H)-CH_3 \longrightarrow H_3C-\dot{C}(CH_3)-CH_3 + \cdot H$	403

4.13.23. Relative free radical stabilities:

(handwritten note: more PE = less stable; larger = more space to hold e⁻)

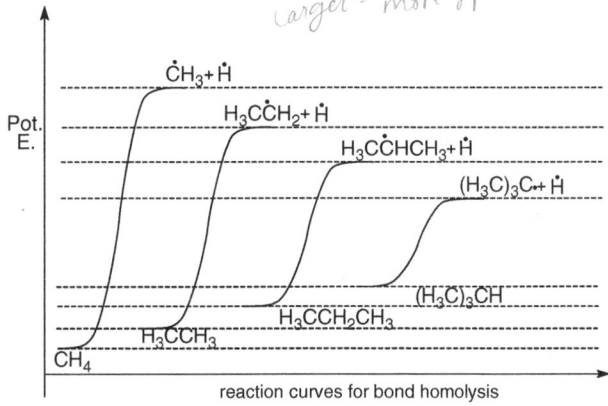

[Diagram: Potential Energy vs reaction curves for bond homolysis, showing curves from CH₄ → ĊH₃+Ḣ (highest), H₃CCH₃ → H₃CĊH₂+Ḣ, H₃CCH₂CH₃ → H₃CĊHCH₃+Ḣ, (H₃C)₃CH → (H₃C)₃Ċ+Ḣ (lowest activation)]

(more stable) 3° > 2° > 1° > methyl (less stable)

- **a.** Problem 8. Using bond dissociation enthalpies (section 4.7.1b) add the following free radicals to the stability series above.

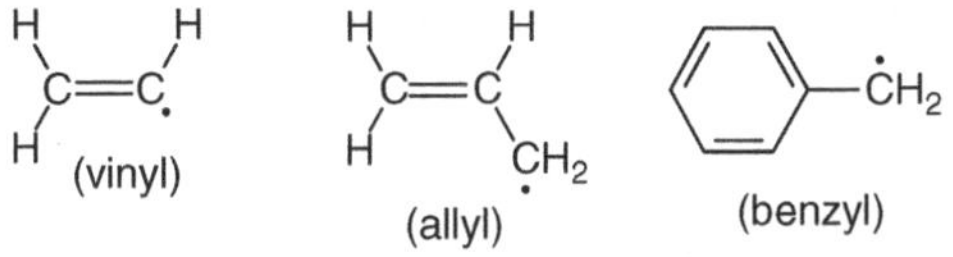

  (vinyl) H₂C=ĊH   (allyl) H₂C=CH–ĊH₂   (benzyl) C₆H₅–ĊH₂

4.13.24. The more stable R· (relative to the alkane from which it was formed), the faster it forms (the lower E_a).

4.13.25. Alkyl free radical stability governs orientation and reactivity.

4.13.26. Transition state structures (‡₂) for higher alkane chlorination *vs.* bromination.

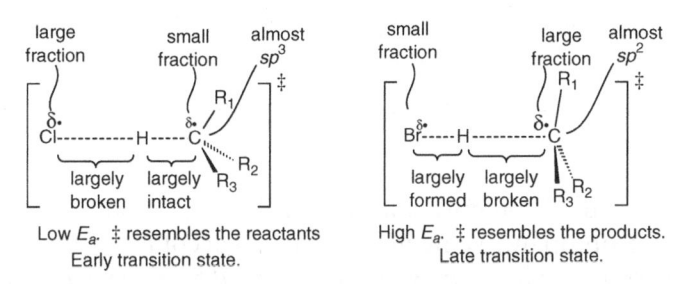

[Diagram of two transition states:
Left: Cl------H------C(R₁)(R₂)(R₃) — large fraction / small fraction, almost sp³, largely broken / largely intact. Low E_a. ‡ resembles the reactants. Early transition state.
Right: Br------H------C(R₁)(R₂)(R₃) — small fraction / large fraction, almost sp², largely formed / largely broken. High E_a. ‡ resembles the products. Late transition state.
R₁, R₂, R₃ = H or alkyl groups]

4.13.27. The larger δ· on C, the more delocalization (section 1.9.3) operates and causes larger rate differences between methane, 1°, 2°, and 3° H atoms.

- **a.** C is almost a full free radical for bromination, accounting for its large rate differences. For chlorination, δ ≈ 0.17 (small fraction), accounting for its small rate differences.

4.13.28. Can a less stable free radical rearrange its C atoms into a more stable free radical (below)?

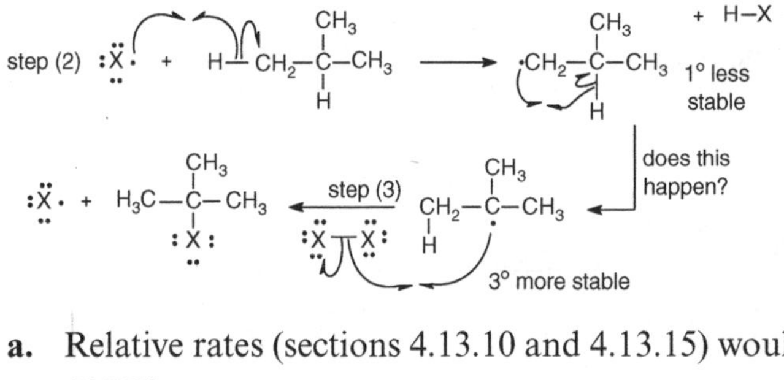

step (2) :Ẍ· + H–CH₂–C(CH₃)(H)–CH₃ ⟶ ·CH₂–C(CH₃)(H)–CH₃ + H–X   1° less stable

:Ẍ· + H₃C–C(CH₃)(:X:)–CH₃ ⟵step (3)⟵ CH₂–C(CH₃)(H)–CH₃ ←does this happen?

3° more stable

- **a.** Relative rates (sections 4.13.10 and 4.13.15) would not be accurate if rearrangement occurs.

4.13.29. H. C. Brown, G. Russell – made 2-deuterio-2-methylpropane:

$$\underset{\underset{CH_3}{|}}{\overset{\overset{CH_3}{|}}{H_3C-C-H}} \xrightarrow[\text{UV light, heat}]{Br_2 \text{ (one mol)}} \underset{\underset{CH_3}{|}}{\overset{\overset{CH_3}{|}}{H_3C-C-Br}} + H-Br \xrightarrow[\text{dry ether (solvent)}]{Mg} \underset{\underset{CH_3}{|}}{\overset{\overset{CH_3}{|}}{H_3C-C-MgBr}} \xrightarrow{D_2O} Mg^{2\oplus} + Br^{\ominus} + OH^{\ominus} + \underset{\underset{CH_3}{|}}{\overset{\overset{CH_3}{|}}{H_3C-C-D}}$$

4.13.30. Some 2-chloro-1-deuterio-2-methylpropane would be obtained if rearrangement occurs. None was actually found.

$$\underset{\underset{D}{|}}{\overset{\overset{CH_3}{|}}{H_3C-C-CH_3}} \xrightarrow[\text{step (2)}]{:\ddot{Cl}\cdot} \xrightarrow[\text{step (3)}]{Cl_2} \underset{\underset{Cl}{|}}{\overset{\overset{CH_3}{|}}{H_3C-C-CH_2-D}} ? + H-Cl$$

  a. The H-Cl : D-Cl ratio was the same as the 1-chloro-2-deuterio-2-methylpropane : 2-chloro-2-methylpropane ratio, i.e., no rearrangement occurred.
  b. Problem 9. How would the H-Cl : DCl product ratio compare to the 1-chloro-2-deuterio-2-methylpropane : 2-chloro-2-methylpropane product ratio if some isobutyl free radicals had rearranged into *tert*-butyl free radicals?

$$\underset{\underset{H}{|}}{\overset{\overset{CH_3}{|}}{H_3C-C-CH_2\cdot}} \longrightarrow \underset{\underset{\cdot}{|}}{\overset{\overset{CH_3}{|}}{H_3C-C-CH_3}}$$

  2. What results would have been expected if the following occurred?

$$\underset{\underset{H}{|}}{\overset{\overset{CH_3}{|}}{H_3C-C-CH_2\cdot}} \quad \underset{\underset{H}{|}}{\overset{\overset{CH_3}{|}}{H_3C-C-CH_3}} \longrightarrow \underset{\underset{H}{|}}{\overset{\overset{CH_3}{|}}{H_3C-C-CH_3}} + \underset{\underset{H}{|}}{\overset{\overset{CH_3}{|}}{H_3C-\overset{\cdot}{C}-CH_3}}$$

  c. Problem 10. 1-Bromopropane and 2-bromopropane monochlorination both yield the same product, 1-bromo-2-chloropropane. Write mechanisms to account for this. *Hint: a less stable free radical can rearrange into a more stable free radical by halogen migration from one C to another, but not by hydrogen migration.*

4.13.31. Cycloalkanes undergo usual alkane reactions:
  a. Free radical substitution.

$$\text{cyclopentane (excess)} \xrightarrow[\text{300°C}]{Br_2 \text{ (one mol), UV light}} \text{bromocyclopentane} + H-Br$$

4.15.1. Free radical inhibitors – substances that consume (or deactivate) free radicals so that chain reactions become very slow. Two examples are hydroquinone and diphenylamine.

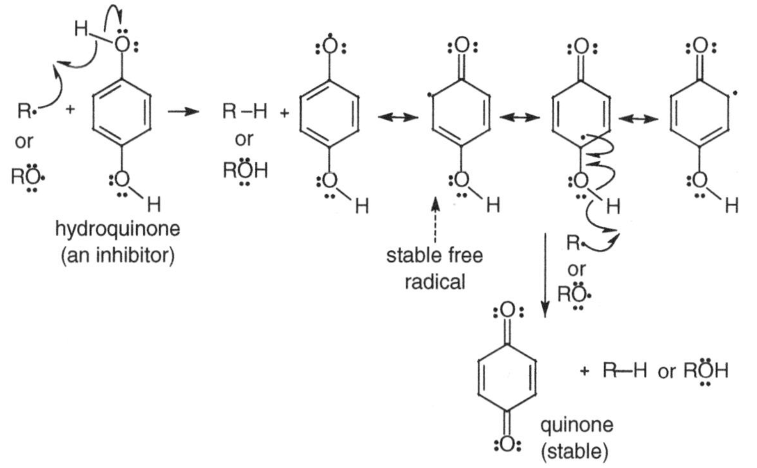

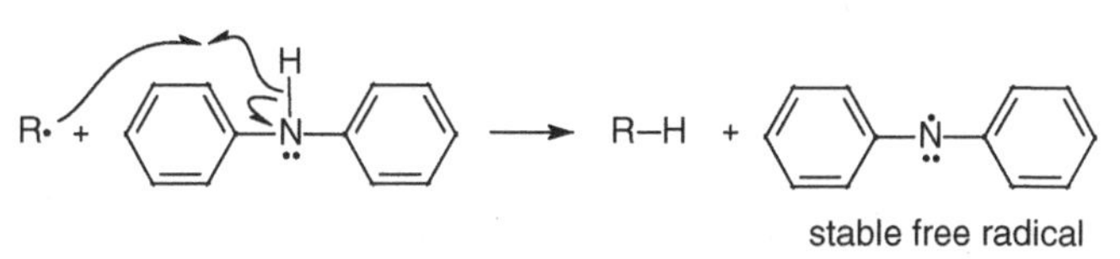

a. These free radicals are extra stable due to lone electron delocalization (through resonance).
b. The reaction curves below show how inhibitors slow chain reactions.

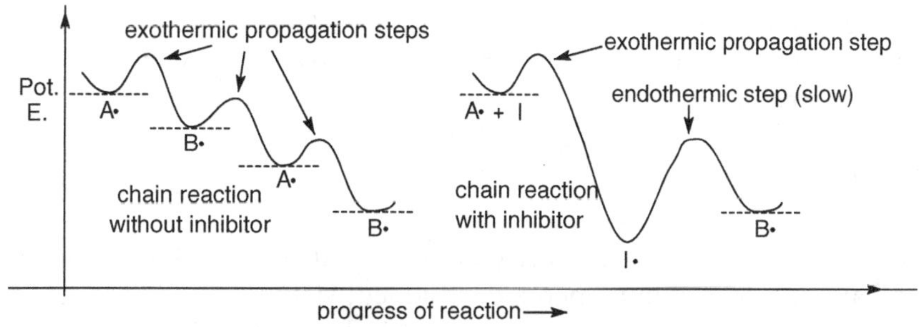

c. Some other common (edible) free radical inhibitors (antioxidants):

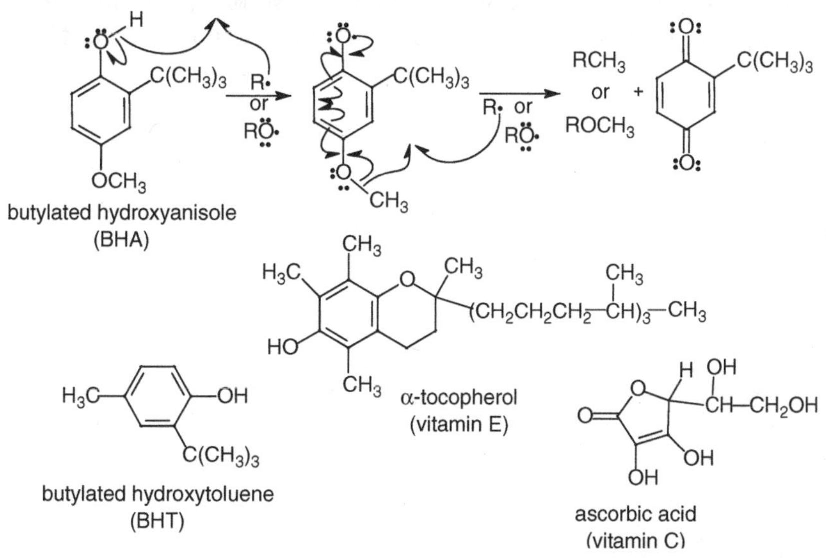

butylated hydroxyanisole (BHA)

butylated hydroxytoluene (BHT)

α-tocopherol (vitamin E)

ascorbic acid (vitamin C)

4.15.2. Free radical initiators. RO-OR' (peroxide) bond is very weak (151 kJ/mol) and easily homolyzes to yield free radicals. *t*-Butyl and benzoyl peroxides and **AIBN** are added as *initiators*, substances that promote (free radical) reactions.

di-*tert*-butylperoxide → 2 *tert*-butoxy radicals, ΔH°_{rxn} +151 kJ/mol

benzoyl peroxide → 2 benzoyloxy radicals (heat or UV light)

azo-(bis)isobutyronitrile (AIBN) → 2 radicals + N≡N (heat or UV light)

a. Example: a free radical from an initiator can react with halogen:

$(CH_3)_3C-O\cdot + :Cl-Cl: \longrightarrow (CH_3)_3C-O-Cl: + :Cl\cdot$

b. Once Cl· forms, it can free radical chlorinate (step 2). Initiators are an alternative to heat or UV light (although all three are often used together).

c. Problem 11. *tert*-Butyl peroxide is a relatively stable, easy to handle liquid that is a convenient free radical source.

$(H_3C)_3C-O-O-C(CH_3)_3 \xrightarrow[\text{or light}]{130°C} 2\ (H_3C)_3C-O\cdot$

A 2-methylpropane / CCl_4 mixture is stable at 130 - 140°C. A reaction occurs when a small amount of *tert*-butyl peroxide is added, yielding main products 2-chloro-2-methylpropane and trichloromethane. 2-Methylpropan-2-ol is also obtained in a stoichiometric amount as the *tert*-butyl peroxide added. Use electron pushing arrows to write possible propagation steps showing how these two products might form. Show all elementary steps (collisions), intermediates, lone pair and unpaired electrons (as dots).

4.16.2. Introduction to reactive intermediates. C almost always forms four bonds to adjacent atoms in stable compounds.

    a. Five bonds to C would mean C shares 10 electrons and violates octet rule – almost never happens.

    b. C sometimes has two or three bonds, but almost all such species are highly unstable and reactive.

4.16.3. An alkyl free radical's central *sp²*-hybridized C atom (intermediate in alkane halogenation) has only seven electrons (section 4.16.1) and three bonds.

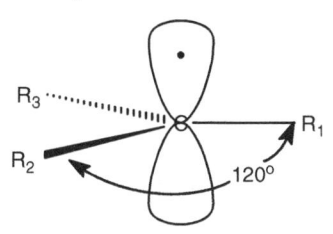

4.16.4. Similar reactive species, *carbocations*, and *carbanions*, will be encountered later.

    a. A carbocation's *sp²*-hybridized central C atom has a + formal charge, only six shared electrons (three bonds), and its $2p_z$-orbital is empty (see section 6.13.7).

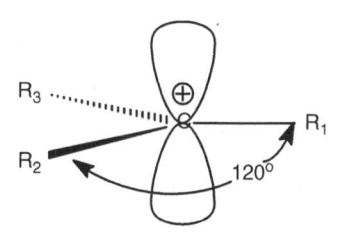

    b. C has intermediate electronegativity (2.5). A carbanion's *sp³*-hybridized central C atom has a − formal charge, only six shared electrons (three bonds), and the fourth *sp³*-orbital contains a lone electron pair.

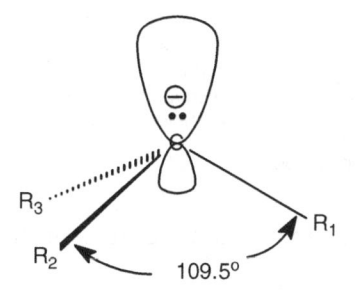

4.16.5. Alkyl free radicals and carbocations are *electron deficient* (electron seeking) and *electrophilic* (Lewis acids). The central C atom withdraws some electron density from adjacent σ-bonds toward the R groups, which helps stabilize it (*inductive effect*).

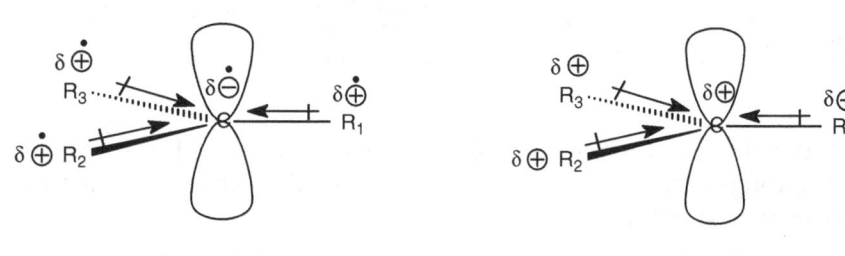

free radical   carbocation

4.13.32. Alternative point of view – lone electron or + charge is delocalized onto adjacent alkyl groups.

ethyl free radical

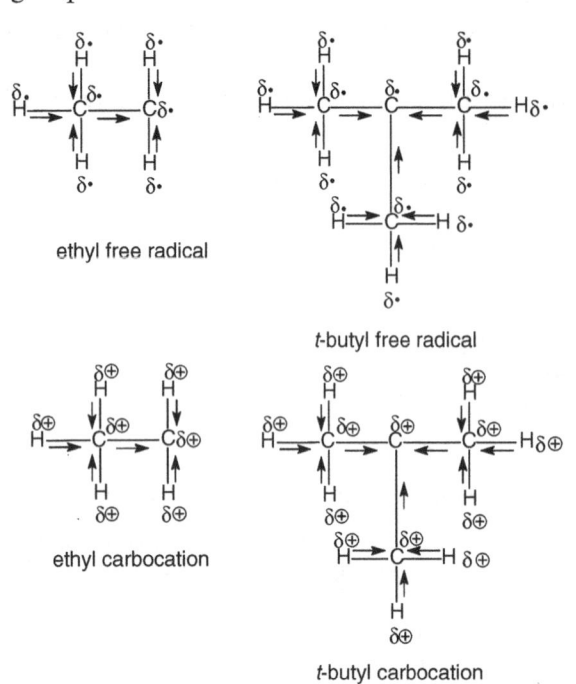

t-butyl free radical

ethyl carbocation

t-butyl carbocation

- **a.** Henry Eyring – "charged particles have claustrophobia." The more the three bonded groups disperse (delocalize) the lone electron or + charge, the more stable (sometimes called the *polar effect*).

4.13.33. Electronegative (electron withdrawing) groups pull electron density away from the already electron deficient C· or C+, destabilizing it.

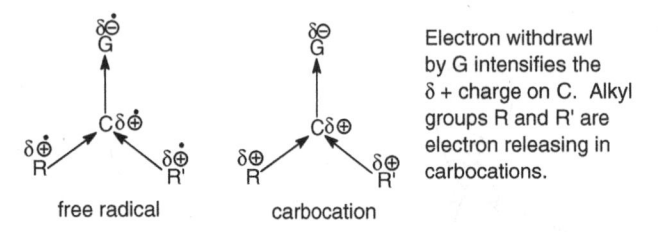

free radical   carbocation

Electron withdrawl by G intensifies the δ + charge on C. Alkyl groups R and R' are electron releasing in carbocations.

- **a.** Electropositive (electron releasing) groups push electron density towards C+, stabilizing it.

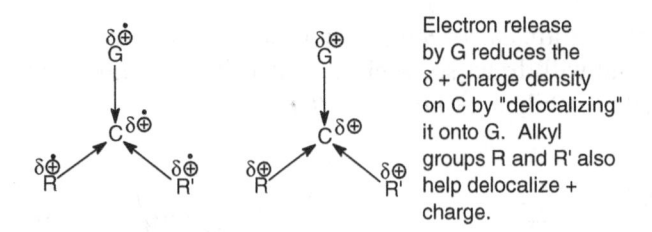

Electron release by G reduces the δ + charge density on C by "delocalizing" it onto G. Alkyl groups R and R' also help delocalize + charge.

- **b.** More alkyl groups bonded to C· or C+ ⇒ more stable. Electronegative C· or C+ induces electron release from adjacent alkyl groups. Alkyl groups sometimes withdraw electron density to stabilize anions.

4.13.34. Two main electron release/withdrawl modes of action exist:
  a. Inductive effect – characteristics:
   1. Electrostatic effect (Coulombic electrical charge attraction or repulsion), most often electron withdrawing.
      a. Alkyl groups are usually weakly electron releasing though.
   2. Operates only through σ-bonds (not through π-bonds) or through space.
   3. Falls off rapidly with (the square of) the distance.
   4. Magnitude is not affected much by other structural features.
   5. Can be stabilizing or destabilizing.
  b. Resonance effect (introduced in section 1.9.2) – characteristics:
   1. Also an electrostatic effect, but can be either electron releasing or withdrawing.
   2. Commonly operates through π-bonds. If molecular orbital geometries are favorable, can operate through σ-bonds (hyperconjugation).
   3. Does not fall off much with increasing distance.
   4. Its magnitude is variable, depending on molecular orbital geometry in a particular structure.
   5. Is usually used to explain stabilizing effects.

4.13.35. Alkyl group electron release. Rules for writing hyperconjugation (σ-bond resonance) forms were given in section 1.9.2.
  a. Example: ethyl free radical.

ethyl free radical electron release by adjacent methyl group (inductive effect)

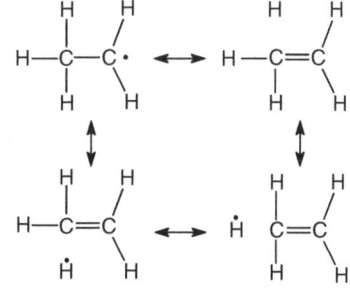

ethyl free radical σ-bond resonance contributing structures

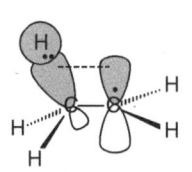

molecular orbital picture -- overlap of empty 2p_z orbital with a C-H σ-bonds on the adjacent methyl group -- stabilizes lone electron

σ-bond resonance hybrid picture -- lone electron delocalized onto adjacent methyl H atoms

b. Example: ethyl carbocation.

ethyl carbocation electron release by adjacent methyl group (inductive effect)

ethyl carbocation σ-bond resonance contributing structures

molecular orbital picture -- overlap of empty $2p_z$ orbital with a C-H σ-bonds on the adjacent methyl group -- stabilizes the + charge

σ-bond resonance hybrid picture -- + charge delocalized onto adjacent methyl H atoms

c. Example: isopropyl free radical hyperconjugation forms:

hybrid

4.13.36. Summary. More important hyperconjugation forms that can be drawn ⇒ more stable. Isopropyl is more stable than ethyl free radical. The same is true for carbocations.

4.13.37. Relative free radical and carbocation stabilities:

more stable 3° > 2° > 1° > methyl less stable

4.13.38. Carbanions are electron rich and are extremely strong bases. Stability order is usually reversed.

more stable methyl > 1° > 2° > 3° less stable

a. Ammonia (normally a weak base) is a very weak acid in this example:

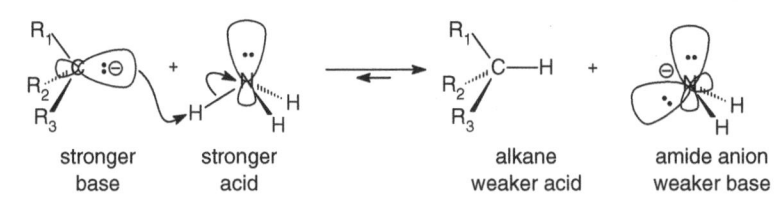

stronger base   stronger acid   alkane weaker acid   amide anion weaker base

b. Alkyl group electron release would localize, intensify, and destabilize a - charge on the central C atom in a carbanion.

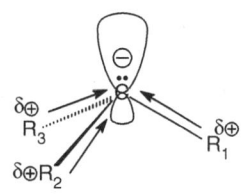

4.13.39. Conventional (π-bond) resonance can stabilize an intermediate. How allyl intermediates form will be described later.

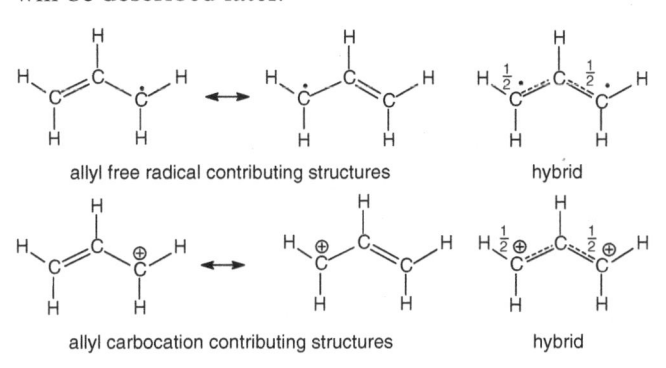

allyl free radical contributing structures   hybrid

allyl carbocation contributing structures   hybrid

a. Resonance can also stabilize a carbanion.

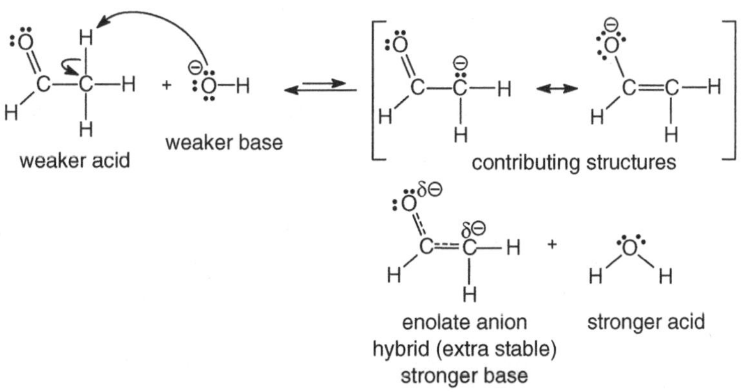

weaker acid   weaker base   contributing structures

enolate anion   stronger acid
hybrid (extra stable)
stronger base

4.16.6. Methylene (or carbene, CH_2, identified spectroscopically in 1959) is a highly reactive reagent, most often used to make cyclopropane rings (section 8.11).

a. Although it has no formal charge, methylene's C atom violates the octet rule, and is electrophilic. Lone electrons can be paired (singlet form) or unpaired (triplet form). It is an extremely toxic and explosive gas.

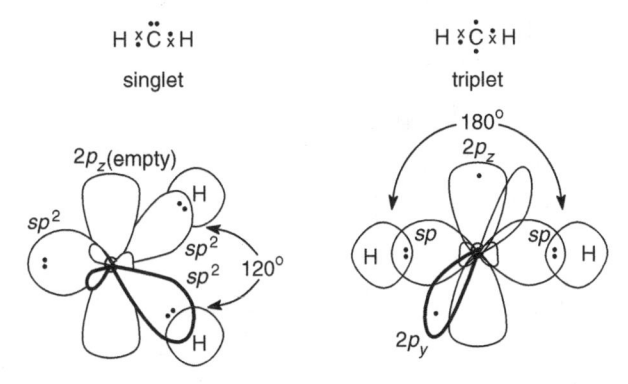

singlet   triplet

4.16.7. Preparation – from diazomethane or ketene photolysis (cleavage with light) or heating.

$$\left[H_2\overset{\ominus}{\underset{..}{C}}-\overset{\oplus}{N}=\overset{..}{N}: \leftrightarrow H_2C=\overset{\oplus}{N}=\overset{\ominus}{\underset{..}{N}}: \leftrightarrow H_2\overset{\ominus}{\underset{..}{C}}-\overset{\oplus}{N}\equiv N: \right]$$
diazomethane

or

$$\left[H_2\overset{\ominus}{\underset{..}{C}}-\overset{\oplus}{C}=\overset{..}{O}: \leftrightarrow H_2C=C=\overset{..}{\underset{..}{O}}: \leftrightarrow H_2\overset{\ominus}{\underset{..}{C}}-C\equiv \overset{\oplus}{O}: \right]$$
ketene

↓ light or heat

:CH$_2$ +   :N≡N: or   :$\overset{\ominus}{C}$≡$\overset{\oplus}{O}$:

carbene
(singlet)

a. Singlet form characteristics.
  1. sp^2-Hybridized C. Lone electron pair is in an sp^2 orbital.
  2. Empty unhybridized $2p_z$ orbital.
  3. 120° H-C-H bond angle.
  4. Often the form first created from photolysis. Conversion to triplet is slow at dry ice temperature (-78°C) in liquid phase.
  5. Less stable than triplet form.
  6. Cycloaddition to a π-bond (usually in liquid solution phase) involves only one elementary step – a *concerted* reaction (bonds break and form at the same time). More substituted alkenes react faster.

    a. *Cis* alkene yields only *cis* product (no *trans*) – a *stereospecific* reaction (section 8.9.3b). Both new σ-bonds form from the same alkene face (*syn*-addition).

b. Triplet form characteristics.
  1. sp-Hybridized C. Unpaired lone electrons; one in $2p_y$ and one in $2p_z$ – a diradical.
  2. 180° H-C-H bond angle.
  3. Forms from singlet (especially in the gas phase by collision with inert gas Ar or N_2).
  4. More stable than singlet.
  5. Cycloaddition to a π-bond is not stereospecific; both *cis* and *trans* products form from *cis* alkene. Two elementary steps (collisions) involved.

4.16.8. Annoying side reactions: (1) (either singlet or triplet form) insertion into C-H bonds, and (2) dimerization.

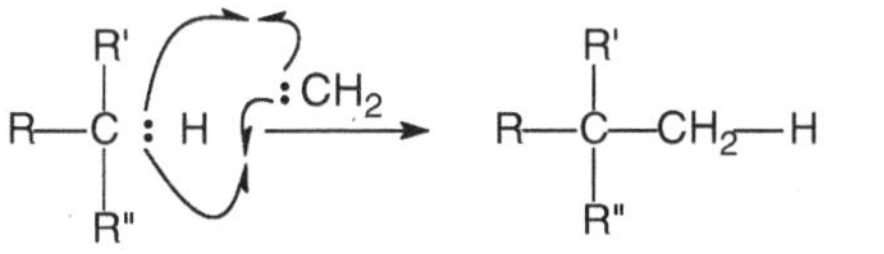
chain lengthened by
one -CH2- unit

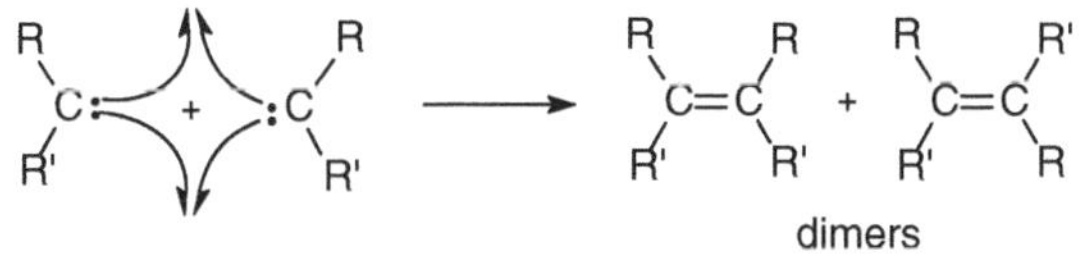
dimers
(cis and trans)

a. Example:

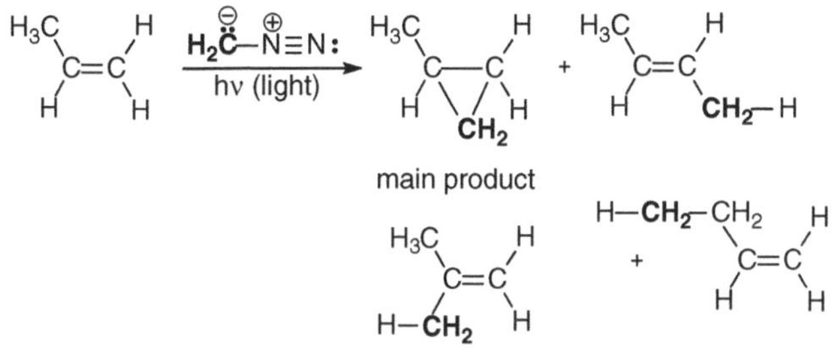
hν (light)
main product

4.16.9. Dichlorocarbene is made by $CHCl_3$ reaction with potassium *tert*-butoxide (KO*t*-Bu, a strong base), a 1,1- (or α-) elimination.

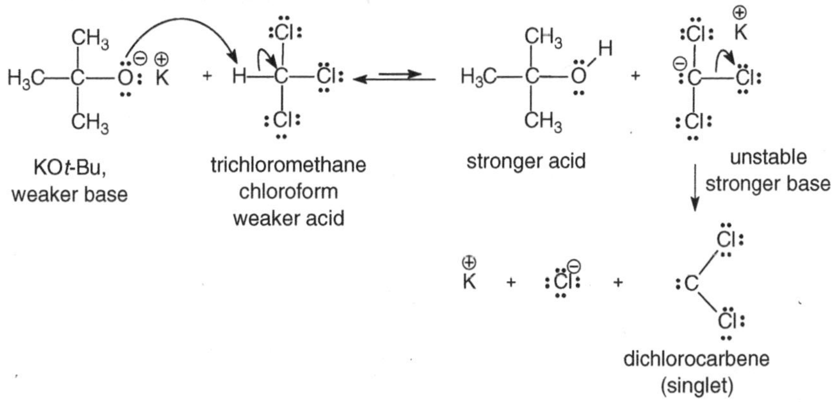
KOt-Bu,
weaker base
trichloromethane
chloroform
weaker acid
stronger acid
unstable
stronger base
dichlorocarbene
(singlet)

a. Singlet form is probably more stable.
b. Stereospecific (*syn*) cycloaddition. :CCl_2 does not insert into C-H bonds.
c. Dichlorocyclopropane product can be dechlorinated or hydrolyzed (section 2.16.5) to a ketone.

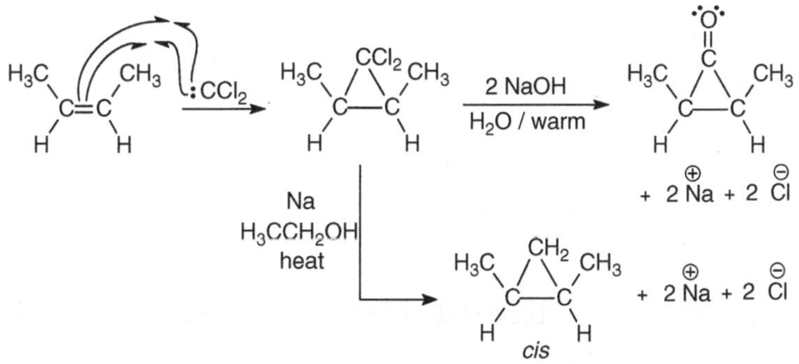
2 NaOH
H2O / warm
Na
H3CCH2OH
heat
cis

d. DuPont chemists H. E. Simmons and R. D. Smith (section 8.11) – an organozinc reagent has "carbene character." It behaves as the singlet, does not insert into C-H bonds, and is safer to handle.

$$H_2Cl_2 + Zn(Cu) \longrightarrow I-CH_2-ZnI + Cu$$

4.16.10. Problem 12. A mixture of singlet and triplet carbenes reacts nonstereospecifically in the gas phase with *cis*-but-2-ene at low concentration yielding both *cis*- and *trans*-1,2-dimethylcyclopropanes. If a little O_2 is added, reaction becomes almost completely stereospecific. Explain the O_2 effect. *Hint*: review section 4.3.2b.

4.16.11. Problem 13. *a.* Draw a likely product structure for :CCl_2 addition to cyclopentene. *b.* Draw two different (stereoisomer) product structures for :CBrCl addition to cyclopentene. Assume the singlet carbene form is more stable in both cases.

4.16.12. Problem 14. The $CHCl_3$ H atom is more acidic (and easily abstracted by base) than a CH_3Cl H atom. Which conjugate base is more stable, ⁻:CCl_3 or ⁻:CH_2Cl?

What main factor influences conjugate base stabilities? *Hint*: review section 4.13.34.

CHAPTER 5

5.1.1. Stereochemistry – three dimensional structure and reactivity.

5.1.2. Stereoisomers (section 1.19.6). Molecules that have the same bonding sequence but different three dimensional atom orientations.

   a. Example: *cis/trans*-isomers (or geometric isomers).

fumaric acid, mp 287°C
essential metabolite

maleic acid, mp 138°C
toxic irritant

5.2.1. All objects have a mirror image. *Superimposable* mirror image – if original object coincides exactly (mirror image is identical to original object).

5.2.2. Chiral object – any geometrical object (example: hands) whose mirror image is not superimposable (Lord Kelvin, 1893; Cahn, Ingold, and Prelog, 1964).

   a. Often a C atom or molecule.
   b. Examples: *cis* and *trans*-1,2-dichlorocyclopentanes.

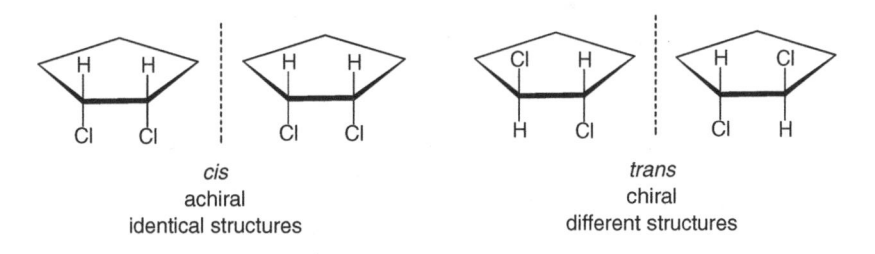

cis
achiral
identical structures

trans
chiral
different structures

5.2.3. Enantiomers – nonsuperimposable mirror image structures.

   a. Example: *trans* isomers above.

5.2.4. Chirality – a necessary and sufficient condition for enantiomerism:

   a. If molecule is chiral, it must have an enantiomer.
   b. If a molecule is achiral, it cannot have an enantiomer.

141

142 SEQUENTIAL ORGANIC CHEMISTRY I

5.2.5. *Chirality center* – a molecular asymmetry point, often an atom, most commonly C, but can be N, P, or S, or even a point in space.

   a. *Stereocenter* - Any point at which interchange of two groups yields a stereoisomer (includes achiral *cis* or *trans* isomer C atoms).

5.2.6. When is an atom chiral?

5.2.7. sp^3 C atom tetrahedral geometry has been directly observed by electron and X-ray diffraction.

5.2.8. Earlier data – J. H. van't Hoff, J. A. LeBel – replace an H with Z. How many isomers are possible?

   a. Since only one H_3C–Z structure exists; all four CH_4 H atoms must be equivalent.
   b. Possible equivalent H atom geometries: (square planar, pyramidal, and tetrahedral).

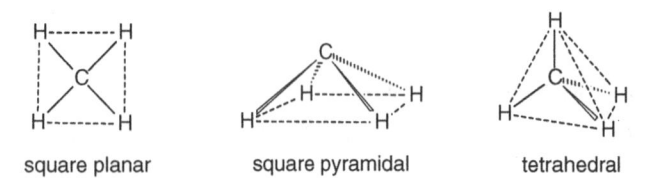

square planar   square pyramidal   tetrahedral

   c. Equivalency test: if two different H atoms yield the same product isomer upon substitution by X, the two H atoms are *equivalent*.

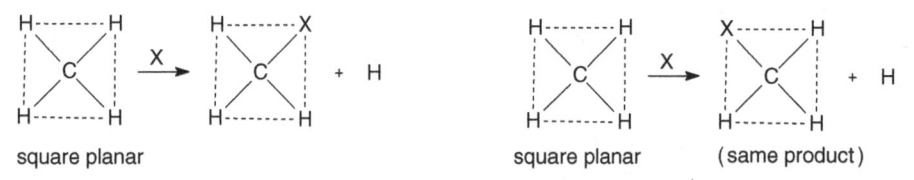

square planar   square planar   (same product)

   d. Superimposability test: if two structure's atoms can be made to coincide by rotations and translations (without breaking any chemical bonds), the structures are superimposable (section 5.2.1).
   e. Problem 1. How many CH_3Y stereoisomers would be possible if methane were a pyramid with a *rectangular* base? Draw them.

   f. Disubstituted isomers (H_2CYZ):

Formula	Number of Stereoisomers Formed
CH_2Cl_2	1
CH_2Br_2	1
CH_2BrCl	1

g. Only tetrahedral geometry is consistent with observations.
   1. A C atom that has two (or more) identical groups bonded to it is usually achiral.
h. Problem 2. How many CH_2YZ stereoisomers would be possible for the following geometries? Draw them.
   1. Rectangular planar (C at center).

   2. Square pyramidal (C at apex).

   3. Rectangular pyramidal (C at apex).

   4. Tetrahedral (C at center).

5.2.9. H_2CClBr has a superimposable (identical) mirror image.

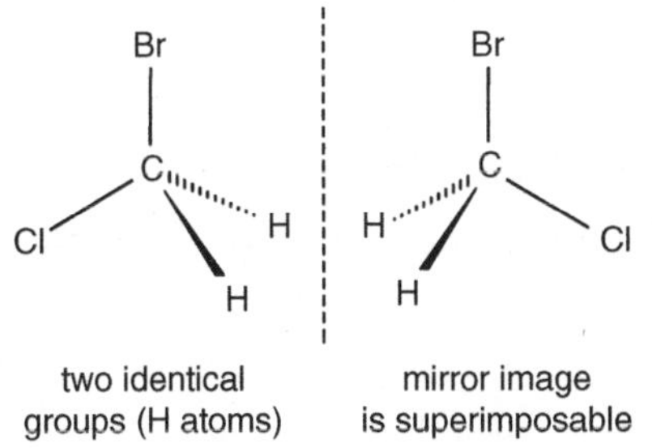

two identical groups (H atoms) | mirror image is superimposable

5.2.10. Tetrasubstituted isomers (CWXYZ):

Formula	Number of stereoisomers found
CWXYZ	2

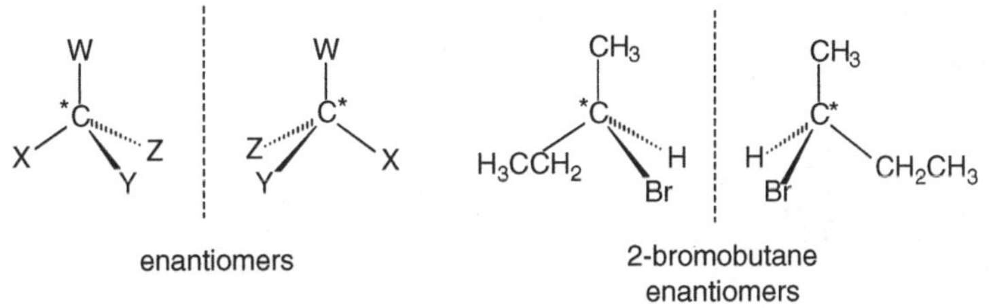

enantiomers

2-bromobutane enantiomers

5.2.11. C must have four different bonded groups to be chiral.

5.2.12. CWXYZ model analysis – nonsuperimposable mirror images.

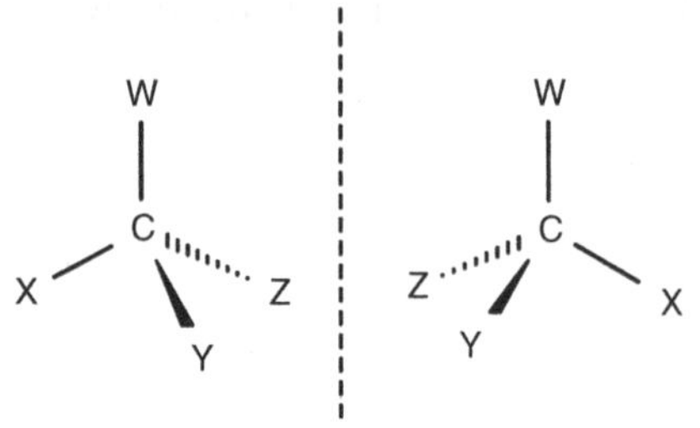

mirror plane

5.2.13. Mirror image lactic acid and 2-methylbutan-1-ol stereoisomers.

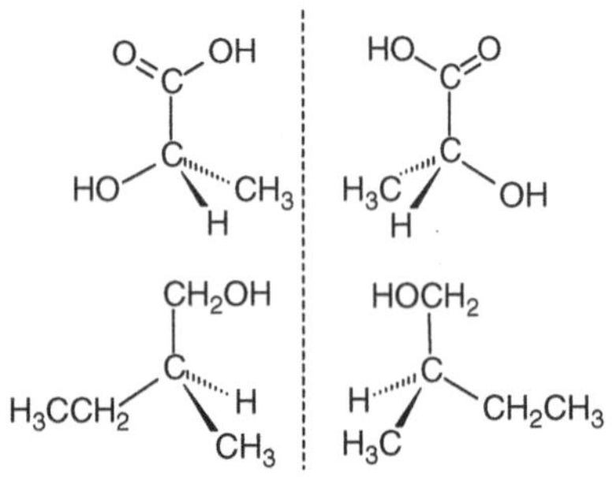

mirror plane

5.2.14. WXYZ can be any four different groups, including isotopes. Example: α-deuterioethylbenzene.

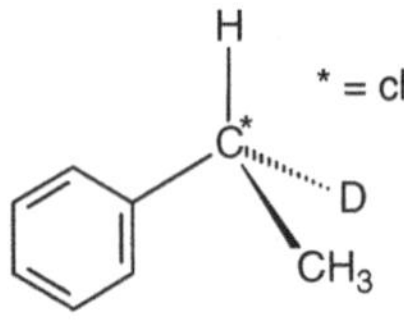

* = chiral atom

α-deuterioethylbenzene

5.2.15. If a molecule has only one chiral center, the molecule will be chiral.

5.2.16. Recognizing chiral centers – look down a chain until a difference is found. Examples:

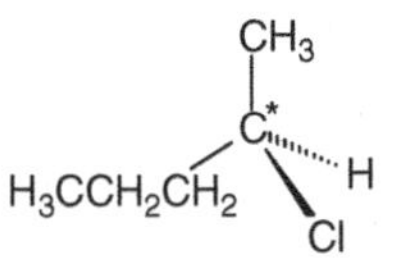

2-chloropentane 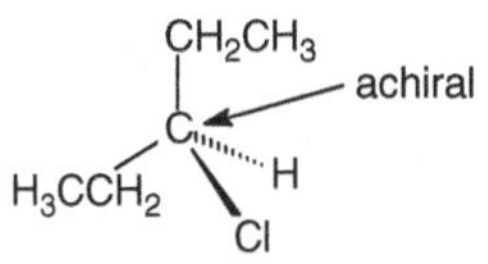   3-chloropentane — achiral

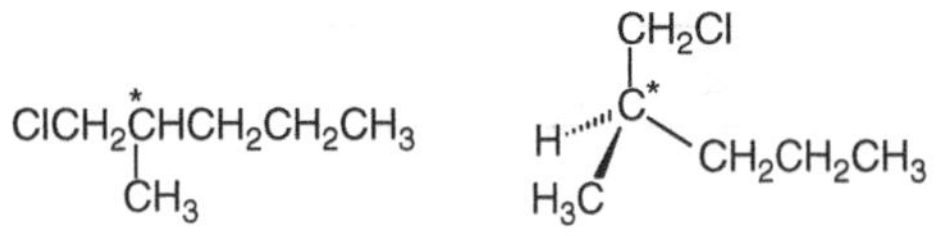

1-chloro-2-methylpentane

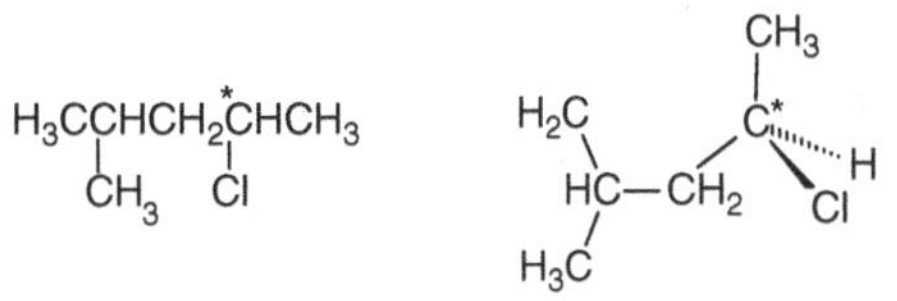

2-chloro-4-methylpentane

a. If a molecule has two (or more) chiral centers, it may or may not be chiral.
   1. A molecule will not be chiral if it has internal symmetry, such as a mirror plane (σ), even if it has chiral C atoms (*cis*–1,2-dichlorocyclopentane has two chiral C atoms). See *meso* compounds, section 5.13.1.

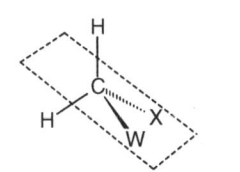

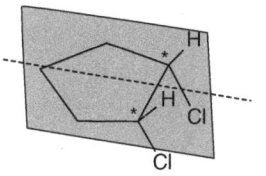

   2. A molecule with chiral atoms may not be chiral if it has an inversion center (*i*).

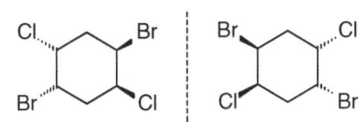

   no internal plane of symmetry
   but superimposable
   (identical structures)

b. Problem 3. Which of the following compounds are chiral? Illustrate chiral atoms as perspective drawings (wedges / dashes).
   1. 1-chloropentane

   2. 2-chloropentane

   3. 3-chloropentane

   4. 1-chloro-2-methylpentane

   5. 2-chloro-2-methylpentane

   6. 2-chloro-3-methylpentane

7. 4-chloro-2-methylpentane

8. 2-bromo-1-chlorobutane

c. Problem 4. Ignoring stereoisomers for the moment, draw all C₃H₆DCl structural isomers. Draw chiral molecules perspectively (wedges / dashes).

5.3.1. R. S. Cahn, C. Ingold, V. Prelog – *R/S* rule system for assigning chiral atom *absolute configuration* (three dimensional group bonding arrangement about a chiral atom). *R* = rectus (right) or *S* = sinister (left).
   a. Assign four bonded groups priorities – in order of decreasing first atom atomic number.
   b. Orient lowest priority group behind paper plane.
   c. Draw a circle from highest to lowest priority for remaining three groups (out of paper plane).
      1. Clockwise = *R* configuration; counterclockwise = *S* configuration.

5.3.2. Example: bromochloroiodomethanes

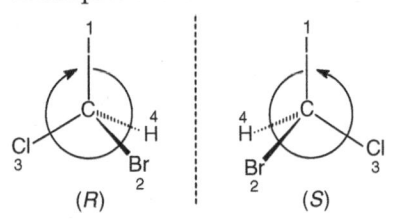

5.3.3. More rules are needed to cover all possible cases.
   a. If two groups have the same atomic number, higher mass number determines priority.
      1. Example: bromodeuterioethane

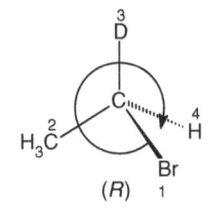

b. If two groups have identical atomic and mass numbers, examine the next atoms out the chains until the first higher atomic number is found – determines group priority.
   1. Example: 2-chlorobutane.

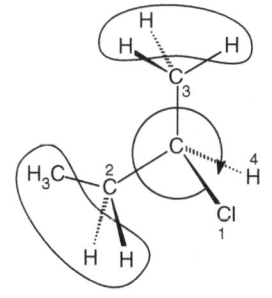

   Compare atoms inside cloud 2 with those inside cloud 3. The C atom has a higher atomic number than one of the H atoms inside cloud 3. Therefore group 2 has higher priority than group 3.

   (R)-2-chlorobutane

   2. All three bonded atoms must be examined before going further out. More examples:

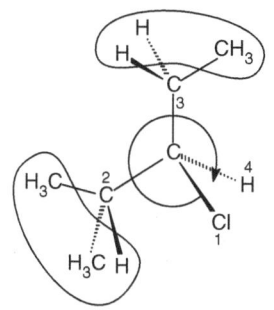

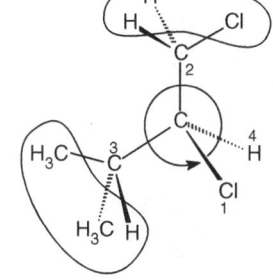

   The chlorine atom in cloud 2 has a higher atomic number than one of the two C atoms in cloud 3. The first difference encountered determines the group priority.

   (R)-3-chloro-2-methylpentane   (S)-1,2-dichloro-3-methylbutane

c. If a chiral C atom's bonded groups have double or triple bonds, respective priorities are assigned by group equivalencies:

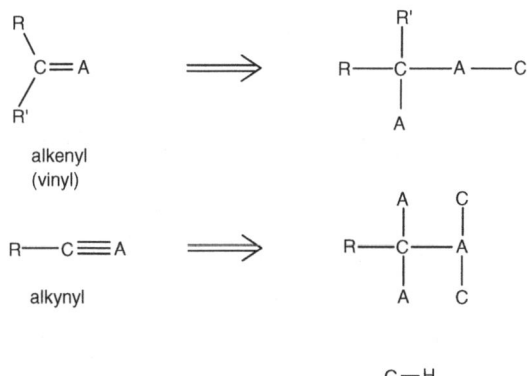

   alkenyl (vinyl)

   alkynyl

   phenyl

d. Example: 3-chlorohex-1-ene – go two atoms out from chiral C atom (clouds) to determine group priorities.

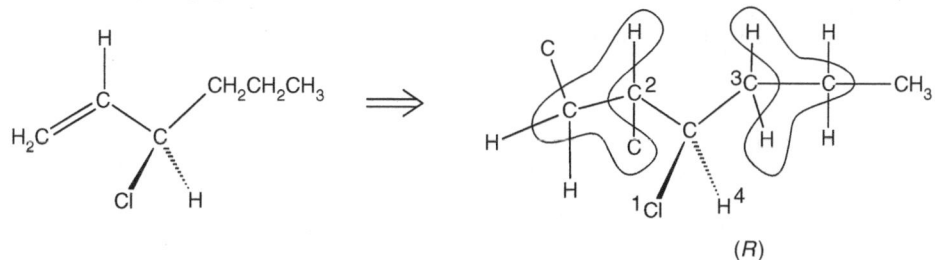

   (R)

e. Example: only (S)-alanine is metabolized by enzymes. Alanine is a common amino acid (protein building block).

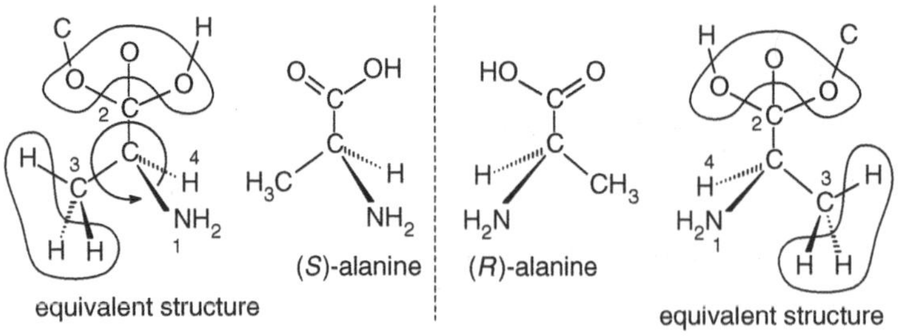

(S)-alanine   (R)-alanine

equivalent structure   equivalent structure

f. Example: the same rules apply to chiral ring C atoms.

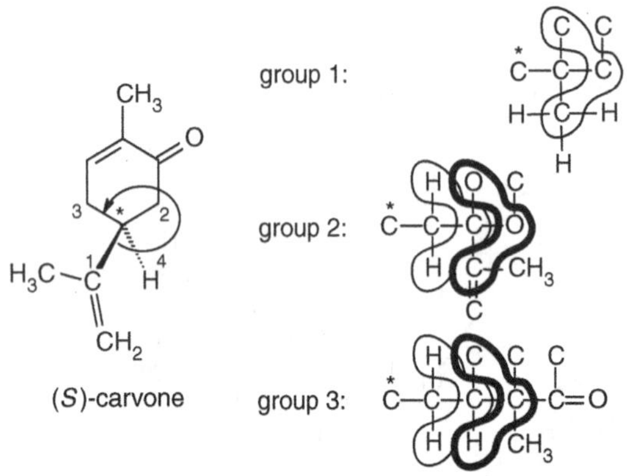

(S)-carvone   group 1:
               group 2:
               group 3:

g. Problem 5. Draw perspectively the other enantiomer and assign absolute configurations to compounds in problem 3.

h. Problem 6. Arrange the following alkyl groups in order of increasing Cahn-Ingold-Prelog priority: methyl, 1°, 2°, 3°.

i. Problem 7. Draw perspectively and designate as R or S enantiomers (where they exist).
   1. 3-chloropent-1-ene

   2. 3-chloro-4-methylpent-1-ene

3. HOOCCH₂CH(OH)COOH (malic acid)

4. C₆H₅CH(CH₃)NH₂

5. 3-ethyl-2,3-dimethylhexane

6. C₆H₅CH(OH)COOH (mandelic acid)

7. H₃CCH(NH₂)COOH (alanine)

j. Problem 8. What are the two lowest molecular mass (enantiomeric) alkanes that are chiral. Make perspective drawings and label each as *R* or *S*.

5.10.1. Cross formulas (Fischer projections) are most useful for sugars (Chapter 23).
    a. Stretch out longest carbon chain vertically.
        1. Position most oxidized functional group (oxygen or other heteroatom containing group) nearest to top.
    b. Rotate around σ-bonds until horizontal bonds at a chiral C are out of plane (most symmetrical conformation, section 5.8.2i).

c. Vertical bonds will be either in plane, or behind plane.

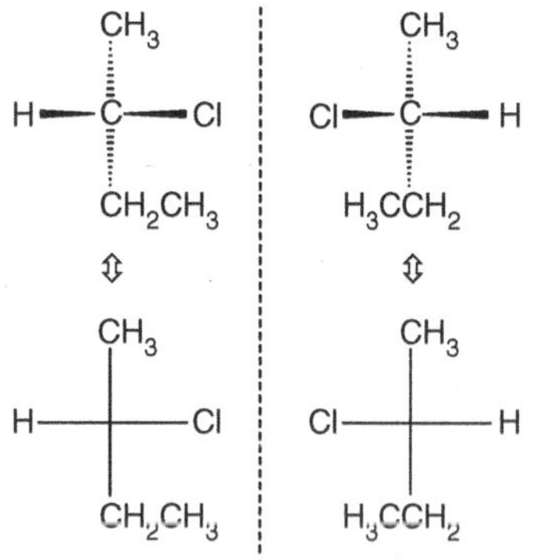

d. Examples:

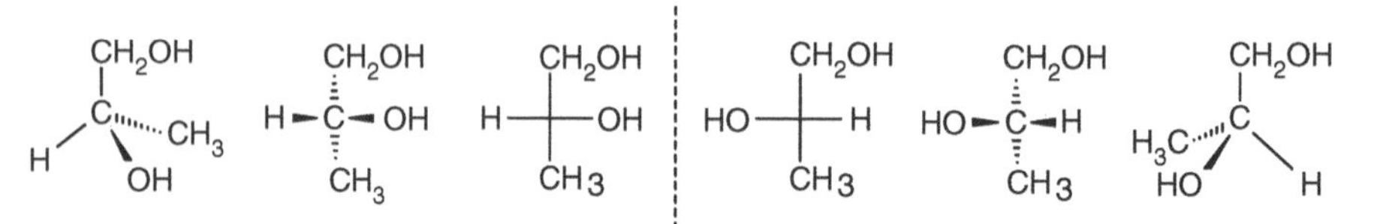

(*R*)-propane-1,2-diol

(*S*)-propane-1,2-diol

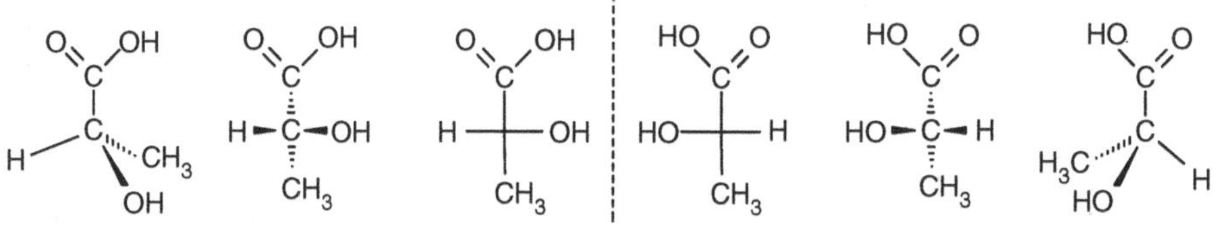

(*R*)-2-hydroxypropanoic acid
(*R*)-lactic acid

(*S*)-2-hydroxypropanoic acid
(*S*)-lactic acid

e. If configuration is assigned from standard Fischer projection form, the arrow around the first three groups is reversed. (*R*) is counterclockwise. Confusing!

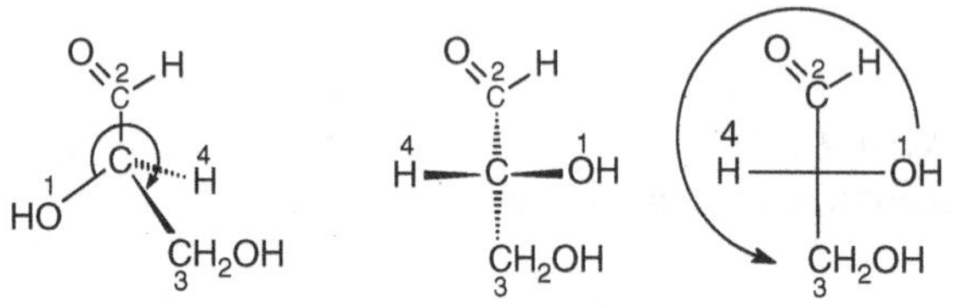

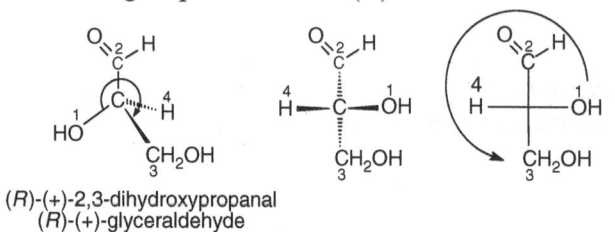

(*R*)-(+)-2,3-dihydroxypropanal
(*R*)-(+)-glyceraldehyde

5.4.1. Polarimetry – measures a stereoisomer's plane polarized light rotation (*optical activity*).

5.4.2. Background: light can be viewed as electromagnetic waves.

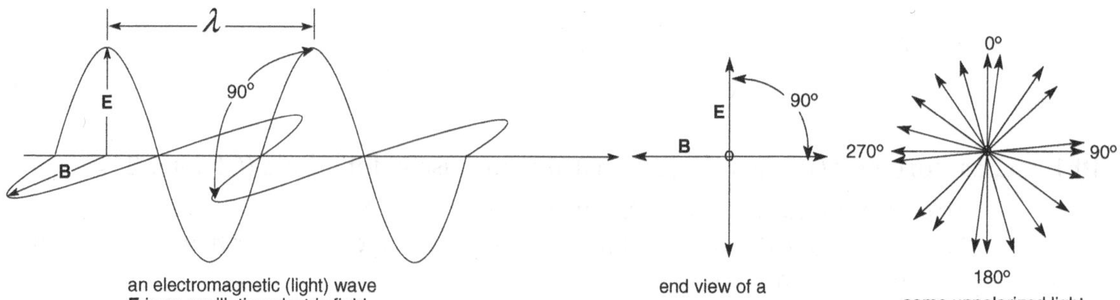

an electromagnetic (light) wave
E is an oscillating electric field
B is an oscillating magnetic field

end view of a single light wave

some unpolarized light waves. **E** has any angle, 0° to 360°.
B is not shown.

4.3. Passing ordinary light through a polarizing filter (Polaroid, Nicol (special plastic) or calcite (a form of $CaCO_3$) yields plane polarized light.

   a. Light waves whose **E** axes are parallel to a filter crystal axis pass through – others are filtered out (converted into heat).

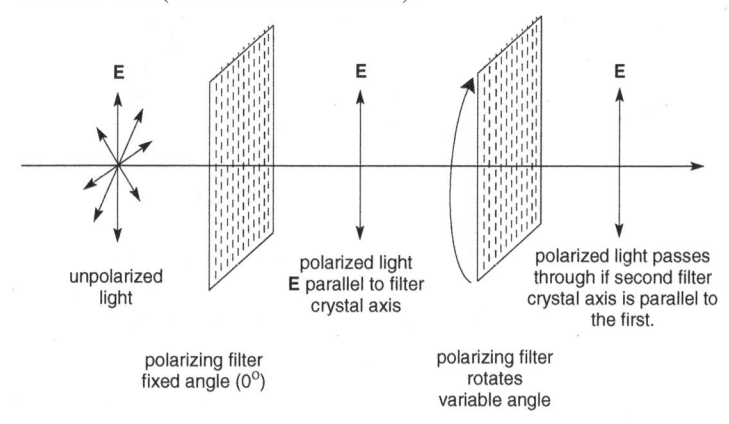

Polarimeter Schematic Diagram

5.4.4. An optically active substance rotates the polarized light plane.

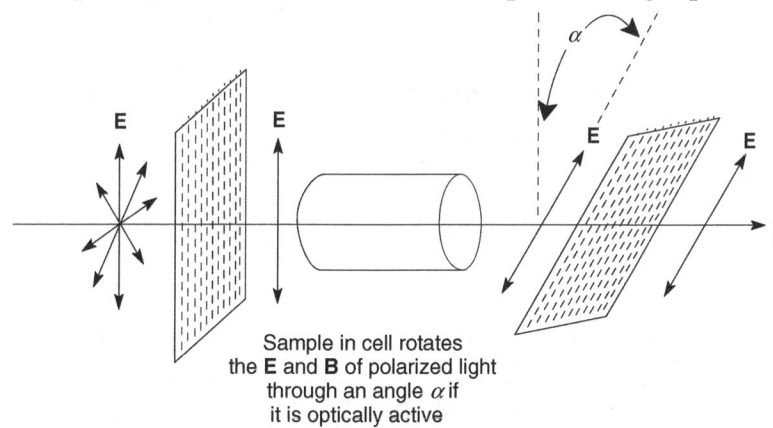

5.4.5. Polarimeter – a monochromatic (single color and wavelength) light source and two polarizing filters – one attached to a protractor (so that angle α can be measured) – rotates about oncoming light axis.

   a. Maximum light passes through when $\alpha = 0$ or 180°, minimum light when $\alpha = 90$ or 270° for an empty cell.
   b. *dextrorotatory* – substance rotates polarized light plane clockwise (+) or *d* (Greek *dexios* "to the right") from zero.
   c. *levorotatory* – substance rotates plane polarized light counterclockwise (–) or *l* (Latin *laevus* "to the left") from zero.
   d. Examples: (+)-lactic acid, (–)-2-methylbutan-1-ol.

(S)-(+)-lactic acid

(S)-(+)-lactic acid Fischer projection drawing

(S)-(–)-2-methylbutan-1-ol

(S)-(–)-2-methylbutan-1-ol Fischer projection drawing

5.4.6. Louis Pasteur (1848) – first noticed mirror image sodium ammonium tartrate crystals (NaNH$_4$C$_4$H$_4$O$_6$), a solid residue in wine bottles.

    a. He separated right and left handed crystals with a hand lens and tweezers.

*Taken from Jaques, J.; Collet, A., Wilen, S. H., "Enantiomers, Racemates, and Resolutions," Wiley-Interscience, John Wiley and Sons, New York (1981), p 6.

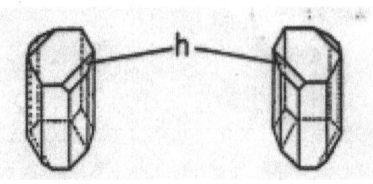

Figure 1 Dextro and levo sodium ammonium tartrates as described by Pasteur. In the absence of hemihedral facets such as *h*, the enantiomorphism of the crystals is not detectable from their aspect. L.Pasteur, *Ann. Chim. Phys.*, 3rd series, 1850, 28,56. Reproduced by permission of Massons, S.A., Paris.

    1. Original mixture in water ⇒ *optically inactive* – does not rotate polarized light plane.
    2. Left handed crystals in water ⇒ *optically active* – rotate polarized light plane.
    3. Right handed crystals in water ⇒ *optically active* – rotation equal but opposite to left handed.

    b. All other "left" and "right" crystal physical properties are identical.

        1. Example: 2-bromobutanes.

property	(*R*)-2-bromobutane	(*S*)-2-bromobutane
boiling point (°C)	91.2	91.2
melting point (°C)	–112	–112
refractive index	1.436	1.436
density	1.253	1.253

        2. Example: (specific rotation is defined in section 5.4.9).

property	(+)-2-methylbutan-1-ol	(–)-2-methylbutan-1-ol
specific rotation	+5.90°·mL·dm^{-1}·g^{-1}	–5.90°·mL·dm^{-1}·g^{-1}
boiling point	128.9°C	128.9°C
relative density	0.8193	0.8193
refractive index	1.4107	1.4107

    c. Pasteur realized that mirror image crystals ⇒ mirror image molecules, and that chiral molecules ⇒ optical activity.

5.4.7. How do molecules rotate the polarized light plane physically?

    a. Any single molecule (optically active or not) held fixed in a given orientation can have an unsymmetrical charged particle arrangement.

    b. Polarized light rotates slightly as it interacts with molecular charged particles.

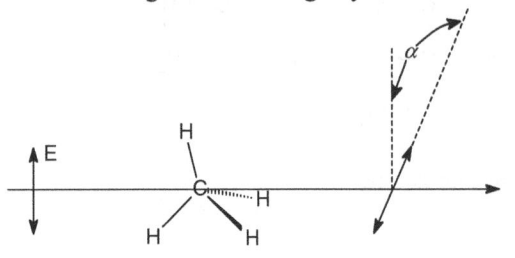

c. For every molecular orientation that rotates polarized light +, another opposite orientation exists rotating polarized light – an equal amount. All interactions cancel statistically for a large sample ⇒ optically inactive.
d. A one enantiomer sample is inherently unsymmetrical – rotations do not all cancel ⇒ optically active.

5.4.8. A given sample's rotation angle α is affected by:
1. molecular structure.
2. temperature (usually 20°C).
3. sample concentration (number of molecules in the light pathway) and cell length.
4. light wavelength (color) (usually 5893 Å, orange sodium D emission line).
5. solvent (if optically active).

5.4.9. "Specific rotation" = rotation angle α for 1.00 g compound/1.00 mL solution and a 10.0 cm (1.00 dm) pathlength. Rotations α not measured under these conditions are converted into specific rotations $[\alpha]$ by the equation:

$$[\alpha] = \frac{observed\ rotation\ \alpha\ (degrees)}{pathlength\ (dm) \times concentration\ (\frac{g}{mL})}$$

g = grams optically active solute

mL = total solution volume

a. Specific rotation – a physical constant.
b. Example: (S)-2-methylbutan-1-ol (2.00 g) was dissolved in enough ethyl alcohol (optically inactive solvent) to make 10.0 mL of solution. The observed rotation $\alpha = -1.18°$. What is the specific rotation $[\alpha]$? The cell is 1.00 dm long.

$$[\alpha] = \frac{\alpha}{d \times \frac{g}{mL}} = \frac{-1.18°}{1.00\ dm \times \frac{2.00\ g}{10.0\ mL}}$$

$$[\alpha]_D^{25} = +5.90°\ mL \cdot g^{-1} \cdot dm^{-1}$$

1. Subscript and superscript, respectively, indicate sodium D line light and temperature 25°C.
2. The (R)-(+)-enantiomer would have $[\alpha]_D^{25} = +5.90°.\ mL \cdot g^{-1} \cdot dm^{-1}$.

c. Problem 9. 6.15 g of cholesterol are dissolved in enough trichloromethane (solvent) to make 100. mL solution. The observed rotation is −1.2° for this solution in a 5.00 cm long cell. Calculate specific rotation $[\alpha]$. What would be the observed rotation angle α for the same solution in a 10.0 cm long cell? What would be the observed rotation α if 10.0 mL of solution were diluted to 20.0 mL, and placed in the 5.00 cm long cell?

d. Problem 10. $\alpha = +45°$ for a pure liquid sample in a 10.0 cm polarimeter cell. How could you tell that the true reading is not −315°, +405°, or +765°?

154 SEQUENTIAL ORGANIC CHEMISTRY I

5.4.10. [α] is small when four groups are very similar: for (S)-n-ethyl-n-propylundecane calculated [α] = 0.00001°·mL·dm⁻¹·g⁻¹. Practically inactive.

(S)-n-ethyl-n-propylundecane

5.4.11. Which 2-chlorobutane three dimensional structure (enantiomer) rotates polarized light (+), and which rotates (−)?

(R)-(−)-2-chlorobutane   (S)-(+)-2-chlorobutane   (R,R)-(+)-tartaric acid

5.14.1 J. M. Bijvoet (1951) - Anomalous X-ray scattering – determined (R,R)-(+)-tartaric acid salt absolute configuration.

   a. Many chiral compound *absolute configurations* have been related back to compounds such as tartaric acid whose absolute configurations are known from X-ray crystallography. The relationship is established by chemical reactions known not to break any bonds to a chiral carbon atom.

   b. Example: the configuration of (S)-2-methylbutan-1-ol has been determined by conversion into (+)-tartaric acid whose absolute configuration is known. The alcohol functional group can be replaced by a chloro or bromo group by reaction with dry HCl or PBr₃, respectively (sections 11.7 and 11.8). No bonds to the chiral C atom are broken in the process.

(S)-(−)-2-methylbutan-1-ol
[α] = −5.90°·mL·dm⁻¹·g⁻¹

(S)-(+)-1-chloro-2-methylbutane
[α] = +1.67°·mL·dm⁻¹·g⁻¹

(S)-(+)-2-methylbutan-1-ol
[α]$_D^{25}$ = +4.0°·mL·dm⁻¹·g⁻¹

   1. Subtlety: although the products still have the *S* configuration, they rotate polarized light (+).
   2. How do we know that the chiral C-CH₂OH bond doesn't break and reform during the reaction? The (S)-(+)-1-halo-2-methylbutane can be converted back into (S)-(−)-2-methyl-1-butanol (by reaction with aqueous NaOH) with no loss of optical activity ([α] is still −5.90°).

5.14.2. E. Fischer (1891) guessed (50:50 chance) that the glucose C5–OH group was on the right (*R*-configuration) in the Fischer projection drawing (Chapter 23), which is correct.

   a. Glucose was degraded into glyceraldehyde. M. A. Rosanoff (1906) guessed that the natural glyceraldehyde C2–OH group is on the right (*R*-configuration) in the Fischer projection drawing, consistent with Fischer's glucose assignment. Natural glyceraldehyde and that obtained from glucose rotate polarized light in the (+) direction. All configurations were *relative* before 1951, since they were based on Fischer's and Rosanoff's guesses.

   b. Fischer-Rosanoff convention – D/L system – for relative configurations – developed before *R/S* system – sometimes ambiguous, but unfortunately, still in common use, especially in biochemistry.

   c. If the heteroatom (O or N) bonded to the last chiral C atom (farthest from top, Fischer projection drawing) is on the right ⇒ D-series. On the left ⇒ L-series. Examples:

5.4.12. No simple relationship exists between *R* and *S* (or D and L) and (+) and (−) (X-ray crystal data).

   a. Some *R* compounds rotate (+), and some rotate (−).

   b. Some *S* compounds rotate (+), and some rotate (−).

5.5.1. A chiral enzyme can distinguish each substrate enantiomer, like a glove on a hand (a *chiral probe*).

   a. Stephen Kent (1992) – synthesized both enantiomers of a peptidase (peptide cleaving) enzyme. One enzyme is specific for just one of the two enantiomeric peptides.

   b. Example: (*R*)-(−)- and (*S*)-(+)-epinephrine.

c. Example: ketamine ("K").

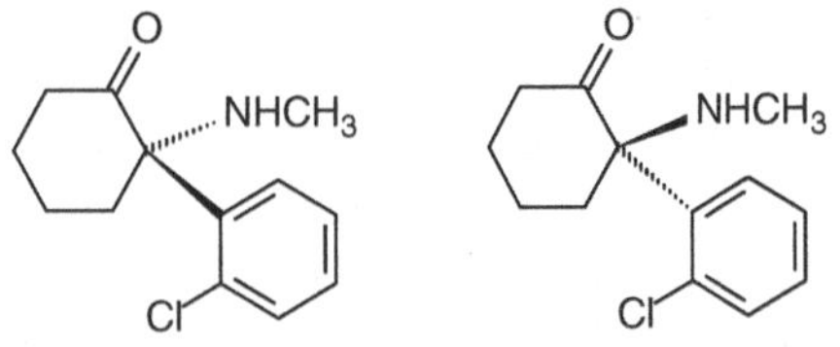

(S)-ketamine
an anesthetic

(R)-ketamine
a hallucinogen

d. Problem 11. Which of the following reactions could safely be used to relate product configuration to reactant configuration? *Hint*: during which of the reactions is a bond broken to a chiral center?

a. (+)-C$_6$H$_5$CH(OH)CH$_3$ + PBr$_3$ ⟶ C$_6$H$_5$CHBrCH$_3$

b. (+)-CH$_3$CH$_2$CHClCH$_3$ + C$_6$H$_6$ + AlCl$_3$ ⟶ C$_6$H$_5$CH(CH$_3$)CH$_2$CH$_3$

c. (−)-C$_6$H$_5$CH(OC$_2$H$_5$)CH$_2$OH + HBr ⟶ (−)-C$_6$H$_5$CH(OC$_2$H$_5$)CH$_2$Br

d. (+)-CH$_3$CH(OH)CH$_2$Br + NaCN ⟶ (+)-CH$_3$CH(OH)CH$_2$CN

e. (+)-CH$_3$CH$_2$−C(=O)−^{18}O−CH(CH$_3$)C$_2$H$_5$ + NaOH ⟶ (+)-CH$_3$CH$_2$−C(=O)−O$^⊖$ Na$^⊕$ + CH$_3$CH$_2$CH(^{18}OH)CH$_3$

f. (−)-CH$_3$CH$_2$CHBrCH$_3$ + CH$_3$CH$_2$O$^⊖$ Na$^⊕$ ⟶ CH$_3$CH$_2$−O−CH(CH$_3$)CH$_2$CH$_3$

g. (+)-CH$_3$CH$_2$CH(OH)CH$_3$ + Na ⟶ (+)-CH$_3$CH$_2$CHCH$_3$ (O$^⊖$ Na$^⊕$) —C$_2$H$_5$Br→ CH$_3$CH$_2$−O−CH(CH$_3$)CH$_2$CH$_3$

5.6.1. Racemic mixture – an equimolar two enantiomer mixture, optically inactive, indicated by (±) prefix or by (R)(S) in separate parentheses.
  a. Plane polarized light rotation by one enantiomer (in a certain orientation) is statistically cancelled by rotation from the other enantiomer in opposite orientation.
  b. Example: (R)(S)-(±)-butan-2-ol. (S)-enantiomer rotates polarized light $[\alpha] = +13.5° \cdot mL \cdot g^{-1} \cdot dm^{-1}$, while (R)-enantiomer rotates plane polarized light $[\alpha] = -13.5° \cdot mL \cdot g^{-1} \cdot dm^{-1}$. An equimolar racemic solution is optically inactive.

5.6.2. Chirality – a necessary but insufficient condition for optical activity:
  a. A sample containing chiral molecules can be optically inactive if it is an equimolar enantiomeric mixture.
  b. A sample must contain chiral molecules to be optically active. However, all samples containing chiral molecules are not optically active. They can be racemic.

5.7.1. Enantiomeric excess (= optical purity) of a mixture:

$$optical\ purity = \frac{observed\ rotation}{pure\ enantiomer\ rotation} \times 100 = \frac{[\alpha]}{[\alpha_+]\ or\ [\alpha_-]} \times 100$$

$$optical\ purity = enantiomeric\ excess = \frac{d-l}{d+l} \times 100 = \frac{excess\ of\ one\ over\ the\ other}{entire\ mixture} \times 100$$

d and l = each enantiomer amount, respectively. Various units: mol, gram, or concentration.

5.7.2. A two enantiomer mixture's specific rotation can be found from its optical purity. Conversely, enantiomeric mol fraction can be calculated from a mixture's specific rotation.

    a. Example: calculate $[\alpha]$ for 2-methylbutan-1-ol mixtures, whose pure (*R*)-(+)-enantiomer specific rotation is $[\alpha_+] = +5.90° \cdot \text{mL} \cdot \text{g}^{-1} \cdot \text{dm}^{-1}$.

        1. 100% (−)-enantiomer – optically pure sample has specific rotation:

$$[\alpha] = [\alpha_+] \times \frac{d-l}{d+l} = 5.90° \cdot \text{mL} \cdot \text{g}^{-1} \cdot \text{dm}^{-1} \times \frac{0.00 - 1.00}{0.00 + 1.00} = -5.00° \cdot \text{mL} \cdot \text{g}^{-1} \cdot \text{dm}^{-1}$$

        2. Racemic mixture – 0% optically pure sample:

$$[\alpha] = [\alpha_+] \times \frac{d-l}{d+l} = 5.90° \cdot \text{mL} \cdot \text{g}^{-1} \cdot \text{dm}^{-1} \times \frac{0.50 - 0.50}{0.50 + 0.50} = 0.00° \cdot \text{mL} \cdot \text{g}^{-1} \cdot \text{dm}^{-1}$$

        3. 0.75 mol fraction + and 0.25 mol fraction − is 50% optically pure + (rotates light half as far as pure +).

$$[\alpha] = [\alpha_+] \times \frac{(d-l)}{d+l} = 5.90° \cdot \text{mL} \cdot \text{dm}^{-1} \cdot \text{g}^{-1} \times \frac{(0.75 - 0.25)}{0.75 + 0.25} = 2.95° \cdot \text{mL} \cdot \text{dm}^{-1} \cdot \text{g}^{-1}$$

    b. Example: an enantiomeric mixture whose specific rotation $[\alpha] = -2.00° \cdot \text{mL} \cdot \text{dm}^{-1} \cdot \text{g}^{-1}$. What is mol fraction ($X_+$)? Mol fraction definition:

$$1.00 = X_+ + X_-$$

and

$$-2.00 = [\alpha] = [\alpha_+] \times (X_+ - X_-)$$

$$-2.00 = [\alpha_+] \times (X_+ - (1.00 - X_+))$$

$$-2.00 = 5.90° \cdot \text{mL} \cdot \text{dm}^{-1} \cdot \text{g}^{-1} \times (X_+ - (1.00 - X_+))$$

solve for X_+.

$$X_+ = 0.331 \text{ and } X_- = 0.669$$

    c. Problem 12. $[\alpha] = +5.90$ and $-1.67° \cdot \text{mL} \cdot \text{g}^{-1} \cdot \text{dm}^{-1}$ for (*R*)-2-methylbutan-1-ol and (*R*)-1-chloro-2-methylbutane, respectively. When an optically impure 2-methylbutan-1-ol sample whose $[\alpha]=+3.54° \cdot \text{mL} \cdot \text{g}^{-1} \cdot \text{dm}^{-1}$ reacts with HCl(conc) / ZnCl$_2$, the product is 1-chloro-2-methylbutane. What will the 1-chloro-2-methylbutane product specific rotation be?

158 SEQUENTIAL ORGANIC CHEMISTRY I

5.9.1. Chiral molecules – can contain:
  a. no chiral atoms (unusual). Examples: allenes, spiro molecules.

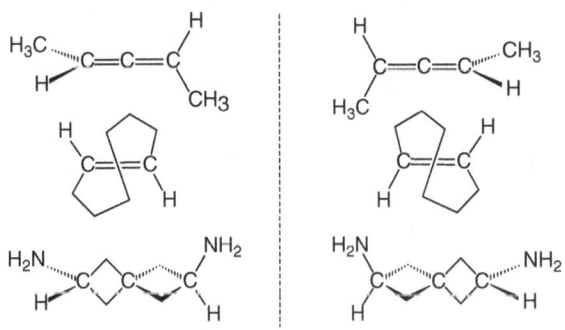

  1. Each molecule above has a chiral center – not at an atom for *trans*-cyclooctadiene.
  2. Allene is an achiral 1,2-diene – central C atom is *sp*-hybridized.

  b. One or more chiral centers.

  1. Some molecules with two or more chiral centers are achiral if they are meso (section 5.13.1).

5.8.1. Configurational isomer interconversion (*R* into *S* or vice versa) – break two σ-bonds and reassemble in reverse order – ordinarily negligibly slow or impossible at 25°C.
  a. Slow interconversion allows stereoisomer isolation (separation).
  b. Stereoisomer interconversion is easy with molecular models. Interchange any two groups.

5.11.1. Diasteriomers – stereoisomers that are not mirror images and are not superimposable. Two main types:
  a. Geometric (*cis/trans*) isomers (sections 1.19.5 and 3.11.1).
  b. Molecules with two or more chiral centers.
  1. Example: 2,3-dichloropentane:

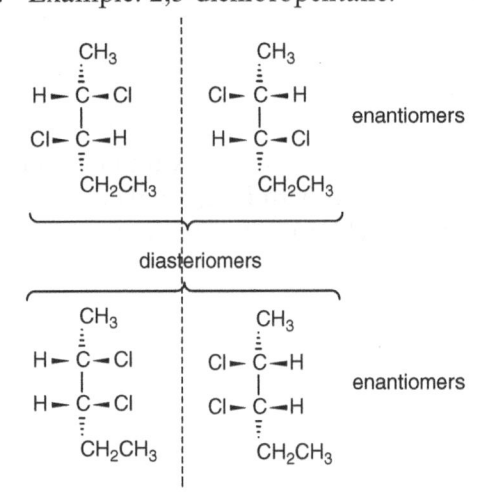

5.15.1. Diasteriomers have similar (but not identical) chemical properties, and different physical properties.

$$\underset{\substack{\mu = 0 \\ \textit{trans}\text{-but-2-ene} \\ \text{bond dipoles cancel} \\ \text{bp} = 0.9°C}}{\overset{H_3C}{\underset{H}{>}}C=C\overset{H}{\underset{CH_3}{<}}} \qquad \underset{\substack{\mu = 0.33\ D \\ \textit{cis}\text{-but-2-ene} \\ \text{bp} = 3.7°C}}{\overset{H_3C\ \text{net dipole}\ CH_3}{\underset{H\qquad H}{>}}C=C}$$

```
   COOH        COOH        COOH
H──┼──Br   Br──┼──H    H──┼──Br
Br──┼──H    H──┼──Br   H──┼──Br
   COOH        COOH        COOH

     mp (either) 158°C      mp (meso) 256°C
```

```
     O  H              O  H
      \\//              \\//
       C                 C
   H──┼──OH          H──┼──OH
  HO──┼──H          HO──┼──H
   H──┼──OH         HO──┼──H
   H──┼──OH          H──┼──OH
     CH₂OH            CH₂OH

  D-(+)-glucose mp 148°C   D-(+)-galactose mp 167°C
```

    a. They can be separated by physical methods (crystallization, fractional distillation, chromatography).

5.12.1. Maximum number of possible stereoisomers = 2^n, where n is the number of chiral C atoms.

    a. Fewer than 2^n stereoisomers sometimes exist due to meso compounds (section 5.13.1).

    b. Sugars often contain several chiral C atoms (D-glucose has four).

5.12.2. Example: 2,3-dichlorobutane has two chiral C atoms.

```
      CH₃           CH₃
   H─C─Cl        Cl─C─H
   Cl─C─H        H─C─Cl       enantiomers
      CH₃           CH₃

         diasteriomers

      CH₃           CH₃
   H─C─Cl        Cl─C─H
   ─────────    ─────────     internal mirror
   H─C─Cl        Cl─C─H       plane
      CH₃           CH₃
       meso compound
       superimposable
       (two different orientations
       of the same compound)
```

5.13.1. Meso (Greek, "middle") compound – contains at least two chiral C atoms, but the mirror image is superimposable. Characteristics:
  a. Internal symmetry plane (when groups are aligned in the most symmetrical eclipsed conformation).
  b. Achiral molecule.
  c. Optically inactive.

5.13.2. If a molecule contains no symmetry plane is it chiral? Not always. Example: 1-chlorobutane.

Is there a horizontal plane of symmetry? No. Is the molecule chiral? No. (No chiral C atoms.)

5.13.3. (2R,3S)-, (2R,3R)-, (2S,3R), (2S,3S)-2,3-dichloropentanes R and S assignments. Work on one C atom at a time.

(2R,3S)   (2R,3R)   (2S,3R)   (2S,3S)

5.13.4. (R,R)-, (S,S)-, (R,S)-2,3-dichlorobutane – carbon numbers are dropped – numbering is equivalent from either end. (S,R)-configuration is the same as (R,S)- since it is a *meso* compound.

(R,S)   (R,R)   (S,S)

5.13.5. Other meso compound examples:

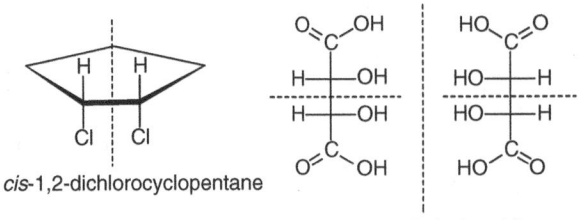

cis-1,2-dichlorocyclopentane

meso-tartaric acid

a. Problem 13. Draw perspectively all possible stereoisomers. Label enantiomer pairs and *meso* compounds. Which ones will be optically active? Assign chiral C atoms as *R* or *S*. Identify diasteriomers.
   1. 1,2-dibromopropane

   2. 3,4-dibromo-3,4-dimethylhexane

   3. 2,4-dibromopentane

   4. 2,3,4-tribromohexane

   5. 1,2,3,4-tetrabromobutane

   6. 2-bromo-3-chlorobutane

   7. 1-chloro-2-methylbutane

   8. 1,3-dichloro-2-methylbutane

5.8.2. Conformers (conformational isomers) – structures that can be interconverted by rotations about σ-bonds.

  a. Example: butane *anti-* and *gauche-*conformations.

  gauche   *anti*

  b. The two *gauche* forms are conformational enantiomers.
  c. Achiral molecules have some chiral conformations, but they interconvert with their enantiomers.
  d. *Anti* form – a conformational diasteriomer of gauche forms.
  e. Conformational stereoisomers normally interconvert rapidly at 25°C and cannot be separated.
  f. Example: *cis*-1,2-dibromocyclohexane.

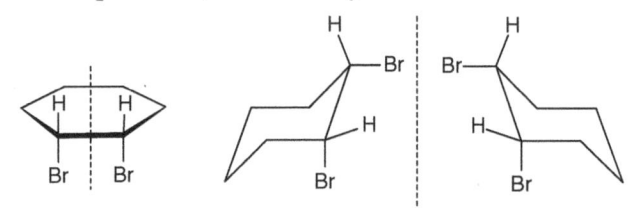

  1. Molecules can be treated as achiral when stereoisomers equilibrate with each other, or with a structure having a mirror symmetry plane.
  2. Examine the most symmetrical conformation when looking for a symmetry plane.

  g. Example: 1,2-dimethylcyclohexane.
  1. Two *cis*-chair forms (conformational enantiomers) interconvert rapidly at 25°C.
  2. Structures I and II (and III and IV) are conformational diasteriomers. Structures I and IV are enantiomers, and do not interconvert. Similarly, structures II and III are enantiomers.

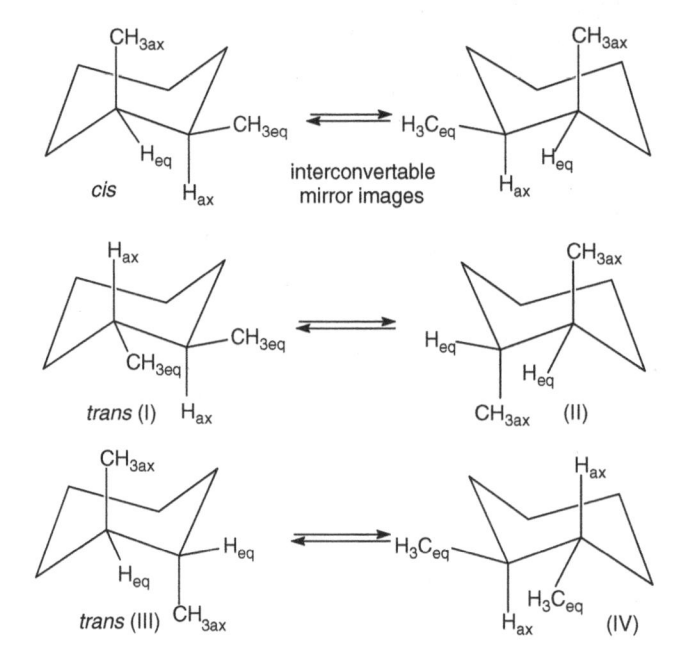

h. "Stereoisomers" from now on will exclude conformers, but includes enantiomers, diasteriomers, and geometrical isomers.
   1. This exclusion is only for learning purposes. Some conformational isomers can be isolated and studied.
   2. The chiral center in the left or right structures is midway between the center C-C atoms.

i. All possible conformations must be examined when testing for superimposability. The most symmetrical (eclipsed) form is the most revealing.

j. Problem 14. At low temperature, where collision energies are small, two isomeric badly crowded CHBr₂CHBr₂ forms have been isolated by crystallization. Draw (Newman projections) of the two isolable conformers. Which would actually be a racemic mixture (conformational enantiomers)?

k. Isomers summary.

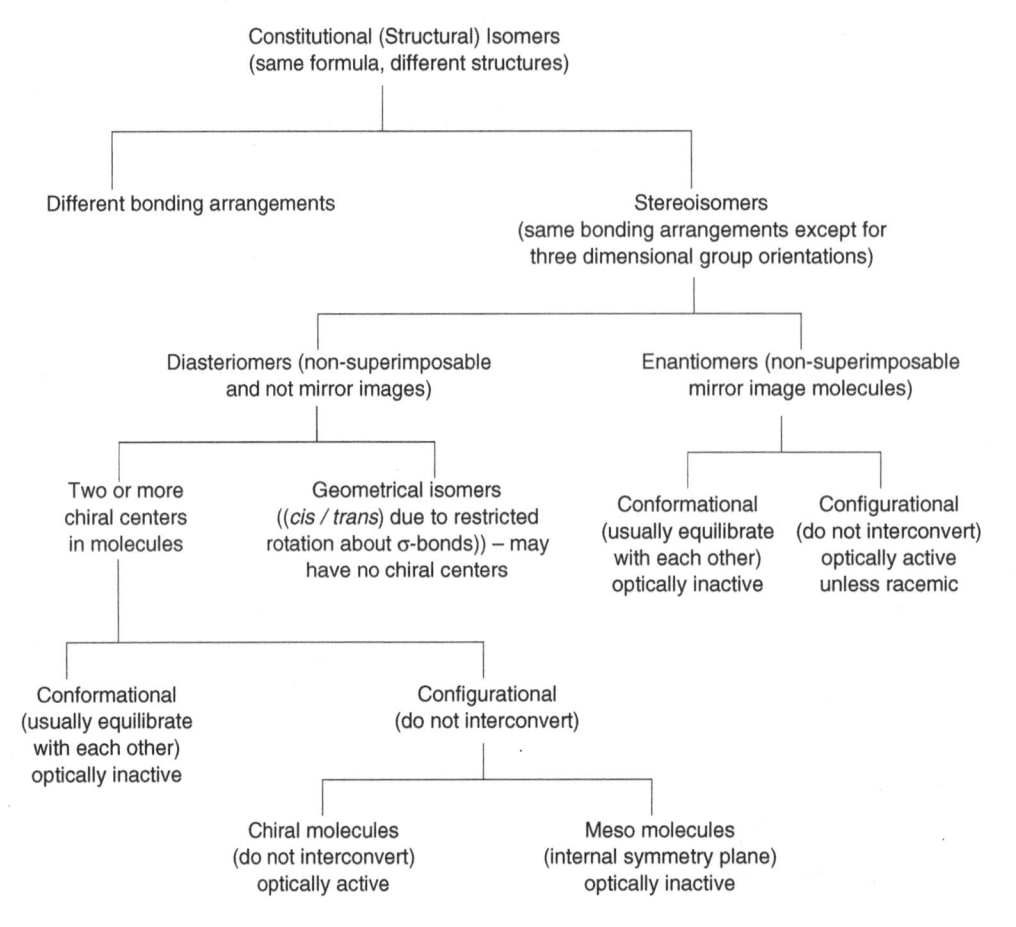

```
                    Constitutional (Structural) Isomers
                    (same formula, different structures)
                                    |
                ┌───────────────────┴───────────────────┐
     Different bonding arrangements              Stereoisomers
                                        (same bonding arrangements except for
                                         three dimensional group orientations)
                                                        |
                                    ┌───────────────────┴───────────────────┐
                    Diasteriomers (non-superimposable       Enantiomers (non-superimposable
                        and not mirror images)                  mirror image molecules)
                                    |                                       |
                        ┌───────────┴───────────┐             ┌─────────────┴─────────────┐
                  Two or more         Geometrical isomers   Conformational          Configurational
                  chiral centers      ((cis / trans) due to (usually equilibrate    (do not interconvert)
                  in molecules        restricted             with each other)       optically active
                                      rotation about σ-bonds)) – may  optically inactive  unless racemic
                                      have no chiral centers
                        |
            ┌───────────┴───────────┐
      Conformational          Configurational
      (usually equilibrate    (do not interconvert)
      with each other)
      optically inactive              |
                          ┌───────────┴───────────┐
                    Chiral molecules         Meso molecules
                    (do not interconvert)    (internal symmetry plane)
                    optically active         optically inactive
```

5.16.1. How are optically pure enantiomers obtained?

    a. Isolation from living sources; examples: (S)-(−)-2-methylbutan-1-ol (from starch fermentation), (+)-tartaric acid from grape juice fermentation, (+)-glucose from sugar beets or cane, grapes, or honey, (+)-alanine from proteins.

(S)-(-)-2-methyl-butan-1-ol   (2R,3R)-tartaric acid L-(+)-tartaric acid   D-(+)-glucose   (S)-(+)-alanine L-(+)-alanine

    1. Many biological molecules are optically active. Why remains largely a mystery.

5.16.2. Each enantiomer's solubility in a chiral solvent may be different, which allows resolution by fractional crystallization.

5.16.3. Resolution of enantiomers can only be achieved by "tweezers" or interaction with other chiral substances.

5.16.4. Chemical resolution with chiral substances (from biological sources).

a. Example: a racemic acid mixture can be separated by neutralization with an optically active base. The two diasteriomeric salts (different properties, such as solubilities) can be separated by fractional crystallization.

(R)(S)-2-chloropropanoic acid (S)-amphetamine (from a plant source)

salts (inside brackets) are diasteriomeric

1. Strong acid converts each pure salt back into acid, optically pure.

resolved   water soluble

2. Nonacidic racemic mixtures (such as alcohols) can be converted into acidic mixtures or esters for resolution, then back into optically pure alcohols.
3. Enantiomers can be separated by column chromatography if a chiral substance is bonded to a solid packing material's surface. One enantiomer passes through a column faster than the other.

5.16.5. Chiral enzymes often metabolize exclusively one enantiomer. The other may even be toxic. Examples:
  a. Only (+)-glucose metabolizes yielding energy.
  b. Penicillium glaucum metabolizes only (+)-tartaric acid.
  c. (−)-Adrenaline is a much more potent stimulant than (+)-adrenaline.
  d. Only (+)-chloromycetin is an antibiotic.
  e. Only (−)-ephedrine is a decongestant. (+)-Ephedrine interferes.
  f. One asparagine and leucine are sweet.
  g. One glutamic acid is a food (meat) flavor enhancer.
  h. (−)-Carvone smells like spearmint, (+)-carvone smells like caraway seeds (rye bread).

5.16.6. Some subtleties can arise during reactions that involve stereocenters. Examples:

5.16.7. Why are racemic (*R*)(*S*)-2-chlorobutanes (along with 1-chlorobutane) produced in butane monochlorination (section 4.13.2)?

  a. The flat intermediate 2° butyl free radical (section 4.16.3) can attack Cl_2 (step (3)) from either face.

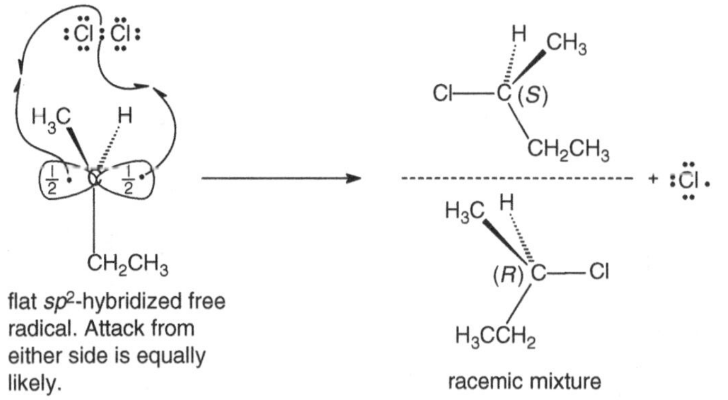

  b. Problem 15. Show why *racemic* 2-chlorobutane would be obtained from butane chlorination if (a) the *sec*-butyl free radical intermediate were not flat, but pyramidal, and (b) chlorination did not involve a free *sec*-butyl free radical at all, but proceeded by a mechanism in which a chlorine atom displaced a hydrogen atom, taking the position on the carbon atom formerly occupied by that hydrogen.

  c. Problem 16. How many separate (different bp) fractions would you expect to obtain from 2-methylbutane monochlorination product fractional distillation? Draw (perspectively if needed) product structures contained in each fraction, and label chiral C's as *R* or *S*. Which (if any) fractions would be optically active? Explain.

d. Problem 17. (S)-2-chlorobutane monochlorination products are separated by fractional distillation. Draw perspectively the 1,2-, 2,2-, and 1,3-dichlorobutane products. Label chiral C atoms R or S. Which (if any) would be optically active or inactive? Explain.

5.16.8. Fractional distillation separates (R)(S)-2-chlorobutanes from 1-chlorobutane. Enantiomer resolution involves interaction with chiral reagents (section 5.16.1).

a. Problem 18. Four dichloro products (**A** - **D**, formula $C_3H_6Cl_2$) are isolated from propane chlorination. What structures are possible?

1. **A** - **D** were each further chlorinated to trichloro products ($C_3H_5Cl_3$) **A** yielded only one product, **B** yielded two, while **C** and **D** each yielded three. Which structure above was **A**? **B**? Which ones are **C** or **D**?

2. By an independent method, an enantiomer of compound **C** was made that was optically active. Which structure is **C**? **D**?

3. When a pure **C** enantiomer was chlorinated, one of the trichloropropanes obtained was optically active, but the other two were optically inactive. Draw the optically active structure. Draw the optically inactive ones.

5.16.9. Generally, optically active products cannot be made from optically inactive reactants, or any chiral products will always be racemic.

  a. Example: butan-2-one hydrogenation yields (R)- and (S)-butan-2-ol, equal amounts.

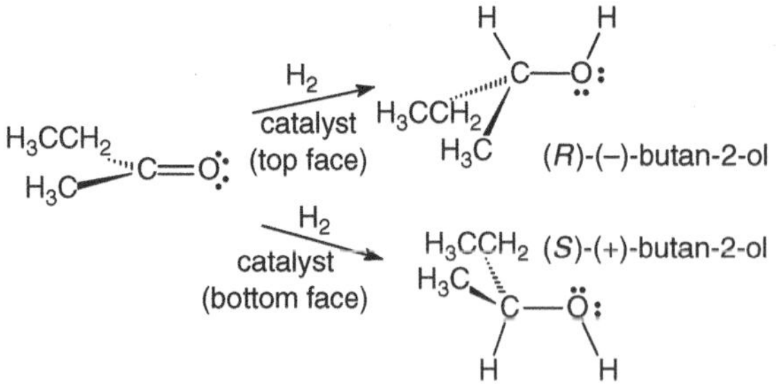

5.16.10. One product from the second (S)-(+)–2-chlorobutane chlorination is (R)–1,2-dichlorobutane, but no (S)–1,2-dichlorobutane is formed.

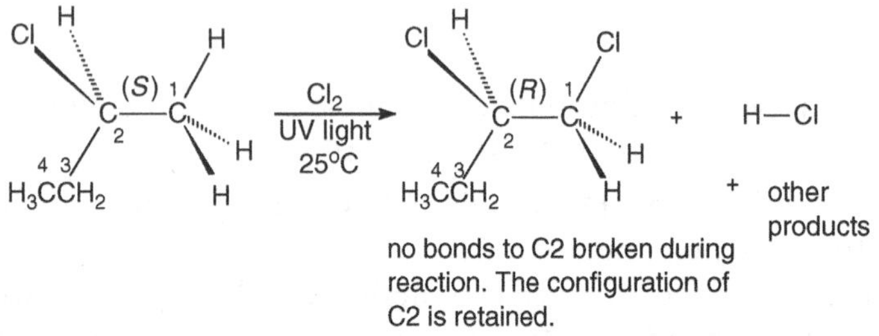

  a. Axiom: configuration is retained if no chemical bond to a chiral C is broken during a reaction.

  b. Subtlety: although C2 configuration is retained (no bonds to C2 are broken), product configuration changes to (R) because Cahn-Ingold-Prelog group priorities have changed. Product –CH$_2$Cl group now has higher priority than the –CH$_2$CH$_3$ group.

5.16.11. Corollary: configuration could be changed if a bond to a chiral C is broken during a reaction.

  a. Example:

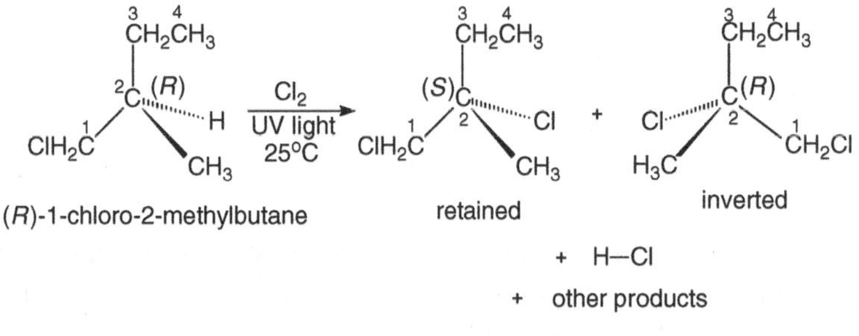

  b. The C2–H bond (to chiral C atom) is broken during reaction, yielding (R)– and (S)–1,2-dichloro-2-methylbutane in equal amounts. Half of reactant molecules invert during reaction.

5.16.12. Existing chiral center influence. Example: (S)-(+)-2-chlorobutane chlorination – 2,3-Dichlorobutane products are formed in *unequal* amounts.

(3)

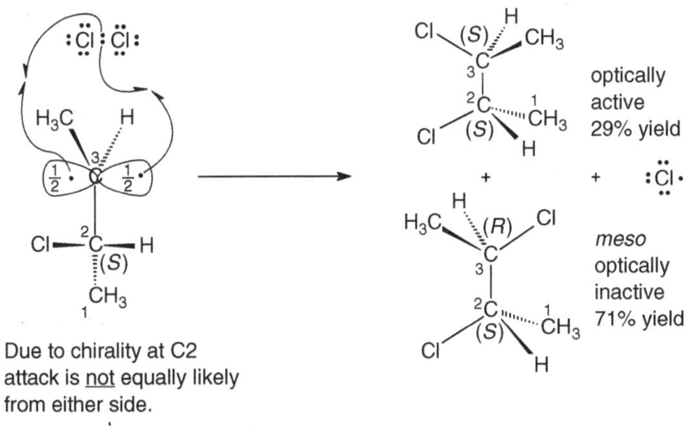

Due to chirality at C2 attack is not equally likely from either side.

→ optically active 29% yield (S,S)

+ meso optically inactive 71% yield

+ :Cl·

a. Left or right attack of the flat free radical at C3 yields (S,S)– and (S,R) (meso) products (diasteriomers).
b. Inherent C2 asymmetry ⇒ left or right attack are unequally likely.
c. C2 configuration is retained ⇒ no (R,R)-product forms – reaction is *stereospecific* (sections 4.16.7a6a, 8.9.3b).

5.16.13. Example: – product stereochemistry – a tool for distinguishing between alternative mechanisms.

a. H. C. Brown, M. S. Kharasch, T. H. Chao – Racemic 1,2-dichloro-2-methylbutane product from (S)-(+)-1-chloro-2-methylbutane chlorination is inconsistent with alternative propagation steps (2a) and (3a) (section 4.7.3).
b. Alternative steps would likely yield retained or inverted products, but not racemization.

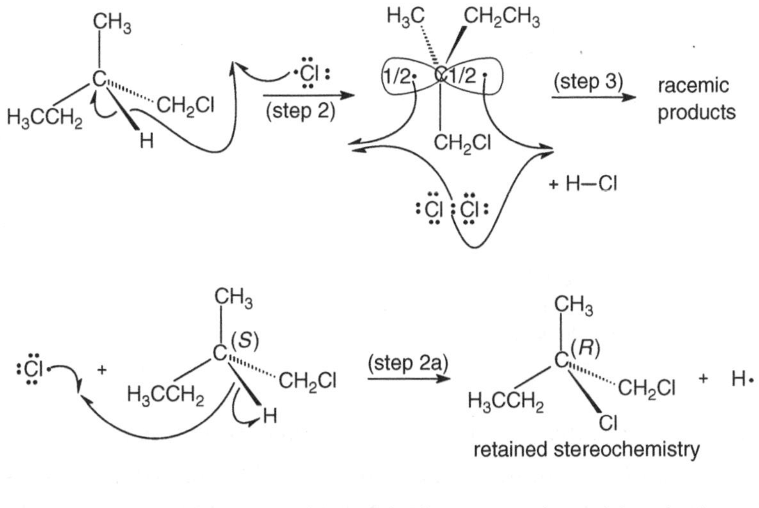

(step 2) → (step 3) racemic products

+ H–Cl

(step 2a) → retained stereochemistry

5.17.1. Reactions of enantiomers with achiral reagents have identical rates and product yields.

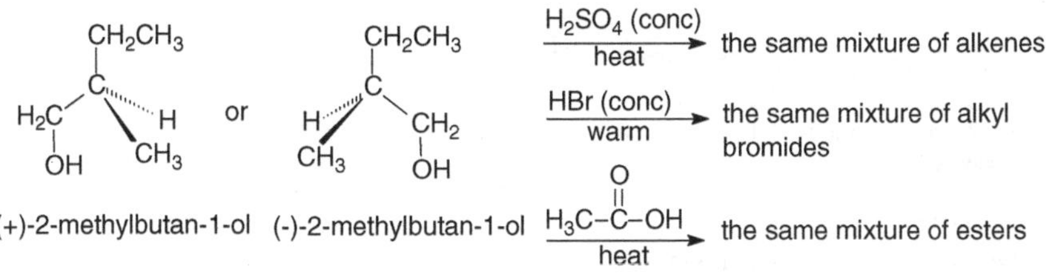

(+)-2-methylbutan-1-ol   (-)-2-methylbutan-1-ol

H_2SO_4 (conc) / heat → the same mixture of alkenes

HBr (conc) / warm → the same mixture of alkyl bromides

$H_3C-C(=O)-OH$ / heat → the same mixture of esters

5.17.3. Analogy: hammering a nail, screwing a right handed screw. Does a right or left hander have an easier time?

5.17.4. When two enantiomers react with a chiral reagent, the respective transition states are diasteriomeric, and reactions have different E_a's, rates, and product yields.

5.17.5. Reactions involving chiral molecules. Reaction types:
   a. Achiral molecule + achiral reagents → chiral molecule. This yields racemic mixtures.
   b. Chiral molecule + achiral reagents and no bond to an existing chiral center is broken → configuration is *retained* (unchanged).
   c. Chiral molecule + achiral reagent and another (second) chiral center is made → enantiomeric and/or diasteriomeric products.
   d. Chiral molecule + chiral reagent → chiral products. A reagent that is already chiral itself can distinguish each different stereoisomer and react at a different rate with each or even yield a different product with each.
   e. Chiral molecule + achiral reagent and a bond to an existing chiral center is broken → chiral products (configurations inverted or retained). Product configurations can help distinguish between possible mechanisms.

5.16.14. Chirality plays huge role in biochemistry. Example – alanine enantiomer enzyme catalyzed reactions. One enzyme incorporates (*S*)-enantiomer into proteins, but reaction with (*R*)-enantiomer is very slow. Another enzyme converts (*R*)-enantiomer into pyruvic acid but reaction with (*S*)-enantiomer is very slow. Both reactions are stereospecific.

5.16.15. *Diasteriospecific reaction* – outcome depends on which diasteriomer is used as reactant.
   a. Examples – pheromones (communication compounds.) Usually only one of several possible diasteriomers is active. Other diasteriomers often inactivate the active one. Enzyme catalyzed reactions are often stereoselective or stereospecific.

(10*trans*,12*cis*)-hexadeca-10,12-dien-1-ol
silk worm sex attractant

(10*cis*,12*cis*)-
(10*trans*,12*trans*)- } all are inactive
(10*cis*,12*trans*)-

b. Problem 19. Draw three dimensional structures and indicate absolute configurations for chiral C atoms in reactants and products **A** - **G**.

1.
　　　　　　　　　　　　+ KMnO₄ ⟶
　　　　　　　　　　　　　(dilute,
　　　　　　　　　　　　　0°C)

_____?
(*R*)-HOCH₂CH(OH)CH=CH₂

_____?
A (optically active)
(HOCH₂CH(OH)CH(OH)CH₂OH)

+

_____?
B (optically inactive)
(HOCH₂CH(OH)CH(OH)CH₂OH)

2. 2　　　　　　　　　⟶ 4 Li　⟶ CuI

_____?
(*S*)-1-chloro-2-methylbutane

_____?
C

_____?
D　　⟵　(*S*)-1-chloro-2-methylbutane
　　　　　　　(Corey-House reaction)

CHAPTER 6

6.1.1. The alkane family (no functional group) has been studied (section 2.15.3).

    a. *Alkyl halides* can be made by free radical alkane halogenation – commonly named from an alkyl group (section 2.15.3c) and halogen as if ionic. Example: methyl chloride.

6.1.2. Only alkyl halides (halogen bonded to sp^3 C) are studied in Chapter 6.

    a. Halogen is bonded to sp^2 C in *vinyl* and *aryl* halides, whose properties are distinct (Chapter 17).

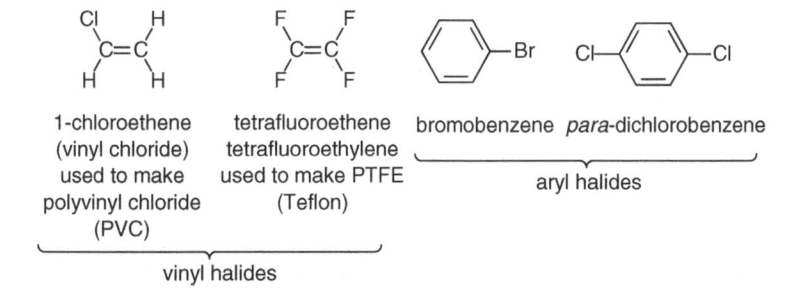

        1-chloroethene (vinyl chloride) used to make polyvinyl chloride (PVC)

        tetrafluoroethene tetrafluoroethylene used to make PTFE (Teflon)

        vinyl halides

        bromobenzene *para*-dichlorobenzene

        aryl halides

6.1.3. Alkane free radical halogenation often yields hard to separate alkyl halide (R-X) mixtures. X = F, Cl, Br, or I. Section 6.6. introduces some better R-X making methods.

6.1.4. X can be replaced by a nucleophile (*substitution*) or it may be lost during π-bond formation (*elimination*), two new common organic reaction types.

    a. Alkyl halides are often synthetic reactants – converted to other functional groups. A C-X bond is polar and weak – heterolyzes (ionizes) easily.

6.1.5. Chapter 4 mostly involved homolytic reactions. This chapter concerns heterolytic reactions.

    a. Bond heterolyses (general reactions):

$$A:B \longrightarrow \overset{\oplus}{A} + :\overset{\ominus}{B} \quad \text{endothermic}$$

$$\overset{\oplus}{C} + A:B \longrightarrow C:A + \overset{\oplus}{B}$$

6.1.6. Heterolyses are usually done in liquid (solution) phase; + and – ions are involved.

6.2.1. Alkyl halides are classified as 1°, 2° or 3° (section 3.3.8 and 4.13.1 in Chapter 3).

$$R-\underset{H}{\overset{H}{C}}-X \qquad R-\underset{H}{\overset{R'}{C}}-X \qquad R-\underset{R''}{\overset{R'}{C}}-X$$

primary (1°)   secondary (2°)   tertiary (3°)

a. Different classes undergo similar reactions, but at different rates.

b. Problem 1. Draw and assign class for the eight isomers having formula $C_5H_{11}Cl$. Ignore stereoisomers.

c. Problem 2. Draw three dimensionally all isomeric structures (including stereoisomers), for:
  1. dichlorocyclopropanes

  2. dichlorocyclobutanes

  3. dichlorocyclopentanes

6.2.2. Alkyl halide names – (sections 2.17.4 and 3.3.7). More examples: (IUPAC name on top, common name below)

H_3CCl	H_2CCl_2	$HCCl_3$	CCl_4
chloromethane	dichloromethane	trichloromethane	tetrachloromethane
methyl chloride	methylene chloride	chloroform	carbon tetrachloride
	(a geminal dihalide)		

1°
$H_3C(CH_2)_2CH_2Br$
1-bromobutane
n-butyl bromide

$\overset{Cl}{\underset{}{|}}$ 2°
$H_3C-CH-CH_3$
2-chloropropane
isopropyl chloride

H_3CCH_2-F
fluoroethane
ethyl fluoride

$\overset{Br}{\underset{}{|}}$
$H_3C-CH-CH_3$
2-bromopropane
isopropyl bromide

$\overset{CH_3}{\underset{}{|}}$ 1°
$H_3C-CH-CH_2Cl$
1-chloro-2-methylpropane
isobutyl chloride

$\overset{CH_3}{\underset{CH_3}{|}}$ 3°
$H_3C-\overset{|}{C}-CH_3$
2-iodo-2-methylpropane
tert-butyl iodide

iodocyclohexane
cyclohexyl iodide

trans-1-chloro-3-methylcyclopentane

$\overset{CH_2-I}{\underset{}{|}}$
$H_3CCH_2-CH-CH_2CH_3$
3-(iodomethyl)pentane?
2-ethyl-1-iodobutane?

$\overset{Cl}{\underset{}{|}}\ \overset{Cl}{\underset{}{|}}$
H_2C-CH_2
1,2-dichloroethane
(a vicinal dihalide)

a. Prefixes (such as iso-) do not help specify class – isopropyl chloride is 2°, but isobutyl chloride is 1°.
b. Problem 3. Write condensed structural formulas for:
   1. 1-bromo-2,2-dimethylpropane

   2. 2-chloro-2,3-dimethylpentane

   3. 1-bromo-3-methylhexane

   4. 2,2-dichloropropane

   5. 1,2-dichloro-3-methylbutane

   6. 3-bromo-2,4-dimethylpentane

c. Problem 4. Draw expanded structural formulas and give IUPAC names.
   1. $(CH_3)_2CHCH_2I$

   2. $(CH_3)_2CHCHClCH_3$

   3. $CH_3CHBrC(CH_3)_2CH_2CH_3$

   4. $(CH_3)_2CClCBr(CH_3)_2$

6.5.1. Alkyl halide physical properties: often have higher boiling points and densities than alkanes:

Compound	Molecular mass	Boiling point (°C)	Density (g / mL)
methane	16.0	−161.5	0.4
ethane	30.1	−88.5	
propane	44.1	−42	
n-pentane	72.2	36	0.626
H_3CF	34	−78	
H_3CCl	50.5	−24	0.92
H_3CBr	94.9	4	1.68
H_3CI	141.9	42	2.279

(Continued)

Compound	Molecular mass	Boiling point (°C)	Density (g / mL)
H_3CCH_2Cl	64.5	12	0.90
H_3CCH_2Br	109.0	38	1.440
H_3CCH_2I	156.0	72	1.933
$H_3CCH_2CH_2Cl$	78.5	47	0.890
$H_3CCH_2CH_2Br$	123.0	71	1.335
$(H_3C)_2CHCl$	78.5	36	0.859
$H_3C(CH_2)_3Cl$	92.5	78	0.884
$(H_3C)_3CCl$	92.5	52	0.840
H_2CCl_2	84.9	50	1.336
$HCCl_3$	119.4	61	1.489
CCl_4	153.8	77	1.595

6.5.2. Apparent trends:

  a. Higher alkyl halide bp's and densities – mostly due to London and dipole-dipole (smaller) attractions.

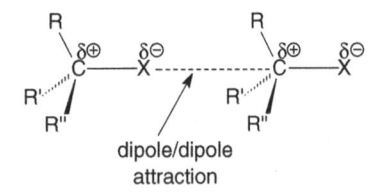

  1. F has about the same surface area as H. Higher bp's (R-F over alkanes) are mostly due to dipole-dipole attractions.
  2. R-F, R-Cl, R-Br, and R-I's bp differences are due to different X surface areas: I > Br > Cl > F. Greater surface area ⇒ greater London attractions.
  3. More isomer branching ⇒ lower bp and density.
  4. Only fluorides and monochlorides are less dense than water (1.00 g/mL at 4°C). More halogen atoms per molecule ⇒ higher bp and density.

6.5.3. Alkyl halides are not very polar:

  a. Almost water insoluble ⇒ slow environmental degradation.
  b. Soluble in nonpolar organic solvents.
  c. Rarely dissolve (polar) ionic compounds.
  d. Smell – like dry cleaning fluid (Cl_3CCH_3).
  e. Problem 5. Three compound's physical properties are listed:

Name	Structure	Dipole moment D	bp °C	Solubility g/100 g H_2O
n-pentane	$CH_3(CH_2)_3CH_3$	0	36	insol.
ethoxyethane	$CH_3CH_2OCH_2CH_3$	1.18	35	8
1-chloro-propane	$CH_3CH_2CH_2Cl$	2.10	47	insol.

What intermolecular forces determine bp's and solubilities in each case?

6.3.1. Alkyl halide uses:
   a. Useful solvents chloromethane, dichloromethane, trichloromethane, and tetrachloromethane (section 4.2.3) are separated by fractional distillation.
   b. Polychlorinated solvents are toxic (liver damage, dermatitis) and carcinogenic (cancer causing).
      1. CCl_4 – was used in dry cleaning and fire extinguishers. Ozone layer depleter. Now replaced by Cl_3CCH_3 and other solvents.
      2. $CHCl_3$ was used as a general anesthetic – also a good degreaser and paint remover. Halothane, $F_3CCHClBr$ is a safer general anesthetic. Chloroethane is a topical anesthetic – numbs partly by freezing (bp 12°C).
      3. CH_2Cl_2 – less toxic than $CHCl_3$ – suspect carcinogen – used in degreasers, paint removers, and as coffee decaffeinator (now replaced by (safer) liquid CO_2). CH_2Cl_2 can be converted by O_2 into deadly phosgene ($Cl_2C=O$).

6.3.2. Chlorofluorocarbon (CFC) refrigerants, and propellants (freons).

Name	Formula
chlorotrifluoromethane (CFC11)– a foaming agent	$FCCl_3$
dichlorodifluoromethane (CFC12) – common refrigerant, use now limited	F_2CCl_2
1,1,1-trifluoro-2,2-dichloroethane (HCFC123) – replacement for CFC11 – destroyed before reaching stratosphere	$HCCl_2CF_3$
dichlorofluoromethane HCFC22 – replacement for CFC12 – destroyed before reaching stratosphere	$HCCl_2F$

   a. Freons were developed to replace (toxic) ammonia. Leaking refrigerators once killed people.

6.3.3. Freon properties:
   a. compressible
   b. inert
   c. nontoxic

6.3.4. When released, CFC's diffuse into Earth's stratosphere (30 –90 km up).

6.3.5. What happens to them? Light energy causes reactions.
   a. Some properties of light were introduced in section 5.4.2.

6.3.6. An ultraviolet (UV) photon has enough energy to break bonds, as seen in step (1):

(1) $X_2 \xrightarrow{h\nu} 2\ X\cdot$

   a. A photon ($h\nu$) is absorbed (disappears) when a bond breaks. The light energy is converted into chemical energy.

6.3.7. N_2 and O_2 absorb wavelengths ≤ 250 nm (in the ultraviolet), but are transparent for λ's ≥ 250 nm.

6.3.8. Ozone (O_3) formation mechanism:

(1) $:\ddot{O}:\ddot{O}: \xrightarrow[\text{short wavelength} < 250\text{ nm}]{h\nu} :\ddot{O}\cdot + \cdot\ddot{O}:$ initiation

(2) $:\ddot{O}\cdot + :\ddot{O}:\ddot{O}: + M \longrightarrow :\ddot{O}:\ddot{O}:\ddot{O}: + M^*$ propagation

(3) $:\ddot{O}:\ddot{O}:\ddot{O}: \xrightarrow[\text{long wavelength} 250-300\text{ nm}]{h\nu} :\ddot{O}\cdot + :\ddot{O}:\ddot{O}:$ propagation

(4) $:\ddot{O}\cdot + \cdot\ddot{O}: + M \longrightarrow :\ddot{O}:\ddot{O}: + M^*$ termination

(5) $:\ddot{O}\cdot + :\ddot{O}:\ddot{O}:\ddot{O}: \longrightarrow :\ddot{O}:\ddot{O}: + :\ddot{O}:\ddot{O}:$ termination

M is an inert gas (N_2 or Ar) that absorbs kinetic energy

6.3.9. Chlorine catalyzes ozone destruction. Reaction (7) destroys O_3 needed for reaction (3); reaction (8) destroys O needed for step (2). Each Cl· formed can destroy 100,000 molecules of O_3 before it undergoes a termination reaction.

(6) $F_3C:\ddot{C}l: \xrightarrow{h\nu} F_3C\cdot + \cdot\ddot{C}l:$ initiation

(7) $:\ddot{C}l\cdot + :\ddot{O}:\ddot{O}:\ddot{O}: \longrightarrow :\ddot{C}l:\ddot{O}\cdot + :\ddot{O}:\ddot{O}:$ propagation

(8) $:\ddot{C}l:\ddot{O}\cdot + \cdot\ddot{O}: \longrightarrow :\ddot{C}l\cdot + :\ddot{O}:\ddot{O}:$ propagation

6.3.10. Insecticides. Dichlorodiphenyltrichloroethane (DDT) was discovered in 1939. Low human toxicity – one ounce kills a human, but protects one acre from insects (mosquitos, fleas, etc.)

   a. DDT is a cumulative poison for species at the top of the food chain (eagles), banned in the U.S. as an agricultural pesticide (1972). It is estimated to have saved over 500 million lives (from malaria, bubonic plague, yellow fever, sleeping sickness) and is still used as a last resort outside the U.S.

   b. Other polychlorinated insecticides (not usually used in agriculture). They have some human toxicity.

DDT

lindane (head lice shampoo)

chlordane (protects wood from termites)

6.4.1. Electronegativity difference and bond length determine bond dipole moments (section 2.1.3).

a.

$$\mu \text{ (in debyes)} = \frac{4.8 \text{ debyes}}{\text{angstrom}} \times \delta \text{ (electron charge)} \times d \text{ (in angstroms)}$$

1. Halogen electronegativities are (section 1.6.1):

    F (4.0) > Cl (3.2) > Br (3.0) > I (2.7)

2. Bond lengths increase as:

    shorter F – C < Cl – C < Br – C < I – C longer

    due to different atomic X radii:

Atom	Van der Waals radius (10^{-8} cm)
H	1.2
F	1.35
Cl	1.8
Br	1.95
I	2.15

3. Opposing factors result in bond dipoles:

    smaller I – C (1.29 D) < Br – C (1.48 D) < F – C (1.51 D) < Cl – C (1.56 D) larger

6.4.2. Structural geometry determines molecular dipole moment (section 2.1.6) – individual bond dipole moment vector sum.

X	H_3CX	H_2CX_2	HCX_3	CX_4
F	1.82 D	1.97 D	1.65 D	0.00 D
Cl	1.94 D	1.60 D	1.03 D	0.00 D
Br	1.79 D	1.45 D	1.02 D	0.00 D
I	1.64 D	1.11 D	1.00 D	0.00 D

6.6.1. Four alkyl halide preparations.
   a. Direct (free radical) alkane halogenation (section 4.13.1).

$$(CH_3)_4C \xrightarrow[25°C]{Cl_2, \; h\nu \text{ (UV light photons)}} (CH_3)_3C-CH_2Cl + H-Cl$$
(excess)

$$\text{(methylcyclohexadiene)} \xrightarrow[127°C]{Br_2/h\nu \text{ (one mol)}} \text{(bromomethyl product)}$$

$$\text{cyclohexane (all H's are equivalent) (excess)} \xrightarrow[\text{or heat}]{Cl_2, \text{ UV light}} \text{chlorocyclohexane, 50\%}$$

$$(CH_3)_3C-H \xrightarrow[\text{and heat}]{Br_2, \text{ UV light}} (CH_3)_3C-Br \quad \text{90\% overall (section 4.13.3)}$$

1. Good yields are obtained only when all H atoms are equivalent, or when one hydrogen class is much more reactive than another (bromination).

$$H_3CCH_2CH_3 \xrightarrow[\text{light}]{Cl_2, \text{ ultraviolet}} H_3CCH_2CH_3 \text{ (unreacted)} + H_3CCH_2CH_2Cl + H_3CCHClCH_2$$

$$+ \; H_3CCHClCH_2Cl \; + \; H_3CCH_2CHCl_2 \; + \; H_3CCCl_2CH_3 \; + \; H_2C(Cl)-CH(Cl)-CH_2Cl$$

+ many others
yield of any one product is poor

2. Example: allyl bromination (section 15.1.11) – driven by allyl free radical intermediate resonance stabilization (section 4.13.39).

$$H_2C=CH-CH_3 \xrightarrow[\text{irreversible}]{\cdot X:} H_2\dot{C}-CH=CH_2 \leftrightarrow H_2C=CH-\dot{C}H_2 \; + \; H-X:$$
allylic attack

$$\downarrow :X-X:$$

$$H_2C=CH-CH_2X \; + \; \cdot X:$$

a. Excess X_2 must be avoided since it adds to double bonds (an unwanted side reaction, section 8.8). *N*-Bromosuccinimide (NBS) is a Br· source – Br_2 forms only as fast as H-Br byproduct.

3. Example:

N-bromosuccinimide, UV light and heat, H–X (trace)

4. Benzylic bromination with NBS works well (section 17.14).

b. Hydrogen halide alkene or alkyne addition (sections 8.3.1 and 9.9.6).

c. Halogen alkene and alkyne addition (sections 8.8.1 and 9.9.5).

182 SEQUENTIAL ORGANIC CHEMISTRY I

d. Alcohol -OH group substitution by halide, -X (section 11.7.1).

$$H_3CCH_2CH_2-OH \xrightarrow[\text{heat}]{\text{HBr(conc)} \text{ or NaBr / H}_2\text{SO}_4\text{(conc)}} H_3CCH_2CH_2-Br$$

[Structure: phenyl-CH(OH)-CH₃ → (PBr₃, pyridine) → phenyl-CH(Br)-CH₃]

6.6.2. Learn reactants, reagents, and products. Know reaction a. mechanism for now.

6.6.3. Many alcohols are commercially available ⇒ reaction d. is most useful.

6.7.1. Alkyl halide heterolytic reactions. Nucleophilic aliphatic substitution – Example: bromomethane + sodium hydroxide – an S_N2 mechanism (section 6.8.5).

[Mechanism diagram: Na⁺ + HO⁻ attacks CH₃-Br with H₂O solvent, warm → transition state [HO---C---Br]‡ → HO-CH₃ + Br⁻ + Na⁺]

  a. Nucleophile and Lewis base HO⁻ forms a coordinate covalent bond to C. Simultaneously, Br takes both bonding electrons as C-Br bond heterolyzes.

6.7.2. Nucleophile – *nucleus lover* – attracted to a (positively charged) atomic nucleus; i.e., a Lewis base (section 2.13.1).

  a. Relative base strength (basicity order) (section 2.6.5) – expanded list:

  $HSO_4^- < I^- < Br^- < Cl^- <$ H₂O $<$ ROH $<$ RCOO⁻ $<$ NH₃ $<$ HCO_3^-
  weaker base

  $< $:C≡N:⁻ $ << $ HO⁻ $<$ RO⁻ $<$:C≡C:⁻ $<$ NH₂⁻ $<$ H:⁻ $<$:CH₃⁻
  stronger base

  b. Nucleophile – electrically neutral or – charged ion – never a + charged ion. Examples:

  $HSO_4^- <$ H₂O $<$ ROH $<$ Cl⁻ $<$ Br⁻ $<$ RCOO⁻ $<$ NH₃
  weaker nucleophile

  $<$ HO⁻ $<$ RO⁻ $<$:C≡N:⁻ $<$ I⁻ $<$ NH₂⁻ $<$ H:⁻ $<$:CH₃⁻
  stronger nucleophile

6.7.3. Nucleophile strength does not exactly parallel base strength due to additional factors besides Lewis basicity.

6.7.4. All strong bases are strong nucleophiles, but some weaker bases (I⁻ and CN⁻) are surprisingly strong nucleophiles.

   a. Base strength – measures equilibrium position as a base reacts *with H⁺ ion* (or another Lewis acid).

   b. Nucleophilicity – measures how fast a nucleophile reacts *with a δ+ C atom*.

   c. Factors that influence nucleophile strength:

   1. Lewis basicity – stronger ⇒ more nucleophilic.

   2. Effective size – smaller ⇒ more nucleophilic.

   3. Polarizability – nucleophile electron orbital (cloud) deformability (attacking δ+ C atom) – more polarizable (softer) ⇒ more nucleophilic.

   a. Higher shell and π-orbitals are more polarizable than second shell orbitals. Example:

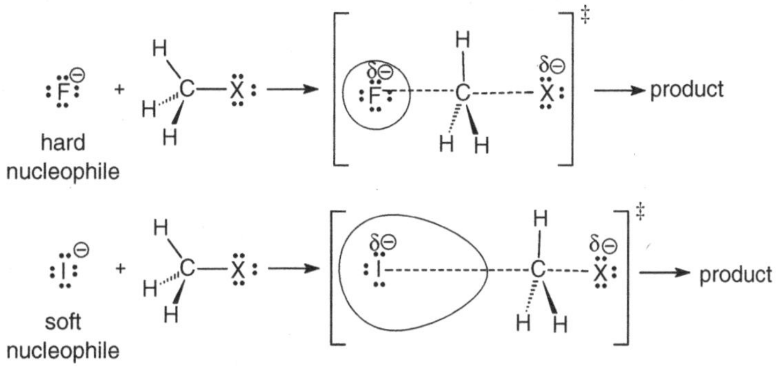

   4. Solvation (section 6.10.1) – less solvated (surrounded by solvent molecules) ⇒ smaller effective size and more nucleophilic. Stripping off solvent molecules is endothermic:

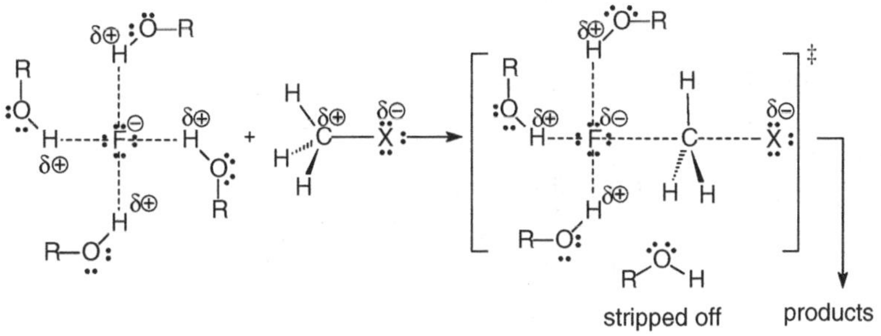

   d. Nucleophilicity trends:

   1. A base is a stronger nucleophile than its conjugate acid.

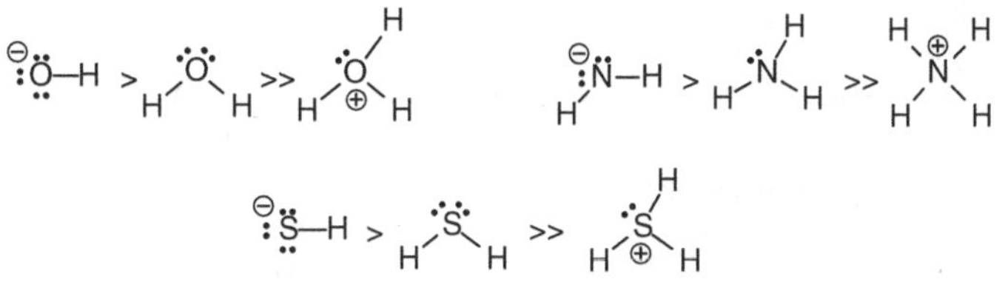

2. Nucleophilicity decreases from left to right in the periodic table for analogous species.

$$H_2\overset{\ominus}{N}-H > H\overset{\ominus}{O}-H > :\overset{..}{\underset{..}{F}}:^{\ominus} \qquad H_3N-H > H_2O-H > H-\overset{..}{\underset{..}{F}}:$$

3. Nucleophilicity increases from top to bottom in the periodic table (paralleling solvated size and polarizability trends).

$$:\overset{..}{\underset{..}{I}}:^{\ominus} > :\overset{..}{\underset{..}{Br}}:^{\ominus} > :\overset{..}{\underset{..}{Cl}}:^{\ominus} > :\overset{..}{\underset{..}{F}}:^{\ominus} \qquad :\overset{..}{Se}-H > :\overset{..}{S}-H > :\overset{..}{\underset{..}{O}}-H$$

$$(H_3CCH_2)_3P: \; > \; (H_3CCH_2)_3N:$$

6.7.5. Substrate – reactant (R-X) attacked by a reagent (a nucleophile in this case).

6.11.1. Leaving group – group departing from carbon with an electron pair (Br⁻ in example above).

   a. Must have an electronegative atom bonded to C, so that C has a δ+ charge and is electrophilic (susceptible to nucleophilic attack). Electronegative atom stabilizes - charge in transition state – commonly X, O, N, S, and P. A few common leaving groups:

anionic: halides, sulfonate, sulfate, phosphate

neutral: H₂O, ROH, :NR₁R₂R₃, :PR₁R₂R₃

   b. A fast (good) leaving group (such as X⁻) becomes a weak Lewis base after it has left, so that products are stable and reverse reaction is slow.

     1. HO⁻ is a poor (slow) leaving group.

$$:Br:^{\ominus} + H_3\overset{\delta+}{C}-\overset{\delta-}{O}-H \; \rightleftharpoons \; :Br-CH_3 + \;^{\ominus}:\overset{..}{O}-H$$

strong base
slow leaving group

     2. An alcohol HO- group usually must be converted into a better leaving group during syntheses. One way is to add acid. Only weakly basic nucleophiles can exist when pH < 7.

$$H_3\overset{\delta+}{C}-\overset{\delta-}{O}-H \;\; \xrightarrow{H-X:} \;\; H_3C-\overset{+}{O}(H)_2 + :X:^{\ominus} \;\; \rightleftharpoons \;\; :X-CH_3 + :\overset{..}{O}(H)_2$$

an oxonium ion   weak base
fast leaving group

3. Convert an HO-group into a sulfonic acid ester, a very good leaving group.

$$\left[\begin{array}{c} H \\ | \\ \overset{H}{\underset{H_3C}{\diagup}} C - \ddot{\underset{\cdot\cdot}{O}} - \overset{\overset{\cdot\cdot}{\ddot{O}}}{\underset{\underset{\cdot\cdot}{\ddot{O}}}{S}} - \!\!\!\bigcirc\!\!\! - CH_3 \end{array} \right]$$

p-toluenesulfonate (tosylate)
fast leaving group

c. Fast leaving groups are polarizable and sustain partial bonding to a δ+ C atom from a longer distance, stabilizing the transition state.
   1. Iodide is both a good nucleophile (in aqueous solution) and good leaving group.
   2. Fluoride is a poor nucleophile (in aqueous solution) and poor leaving group.

6.8.1. Rate law – shows how a reaction rate depends on reactant concentrations (at a given temperature).
   a. The general rate expression (section 4.9.4) collision frequency term includes concentrations.

6.8.2. Example: H₃C-Br with HO⁻ (55°C, solvent 80% ethyl alcohol / 20% water). Rate law:

$$H_3C-\ddot{Br}: \xrightarrow[\substack{80\% \, H_3CCH_2OH \\ 20\% \, H_2O \\ 55°C}]{H-\ddot{O}:^{\ominus} \, Na^{\oplus}} H_3C-OH + :\ddot{Br}:^{\ominus}$$

$$\text{rate} = k[H_3C - Br][HO^-]$$

$$k = \text{(second order) rate constant} = 0.0214 \text{ L} \cdot \text{mol}^{-1} \cdot \text{sec}^{-1}.$$

[ ] brackets mean molarity. Rate law tells how many mols per liter of methanol product are formed per sec.

   a. Since both [H₃C-Br] and [HO⁻] appear in the rate law ⇒ a collision must occur between H₃C-Br and HO⁻ particles in the rate determining (slow) mechanism step.

6.8.3. Another example – (H₃C)₃C-Br with HO⁻ (dil) (55°C, 80% ethyl alcohol / 20% water). Rate law:

$$\underset{\underset{CH_3}{|}}{\overset{\overset{CH_3}{|}}{H_3C-C-\ddot{Br}:}} \xrightarrow[\substack{80\% H_3CCH_2OH \\ 20\% H_2O \\ 55°C}]{H-\ddot{O}:^{\ominus} \, Na^{\oplus}} \underset{\underset{CH_3}{|}}{\overset{\overset{CH_3}{|}}{H_3C-C-OH}} + :\ddot{Br}:^{\ominus}$$

$$\text{rate} = k[(H_3C)_3C - Br]$$

$$k = \text{first order rate constant} = 0.010 \text{ sec}^{-1}$$

   a. [HO⁻] is absent ⇒ HO⁻ must <u>not</u> collide with (H₃C)₃C-Br in the slow mechanism step ⇒ reaction goes by a different mechanism.

6.8.4. Methyl and 1° alkyl halides normally follow 2nd order kinetics. 2° is sometimes 2nd order and sometimes 1st order. 3° is usually 1st order.
   a. Reactivity (with constant [HO⁻]):

6.8.5. E. D. Hughes and C. Ingold (1935) – Crossing reactivity curves ⇒ two competing mechanisms must operate:

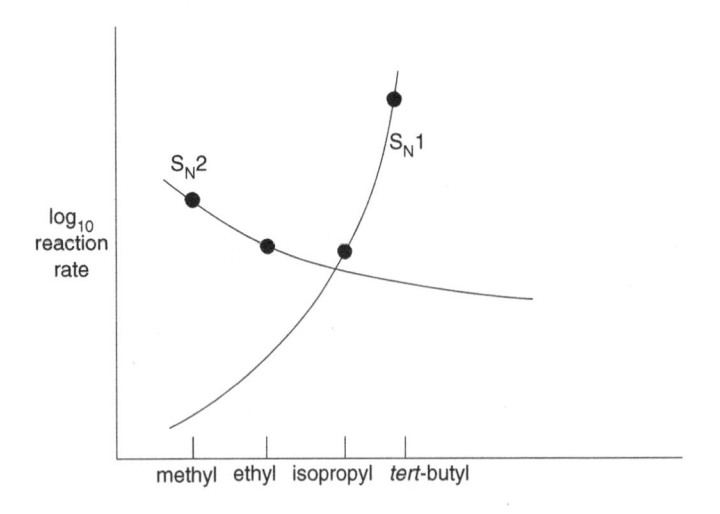

    a. S_N2 operates for methyl, 1°, and sometimes 2°, and S_N1 operates sometimes for 2°, and 3°.

6.8.6. General S_N2 (substitution, nucleophilic, bimolecular) mechanism. M^+ is a metal (spectator) ion, and $:Z^-$ is any negatively charged nucleophile.

6.8.7. S_N2 features:
    a. one elementary step (collision) – concerted bond breaking and formation.
    b. first order in [R-X] and first order in [$:Z^-$] (nucleophile), second order overall.
    c. backside attack – $:Z^-$ stays as far from X as possible during collision with R-X.
    d. complete C inversion.
    e. bimolecular, pentavalent transition state (includes both $:Z^-$ and R-X).
    f. fastest for methyl, slowest for 3° alkyl halides.
    g. $:Z^-$ forms a new σ-bond (coordinate covalent bond, section 1.5.2) to C with its own lone electron pair.

6.8.7. Reaction diagram (for iodomethane reacting with hydroxide) – consistent with S_N2 features:

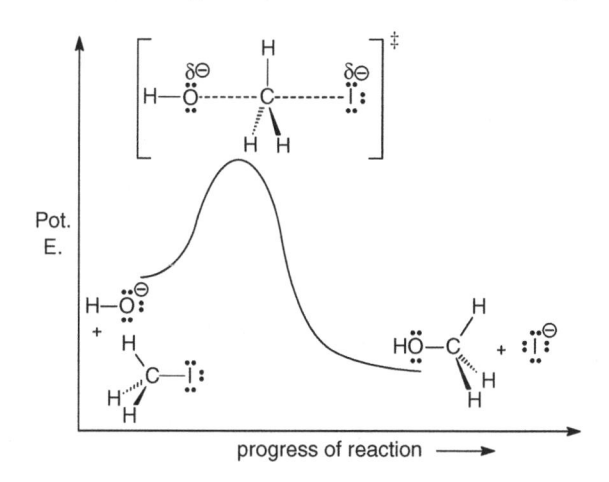

6.8.8. Methyl, 1°, 2°, and 3° alkyl bromides all halogen exchange mainly via S_N2 in dimethylformamide solvent (DMF, section 6.10.5).

compound	methyl	ethyl	isopropyl	t-butyl
relative rate	37	1.0	0.02	0.0008
relative rate	46,000	1,300	25	1.0

6.8.9. Reactants and products are far from the S_N2 transition state on the reaction curve (section 6.8.7), and do not look much like it.

   a. Pentavalent C has little electrical charge (because bond breaking and bond formation are nearly equal) in all four transition states:

      1. The three fully bonded H- or H_3C-groups cannot stabilize the transition state through electrical charge delocalization since the central C has little or no charge. The three substituent group polar effects (section 4.13.32) are small and do not affect reaction rate much.

   b. The more crowded the five groups about the central C atom $\Rightarrow$ higher E_a and slower reaction.

6.11.2. Steric effect – a rate difference resulting from different sized groups (near a reaction site) in one structure *vs.* another.

[Reaction diagram showing nucleophilic attack of HO⁻ on (CH₃)₃C-Br through sterically crowded pentavalent transition state, producing HO-C(CH₃)₃ + Br⁻. Angles labeled 109.5° and 90°.]

sterically crowded pentavalent transition state

6.11.3. Steric effects are largely responsible for S$_N$2 reaction rate differences:

faster methyl > 1° > 2° > 3° slower

   a. Nucleophile steric hindrance – bulky *tert*-butoxide is a stronger base (for abstracting protons) than ethoxide, but smaller ethoxide is a stronger nucleophile (towards C).

[Structures of potassium *tert*-butoxide and sodium ethoxide]

potassium *tert*-butoxide   sodium ethoxide

6.11.4. Example: all C atoms are primary in the alkyl bromide rate comparison:

compound	ethyl	*n*-propyl	isobutyl	neopentyl
	H₃CCH₂Br	H₃CCH₂CH₂Br	(H₃C)₂CHCH₂Br	(H₃C)₃CCH₂Br
relative rate (S$_N$2)	1.0	0.69	0.33	0.000006

   a. Nucleophiles attack the sp^3-C-Br bond back lobe. Nearby large groups block approach.
   b. Problem 6. Draw structures for the following compounds and then rank them in order of increasing S$_N$2 reactivity (slowest to fastest):
   1. bromocyclohexane, 1-bromo-1-methylcyclohexane, 1-bromo-1-cyclohexylmethane

   2. Rank part 1 compound reactivity towards S$_N$1 (slowest to fastest).

6.12.1. Molecular orbital theory – :Z⁻ electron density flows into the backside of the sp^3-C-X antibonding σ*-orbital (since σ-bonding orbital is full already) as :Z⁻ approaches R-X. The added electron density helps push out X⁻.

6.12.2. Stereochemical evidence supports *Walden inversion* in S_N2: (R)-$(-)$-2-bromooctane + NaOH. Product configuration *inverts* from R to S like an umbrella turning inside out.

  a. Example:

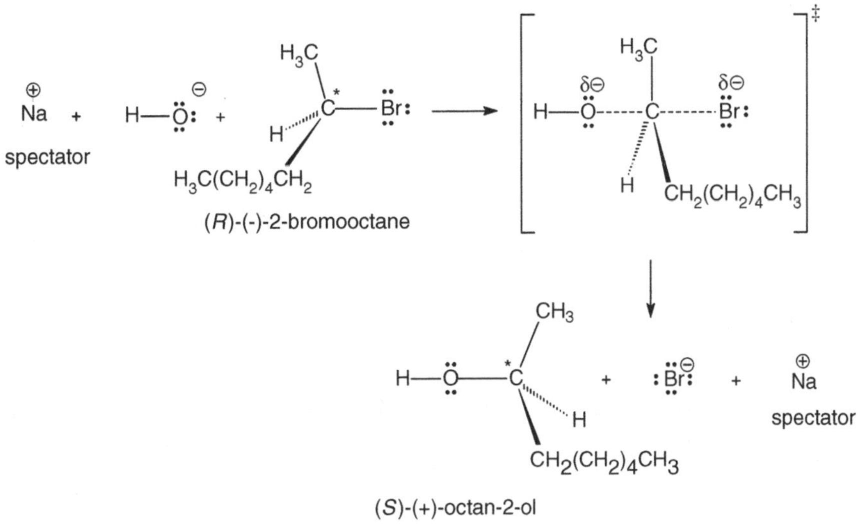

  b. (S)-$(+)$-Octan-2-ol optical purity is 100%.
  c. Example:

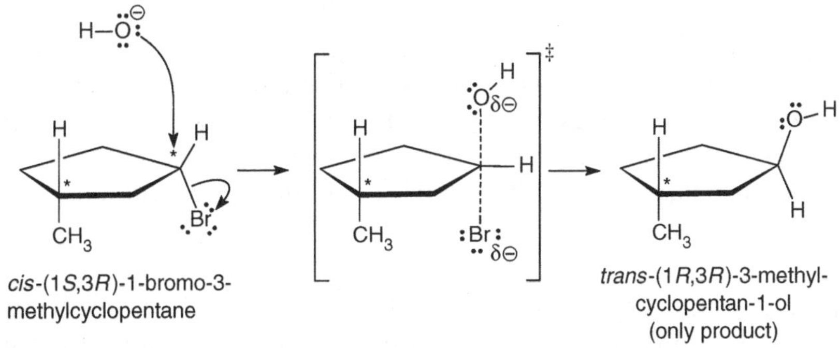

  d. Inversion cannot be proven for optically inactive substrates.
  e. Problem 7. Specific rotation for the optically pure 2-bromooctane enantiomer shown is $+39.6° \cdot dm^{-1} \cdot g^{-1} \cdot mL$, and that of the optically pure octan-2-ol enantiomer is $+10.3° \cdot dm^{-1} \cdot g^{-1} \cdot mL$. What would be the octan-2-ol specific rotation obtained from S_N2 reaction of an optically impure 2-bromooctane sample whose initial $[\alpha] = +24.9° \cdot dm^{-1} \cdot g^{-1} \cdot mL$?

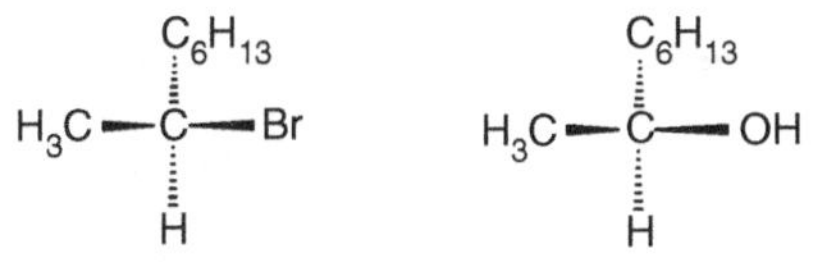

f. Problem 8. In 1923, Henry Phillips (Battersea Polytechnic, London) reported the following experiment:

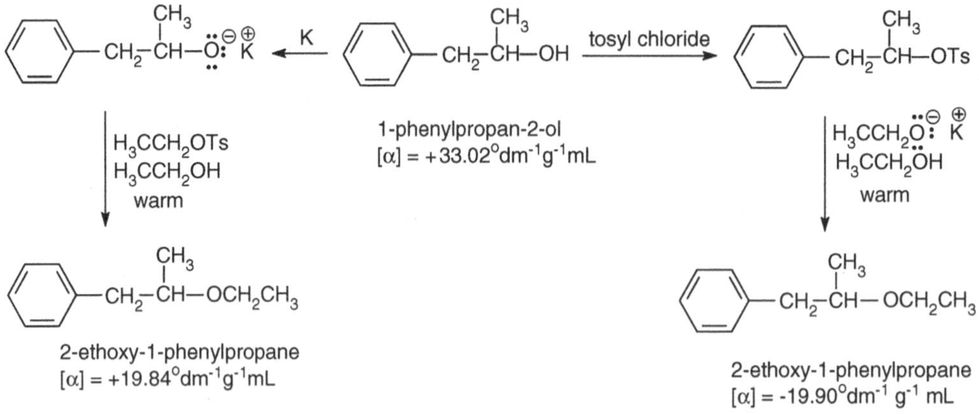

[Reaction scheme: 1-phenylpropan-2-ol ([α] = +33.02°dm⁻¹g⁻¹mL) is converted on the left side to its potassium alkoxide (with K), which reacts with H₃CCH₂OTs / H₃CCH₂OH (warm) to give 2-ethoxy-1-phenylpropane ([α] = +19.84°dm⁻¹g⁻¹mL). On the right side, 1-phenylpropan-2-ol reacts with tosyl chloride to give the tosylate, which then reacts with H₃CCH₂O⁻K⁺ / H₃CCH₂OH (warm) to give 2-ethoxy-1-phenylpropane ([α] = −19.90°dm⁻¹g⁻¹mL).]

Account for the fact that the ethers obtained by the two routes have opposite but equal (within experimental error) optical rotations. In which reaction(s) is carbon configuration inverted?

g. Problem 9. When optically active 2-iodooctane was allowed to stand in acetone (propan-2-one) solution containing Na^{131}I (radioactive iodide), the 2-iodooctane was observed to lose optical activity and to exchange ordinary iodide for radioactive iodide. Each process' rate depended on both [RI] and [^{131}I], but optical activity loss was exactly twice as fast as radioactivity gain. Combining as it does kinetics and stereochemistry, this experiment, reported in 1935 by E. D. Hughes is considered to have established the stereochemistry of the S$_N$2 reaction: that each molecule undergoing substitution suffers configuration inversion. Show what is happening on the molecular level, and justify this conclusion.

6.12.3. S$_N$2 is stereospecific (sections 4.16.7.a6a and 8.9.3b) – starting with one of the two possible reactant configurations, only one of the two possible product configurations is obtained.

a. Example – product enantiomer formed depends on the reactant enantiomer (an *enantiospecific reaction*).

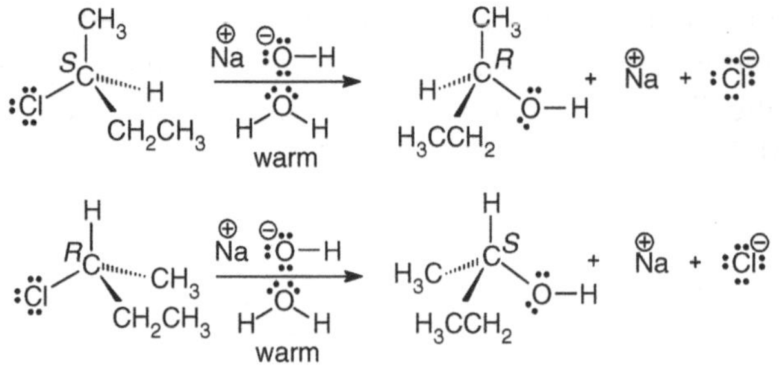

[Two reaction schemes showing: (S)-2-chlorobutane + NaOH (warm) → (R)-2-butanol + Na⁺ + Cl⁻; and (R)-2-chlorobutane + NaOH (warm) → (S)-2-butanol + Na⁺ + Cl⁻]

6.8.10. Summary: evidence supporting S$_N$2 mechanism:
a. kinetic data:
    1. second order reaction rate law.
    2. steric effect.
b. product studies:
    1. stereochemistry (configuration inversion).

6.9.1. Some useful S$_N$2 alkyl halide synthetic reactions:

$$R-X + I^- \longrightarrow R-I + X^-$$
$$R-X + H-O^- \longrightarrow R-OH + X^-$$
$$R-X + R-O^- \longrightarrow R-O-R' + X^-$$
$$R-X + H-S^- \longrightarrow R-S-H + X^-$$
$$R-X + R'-S^- \longrightarrow R-S-R' + X^-$$
$$R-X + {}^-NH_2 \longrightarrow R-NH_2 + X^-$$
$$R-X + NH_3 \longrightarrow R-NH_2 + H-X$$
$$R-X + {}^-N=N=N^- \longrightarrow R-N=N=N^- + X^-$$
$$R-X + {}^-C\equiv C-R' \longrightarrow R-C\equiv C-R' + X^-$$
$$R-X + {}^-C\equiv N \longrightarrow R-C\equiv N + X^-$$
$$R-X + {}^-O-\overset{\overset{O}{\|}}{C}-R' \longrightarrow R-O-\overset{\overset{O}{\|}}{C}-R' + X^-$$
$$R-X + :P(Ph)_3 \longrightarrow R-\overset{+}{P}(Ph)_3 + X^-$$
$$R-X + R'_2CuLi \longrightarrow R-R' + R'-Cu + Li^+ + X^-$$

 a. Corey-House reaction (with R'$_2$CuLi) is probably not S$_N$2, since coupling occurs with vinyl and aryl halides.

6.9.2. S$_N$2 reaction – synthesis "workhorse."

6.9.3. Halogen exchange examples. Good for making iodides and fluorides – hard to make by other methods.

$$\overset{+}{Na} + :I:^- + \overset{H}{\underset{H}{\overset{|}{C}}}^{\delta+}-Br^{\delta-} \xrightarrow[\text{warm}]{H_2O \text{ (solvent)}} \left[:I^{\delta-}\cdots\overset{H}{\underset{H H}{C}}\cdots Br^{\delta-} \right]^{\ddagger}$$

$$\longrightarrow :I-\overset{H}{\underset{H}{\overset{|}{C}}}-H + :Br:^- + \overset{+}{Na}$$

$$\overset{+}{Na} + H_2C=CH-CH_2-Cl: + :I:^- \xrightarrow[\text{(solvents)}]{\text{acetone}} \overset{+}{Na} + H_2C=CH-CH_2-I:$$
spectator ion $\qquad\qquad\qquad\qquad$ H$_2$O $\qquad$ spectator ion $+ :Cl:^-$
$\qquad\qquad\qquad\qquad\qquad$ warm

$$\overset{+}{K} + H_3C-CH_2-Cl: + :F:^- \xrightarrow[\text{H}_3\text{C}-\text{CN}]{\substack{\text{18-crown-6} \\ \text{(phase transfer} \\ \text{catalyst)}}} H_3CCH_2-F: + \overset{+}{K} + :Cl:^-$$
$\qquad\qquad\qquad\qquad\qquad\qquad$ (solvent)

6.9.4. A proton transfer step occurs before product (alcohol) is isolated for electrically neutral nucleophiles (such as H₂O):

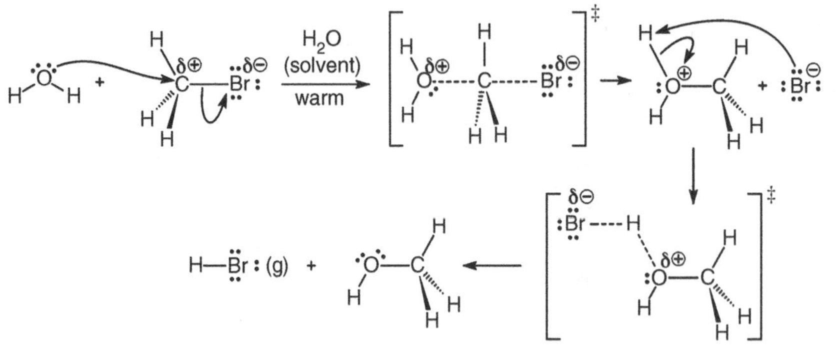

6.10.1. Solvent can exert an even larger rate effect than polar and steric factors in heterolytic chemistry. Solvent is not really an inert medium.
  a. Solute-solvent attractions and entropy must be larger than solute-solute + solvent-solvent attractions and entropy for dissolution to occur (section 2.3.2).
  b. Reactant and transition state energies are related to their structures *which include solvation molecules*.

6.10.2. Weak, secondary attractive forces between molecules (section 2.2.3) determine biomolecule and polymer shapes.
  a. An individual ion-dipole bond is fairly weak, but many such bonds working together can be stronger than a full ionic or covalent bond (> 420 - 1260 kJ/mol).
  b. Example: NaCl (bp 1413° C), easily dissolves in H₂O at 25° C due to ion-dipole attractions.

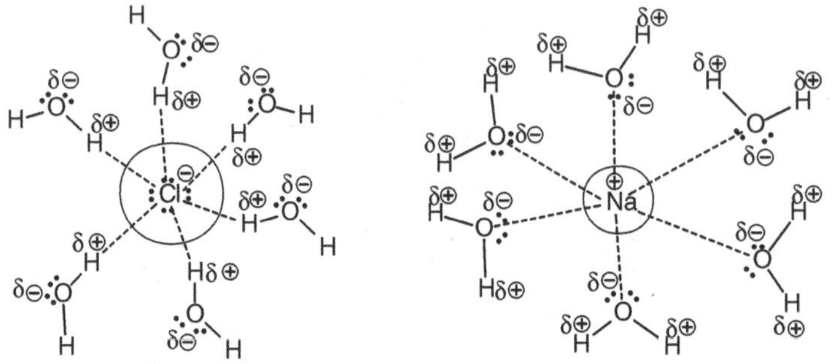

6.10.3. Alkane (nonpolar) and alkyl halide (weakly polar) molecules are attracted to each other only by weak London and dipole-dipole attractions. They have weaker attractions for water molecules than water molecules have for themselves (due to H-bonding).
  a. Alkanes and alkyl halides are practically insoluble in water – *hydrophobic* (water hating) or *lipophilic* (fat loving).

6.10.4. Protic (proton donating) solvent – forms hydrogen bonds – contains at least one -OH or -NHR group (R = H or alkyl).

6.10.5. Polar aprotic solvent – cannot form hydrogen bonds – has a moderately high dielectric constant.
  a. Examples: acetone, dimethyl sulfoxide (DMSO), dimethylformamide (DMF), hexamethylphosphorotriamide (HMPT), and acetonitrile.

propan-2-one   dimethylsulfoxide   N,N-dimethylmethanamide
acetone   (DMSO)   N,N-dimethylformamide
                                   (DMF)

hexamethylphosphordiamide   ethanenitrile
(HMPT)   acetonitrile

1. All have uncrowded O atoms (with lone pair electrons) and strongly solvate cations.
2. They solvate anions weakly. δ+ is sterically crowded and forms only weak dipole-ion attractions.

b. Problem 10. Which of the following solvents would be protic? Aprotic?

a. NH$_3$(l)   b. SO$_2$(l)   c. CH$_2$Cl$_2$ d. CH$_3$CH$_2$OH
   ammonia   sulfur dioxide   dichloromethane   ethanol

e. CH$_3$CH$_2$OCH$_2$CH$_3$ f. CH$_3$–C(=O)–OH   g. CH$_3$–C(=O)–CH$_3$ h. H–C(=O)–NH$_2$
   ethoxyethane   ethanoic acid   propan-2-one   methanamide
   diethyl ether   acetic acid   acetone   formamide

i. H–C(=O)–N(H)–CH$_3$ j. CH$_3$C≡N   k. (oxolane)   l. (sulfolane)
   N-methylmethanamide   ethanenitrile   oxolane   sulfolane
   N-methylformamide   acetonitrile   tetrahydrofuran

6.10.6. Nucleophiles (often polar or ionic) and alkyl halide (nonpolar or covalent) molecules must collide with each other in a medium that dissolves both substrates in order for S$_N$2 to occur. Only protic solvents (such as alcohol / water mixtures) were practical until the advent of polar aprotic solvents.

a. Anions are strongly solvated (stabilized) by polar protic solvents, lowering their nucleophilicity. Small "hard" (nonpolarizable) ions, like F$^-$, are more strongly solvated than larger (polarizable) ions, like I$^-$ and C≡N$^-$.

1. Protic solvent anion stabilization. Some - charge is delocalized onto H$_2$O molecules.

b. Hard anions are only weakly solvated (naked ions) in polar aprotic solvents, and are up to a million times more nucleophilic.

c. Problem 11. Like most inorganic salts, ammonium chloride is insoluble in nonpolar organic solvents. If H atoms are replaced by methyl groups, the salt becomes appreciably soluble in nonpolar organic solvents. How do you account for this? How might the ammonium salt solubility be further increased?

ammonium chloride

tetramethylammonium chloride
a quaternary ammonium salt

6.10.7. Gas phase S_N2 reactions, where nucleophiles are not solvated at all, can be up to 10^{20} times faster than in solution.

a. R-X is only weakly solvated by strongly ionizing (polar) solvents since it is relatively nonpolar. A nucleophile (such as HO⁻) with a full (hard) - charge on oxygen forms ion-dipole bonds with strongly ionizing solvents, which dramatically stabilizes (and weakens) the nucleophile.

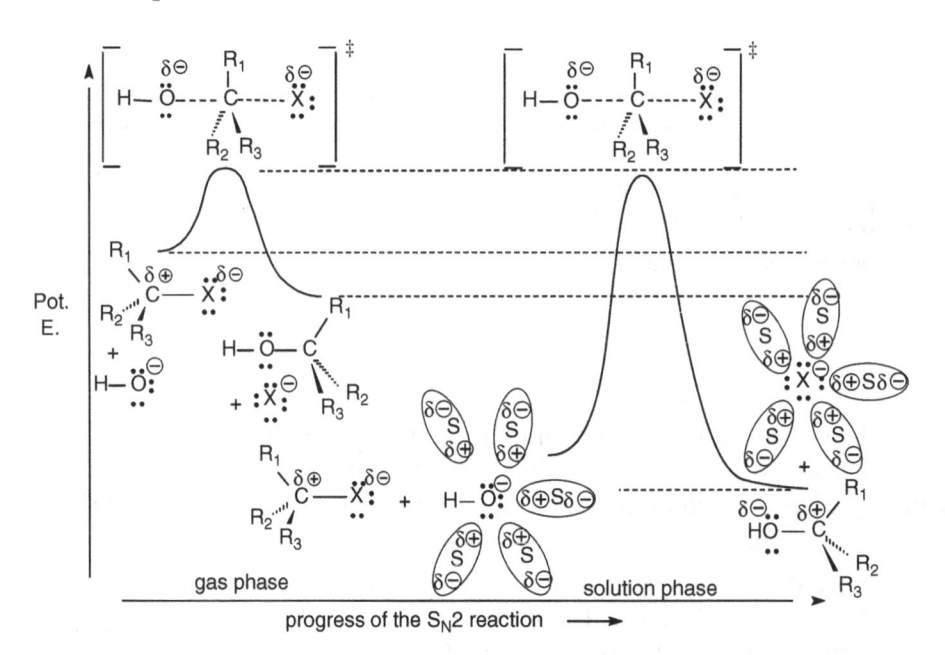

b. Transition state - charge is delocalized between nucleophile and leaving group. This "soft" charge is only weakly solvated and is not stabilized much by a strongly ionizing solvent.

c. Nucleophile (hydroxide ion) is stabilized more than transition state – E_a is higher, and S_N2 reaction rate is slower in strongly ionizing solvents.

d. Example: reaction of H_3C-Br with partially hydrated HO⁻.

$$H_3C - Br + HO^- \cdot (H_2O)_n \rightarrow H_3C\text{-}OH + Br^-$$

n =	0	1	2	3	solution
$k_{relative}$ =	1	0.6	0.002	0.0002	10^{-16}

6.10.8. In the gas phase, nucleophilicity parallels Lewis basicity. For the halide exchange reaction:

$$H_3C-Br + X^- \rightarrow H_3C-X + Br^-$$

stronger $F^- > Cl^- > Br^- > I^-$ weaker

6.10.9. In methanol or water (polar protic solvents) nucleophilicity is reversed.

weaker $F^- < Cl^- < Br^- < I^-$ stronger

    a. F^- is the hardest ion, and the most strongly solvated (stabilized) by H-bonding.

6.10.10. Nucleophilicity in polar aprotic solvents is similar to the gas phase due to weak solvation.

    a. Problem 12. Arrange each set of compounds in order of increasing reactivity toward S_N2 displacement in warm ethanenitrile solvent.
        1. 2-bromo-2-methylbutane, 1-bromopentane, 2-bromopentane

        2. 1-bromo-3-methylbutane, 2-bromo-2-methylbutane, 3-bromo-2-methylbutane

        3. 1-bromobutane, 1-bromo-2,2-dimethylpropane, 1-bromo-2-methylbutane, 1-bromo-3-methylbutane

    b. Problem 13. Arrange the compounds in parts 1 and 2 above in order of increasing S_N1 reactivity in warm ethanol solvent.

    c. Problem 14. Solvent effects above apply to an anionic nucleophile attacking an electrically neutral substrate. Use similar logic to account for the following facts:
        1. Increasing solvent polarity (from a nonpolar solvent to a polar solvent) greatly increases S_N2 reaction rate when ammonia is the nucleophile.

$$R-\ddot{X}: + \ddot{N}H_3 \rightarrow R-\overset{\oplus}{N}H_3 + :\ddot{X}:^{\ominus}$$

        2. Increasing solvent polarity causes a large S_N2 attack rate decrease by hydroxide on trimethylsulfonium ion.

3. Increasing solvent polarity causes a small S$_N$2 attack rate decrease by trimethylamine on trimethylsulfonium ion.

$$H_3C-\overset{CH_3}{\underset{CH_3}{\overset{\oplus}{S}}}\!\!\cdot + :\!\ddot{\underset{\cdot\cdot}{Cl}}\!:^{\ominus} + CH_3-\overset{CH_3}{\underset{CH_3}{N}}\!: \longrightarrow H_3C-\overset{CH_3}{\underset{CH_3}{\ddot{S}}}\!: + CH_3-\overset{CH_3}{\underset{CH_3}{\overset{\oplus}{N}}}\!\!-CH_3 + :\!\ddot{\underset{\cdot\cdot}{Cl}}\!:^{\ominus}$$

trimethylsulfonium ion   trimethylamine   dimethyl sulfide   tetramethyl-ammonium ion

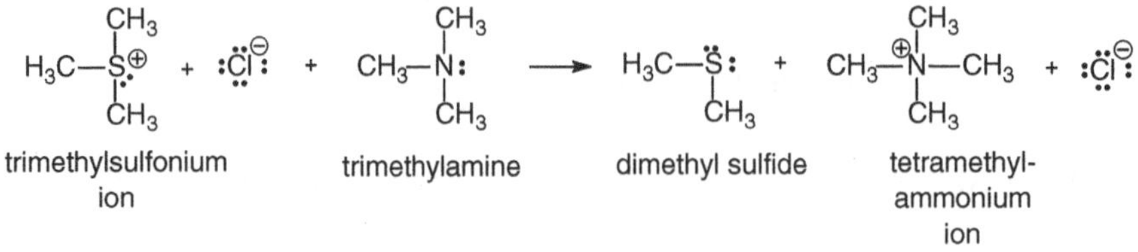

d. **Problem 15.** For each of the following second-order reactions suggest (a) possible explanation(s) for the solvent effects given:
1. Relative rates: 1.0 (H$_2$O) : 16 (methanol) : 44 (ethanol)

$$^{131}I^- + CH_3I \rightarrow CH_3{}^{131}I + I^-$$

2. Relative rates: 1.0 (hexane) : 13,000 (trichloromethane)

$$(n-C_3H_7)_3N + CH_3I \rightarrow (n-C_3H_7)_3N^+CH_3 I^-$$

3. Relative rates: 1.0 (CH$_3$OH) : 10^5 (HMPT)

$$Br^- + CH_3OTs \rightarrow CH_3Br + TsO^-$$

e. **Problem 16.** The following reaction is carried out in the weakly ionizing solvent, acetone. The halide reactivity order depends on the salt that is used as X$^-$ source. If Li$^+$ X$^-$ is used relative rates are: faster I$^-$ > Br$^-$ > Cl$^-$ slower, but if (n-C$_4$H$_9$)$_4$N$^+$ X$^-$ is used, faster Cl$^-$ > Br$^-$ > I$^-$ slower. Suggest factors that might explain this behavior?

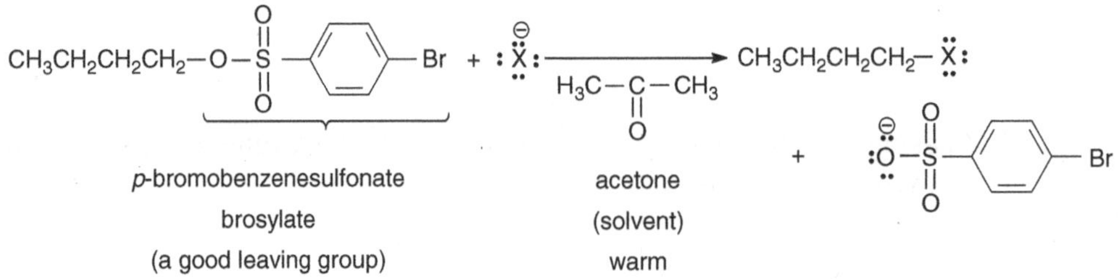

CH$_3$CH$_2$CH$_2$CH$_2$—O—S(=O)(=O)—C$_6$H$_4$—Br + :X:$^{\ominus}$ $\xrightarrow[\text{warm}]{H_3C-C(=O)-CH_3}$ CH$_3$CH$_2$CH$_2$CH$_2$—X: + :O$^{\ominus}$—S(=O)(=O)—C$_6$H$_4$—Br

p-bromobenzenesulfonate
brosylate
(a good leaving group)

acetone
(solvent)
warm

6.13.1. S$_N$1 mechanism (substitution, nucleophilic, unimolecular).

a. Example: 2-bromo-2-methylpropane reaction with HO$^-$ (dil) in 80% ethanol/ 20% water at 55° C (section 6.8.3) yields 2-methylpropan-2-ol. Some 2-methylpropene byproduct also forms by the E2 mechanism (section 6.17.2).

b. Example: 2-bromo-2-methylpropane reacts with boiling methanol (a weak nucleophile) via S$_N$1.

1. *Solvolysis* (solvo from solvent, and lysis means cleavage) – solvent is the nucleophile.

6.13.2. General S_N1 mechanism:

6.13.3. S_N1 reaction features:
  a. At least two and often three elementary steps (collisions); bond breaking and bond making are not simultaneous.
  b. Reaction rate is first order in [R-X] but independent of [:Z-H] (zeroth order). See section 6.8.3. Rate law:

  $$\text{rate} = k[\text{R-X}]$$

  c. Step 1 is rate determining; steps 2 and 3 are fast by comparison.
  d. Unimolecular transition state structure $\ddagger_1$ (for the slow step).
  e. R-X overall reactivity:

  $$\text{slowest methyl} < 1° < 2° < 3° \text{ fastest}$$

  f. Intermediate products are ions: carbocations, oxonium ions, halides, etc.).
  g. Rearranged products form where possible.
  h. Optically active substrates ⇒ at least partially racemized products (R-X).

6.13.4. Reaction (section 6.13.1a) diagram. Compare to section 6.8.7.

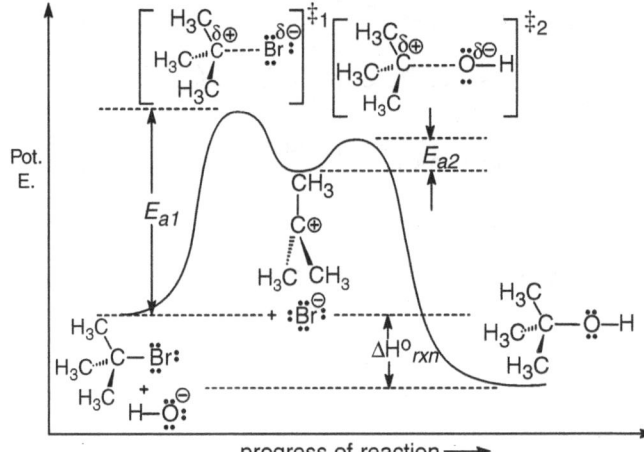

6.13.5. Carbocation (carbenium ion) (section 4.16.4) – a high energy intermediate product formed during S_N1.

6.13.6. Carbocation description – a group of covalently bonded atoms containing a carbon atom that has a positive formal electrical charge, and only six (shared) electrons.

   a. Strictly, carbocations include both and carbenium ions and carbonium ions.

   1. Carbonium ion – has ten electrons around C, as in $^+CH_5$. Such species have very high potential energy and decompose rapidly.

   b. Carbocation (carbenium ion) classification: methyl, 1°, 2°, or 3°:

   methyl   ethyl (1°)   isopropyl (2°)   *tert*-butyl (3°)

   c. Properties:
   1. Highly reactive Lewis acid (electrophile).
   2. Cannot easily be isolated and studied.

   d. Georg Olah (1994 Nobel Prize) in 1963 directly observed *t*-butyl carbocation intermediates in superacid (by NMR, Chapter 13) – a stepwise S_N1 reaction:

6.13.7. A carbocation C atom is similar to the BF_3 B atom (section 1.15.7) – sp^2 hybridized (six valence electrons). Supporting evidence: calculations, infrared and NMR spectroscopy.

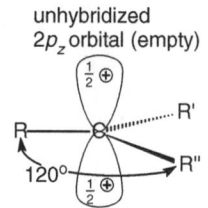

6.13.8. Heterolytic bond dissociation enthalpy – energy (kJ) to convert one mol of alkyl bromide into one mol of carbocations and one mol of halide ions (gas phase).

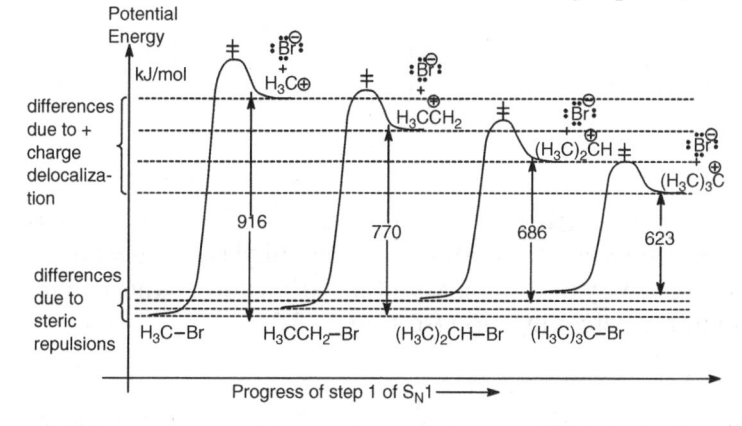

    a. Carbocation stability order (relative to corresponding R-X):

$$\text{more stable } 3° > 2° > 1° > \text{ methyl less stable}$$

6.13.9. Solution $\Delta H^°_{rxn}$ values are similar to those for the gas phase (G. Olah).

    a. Energy differences are larger for carbocations than for free radicals. *t*-Butyl carbocation is 276–301 kJ/mol more stable than methyl carbocation, while *t*-butyl free radical is only 50 kJ/mol more stable than methyl free radical.

    b. Steric effects are responsible for small reactant potential energies differences.

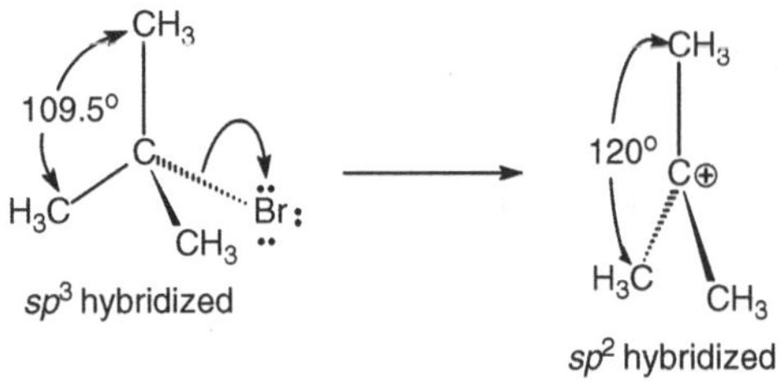

        1. Reactant bond angles are smaller. Steric crowding increases in the order methyl < 1° < 2° < 3°. Carbocation bond angles are larger and crowding is reduced.

6.13.10. Methyl, 1°, 2°, and 3° alkyl halides all react at least partly via S_N1 in the nonnucleophilic solvent CF_3COOH.

Alkyl group	*t*-butyl	isopropyl	ethyl	methyl
relative rate	>10⁶	1.0	<10⁻⁴	<10⁻⁵

    a. Rate differences are larger for S_N1 than for S_N2 (compare with section 6.8.8).

6.13.11. Alkyl halide S_N1 reactivity (section 6.13.3e) parallels carbocation stability.

    a. A more stable carbocation (relative to reactant (R-X) from which it forms) $\Rightarrow$ faster S_N1.

6.13.12. Step 1 transition state structure (section 6.13.1):

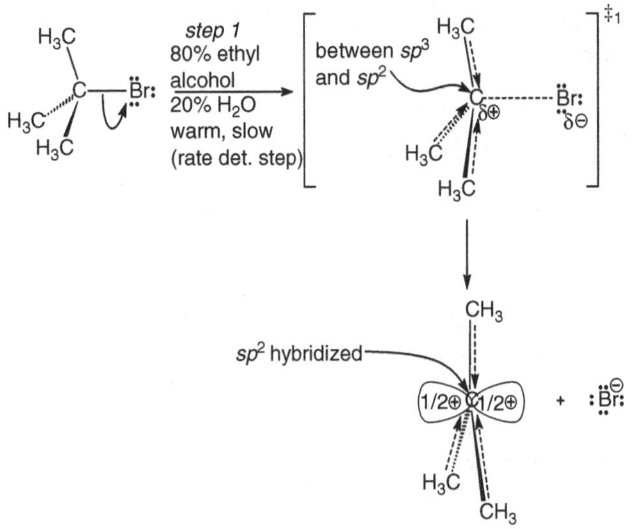

    a. Central C atom has carbocation ($\delta+$) character in $\ddagger_1$, i.e., it has partial (or fractional) + charge. Electron release (from three adjacent alkyl substituents) stabilizes a transition state in a similar way as it stabilizes a full carbocation – through inductive electron release and/or hyperconjugation (section 4.13.35).

1. ‡$_1$ structure is near the carbocation on the reaction curve (see reaction diagram section 6.13.4), since step 1 is very endothermic.
2. If a carbocation is more stable, the nearby ‡$_1$ structure will also be more stable. E_a for step 1 will also be lower, and reaction will be faster.

6.1.7. The product whose yield is highest is also formed fastest for most (but not all) organic reactions.

6.13.13. Many reactions have carbocation intermediates, and are analyzed similarly to S$_N$1.

a. Problem 17. 1-Halo-2,2-dimethylpropanes are notoriously slow to substitute under both S$_N$2 and S$_N$1 conditions.
1. Why is the S$_N$2 reaction slow? What is the main factor that influences S$_N$2 reaction rates?

2. Why is the S$_N$1 reaction slow? What is the main factor than influences S$_N$1 reaction rates?

6.13.14. Resonance stabilizes carbocations (section 4.13.39). Allyl bromide (a 1° R-X) undergoes S$_N$1 about as fast as a 2° R-X.

6.13.15. Similar factors influence leaving group ability in S$_N$1 as in S$_N$2 (section 6.11.1).

6.13.16. S$_N$1 is very slow in the gas phase. Heterolytic bond breaking is very endothermic.

a. S$_N$1 step 1 E_a (unimolecular heterolysis for 2-bromo-2-methylpropane) would be at least 623 kJ/mol (gas phase). E_a = 84 to 126 kJ/mol in solution. The 497 to 539 kJ/mol additional energy to heterolyze the C-Br bond in solution comes from ion-dipole (solvent) attractions. Polar solvent pulls the C-X bond apart.

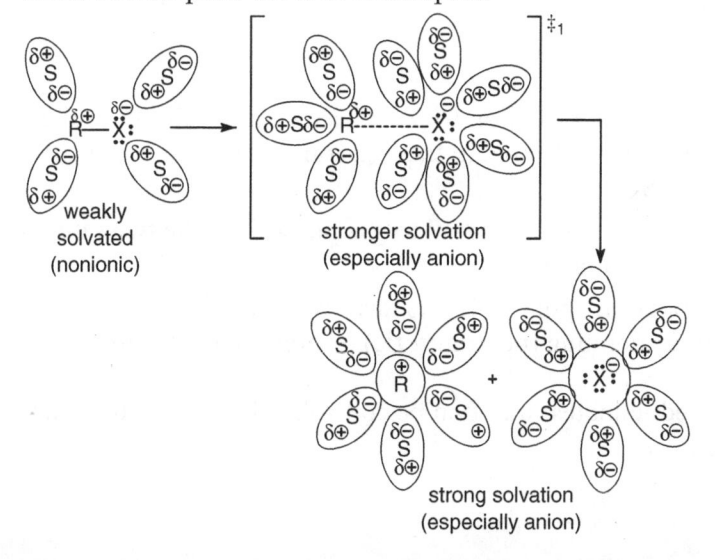

a. Reactant R-X is only weakly solvated by a polar solvent. Stabilization is small compared to a nonpolar solvent.
b. The transition state C—X partial bond is assumed to be strongly polarized, since C and X are on their way to becoming ions. Transition state ‡₁ is highly stabilized in polar *vs.* nonpolar solvents (or gas phase). E_a is much lower, and step 1 is much faster in polar solvents.

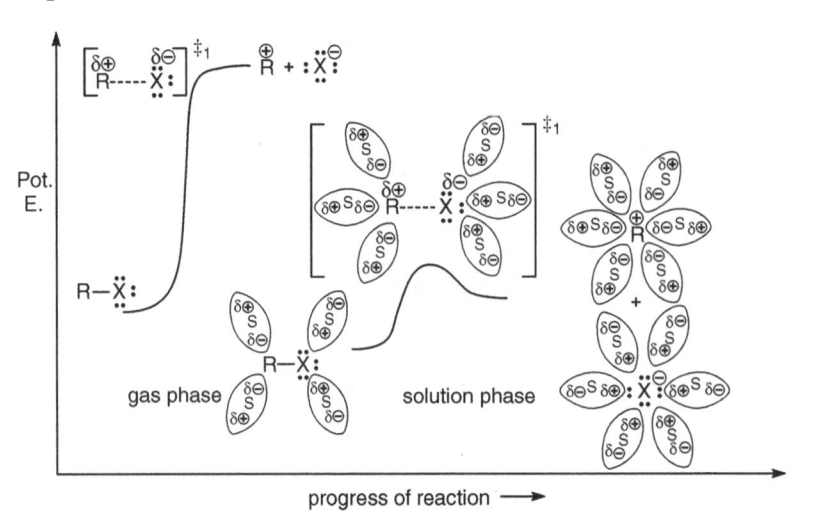

c. Most carbocations are stabilized by delocalization (polar effects), and are soft ions (not as strongly solvated as the anion X⁻). Better solvated X⁻ ⇒ greater stability, lower E_a, and faster step 1. S_N1 is usually faster in protic solvents (*vs.* polar aprotic or nonpolar solvents), where X⁻ is strongly solvated (especially by "hydrogen bonds").
d. More acidic solvents form stronger hydrogen bonds to X⁻. S_N1 reactions are very fast in F_3CCH_2OH, $HCOOH$, F_3CCOOH solvents. (Fluorine atoms inductively stabilize conjugate bases, enhancing acidity.)
e. Problem 18. Solvation effects above apply to electrically neutral substrates. Using similar logic, account for the fact that increasing the solvent polarity causes a modest *decrease* in the following S_N1 reaction rate.

f. By contrast to S_N1, bond formation to :Z⁻ releases energy almost at the same time as it is needed to break the C-X bond in the S_N2 transition state ⇒ solvation has little stabilizing effect on S_N2.

6.13.17. Although gas phase S_N1 is very slow compared to gas phase S_N2, the two rates can be comparable in solution. To summarize:
  a. Strongly ionizing, H-bonding, polar solvents favor S_N1 over S_N2.
  b. Weakly ionizing, non-H-bonding (polar aprotic) or nonpolar solvents favor S_N2 over S_N1.
  c. Which mechanism predominates can often be controlled by nucleophile, substrate, and (above all) solvent choice.

6.14.1. S$_N$1 stereochemical evidence – optically active *n*-halooctane yields partially racemized products.

a. A nucleophile can attack either carbocation face (backside preferred, but not exclusive). (*R*)-product has *retained* configuration, while (*S*)-product has *inverted* configuration.

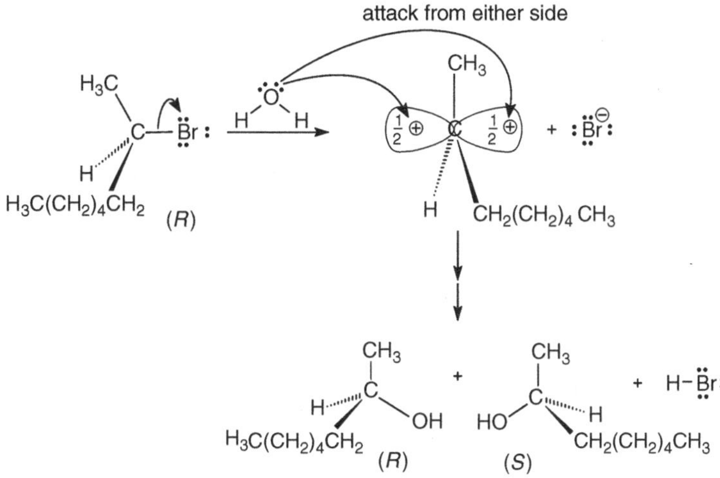

6.14.2. Incomplete racemization (excess inversion) is due partly to ion pair formation after R-X bond cleavage, and partly to leaving group steric hindrance.

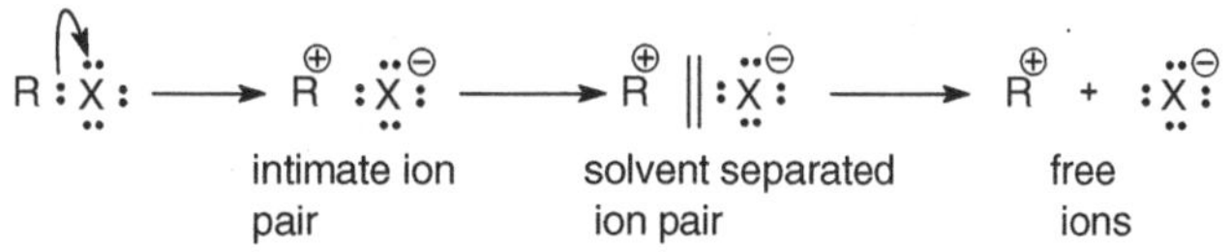

a. Example:

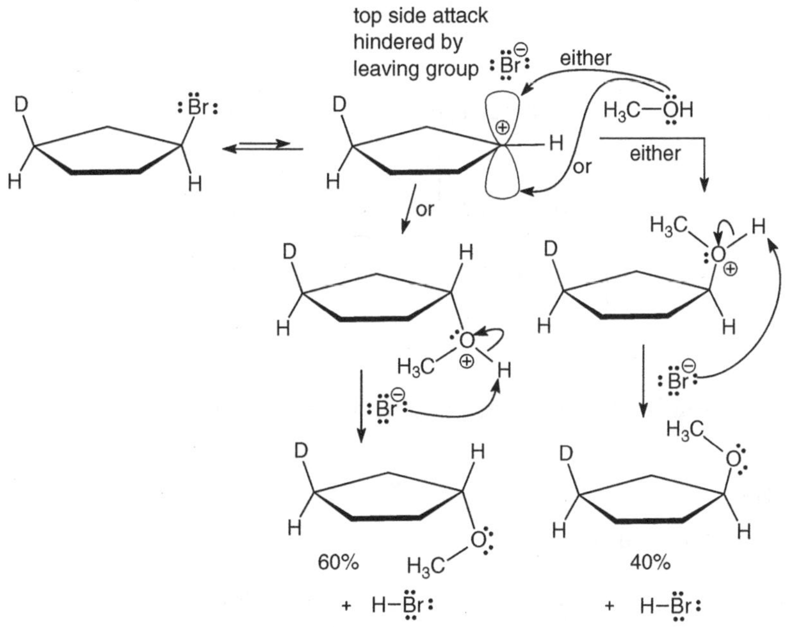

b. Problem 19. Optically pure (*R*)-1-chloro-1-phenylethane has [α] = –109° · dm^{-1} · g^{-1} · mL. Optically pure (*R*)-1-phenylethan-1-ol has [α] = –42.3° · dm^{-1} · g^{-1} · mL. When optically

impure 1-chloro-1-phenylethane ($[\alpha] = -34° \cdot dm^{-1} \cdot g^{-1} \cdot mL$) is treated with dilute NaOH(aq), product alcohol has $[\alpha] = +1.7° \cdot dm^{-1} \cdot g^{-1} \cdot mL$.
1. What is the optical purity of reactant? Product?

2. What is the percentage of retention? Inversion? Racemization?

6.15.1. H. Meerwein (1922) – carbocation rearrangements – nucleophile does not bond to the same C atom from which the leaving group leaves during some S_N1 reactions.
  a. 1-Bromo-2,2-dimethylpropane reactions:

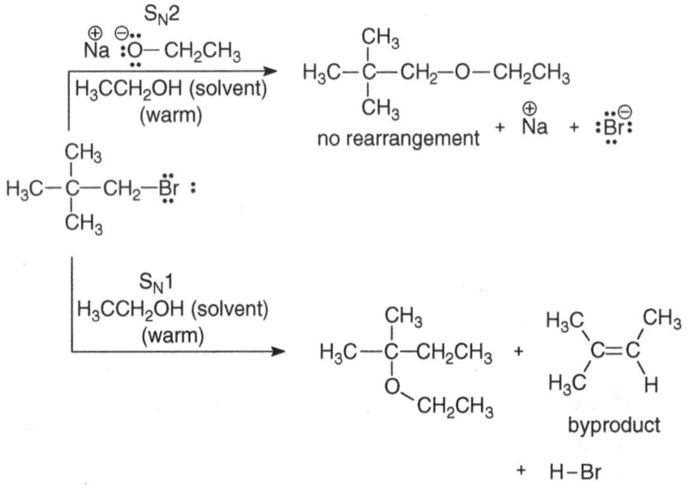

6.15.2. Main pathway – a less stable carbocation rearranges to a more stable carbocation. Rearrangements are fast intramolecular (within the same molecule) Lewis acid/base reactions.
  a. Most commonly occurs by:
  1. a 1,2-hydride shift. (A full hydride ion is <u>not</u> involved.)

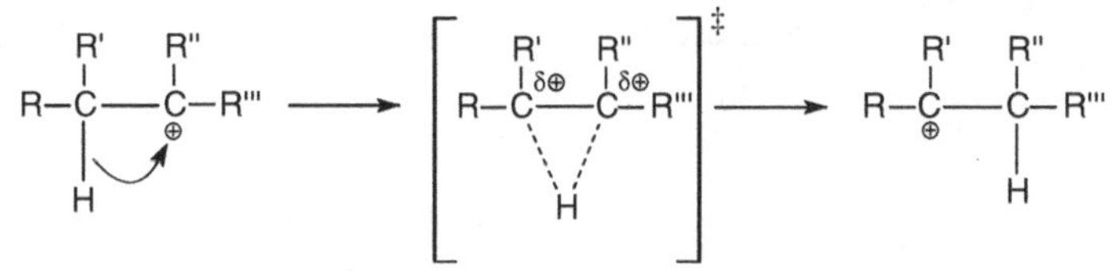

    R, R', R", R''' = H or alkyl

    a. A hydride ion is a hydrogen atom with two electrons and a formal - charge.

    a hydride ion

  2. a 1,2-alkyl (alkide) shift.

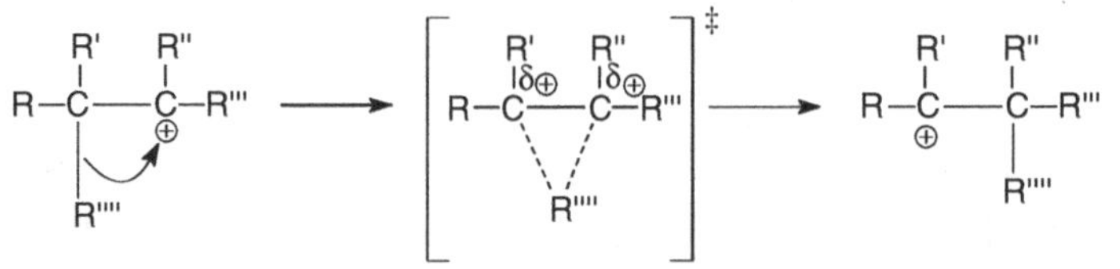

    R, R', R", R''' = H or alkyl
    R'''' = alkyl

6.15.3. Examples.

a. *n*-halopropane:

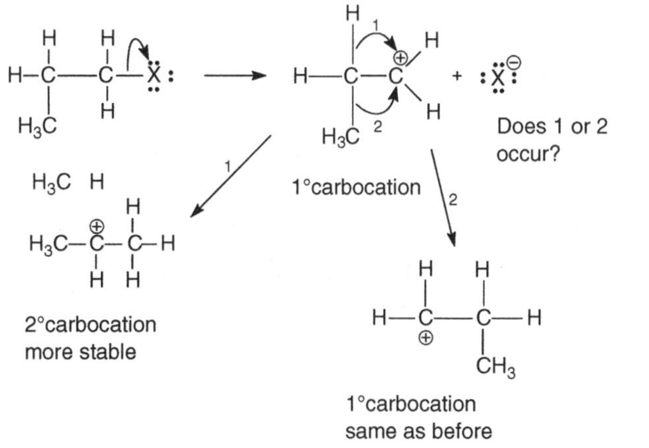

[Reaction scheme showing n-halopropane ionizing to a 1° carbocation, which can rearrange via pathway 1 (hydride shift) to give a 2° carbocation (more stable), or via pathway 2 to give the same 1° carbocation as before.]

1. Rearrangement actually occurs simultaneously with C-X bond breaking for 1° R-X, i.e., is concerted with leaving group loss. A 1° carbocation does not actually form.

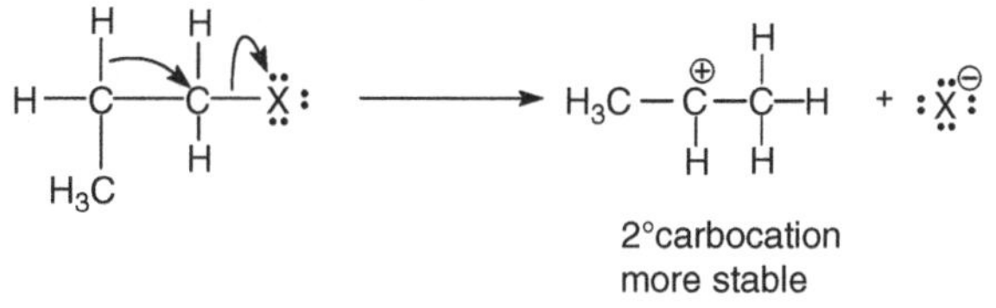

[Concerted mechanism showing hydride shift simultaneous with X⁻ departure to give 2° carbocation directly.]

b. 1-halo-2-methylpropane

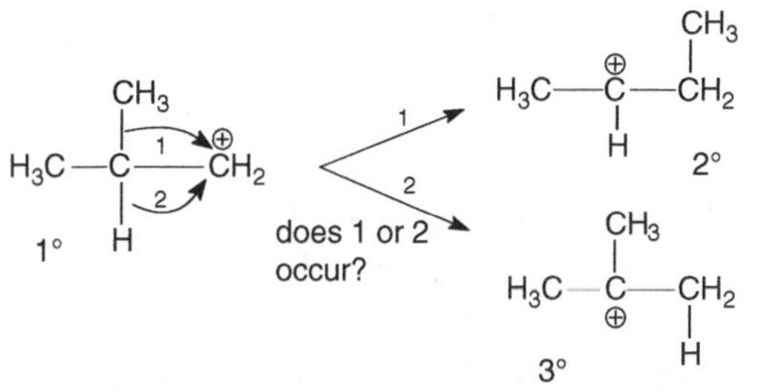

[Scheme showing 1° carbocation from 1-halo-2-methylpropane; pathway 1 gives a 2° carbocation, pathway 2 gives a 3° carbocation.]

c. 1-halo-2,2-dimethylpropane

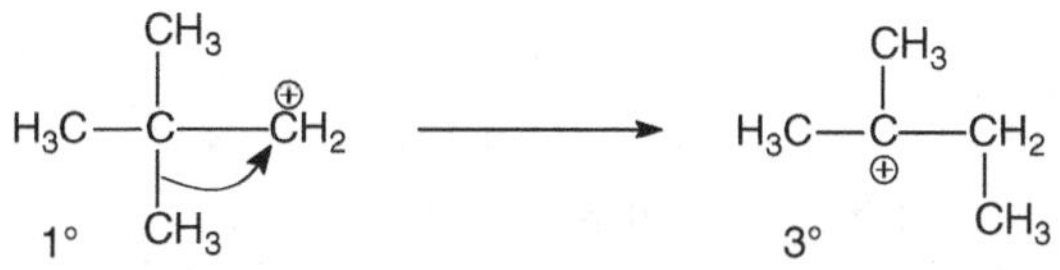

[Scheme showing methyl shift from 1° carbocation to give 3° carbocation.]

d. 3,3-dimethyl-2-halobutane

[Scheme showing 2° carbocation from 3,3-dimethyl-2-halobutane; pathway 1 (methyl shift) gives 3° carbocation, pathway 2 (hydride shift) gives 1° carbocation.]

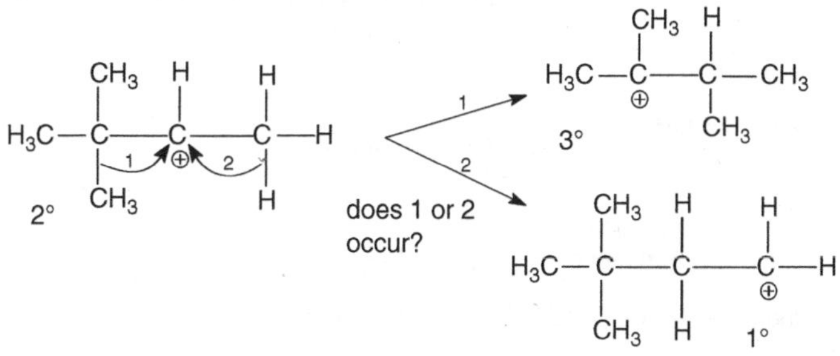

e. Problem 20. When 3,3-dimethylbut-1-ene is treated with HI, a product mixture is obtained:

$$\text{H}_3\text{C}-\underset{\underset{\text{CH}_3}{|}}{\overset{\overset{\text{CH}_3}{|}}{\text{C}}}-\text{CH}=\text{CH}_2 \xrightarrow[\text{(dry gas)}]{\text{H}-\text{I}} \text{H}_3\text{C}-\underset{\underset{\text{CH}_3}{|}}{\overset{\overset{\text{CH}_3}{|}}{\text{C}}}-\underset{\underset{\text{I}}{|}}{\text{CH}}-\underset{\underset{\text{H}}{|}}{\text{CH}_2} + \text{H}_3\text{C}-\underset{|}{\overset{\overset{\text{CH}_3}{|}}{\text{C}}}-\underset{|}{\overset{\overset{\text{CH}_3}{|}}{\text{CH}}}-\underset{\underset{\text{H}}{|}}{\text{CH}_2}$$

3,3-dimethylbut-1-ene   3-iodo-2,2-dimethylbutane   3-iodo-2,3-dimethylbutane

Realizing that π-electrons are basic and that a carbocation intermediate is involved, propose a mechanism for each product formation. Use electron pushing arrows and show all elementary steps (collisions), intermediates, formal electrical charges, and involved lone pair electrons. This is "electrophilic addition," the main Chapter 8 mechanism.

6.15.4. G. Olah has measured some rearrangement rates spectroscopically in super acid solutions. Usually they are much faster than S_N1 step 2.

6.15.5. Carbocation rearrangements are involved in natural product biosynthesis, such as menthol, camphor, and α-pinene (common names).

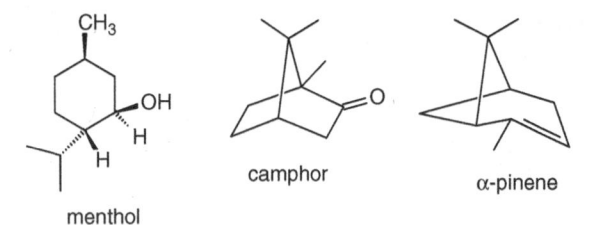

menthol   camphor   α-pinene

6.16.1. Both S_N2 and S_N1 operate to some extent in all R-X reactions. However, one mechanism is usually much faster than others. Factors affecting S_N2 vs. S_N1 rates:
   a. Better (faster) leaving group ⇒ faster both mechanisms.
   b. Alkyl substituents speed up S_N1 (through carbocation + charge delocalization), but slow down S_N2 (through steric hindrance).
      1. *Tertiary* R-X's react mainly by S_N1.
      2. Methyl and 1° R-X's react mainly by S_N2.
      3. 2° R-X's probably react by S_N2.
   c. High :Z⁻ (nucleophile) concentration favors S_N2 over S_N1 for all R-X's.
   d. Strong :Z⁻ (nucleophile) favors S_N2 over S_N1 for all R-X's.
   e. Polar protic (hydrogen-bonding) solvent favors S_N1 over S_N2 for all R-X's, because it solvates ions better.
   f. AgNO₃ addition (which precipitates halide ions) can cause otherwise unlikely ionizations:

$$\text{H}_3\text{C}-\underset{\underset{\text{CH}_3}{|}}{\overset{\overset{\text{CH}_3}{|}}{\text{C}}}-\text{CH}_2-\text{I} + \text{AgNO}_3 \xrightarrow[\text{warm}]{\text{H}_3\text{CCH}_2\text{OH}} \text{H}_3\text{C}-\underset{\underset{\text{CH}_3}{|}}{\overset{\overset{\text{CH}_3}{|}}{\text{C}}}-\overset{\oplus}{\text{CH}}_2 + \text{AgI(s)}$$

1° carbocation (unstable)

products ⟵ ⟵ $\text{H}_3\text{C}-\underset{\oplus}{\overset{\overset{\text{CH}_3}{|}}{\text{C}}}-\text{CH}_2-\text{CH}_3$

g. Problem 21. In 80% ethanol/20% H₂O solvent at 55° C, 2-bromopropane was thought to react with hydroxide ion according to the following rate law:

$$\text{rate} = 4.7 \times 10^{-5} [RX][^-OH] + 2.4 \times 10^{-4} [RX]$$

Hint: The first term is the S_N2 rate contribution, the second term is the S_N1 rate contribution.
  1. What 2-bromopropane percentage reacts via S_N2 when $[^-OH]$ = 0.001 M? 0.01 M? 0.1 M? 1.0 M? 5.0 M?

h. Problem 22. Fill in the table, referring to a reaction between an alkyl halide and hydroxide ion in 80% ethanol / 20% H₂O / 55° C (see section 6.8.2).

feature	S_N2	S_N1
stereochemistry		
reaction order		
rearrangements?		
relative rates by class		
relative rates by leaving group Cl, Br, and I		
relative rates by protic vs. aprotic solvent		
effect on rate of a temperature rise		
effect on rate of doubling [RX]		
effect on rate of doubling [⁻OH]		
effect on rate of increasing solvent H₂O content		
effect on rate of increasing solvent alcohol content		
effect on rate of changing solvent to HMPT		

6.8.12. Review: other alkyl halide reactions (non-S_N2):
  a. Grignard reagent preparation (section 3.5.7).

$$R-X \xrightarrow[\text{dry ether}]{Mg} R-MgX \xrightarrow{\text{various reagents}} \text{many products}$$

  b. Reduction (to a hydrocarbon). D is heavy hydrogen (D = ^2_1H).

$$(H_3C)_3C-Cl \xrightarrow[\text{dry ether}]{Mg} (H_3C)_3C-MgCl \xrightarrow{D_2O} (H_3C)_3C-D + Mg(OD)Cl$$

6.16.2. R-X analysis:
  a. Insoluble in H₂SO₄(conc).
  b. Do not react with Br₂ (cold, dark). Do not react with oxidizing agents – KMnO₄/KOH (cold/dil), or CrO₃.

208 SEQUENTIAL ORGANIC CHEMISTRY I

---

c. Sodium fusion (section 1.11.1c1), then AgNO₃(aq) precipitates AgX. Reaction with AgNO₃ can often be done directly, reactivity: R-I > R-Br > R-Cl

d. Spectroscopic methods (Chapters 12 and 13).

7.9.1. Elimination – a σ-bond is broken to each of two adjacent bonded atoms (often sp^3 C's), and a double (π) bond forms in between.

7.9.2. Examples – alkyl halide dehydrohalogenation.

7.12.1. Basic H and X 1,2-elimination. R-X reactivity: faster 3° > 2° > 1° slower.

a. Example: E2 mechanism (elimination, bimolecular):

b. Base (:OH⁻) removes a β-H⁺ ion (proton) from R-X.
  1. The non-IUPAC α, β, γ, δ, ε, ω system.

c. X (leaving group) takes a bonding electron pair; H leaves its bonding electron pair behind which becomes the π-electron pair.
d. An alkene product mixture can form from nonequivalent β-H atoms.

[reaction scheme showing E2 elimination of 2-bromo-2-methylbutane (or similar) with KOH/ethanol yielding three alkene products in 53%, 28%, and 19% ratios]

e. KOH is often used as base – more soluble than NaOH in ethanol solvent. Other bases: NaOCH$_3$, NaOCH$_2$CH$_3$, KOC(CH$_3$)$_3$. Water or alcohol solvents sometimes act as (weak) bases.
f. E2 does not yield rearranged alkene products because carbocations are not involved.

7.12.2. Elimination competes with S$_N$2 because strong nucleophiles are often also strong bases. E2 predominates and S$_N$2 is very slow for 3° alkyl halides. Why?

[reaction scheme showing 3° alkyl bromide with KOCH$_3$/HOCH$_3$ undergoing E2 to give alkene + H$_3$C—OH + :Br:⁻; S$_N$2 too sterically hindered]

7.12.3. Hughes and Ingold (1935) proposed two elimination mechanisms, E2 (elimination bimolecular) and E1 (elimination unimolecular), similar to S$_N$2 and S$_N$1. Mechanism evidence:

a. With concentrated base, elimination is first order in [R-X] and first order in base [:Z⁻], second order overall. Rate law:

$$\text{rate} = k[\text{R}-\text{X}][:Z^-]$$

1. If :Z⁻ strength or concentration is reduced, sometimes elimination rate is independent of base concentration (first order overall) for 2° and 3° R-X. Rate law:

$$\text{rate} = k[\text{R}-\text{X}]$$

7.12.4. General E2 mechanism:

7.12.5. E2 features:
  a. Second order rate law.
  b. Only one elementary step (bond breaking and bond formation are simultaneous, or concerted).
      1. Bimolecular transition state structure (containing both R-X and :Z⁻).
  c. No rearranged products.
  d. Large deuterium isotope effect (section 7.12.7).
  e. No hydrogen exchange occurs.
  f. Large element effect.
  g. H$_\beta$ and X are either *syn*- or *anti*-coplanar in transition state structures.

7.12.6. E2 reaction diagram:

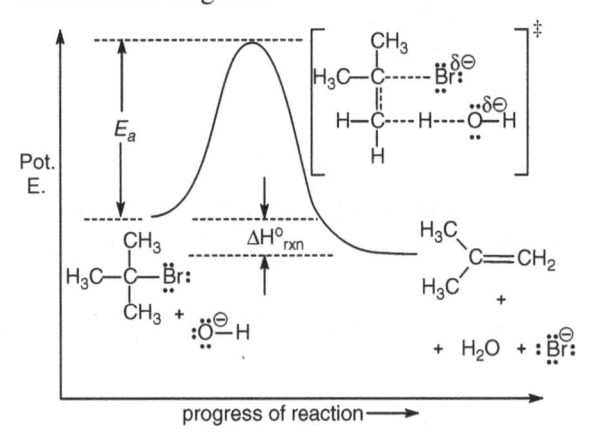

7.12.7. Isotope effect – a reaction rate (or equilibrium position) difference caused by a different atomic isotope presence. If a particular atom is not bound as tightly in a transition state as it was in a reactant, the same reaction involving a heavier atomic isotope will be slower.
  a. Primary isotope effect – largest isotope effect type – if a bond to an atom is broken in the rate determining step, the overall reaction will be slower with a heavier atomic isotope.

b. A given element's isotopes have similar (but not identical) chemical properties. They often undergo the same reactions, but at different rates.
   1. Hydrogen isotopes have the greatest proportional mass differences of any isotopes ($_1^2H$ mass ~ 2 × $_1^1H$ mass). Hydrogen isotopic rate differences are the largest observed. Primary isotope effects (25° C) are (k^H/k^D = 5–8) for reactions such as:

c. Two isotopic reaction rates can be measured separately, or from competition experiments using a limited :Z⁻ reagent amount with a hydrogen and deuterium labeled substrate mixture. Rates can be calculated by comparing relative H-Z vs. D-Z yields. The higher yield product is assumed to form faster.

d. Example: 2-phenyl-1-bromoethane vs. 2,2-dideuterio-2-phenyl-1-bromoethane dehydrobromination with NaOCH₂CH₃ in ethanol.

   1. Observed k^H/k^D = 7 means that a C-H bond breaks in the rate determining step – consistent with a one step E2 mechanism written above.

e. Problem 23. When excess toluene-α-d was photochemically monochlorinated at 80° C with 0.1 mol of Cl₂, 0.0212 mol DCl and 0.0868 mol HCl were obtained. What is the deuterium isotope effect (k^H/k^D) per methyl hydrogen atom? How many mols of DCl and HCl would be expected from toluene-α-d₂ monochlorination under the same conditions?

7.12.8. All section 7.12.5a-d features are consistent with a one step (E2) mechanism. However, a two step carbanion mechanism (E1cB, elimination, unimolecular, conjugate base) is equally consistent.

    a. A carbanion is the conjugate (very strong) base of a (very weak) carbon acid (section 4.16.4b).

    b. A "deuterium exchange" experiment was done to see if E1cB actually operates – 2-phenyl-1-bromoethane reaction with sodium ethoxide in H_3CCH_2OD (solvent).

        1. A phenyl group makes the benzylic H-atoms more acidic than normal, which should promote E1cB.

        2. After about half of the 2-phenyl-1-bromoethane had been converted into alkene, reaction was interrupted and unreacted 2-phenyl-1-bromoethane was analyzed by mass spectrometry. It contained no deuterium. Reversible carbanion formation (E1cB step 1) can therefore be ruled out.

7.12.9. Irreversible carbanion formation cannot be ruled out by the above experiment. k_2 could be much larger than k_{-1} and k_1, and forward step 1 could be rate limiting. If step 2 were rate limiting, deuterium exchange should be observed. If step 1 is rate limiting, overall reaction rate would be independent of the C-X bond breaking rate, since this occurs in the second (fast) step. Can this be tested?

    a. A halogen isotope effect would be difficult to measure since the mass difference between ^{35}Cl and ^{37}Cl (or other halogen isotopes) is only a few percent. No rate difference would be observable anyway if the C-X bond breaks in a fast step 2.

    b. Reaction rates should be the same for R-F, R-Cl, R-Br, and R-I if the C-X bond breaks in a fast step 2.

    c. Experimental R-X elimination reactivity:

$$R-I > R-Br > R-Cl > R-F$$

Rate differences are similar to those of S_N2 and S_N1, where the C-X bond breaks in the rate determining step – R-I reacts 25,000 times faster than R-F. These large differences are called the "element" effect (J. Bunnett).

d. A large element effect is consistent only with E2, and not E1cB.

e. Problem 24. Why doesn't $CHCl_3$ undergo β-elimination when warmed with NaOH / H_2O?

7.12.10. For alkyl halides with nonequivalent β-hydrogen atoms, alkene mixtures form (section 7.12.1d). Which product predominates?

7.12.11. 2-Bromobutane yields mainly *cis*- and *trans*-but-2-enes, although only two β-hydrogen atoms are bonded to C3 *vs.* three bonded to C1.

7.8.1. A. Zaitsev's (Saytzeff's) rule (1875) – "Dehydrohalogenation's preferred alkene product has the greater number of alkyl groups bonded to the (sp^2) C atoms. More substituted alkene ⇒ more stable and forms faster."

a. Alkene stabilities (see section 7.8):

b. Explanation 1 – alkyl groups stabilize alkenes through hyperconjugation effects (section 4.13.35).

c. Explanation 2: adjacent alkyl groups donate electron density to the π-bond, strengthening it. (Do we believe this?)
d. Explanation 3: more substituted alkene alkyl groups have less steric interaction.

monosubstituted (less stable)   trisubstituted (more stable)

e. Once formed, alkene products do not normally revert back to reactants under the reaction conditions. But-2-ene predominates over but-1-ene because it forms (irreversibly) faster.
   1. But-2-ene is more stable than but-1-ene. Often (but not always) a more stable product will form faster.
   2. The E2 transition state structure has "alkene character." The same factor that stabilizes product alkenes (R-group number bonded to sp^2 C atoms) also stabilizes the transition state structure. If a transition state is more stable (more substituted), E_a will be lower, and E2 will be faster.

   3. Zaitsev's rule exceptions – when -X is fluoride or a quaternary ammonium salt, section 19.14.2. Factors besides hyperconjugation (in the alkene-like transition state) come into play.

7.8.2. Transition state alkene character explains reactivity order. Examples:

Reaction	Relative rate	Relative rate per H$_\beta$
H$_3$CCH$_2$Br $\xrightarrow[\text{heat}]{\text{KOH} \atop \text{HOCH}_2\text{CH}_3}$ H$_2$C=CH$_2$	1.0	1.0
H$_3$CCH$_2$CH$_2$Br $\xrightarrow[\text{heat}]{\text{KOH} \atop \text{HOCH}_2\text{CH}_3}$ H$_2$C=CHCH$_3$	3.3	5.0
H$_3$CCHCH$_3$ with Br $\xrightarrow[\text{heat}]{\text{KOH} \atop \text{HOCH}_2\text{CH}_3}$ H$_2$C=CHCH$_3$	9.4	4.7
H$_3$CCCH$_3$ with CH$_3$ and Br $\xrightarrow[\text{heat}]{\text{KOH} \atop \text{HOCH}_2\text{CH}_3}$ H$_2$C=C(CH$_3$)$_2$	120	40

6.18.3. At least two other factors influence E2 rates:

  a. More β-hydrogen atoms in a substrate ⇒ higher collision probability with base.

  b. Weaker C-H$_\beta$ bond ⇒ easier to break, and faster E2. Bond strength order:

  $$\text{stronger } 1° \text{ C} - \text{H} > 2° \text{ C} - \text{H} > 3° \text{ C} - \text{H} \text{ weaker}$$
  $$\phantom{\text{stronger }} 423 413 403 \text{ kJ/mol}$$

  c. These two factors work in the same direction with alkyl substituent hyperconjugation stabilization (in the alkene-like transition state). Hyperconjugation is usually dominant, however.

  d. Problem 25. Draw the E2 reaction (using KOH / ethanol / heat) for the following molecules: bromoethane, bromopropane, 1-bromo-2-methylpropane, 1-bromo-2,2-dimethylpropane. Arrange them in order of increasing reaction rate.

  Why is this reactivity order observed? Explain.

  e. Problem 26. 2-Bromopropane dehydrohalogenation normally requires refluxing several hours in KOH/CH$_3$CH$_2$OH/78° C. The reaction occurs in less than a minute at 25° C with K$^+$ $^-$OC(CH$_3$)$_3$ base in DMSO (polar aprotic solvent). Explain why reaction is so much faster in DMSO.

7.14.1. E2 occurs from an *anti*-conformation whenever an alkyl halide H$_\beta$ and X atom can achieve it.

    a. Maximum orbital overlap occurs when H$_\beta$ and X are coplanar – allows C-H$_\beta$ σ-electron "flow" into forming π-orbital.

7.14.2. Alternative view – E2 is similar to S$_N$2 where the "nucleophile" is the β-C atom lone pair (after H$^+$ is removed).

7.14.3. E2 is stereospecific (sections 4.16.7.a6a and 8.9.3b). Example: (1*R*,2*R*)- or (1*S*,2*S*)-*erythro*-1-bromo-1,2-diphenylpropane and (1*R*,2*S*)- or (1*S*,2*R*)-*threo*-1-bromo-1,2-diphenylpropane.

a. *Erythro* – like groups on the same side of Fischer projection drawing.
b. *Threo* – like groups on opposite sides in Fischer projection drawing.

7.15.1. E2 occurs from the *anti*-conformation in which Br is axial in bromocyclohexane.

7.15.2. *Syn*-eliminations can occur in cyclic compounds where an *anti* configuration is not possible.

a. Problem 27. Of the several 1,2,3,4,5,6-hexachlorocyclohexanes, which one isomer undergoes basic dehydrohalogenation much more slowly than the others? Why?

b. Problem 28. Which E2 elimination would you expect to be faster, *cis*- or *trans*-2-bromo-1-methylcyclohexane? Why?

218 SEQUENTIAL ORGANIC CHEMISTRY I

7.10.1. A different mechanism (E1, elimination, unimolecular) operates when first order kinetics are observed (section 7.12.3). Example:

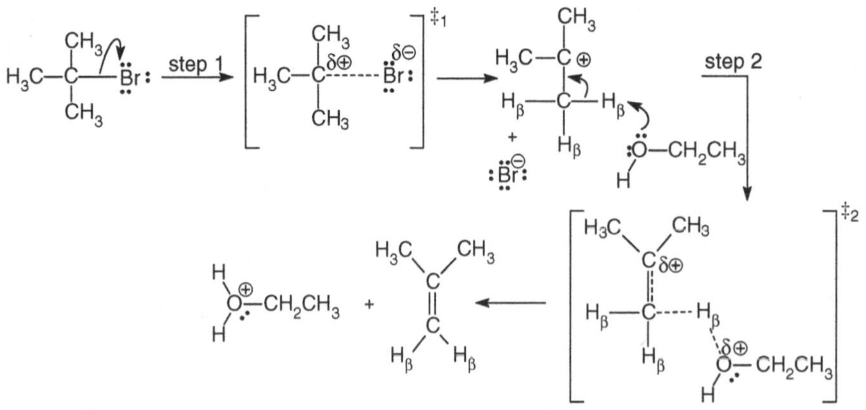

7.10.2. General E1 mechanism:

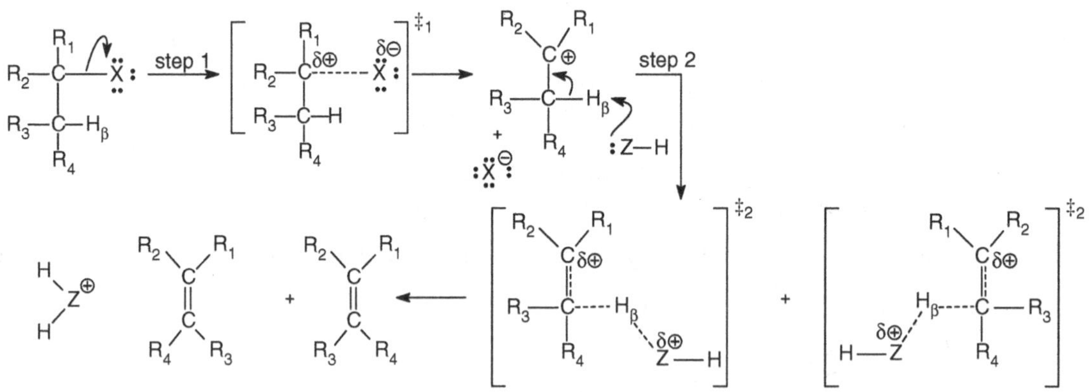

7.10.3. E1 features:
- **a.** Reaction rate depends on [R-X], but not [:Z-H] (first order overall).
- **b.** Two elementary steps (bond breaking and bond making are not simultaneous).
   1. Unimolecular transition state ($\ddagger_1$) structure (containing only R-X and not :Z-H) for rate determining step.
- **c.** Step 1 determines overall reaction rate; step 2 is fast by comparison.
- **d.** R-X reactivity order is the same as for S_N1 – rate is determined by carbocation stability:

    slowest 1° < 2° < 3° fastest

- **e.** Intermediate products are ions (carbocation and halide).
- **f.** Rearranged products form, where possible.
- **g.** No deuterium isotope effect.
- **h.** No hydrogen exchange.
- **i.** Large element effect.
- **j.** Not stereospecific.

7.10.4. E1 reaction diagram.

7.10.5. E1 and S_N1 have the same first step. :Z-H is a base in the second step, (attacks δ+ H) rather than a nucleophile (attacks R+).

  a. Carbocations can
     (1) combine with a nucleophile (S_N1).
     (2) rearrange to a more stable carbocation through a 1,2-alkyl or 1,2-hydride shift.
     (3) eliminate a proton to form an alkene (E1), i.e., they are Brønsted-Lowry acids.
        1. S_N1 and E1 are not usually synthetic reactions because product mixtures are often obtained. They are used in carbocation studies.

7.10.6. C-H bond breaking in the second (fast) step is consistent with no deuterium isotope effect.

7.10.7. The C-H$_\beta$ σ-orbital must be coplanar with the adjacent $2p_z$-orbital.

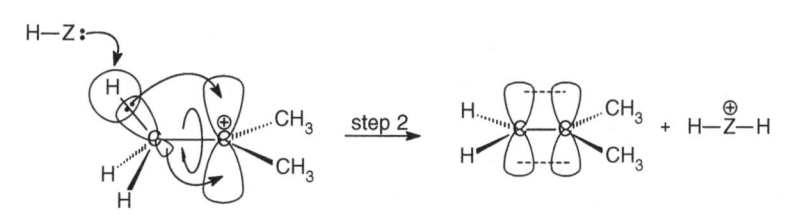

a. However, rotation about a C-C⁺ bond can occur before step 2 – leads to stereospecificity loss.

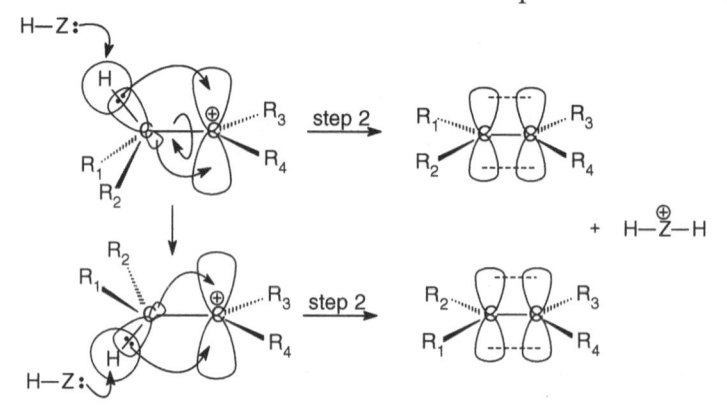

b. Problem 29.
   a. Menthyl chloride yields only 2-menthene upon heating with sodium ethoxide (strong base) in ethanol solvent. Why is 2-menthene the only alkene product?

   b. When menthyl chloride is heated in ethanol (without sodium ethoxide), it yields both 3-menthene (68%) and 2-menthene (32%). Explain this behavior.

7.10.7. Rearranged (via 1,2-shifts) alkene products are consistent with carbocation intermediates. Examples:

Ts is tosylate
a fast leaving group

a. The 2° carbocation (structure **1**) can also rearrange to a 3° carbocation (structure **4**), which eliminates to either structure **5** or **3**.

7.11.1. Zaitsev's rule applies to E1 as well as E2 – the more substituted alkene is preferred.

a. Example: 2-tosylpentane. Both alkene products form (in the rate determining step) from the same carbocation. However, the second (fast) step for C3-H bond heterolysis must be faster than for C1-H bond heterolysis.

1. The more substituted transition state (which has alkene character) is more stable, E_a is lower, and reaction rate is faster.
2. Zaitsev's rule is observed for rearranged carbocations.

7.21.1. Which mechanism operates, E2 or E1?

a. Both mechanisms are fastest for 3° R-X.
b. More concentrated and stronger base :Z⁻ ⇒ faster E2 vs. E1 for all R-X's.
c. E1 occurs only for 2° and 3° R-X's (which can form stable carbocations) with low concentration weak bases (:Z-H solvent, and no :Z⁻).
d. Like S_N2, E2 is not very sensitive to solvent polarity. Like S_N1, E1 is faster in polar protic solvents.

7.17.1. All four elimination (E1 and E2) and substitution (S_N1 and S_N2) mechanisms compete in a given R-X solution (X = halogen or sulfonate). One mechanism can often be made to predominate over others by appropriate reaction condition choices, but outcomes cannot always be predicted. General guidelines:

  a. S_N2 and E2 predominate over S_N1 and E1 for all R-X's when a strong nucleophile (usually also a strong base) is present (bimolecular conditions).

    1. S_N2 R-X reactivity:

$$\text{faster methyl} > 1° > 2° > 3° \text{ slower}$$

    2. E2 R-X reactivity:

$$\text{faster } 3° > 2° > 1° \text{ slower}$$

    3. Methyl R-X's cannot eliminate, and 3° R-X's eliminate almost exclusively under bimolecular conditions. 1° R-X's are difficult to eliminate, while 2° R-X's usually give both substitution and elimination.

    4. Substitution is favored with weakly basic strong nucleophiles (for examples, I^-, HS^-, or CN^-).

    5. A less polar solvent favors elimination.

    6. E_a is often higher for elimination than for substitution. A higher temperature yields more elimination.

      a. $\Delta S°_{rxn}$ for E2 or E1 is positive, since each R-X molecule yields two separate particles (alkene and X^-). The $-T\Delta S°_{rxn}$ term makes $\Delta G°_{rxn}$ more negative (favorable) at higher T (section 4.5.1).

    7. E2 elimination is promoted over S_N2 by sterically hindered strong bases, such as *tert*-butoxide ion (section 6.11.3).

  b. If strong nucleophiles are absent (unimolecular conditions), S_N1 and E1 predominate for 2° and 3° R-X's.

    1. Adding $AgNO_3$ or raising temperature can force a 1° R-X to ionize, usually with rearrangement. 1° alkyl halides do not easily ionize unless R^+ is resonance stabilized.

    2. More substitution occurs for 2° R-X's than for 3° R-X's, due to greater steric hindrance as a nucleophile approaches a 3° carbocation. Example: 2-bromo-2-methylpropane *vs.* 2-bromopropane in ethanol at 80° C:

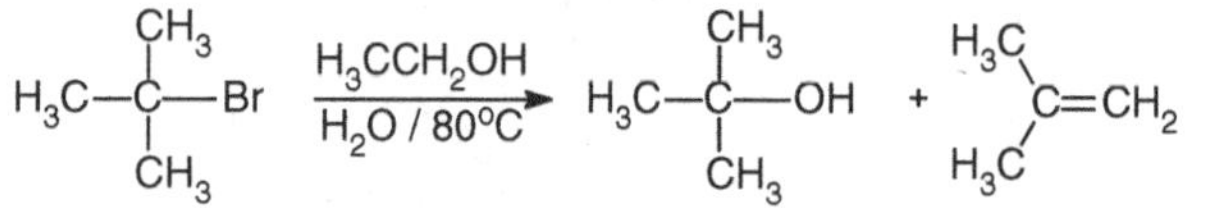

$$\underset{3° \text{ R-X}}{(CH_3)_3C-Br} \xrightarrow[H_2O\,/\,80°C]{H_3CCH_2OH} \underset{81\%\,(S_N1)}{(CH_3)_3C-OH} + \underset{19\%\,(E1)}{(CH_3)_2C=CH_2}$$

$$\underset{}{(CH_3)_2CH-Br} \xrightarrow[H_2O\,/\,80°C]{H_3CCH_2OH} \underset{95\%\,(S_N1)}{(CH_3)_2CH-OH} + \underset{5\%\,(E1)}{CH_3CH=CH_2}$$

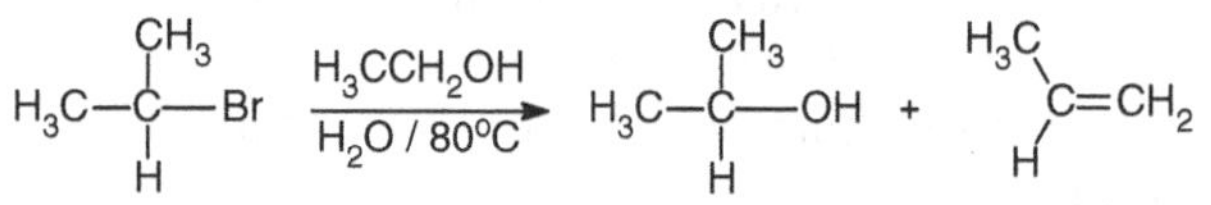

    3. E1 is often a nuisance that interferes with S_N1. It is usually better to use E2 conditions (less polar solvent, strong sterically hindered base, and heat) to synthesize an alkene from R-X.

c. Problem 30. Write the chief product structures and IUPAC names expected from 1-bromobutane reaction with:
  1. NaOH / H$_2$O / warm

  2. Zn / HCl(dil) / H$_2$O (*Hint*: initially forms organozinc bromide)

  3. 1. Li (2 equivalents), 2. CuI, 3. bromoethane

  4. Mg (dry ether solvent)

  5. Product from 4 + D$_2$O

  6. NaI / acetone (dry solvent) / warm

  7. Br$_2$ / hexane (solvent) / 25° C / dark

  8. NH$_3$ / H$_2$O / warm

  9. NaOCH$_2$CH$_3$ / HOCH$_2$CH$_3$ (solvent) / warm

  10. NaCN (pH 9) / H$_2$O / warm

  11. NaSCH$_3$ / H$_2$O / warm

d. Problem 31. What reagents and conditions would you need to convert bromobutane into:
  1. iodobutane

  2. chlorobutane

  3. methoxybutane

  4. butanol

5. *n*-butylmagnesium bromide

6. lithium di-*n*-butylcopper

e. Problem 32. Substitution and elimination products are obtained from the following primary alkyl bromides with strongly basic bimolecular conditions, $CH_3CH_2O^- Na^+$ / CH_3CH_2OH (solvent) / heat. Account for the substitution/elimination percentages.

alkyl bromide	% substitution	% elimination
bromoethane	99	1
bromopropane	91	9
1-bromo-2-methylpropane	40	60

CHAPTER 7

7.1.1. Alkene family – hydrocarbons containing a π-bond functional group (section 1.17.1).
    a. Hydrocarbon – a compound containing only C and H.

7.1.2. *Olefin* – an older alkene name – from *olefiant* (oil-forming) *gas*. Alkenes (nonpolar) have an oily appearance (do not mix with water).

7.2.3. Ethene Lewis dot, LCAO and MO pictures were drawn in section 1.17.

7.3.1. *Unsaturated hydrocarbon* – can absorb H_2 (when a catalyst is present).

$$\underset{\text{unsaturated}}{H_2C=CH_2} \xrightarrow{H_2 \text{, Ni (catalyst)}} \underset{\text{saturated}}{H_3C-CH_3}$$

    a. Problem 1. Write molecular formulas for each compound below. How many carbon and hydrogen atoms does each contain?
        1. *n*-hexane and cyclohexane

        2. *n*-pentane and cyclopentane

        3. hex-1-ene and cyclohexene

        4. dodecane and *n*-hexylcyclohexane and cyclohexylcyclohexane

Generally, how can you deduce the number of rings in a compound if you know its molecular formula and number of π-bonds?

7.3.2. Open chain (no ring) alkenes have formulas C_nH_{2n}, two fewer H atoms than nonring alkanes (C_nH_{2n+2}). A ring also decreases the number of H atoms.

$$H_3C-CH_2-CH_3 \qquad H_2C=CH-CH_3 \qquad \text{cyclopropane}$$

propane propene cyclopropane
C_3H_8 C_3H_6 C_3H_6

a. Examples: propene and cyclopropane each have six H atoms, two fewer than *n*-propane's eight.

b. Element of unsaturation – *either* a double bond *or* a ring. Examples: C_4H_8 isomers:

1. *"Rings or double bonds"* is more descriptive than *"elements of unsaturation."* Larger cycloalkanes cannot easily be hydrogenated, although cyclopropane and cyclobutane can.

2. Rings or double bonds are calculated from:

$$\text{rings or double bonds} = \frac{1}{2}(2C + N + 2 - H)$$

where *C* is the number of C atoms,
H is the number of H or X atoms,
N is the number of nitrogen atoms,
and *O* atoms are ignored.

3. Examples: each isomer below has one ring or double bond.

7.3.3. More alkene isomers are possible than for alkanes with the same number of C atoms (because the double bond position is an additional variable).

a. Problem 2. Draw six isomeric C_5H_{10} alkene structures and four isomeric C_3H_5Cl alkene structures, and give IUPAC names. Assign *E* or *Z* geometry where necessary (see section 7.5.3.).

d. Problem 3. Write molecular formulas for cyclohexane, methylcyclopentane, and 1,2-dimethylcyclobutane. Does the molecular formula give any information about ring size?

e. Problem 4. The yellow plant pigments α-, β-, and γ-carotene and the red tomato pigment lycopene are converted into vitamin A_1 in the liver. All four have the molecular formula $C_{40}H_{56}$. Upon catalytic hydrogenation, α- and β-carotene yield a product whose formula is $C_{40}H_{78}$, while γ-carotene yields $C_{40}H_{80}$, and lycopene yields $C_{40}H_{82}$. How many rings, if any, does each compound contain?

7.3.4. Lewis dot propene structure (propylene).

7.3.5. More *s*-character ⇒ stronger electron attraction to nucleus, and shorter (and stronger) σ-bond. Hybridization and bond distances:*

C-H bond	Distance (Å)	C-C bond	Distance (Å)
sp³-1s	1.09	*sp³-sp³*	1.54
sp²-1s	1.08	*sp³-sp²*	1.501
sp-1s	1.06	*sp²-sp²*	1.48
		sp³-sp	1.459
		sp²-sp	1.43
		sp-sp	1.38
		sp²=sp²	1.33
		sp≡sp	1.20

*Some data in this table were taken from Morrison and Boyd, "Organic Chemistry," 6th ed.

a. An *sp* orbital is ½ *s* and ½ *p*; an *sp²* orbital is ⅓ *s* and ⅔ *p*; while an *sp³* orbital is ¼ *s* and ¾ *p*.

b. Example: bonds ranked in order of increasing length.

7.3.6. Bond strength parallels length. Some bond energies:

bond type	bond energy (kJ/mol)
sp^3-$1s$	423
sp^2-$1s$	463
sp^3-sp^3	377
sp^3-sp^2	385
sp^2=sp^2 (σ-bond)	407

7.3.7. Four butene (butylene) isomer Lewis dot structures:

but-1-ene   trans-but-2-ene   cis-but-2-ene   2-methylprop-1-ene
1-butene   trans-2-butene   cis-2-butene   isobutene

7.3.8. Isomeric butene physical properties are different:

Compound	bp °C	mp °C	d (−20°C)	R.I. (−12.7°C)
isobutene	−7	−141	0.59	1.3727
but-1-ene	−6	<−195	0.59	1.3711
trans-but-2-ene	+1	−106	0.60	1.3778
cis-but-2-ene	+4	−139	0.62	1.3868

a. But-1-ene and but-2-ene (*cis* or *trans*) hydrogenation (see section 7.8.1) with H_2 / Pt (catalyst) yields *n*-butane, and isobutene (bp −7°C) yields isobutane.
    1. Double bond position in but-1- and but-2-enes was verified by double bond cleavage (section 8.15) and product analysis.
b. *Cis* and *trans* isomers do not interconvert at 25°C due to a 293 kJ/mol E_a barrier.

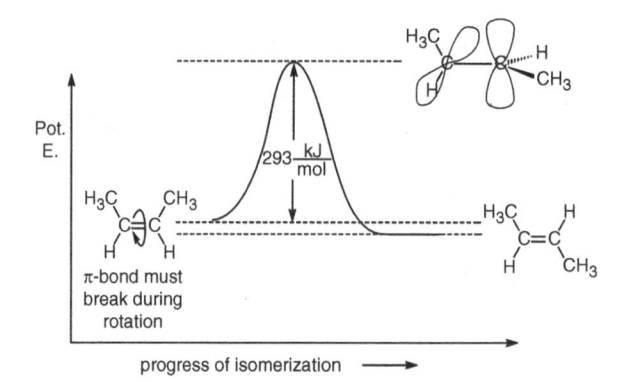

π-bond must break during rotation

progress of isomerization ⟶

7.5.1. *Cis* (same side) and *trans* (across) but-2-enes (section 1.19.5) are diasteriomers – stereoisomers that are not mirror images and are not superimposable. More specifically, they are geometrical isomers, section 5.11.1).

    a. Geometric isomers are possible only if each sp^2 C atom has two different groups bonded to it.

    b. *Cis* is usually less stable than *trans*, due to steric (adjacent group) crowding.

7.5.2. *Cis* and *trans* do not cover all possible geometrical cases.

7.5.3. *E / Z* designation – assign groups bonded around C=C their Cahn-Ingold-Prelog priorities (section 5.3.1). If a higher priority group on each C atom is on the same side of a line joining the two C atoms, (*cis*-like), configuration is **Z** (zusammen, together). *Cis* is a special case of **Z**. If higher priority groups on each C atom are on opposite sides (*trans*-like), configuration is **E** (entgegen, opposite). *Trans* is a special case of **E**.

    a. Examples: (*E*)- and (*Z*)-1-bromo-1-chloropropene; (*E*)- and (*Z*)-2-bromo-1-chloropropene; (*E*)- and (*Z*)-1-bromo-1,2-dichloroethene

★ higher priority

    b. Cyclic double bonds are always *cis* (or *Z*) unless a ring contains seven or more atoms.

cis or (*Z*)-cyclohexene   (*Z*)-cyclooctene   *trans* or (*E*)-cyclodecene   *cis* or (*Z*)-cyclodecene

c. Multiple π-bonds – treat each one separately.

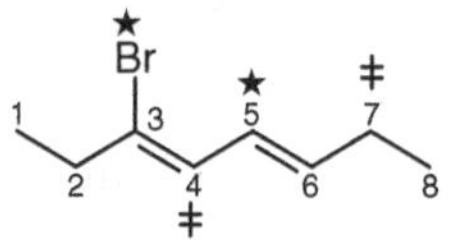

(3Z,5E)-3-bromoocta-3,5-diene
★priorities for the 3,4 double bond
‡priorities for the 5,6 double bond

d. Problem 5. Write condensed structural formulas for
1. 2,3-dimethylbut-2-ene

2. 3-bromo-2-methylpropene

3. *cis*-2-methylhept-3-ene

4. (*E*)-2-chlorobut-2-ene

5. 3,6-dimethyloct-1-ene

6. 2,4,4-trimethylpent-2-ene

7. *trans*-3,4-dimethylhex-3-ene

8. (*Z*)-3-chloro-4-methylhex-3-ene

9. (*E*)-1-deuterio-2-chloropropene

10. (*R*)-3-bromobut-1-ene

11. (*S*)-*trans*-4-methylhex-2-ene

12. (*E*,*Z*)-hexa-2,4-diene

c. Problem 6. Give IUPAC names for the following:

d. Problem 7. Which of the following are different for *cis*-hex-3-ene vs. *trans*-hex-3-ene?
   a. bp
   b. mp
   c. dipole moment
   d. refractive index
   e. hydrogenation rate
   f. hydrogenation product
   g. ethanol solubility
   h. density
   i. gas chromatography retention time

7.5.4. Application. The first step in vision is 11-*cis*-retinal (covalently bonded to rhodopsin in a lipophilic pocket) photochemical transformation into 11-*trans*-retinal. Retinal is made from vitamin A_1 (retinol), which comes from β-carotene (carrots).

a. Rhodopsin changes conformation to accommodate the *trans* form, which allows Ca^{2+} to flow through a membrane, triggering a nerve impulse to the brain.

b. Enzymes then convert *trans* form back into *cis* form (an endothermic process).

7.4.1. Parent and substituent alkene names were given in section 2.15.6. Alkene IUPAC rules:
   a. Find longest carbon chain containing C=C and assign parent alkane name.
   b. Drop alkane "ane" ending and add "ene," alkene parent suffix.
   c. If a higher priority functional group is present and C=C is in the parent chain, use an "en" suffix (see examples below).
   d. Number C=C position at the first sp^2 C encountered, counting from the end nearest the double bond.
      1. IUPAC (1993) – put the number just before the "ene" suffix rather than before the parent chain name. In examples below, older names are on top; 1993 names are below, and common names are last.
   e. If more than one C=C is present, number positions (as in d), drop alkane name "ne", and add "diene, triene, or tetraene, etc." as parent suffix.
   f. Designate geometrical isomers with prefixes *cis-* or *trans-*, or *(E)-* or *(Z)-*.
   g. Common alkene containing branch names are vinyl, allyl, and methylene.

vinyl   allyl   methylene

7.4.2. More examples:

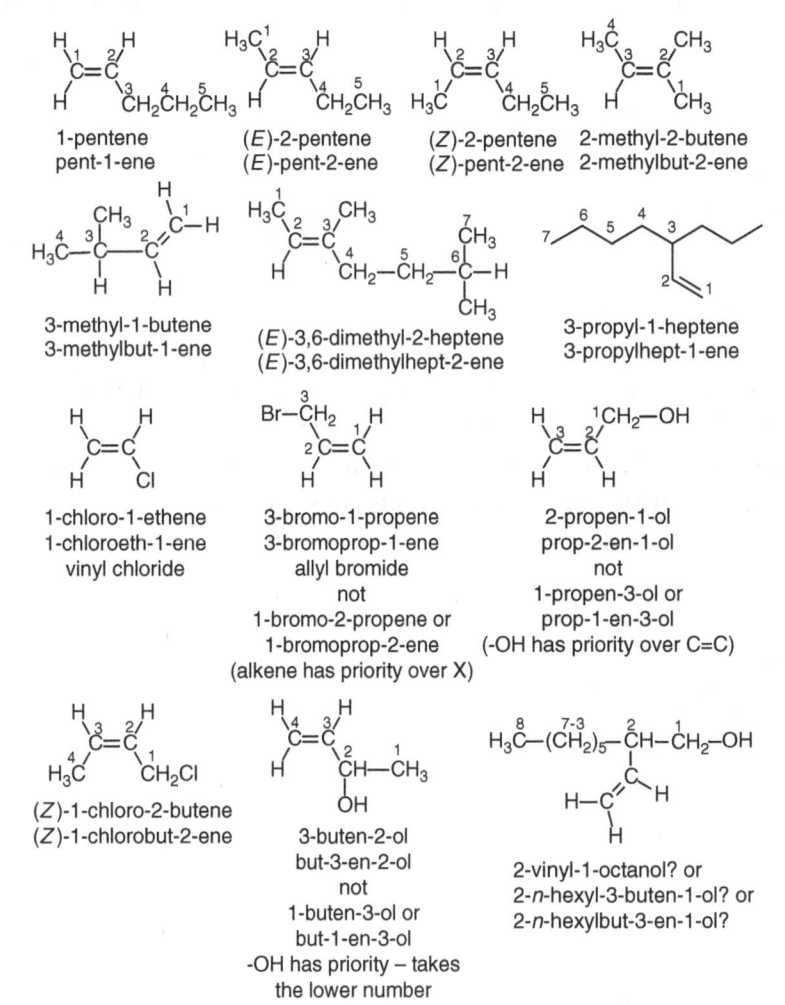

1-pentene / pent-1-ene

(*E*)-2-pentene / (*E*)-pent-2-ene

(*Z*)-2-pentene / (*Z*)-pent-2-ene

2-methyl-2-butene / 2-methylbut-2-ene

3-methyl-1-butene / 3-methylbut-1-ene

(*E*)-3,6-dimethyl-2-heptene / (*E*)-3,6-dimethylhept-2-ene

3-propyl-1-heptene / 3-propylhept-1-ene

1-chloro-1-ethene / 1-chloroeth-1-ene / vinyl chloride

3-bromo-1-propene / 3-bromoprop-1-ene / allyl bromide / not 1-bromo-2-propene or 1-bromoprop-2-ene (alkene has priority over X)

2-propen-1-ol / prop-2-en-1-ol / not 1-propen-3-ol or prop-1-en-3-ol (-OH has priority over C=C)

(*Z*)-1-chloro-2-butene / (*Z*)-1-chlorobut-2-ene

3-buten-2-ol / but-3-en-2-ol / not 1-buten-3-ol or but-1-en-3-ol / -OH has priority – takes the lower number

2-vinyl-1-octanol? or 2-*n*-hexyl-3-buten-1-ol? or 2-*n*-hexylbut-3-en-1-ol?

7.4.3. Cyclic alkenes:

1-methyl-1-cyclopentene 3-methyl-1-cyclohexene (3R,6S)-3,6-dibromo-1-cyclohexene
1-methylcyclopent-1-ene 3-methylcyclohex-1-ene (3R,6S)-3,6-dibromocyclohex-1-ene
or
(Z)-3,6-dibromo-1-cyclohexene
(Z)-3,6-dibromocyclohex-1-ene

7.4.4. Polyenes.

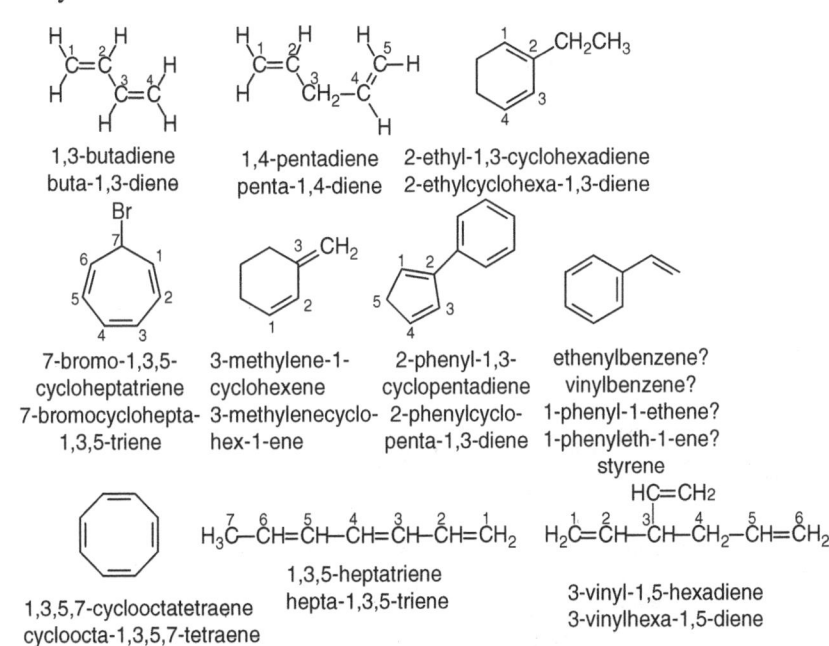

1,3-butadiene   1,4-pentadiene   2-ethyl-1,3-cyclohexadiene
buta-1,3-diene   penta-1,4-diene   2-ethylcyclohexa-1,3-diene

7-bromo-1,3,5-   3-methylene-1-   2-phenyl-1,3-   ethenylbenzene?
cycloheptatriene   cyclohexene   cyclopentadiene   vinylbenzene?
7-bromocyclohepta- 3-methylenecyclo- 2-phenylcyclo-   1-phenyl-1-ethene?
1,3,5-triene   hex-1-ene   penta-1,3-diene   1-phenyleth-1-ene?
                                                         styrene
                                                         HC=CH₂

1,3,5,7-cyclooctatetraene   H₃C–CH=CH–CH=CH–CH=CH₂   H₂C=CH–CH–CH₂–CH=CH₂
cycloocta-1,3,5,7-tetraene 1,3,5-heptatriene
                            hepta-1,3,5-triene   3-vinyl-1,5-hexadiene
                                                       3-vinylhexa-1,5-diene

7.19.1. Catalytic and steam petroleum cracking (section 3.6.8b) at high temperatures yields alkenes industrially.

7.6.1. Mixtures are always obtained, but can be separated by fractional distillation. Which alkenes form depends on temperature, catalyst, and H₂ pressure.

7.6.2. The petroleum industry sells pure alkenes with four or fewer C atoms. Some (but not all) higher molecular mass isomers are also available.

7.6.3. Cracking is endothermic but has a positive entropy factor. The $-T\Delta S^{\circ}_{rxn}$ term must be larger (negative) at higher temperatures than $-\Delta H^{\circ}_{rxn}$ so that ΔG°_{rxn} becomes negative and cracking becomes possible.

7.1.3. Alkenes are common in nature. Ethene causes fruit to ripen. Examples:

ethene   α-pinene   cis-9-tricosene
ethylene   (in terpentine) muscalure
                                 (house fly sex
                                 attractant)

7.6.5. Ethene is the fifth largest produced industrial chemical (and most produced organic chemical) in the U.S.A. Polyethylene production (alkene polymerization, section 8.16) is its largest use. Some useful products made from ethene:

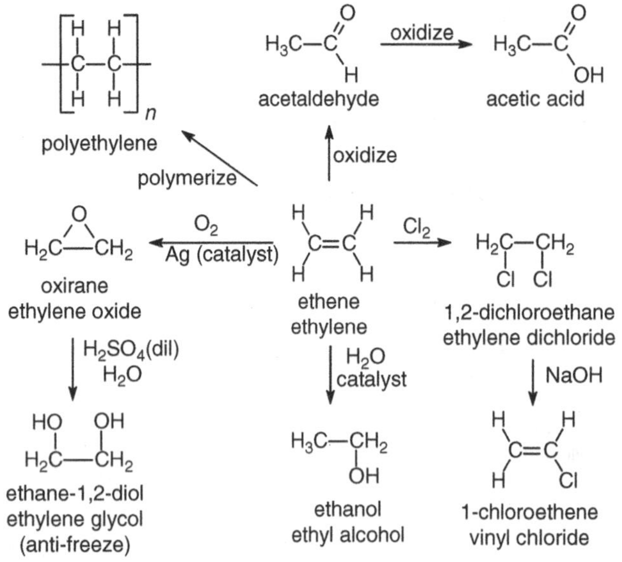

7.6.6. Similar products are made from propene.

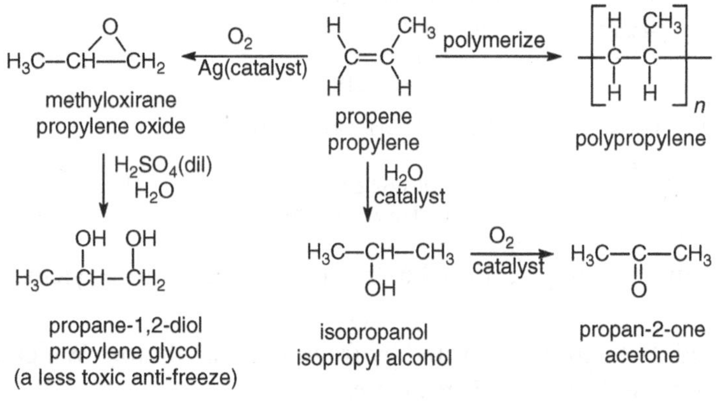

7.8.3. Evidence for alkene stability order and Zaitsev's rule (section 7.8.1) – heat of hydrogenation data (section 7.8.4).

  a. Heat of combustion data also support the alkene stability series, but are harder to measure accurately because heats evolved are large.

7.8.4. C=C double bond reaction types (Chapter 8):

  a. Addition – C=C π-bond (but not σ-bond) breaks and two new σ-bonds form in its place. Product has no π-bond.

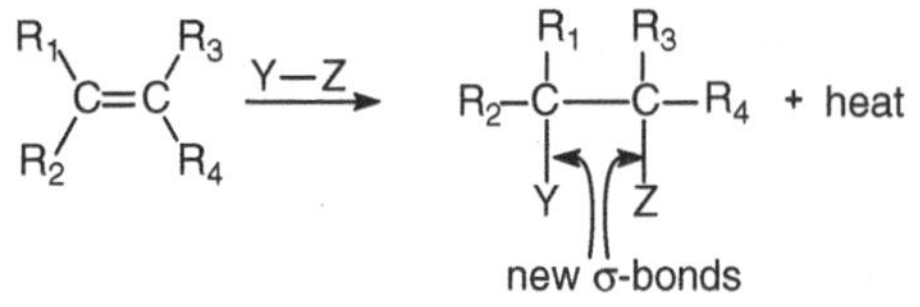

1. Two (or more) molecules combine to yield a single product molecule; the opposite of elimination; limited to compounds that have π-bonds.
   b. Allylic substitution – C=C influences reactivity at an allylic position but C=C remains intact. Example: allylic bromination (sections 6.6.1a2 and 15.7.3).
8.10.1. Hydrogenation – two new σ-bonded H atoms add and replace a π-bond.
8.10.2. Heterogeneous (two phase) hydrogenation – an alcohol, alkane, or ethanoic acid alkene solution is agitated under slight $H_2(g)$ pressure at 25°C with a finely divided, insoluble, catalyst (such as Pt, Pd, or Ni). Catalyst is filtered out and solvent distilled leaving a (higher bp) saturated (alkane) product.
   a. π-Bond is broken and H-H σ-bond is at least partially broken on a catalyst (such as Pt) surface.

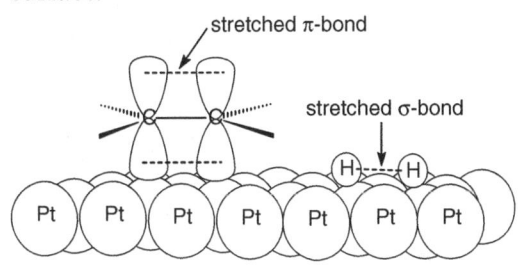

   b. Mechanism:

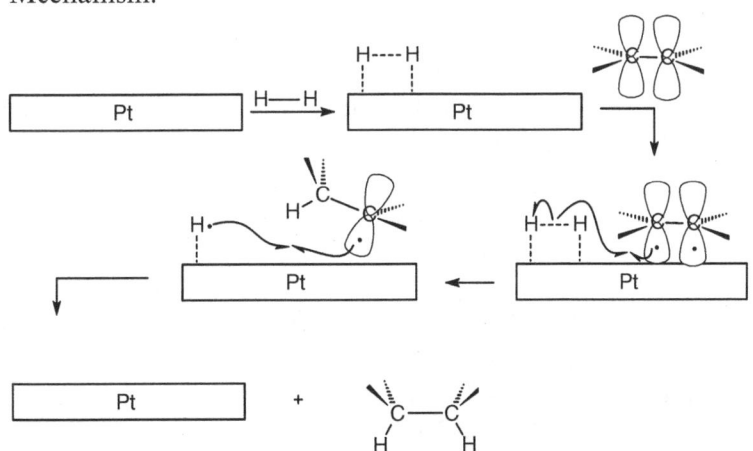

      1. Evidence – an H_2 / D_2 mixture forms H-D on the Pt surface (without alkene).
   c. Hydrogenation is stereospecific (*syn*-addition – both H atoms add from the same face), although more than one step is involved.

   d. Homogeneous (one phase) hydrogenation – process similar to above, but catalyst (an organic rhodium, ruthenium or iridium complex) is soluble.

e. **Problem 8.** Butenedioic acid homogeneous hydrogenation with D_2 yields saturated (2,3-d_2)-butanedioic acid. *cis*-Butenedioic acid yields only the *meso* product, while *trans*-butenedioic acid yields only the racemic product. Assuming these results are typical (they are), what is the homogeneous hydrogenation stereochemistry?

f. Optically active catalysts can produce optically active products from optically inactive reactants (R. Noyori and W. Knowles, Nobel Prize 2001).
   1. The two different hydrogenation transition states are diasteriomeric (different energies). The one leading to the *R*-enantiomer has the lower energy so *R*-product forms faster in this reaction:

g. Example: *asymmetric induction*. An optically active catalyst causes hydrogenation from one face to be faster than from the other.

h. (*S*)-(-)-Dopa is converted by brain enzymes into dopamine, which helps increase levels in Parkinson's disease patients.

i. Catalysts can be attached to an (insoluble) polymer surface to facilitate separation (via filtration) from products.

8.10.3. H₂(g) amount (mols, n) consumed is easy to calculate from the volume decrease (at constant P and T), using the ideal gas law ($PV = nRT$).

   a. If an unsaturated unknown compound's molecular mass is known, the number of multiple bonds can be determined.

   1. Example: a certain compound has a mol. mass = 54.1 g/mol. If 10.0 g of this compound consumes 8.29 L of H₂ at STP (273.15 K, 1 atm), how many π-bonds does the molecule contain? R (gas constant) = 0.0821 L · atm / mol · K)

$$10.0 \text{ g unk} \times \frac{1 \text{ mol unk}}{54.1 \text{ g unk}} = 0.185 \text{ mol unk}$$

$$n = \frac{PV}{RT} = \frac{(1 \text{ atm})(8.29 \text{ L } H_2)}{\left[\frac{0.0821 \text{ L} \cdot \text{atm}}{\text{mol} \cdot \text{K}}\right](273 \text{ K})}$$

$$n = 0.370 \text{ mol } H_2$$

The unknown consumes 2 × 0.185 = 0.370 mol of H₂. Each molecule must have two π-bonds. Some possibilities:

$$H_2C=C=CH-CH_3 \qquad H_3C-C\equiv C-CH_3$$
$$H_2C=CH-CH=CH_2 \qquad H-C\equiv C-CH_2CH_3$$

7.8.5. Heat of hydrogenation – heat (ΔH^o_{rxn}) evolved when one mol of an unsaturated compound is hydrogenated.

   a. Hydrogenation is usually mildly exothermic (110 - 136 kJ/mol) because two (strong) C-H σ-bonds replace one (weaker) π-bond and one H-H σ-bond. Alkanes are more stable than alkenes.

   1. Heats of hydrogenation can be accurately measured because they are not overly exothermic.

   b. Uncatalyzed hydrogenation is very slow (high E_a) even at high temperatures.

   c. A catalyst provides an alternative pathway (with lower E_a), but does not affect ΔH^o_{rxn}.

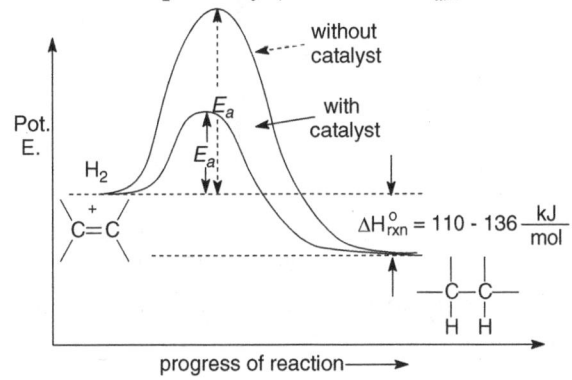

   1. Catalyst also speeds up reverse reaction (dehydrogenation), but does not affect equilibrium reactant vs. product concentrations.

7.8.6. Heat of hydrogenation data:

Structure	Type	ΔH^o_{rxn} (kJ/mol)
H₂C=CH₂	unsubstituted	−136.0
H₃CCH=CH₂	monosubstituted	−123.4

(Continued)

Structure	Type	ΔH^0_{rxn} (kJ/mol)
H₃CCH₂CH=CH₂	monosubstituted	−125.9
H₃C\C=C/CH₃ (H, H)	cis-disubstituted	−118.5
H₃C\C=C/H (H₃C, H)	geminal-disubstituted (terminal)	−117.8
H₃C\C=C/H (H, CH₃)	trans-disubstituted	−114.6
H₃C\C=C/CH₃ (H₃C, H)	trisubstituted	−111.6
H₃C\C=C/CH₃ (H₃C, CH₃)	tetrasubstituted	−110.4

    a. Example compound values above are typical of un-, mono-, di-, tri- and tetrasubstituted alkenes.

7.8.7. *Cis-* and *trans*-but-2-ene both yield *n*-butane when hydrogenated. *trans*-But-2-ene must initially contain 3.9 kJ/mol less energy than *cis*-but-2-ene – *trans* must be more stable than *cis*.

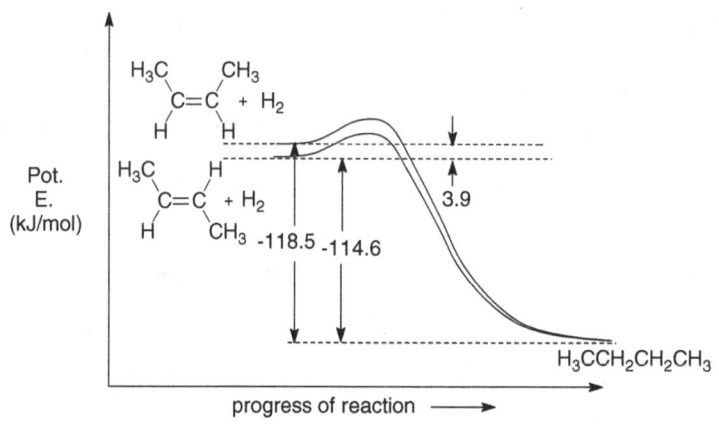

7.8.8. General trend – more substituted an isomeric alkene ⇒ lower heat of hydrogenation, and greater stability.

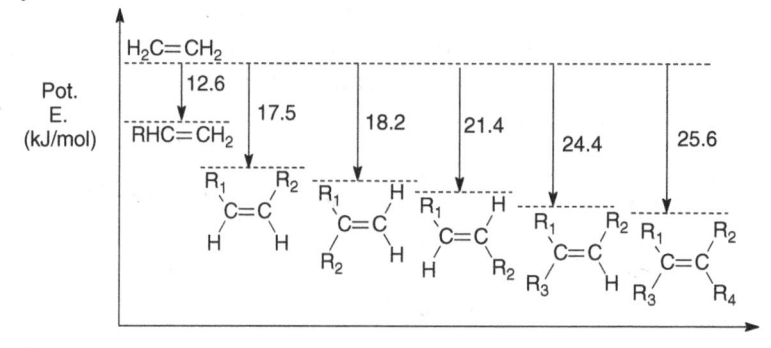

    a. Terminal alkenes are not always between *cis* and *trans*.

7.8.9. Cyclobutene bond angles are near 90°, far from ideal 109.5° for sp^3 C or 120° for sp^2 C. Cyclobutane has less angle strain than cyclobutene. Cyclopentene and cyclopentane have little angle strain. The extra cyclobutene hydrogenation heat (17.0 kJ/mol) is due to (angle) strain relief.

$\Delta H^{\circ}_{rxn} = -129.7 \frac{kJ}{mol}$

$\Delta H^{\circ}_{rxn} = -112.7 \frac{kJ}{mol}$

a. Cyclopropene (stable if kept cold) contains even more angle strain (bond angles near 60°).

sterculic acid (in a tropical tree oil)

b. *Trans* double bonds are only stable at 25°C in rings with seven or more C atoms.

trans-cyclooctene   *cis*-cyclooctene

c. Bredt's rule (a *trans* cyclic double bond consequence) – a bridged bicyclic compound cannot have a double bond at a bridgehead position (section 3.16.1b) unless one ring has seven or more C atoms.

unstable   stable

★ groups are *cis*
‡ groups are *trans*

    a. A few seven membered ring bridgehead alkenes have been made at low temperatures.

7.11.4. Alkanes can be dehydrogenated at high temperatures – similar to cracking (section 3.6.8 and 7.19.1).

a. Spontaneous dehydrogenation requires that $\Delta G^{\circ}_{rxn} < 0 \Rightarrow -T\Delta S^{\circ}_{rxn}$ term must be larger (negative) than $\Delta H^{\circ}_{rxn} \approx +110$ to $+136$ kJ/mol. ΔG°_{rxn} becomes negative only at high temperatures for ordinary alkanes ($\Delta S^{\circ}_{rxn} \approx +126$ J·°C^{-1}). A catalyst lowers E_a and speeds up reaction.

7.7.1. Alkene physical properties:

Alkene	bp °C	Density
H$_2$C=CH$_2$	−104	
H$_3$CCH=CH$_2$	−47	0.52
(H$_3$C)$_2$C=CH$_2$	−7	0.59
H$_3$CCH$_2$CH=CH$_2$	−6	0.59
trans-CH$_3$CH=CHCH$_3$	1	0.60
cis-CH$_3$CH=CHCH$_3$	4	0.62
(H$_3$C)$_2$CHCH=CH$_2$	25	0.65
H$_3$CCH$_2$CH$_2$CH=CH$_2$	30	0.64
trans-CH$_3$CH=CHCH$_2$CH$_3$	36	0.65
cis-CH$_3$CH=CHCH$_2$CH$_3$	37	0.66
(H$_3$C)$_2$C=CHCH$_3$	39	0.66
H$_3$CCH$_2$CH$_2$CH$_2$CH=CH$_2$	64	0.68
(H$_3$C)$_2$C=C(CH$_3$)$_2$	73	0.71
H$_3$C(CH$_2$)$_4$CH=CH$_2$	93	0.70
H$_3$C(CH$_2$)$_5$CH=CH$_2$	122	0.72
H$_3$C(CH$_2$)$_6$CH=CH$_2$	146	0.73
H$_3$C(CH$_2$)$_7$CH=CH$_2$	171	0.74

7.7.2. Alkene properties are similar to alkanes.
  a. Almost insoluble in H$_2$O, soluble in nonpolar solvents hexane, gasoline, chlorinated solvents, or ethers.
  b. Less dense than H$_2$O (0.6 - 0.7 g/mL).
  c. Low bp's (similar to alkanes), increasing 20-30 °C for each additional -CH$_2$-. More branching for isomers ⇒ usually lower bp.
  d. Smaller dipole moments than alkyl halides, but a little larger than alkanes. Alkyl groups are (apparently) electron releasing (inductively), and sp^2 C atoms are more electronegative than sp^3 C atoms.
     1. Examples: propene, but-1-ene, *cis*-but-2-ene, *trans*-but-2-ene. Symmetrical compounds have only London attractions between molecules. Unsymmetrical molecules also have dipole-dipole attractions.

$\mu = 0.35\ D$ $\mu = 0.33\ D$ $\mu = 0.00\ D$ $\mu = 2.4\ D$ $\mu = 0.00\ D$
bp = -47°C   bp = 4°C   bp = 1°C   bp = 60°C   bp = 48°C
　　　　　　mp = -139°C   mp = -106°C   mp = -80°C   mp = -50°C

  2. Symmetrical isomers pack more efficiently in the solid state and have higher melting points.
 e. *Cis*-isomers – often have higher bp's due to their dipole moment > 0.
    But they have lower mp's than *trans* due to lower symmetry.
 f. Smell – very stinky if volatile, skin and mucous membrane irritants.
 g. Problem 9a. Assuming that methyl groups release electron density towards chlorine, would *cis*-2,3-dichlorobut-2-ene have a larger or smaller dipole moment than *cis*-1,2-dichloroethene? 9b. Draw the net molecular dipole moment for *cis*-1,2-dibromo-1,2-dichloroethene. Is it larger or smaller than the *cis*-1,2-dichloroethene dipole moment? Explain.

7.9.1. Alkene preparations: alkyl halide E2 or E1 elimination (sections 7.9.1 and 7.14.1). Review:
   a. E2 usually has fewer side reactions than E1.
   b. E2 reaction is fastest for 3° R-X's, especially those with bulky R groups.
   c. A bulky base and higher temperature promotes E2 over S_N2.

  1. A bulky base (such as diisopropylamine) promotes the less substituted (nonZaitsev, or Hofmann) alkene as major product.

d. Review: E2 is an *anti*-elimination.

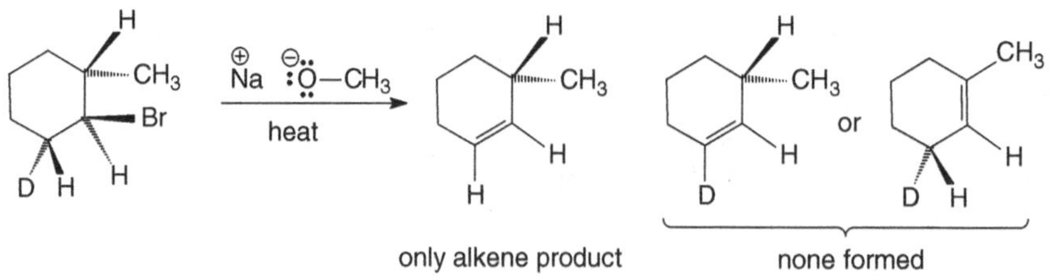

only alkene product   none formed

1. Problem 10. Diasteriomer I gave *cis*-but-2-ene without deuterium loss and *trans*-but-2-ene with deuterium loss when heated with $H_3CCH_2O^-\ K^+$ / $HOCH_2CH_3$. Diasteriomer II gave *cis*-but-2-ene with deuterium loss and *trans*-but-2-ene without deuterium loss. Explain. What is the reaction stereochemistry?

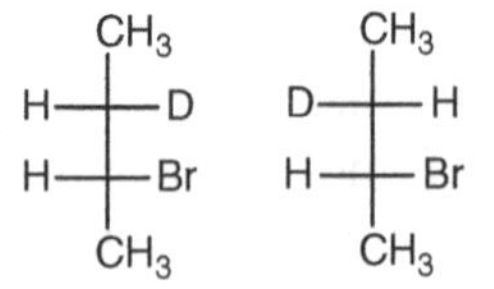

e. E1 occurs for 2° and 3° R-X's in polar, ionizing solvents (such as H_2O or ROH), and weak bases (section 7.16.1). Higher temperature promotes E1 relative to S_N1.

f. Problem 11. Draw all alkene structures that would be obtained from the R-X dehydrohalogenations (with strong base and heat). Which is the major product?
   1. 1-chloropentane

   2. 2-chloropentane

   3. 3-chloropentane

   4. 2-chloro-2-methylbutane

   5. 3-chloro-2-methylbutane

CHAPTER 7   243

   6. 2-chloro-2,3-dimethylbutane

   7. 1-chloro-2,2-dimethylpropane

g. Problem 12. What alkyl halide (if any exists) would you need to make each of the following pure alkenes from dehydrohalogenation (with strong base and heat)?
   1. 2-methylpropene

   2. pent-1-ene

   3. pent-2-ene

   4. 2-methylbut-1-ene

   5. 2-methylbut-2-ene

   6. 3-methylbut-1-ene

h. Problem 13. When 2-methyl-3-tosylpentane is heated in *n*-butanol (solvent, no base added), the following alkene products formed: 2-methylpent-2-ene (80%), 4-methylpent-2-ene (11%), and 2-methylpent-1-ene (9%). Write a mechanism that accounts for these alkene products. Why are these relative amounts formed? Why is the 4-methylpent-2-ene entirely *trans* isomer?

i. Problem 14. Quaternary ammonium ions undergo elimination upon treatment with strong base and heat – for example:

H₃C—CH—CH₃ + ⁻:Ö—H —125°C→ H₂C=C(CH₃)(H) + :N(CH₃)₃ + H-Ö-H
       |
       N(CH₃)₃⁺

N,N,N-trimethylpropan-2-ammonium ion   propene   weak base (good leaving group)

1°, 2°, and 3° ammonium ions R-⁺NH₃, R₂-⁺NH₂, and R₃-⁺NH ions do not eliminate with strong base, although corresponding amine basicity is similar to :N(CH₃)₃. What other reaction occurs instead? Draw with electron pushing arrows.

7.9.2. Vicinal 1,2-dihalide reduction. 1,2-Dihalides are usually made from alkenes, rather than the reverse. Reaction is still sometimes useful.

   a. With NaI / acetone.

   [Mechanism diagram showing anti-coplanar dibromide (meso-stilbene dibromide) reacting with NaI in acetone solvent via E2 to give trans-stilbene + I-Br + Na⁺ + Br⁻, 89% (no cis)]

   bromine atoms are *anti*-coplanar
   solvent: H₃C-C(=O)-CH₃
   E2   89% (no cis)

   b. With Zn in acetic acid (solvent) – a redox reaction occurs on the Zn surface (a *heterogeneous reaction* – part solid, part liquid). Zn is oxidized (from 0 to +2).

   H₃C-CH(Br)-CH(Br)-CH₃ —Zn / H₃C-C(=O)OH→ H₃C-CH=CH-CH₃ + ZnBr₂
                                              cis and *trans* but no but-1-ene

   [Diagram: H₂C=CH-CH₂OH —Br₂, hexane (solvent), 0°C / dark→ H-C(Br)(H)-C(Br)(H)-CH₂OH —KMnO₄/KOH (conc, heat)→ H-C(Br)(H)-C(Br)(H)-C(=O)OH —H₂SO₄(dil), H₂O/25°C (neutralize pH)→ ... —NaI, acetone→ H₂C=CH-C(=O)OH + ZnBr₂]

   would oxidize — protect alkene — deprotect alkene

   1. A vicinal dihalide is sometimes used as an alkene *protecting group* during synthesis. Addition of Br₂ to an alkene is in section 8.8, and oxidation of an alcohol to a carboxylic acid is in section 11.2.

   c. Problem 15. When 1-bromocyclohexene reacts with radioactive Br₂, and the resulting tribromide is heated with NaI/acetone, 1-bromocyclohexene is recovered that contains less than 0.3% radioactive bromine. Write mechanism steps that explain this result.

7.18.1. Alkyl halides (E2 substrates) are often made from alcohols (sections 6.6.1d and 11.7 - 11.8). Alkenes can be made directly from alcohols by water removal (*dehydration*).

7.18.2. Alcohol dehydration is acid catalyzed: H_2SO_4(conc), H_3PO_4(conc), or Al_2O_3 (a Lewis acid).

   a. E2 dehydrohalogenation is often the better method, despite the extra step.

$$R-OH \begin{cases} \xrightarrow{PX_3} R-X \xrightarrow[E2]{base} alkene \\ \xrightarrow{Ts-Cl} R-OTs \xrightarrow[E2]{base} alkene \\ \phantom{\xrightarrow{Ts-Cl}} + H-Cl \\ \xrightarrow{H_2SO_4(conc)} alkene \end{cases}$$

   b. R-OH reactivity is in the order: faster 3° > 2° >> 1° slower. Dehydration yields are high for many 2° and 3° alcohols.

     1. Example: 1° alcohols usually react via an E2-like mechanism.

     2. Examples: 2° and 3° alcohols react via an E1-like mechanism. The reverse of step 3 is known to be about as slow as forward step 2.

c. –OH is a slow leaving group (would leave as strongly basic ⁻OH). Fast, reversible protonation converts –OH into R-⁺OH₂. Weakly basic H₂O then leaves (a much faster leaving group).

d. Unlike dehydrohalogenation, dehydration is reversible and can be driven to completion by distilling out the (low boiling) alkene product, or by H₂O byproduct removal (by reaction with excess acid).

[reaction diagram: H₂O (stronger base) + H₂SO₄ (stronger acid) ⇌ H₃O⁺ (weaker acid) + ⁻OSO₃H (weaker base)]

1. H₂O removal shifts equilibrium to the right so that more alkene continues to form (Le Chatelier's principle). If the reverse reaction is desired (R-OH product from an alkene), excess H₂O is added (as dilute H₂SO₄), and the mixture is not distilled, so that equilibrium is driven back to reactants.

e. Reaction diagrams.

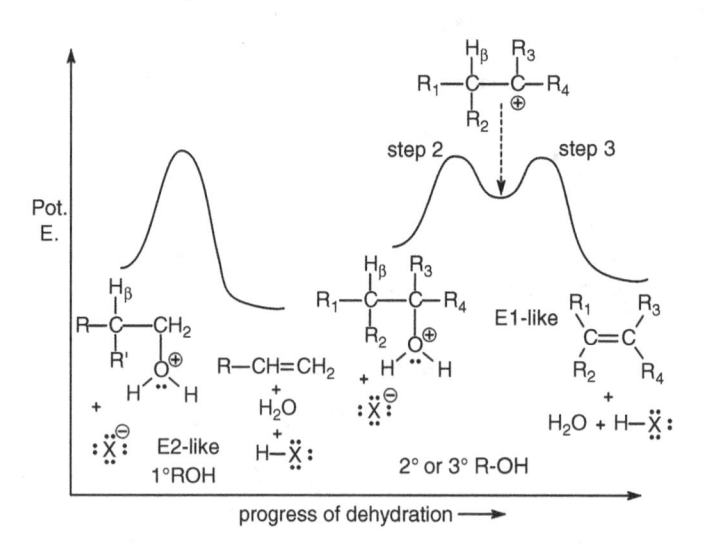

[potential energy diagram showing E2-like pathway for 1° ROH and E1-like pathway (step 2, step 3) for 2° or 3° R-OH vs. progress of dehydration]

f. Acid consumed in the first step is regenerated at the end. Only a trace of acid catalyst is needed, while in dehydrohalogenation one mol of base is consumed per mol of R-X.

g. The base in the last step is weak (H₂O or ⁻OSO₃H) in acidic solution.

h. **Problem 16.** Write a mechanism for alkene hydration, the reverse of dehydration – that is, acid catalyzed water addition to alkenes yielding alcohols. Include all elementary steps (collisions), intermediates, formal electrical charges, and involved lone pair electrons.

i. **Problem 17.** 2-Methylpropan-2-ol was heated with H₂SO₄ in H₂¹⁸O enriched water. Samples were withdrawn at intervals and analyzed for 2-methylpropene and (CH₃)₃C-¹⁸OH. ¹⁸O labeled alcohol formed 20 to 30 times faster (isotope exchange) than alkene. Write mechanism steps consistent with these results. What can be concluded about relative rates of the steps?

j. Problem 18. Write balanced reactions (not mechanisms) showing how propene would be made from:
1. CH₃CH₂CH₂OH

2. CH₃CHOHCH₃

3. CH₃CHClCH₃

4. CH₃CH₂CH₂OTs

5. 1,2-dibromopropane

k. Problem 19. Which alcohol in each pair would be easier to dehydrate?
1. CH₃CH₂CH₂CH₂CH₂OH or CH₃CH₂CH₂CH(OH)CH₃
2. (CH₃)₂C(OH)CH₂CH₃ or (CH₃)₂CHCH(OH)CH₃
3. (CH₃)₂CHC(OH)(CH₃)₂ or (CH₃)₂CHCH(CH₃)CH₂OH

l. 1,2-Shifts yield rearranged products (section 7.10.11).

$$\text{H}_3\text{CCH}_2\text{CHCH}_2-\overset{..}{\text{O}}: \;\; \overset{\text{H}-\overset{..}{\text{O}}-\text{SO}_3\text{H}}{\rightleftarrows} \;\; \text{H}_3\text{CCH}_2\overset{\oplus}{\text{C}}-\text{CH}_2-\overset{+}{\text{O}}\!\!\diagup\!\text{H}$$

(with CH₃ branch, H on central C)

Hofmann product:
$$\text{H}_3\text{C}-\underset{\text{H}_\beta}{\overset{\text{H}_\beta}{\text{C}}}-\underset{\text{CH}_2}{\overset{\text{CH}_3}{\text{C}}}$$

or

$$\text{H}_3\text{C}-\underset{\text{H}_\beta}{\overset{\text{H}_\beta}{\text{C}}}-\underset{\text{H}_\beta}{\overset{\text{CH}_3}{\overset{\oplus}{\text{C}}}}-\text{CH}_2 + \;\; :\!\overset{..}{\text{O}}\!\!\diagup\!\text{H}$$

rearranged Zaitsev:
$$\underset{\text{H}_3\text{C}}{\overset{\text{H}_\beta}{\text{C}}}=\underset{\text{CH}_3}{\overset{\text{CH}_3}{\text{C}}}$$

7.18.3. Because dehydration is reversible, an equilibrium can exist between two different alkene products and reactant R-OH at a given temperature. When this equilibrium exists, the more stable alkene product will predominate.

    a. However, at a lower temperature, a product alkene may not have enough thermal energy to overcome E_a for the reverse reaction, and may not be able to hydrate back to the alcohol R-OH. In this case the less stable product alkene can predominate over the more stable alkene if it forms faster (i.e., if the path to the less stable alkene has the lower E_a).

    b. Although such cases are known, they are not the norm. Usually the more stable product is also the one that forms the fastest.

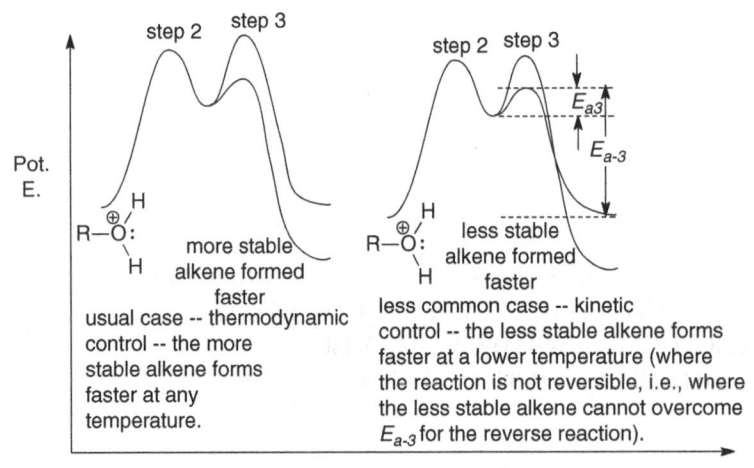

[Energy diagram 1: Pot. E. vs reaction coordinate; step 2 and step 3 barriers; starting from R-Ö(+)H-H; labeled "more stable alkene formed faster — usual case -- thermodynamic control -- the more stable alkene forms faster at any temperature."]

[Energy diagram 2: step 2 and step 3 barriers with E_{a3} and E_{a-3} indicated; "less stable alkene formed faster — less common case -- kinetic control -- the less stable alkene forms faster at a lower temperature (where the reaction is not reversible, i.e., where the less stable alkene cannot overcome E_{a-3} for the reverse reaction)."]

7.18.4. Alkene making methods from alkynes, aldehydes, and ketones will be learned in later chapters.

7.18.5. Biochemical hydration and dehydration reactions are common. Example:

sodium fumarate $\;\;\overset{\text{H}_2\text{O}}{\underset{37°\text{C}}{\overset{\text{fumarase enzyme}}{\rightleftarrows}}}\;\;$ sodium (S)-malate

CHAPTER 8

8.1.1. Alkene addition reaction (section 7.8.4) – two (or more) molecules combine to yield a single product molecule – opposite of elimination.

$$\underset{R_2}{\overset{R_1}{\diagdown}}C=C\underset{R_4}{\overset{R_3}{\diagup}} \xrightarrow{Y-Z} R_2-\underset{\underset{Y}{|}}{\overset{R_1}{\overset{|}{C}}}-\underset{\underset{Z}{|}}{\overset{R_3}{\overset{|}{C}}}-R_4 \;+\; \text{heat}$$

new σ-bonds

    a. σ-Electrons are more strongly attracted to a C nucleus than "exposed" (weakly Lewis basic) π-electrons.

8.1.2. Electrophile (*electron lover*) – opposite of nucleophile (section 2.13.1) – attracted to (negatively charged) electrons, i.e., a Lewis acid. Represented by general symbol Y^+ (or $Y^{\delta+}$ in some cases).

    a. Electrophile – electrically neutral or + charged ion, but never a – charged ion.
    b. Free radicals are electrophilic – seeking one electron.

8.2.1. Alkenes add dry H-X(g) (hydrohalogenation). Water is excluded to avoid competition with X^-. Dry (glacial) ethanoic (acetic) acid can be used as a moderately polar, weakly nucleophilic solvent (dissolves both alkene and H-X).

    a. Problem 1. What is the actual acid for H-Br(aqueous)? For H-Br(dry gas)? Which is the stronger acid – better able to transfer H^+ ion to an alkene?

8.2.2. Example: ethene and HI – electrophilic addition (EA) mechanism:

249

8.2.3. General electrophilic addition (EA) mechanism:

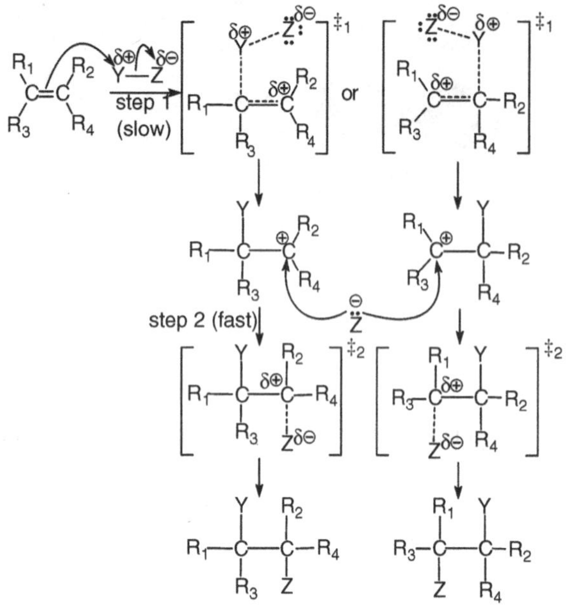

[mechanism diagram showing step 1 (slow) forming carbocation intermediate via transition state ‡1, then step 2 (fast) with Z⁻ attack via transition state ‡2, giving two possible regiochemical products with R₁,R₃ and R₂,R₄ substituents and Y, Z added across the former C=C]

8.2.4. EA features:
   a. Two elementary steps (C-Y and C-Z bond formation are not simultaneous).
   b. Reaction rate is first order in [alkene] and first order in electrophile [Y⁺], second order overall (bimolecular transition state).

$$\text{rate} = k[\text{alkene}][Y^+]$$

   c. Reaction occurs only under acidic conditions.
   d. Step 1 is rate determining; step 2 is relatively fast.
   e. Carbocation intermediate.
      1. Rearranged products form where possible.
      2. Products are racemic where chiral C atoms are created from achiral substrates.
   f. More substituted (with alkyl groups) sp^2 C=C atoms ⇒ faster reaction.
   g. C=C π-electron pair forms new C-Y σ-bond. Z:⁻ lone electron pair forms new C-Z bond to carbocation.
   h. Problem 2. The following gas phase "heat of protonation" (ΔH^o_{rxn}) values have been measured:

Reaction	ΔH^o_{rxn} (kJ/mol)
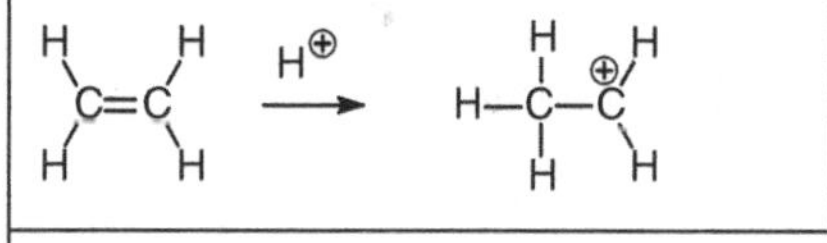 H₂C=CH₂ + H⁺ → H₃C–CH₂⁺	−672
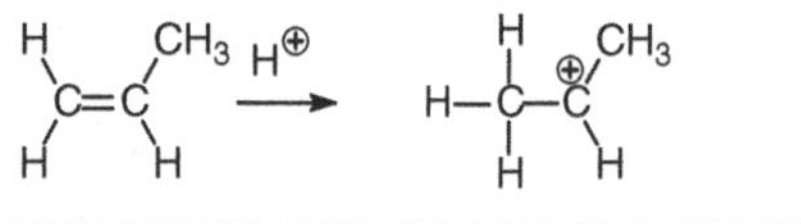 H₂C=CH(CH₃) + H⁺ → H₃C–CH⁺(CH₃)	−755
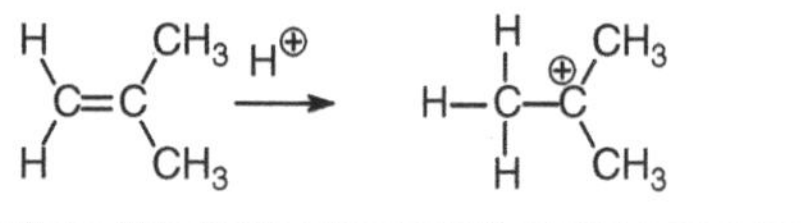 H₂C=C(CH₃)₂ + H⁺ → H₃C–C⁺(CH₃)₂	−810

How do these energy differences compare with those from alkyl halides (unimolecular conditions)? Fill in the table.

Standard	Energy difference (kJ/mol) Et⁺ – iPr⁺	Energy difference (kJ/mol) Et⁺ – t-Bu⁺
R-Cl	87.7	142
R-Br	83.7	146
R-I	83.7	151
alkenes	?	?

Make a general statement concerning carbocation stability relative to reactant.

8.3.1. V. Markovnikov's rule (1869) – during alkene H-X addition, H⁺ ion bonds to whichever C atom already has the most H atoms. Examples:
least substituted

a. *Regiospecific reaction* – yields only one of two (or more) possible isomeric products.
  1. *Regioselective reaction* – all possible isomeric products form detectably, but one (or more) product(s) is (are) preferred over others.

b. Examples:

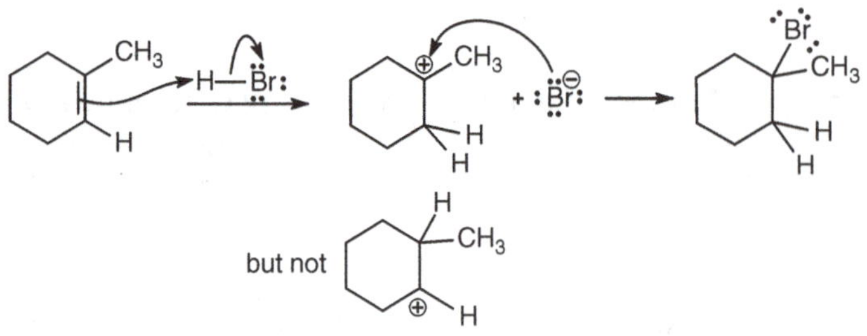

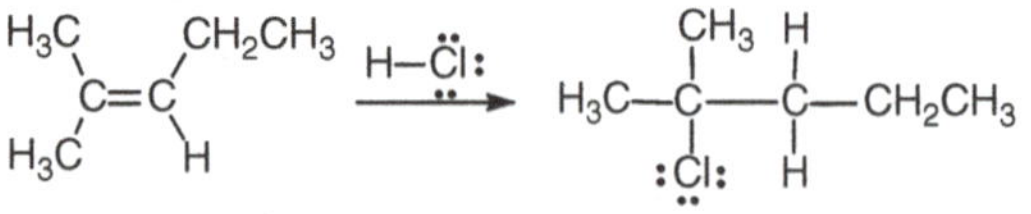

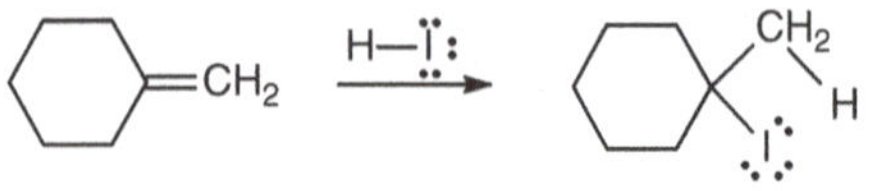

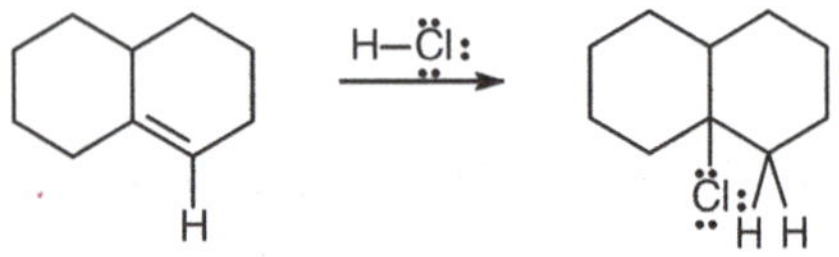

[Mechanism diagrams showing electrophilic addition of H-Br to 1-methylcyclohexene giving 1-bromo-1-methylcyclohexane via tertiary carbocation, "but not" the secondary carbocation alternative.]

[2-methyl-2-pentene + H-Cl → 2-chloro-2-methylpentane (CH₃ and Cl on same carbon)]

[Methylenecyclohexane + H-I → 1-iodo-1-methylcyclohexane]

[Octahydronaphthalene alkene + H-Cl → chloro-decalin product]

8.3.2. Markovnikov's rule (restated) – more stable (substituted) carbocation intermediate determines electrophilic attack orientation.

  a. + Charge delocalization by adjacent alkyl groups is the main factor affecting carbocation stability (3° > 2° > 1° > methyl).
  1. Fundamental point – carbocation stability controls orientation and reactivity in many mechanisms.

  b. Alkene relative EA reactivity parallels carbocation stability.

  c. Problem 3. H-Cl(dry gas) addition to 3-methylbut-1-ene yields a mixture of two alkyl chlorides. Use electron pushing arrows and draw a mechanism that explains these product's formation. Include all elementary steps (collisions), intermediates, formal electrical charges, and involved lone pair electrons. Which one should be formed in higher amount?

  d. Problem 4. Which alkene in each pair would react faster with H_2SO_4(dil)/H_2O ?
  1. ethene or propene
  2. propene or but-2-ene
  3. ethene or bromoethene
  4. but-2-ene or 2-methylpropene
  5. chloroethene or 1,2-dichloroethene
  6. pent-1-ene or 2-methylbut-1-ene
  7. ethene or prop-2-en-1-oic acid ($H_2C=CHCOOH$)
  8. propene or 3,3,3-trifluoropropene

e. **Problem 5.** Draw and name the main product(s) expected from HI(dry gas) reaction with:

1. but-2-ene

2. pent-2-ene

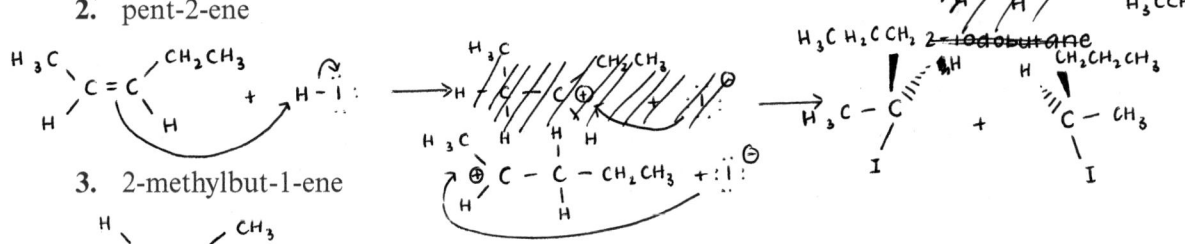

3. 2-methylbut-1-ene

4. 2-methylbut-2-ene

5. 3-methylbut-1-ene (2 products)

6. bromoethene

7. 2,3-dimethylbut-1-ene

8. 2,2,4-trimethylpent-2-ene

f. **Problem 6.** HCl(dry gas) addition to 3,3-dimethylbut-1-ene yields a mixture of 3-chloro-2,2-dimethylbutane and 2-chloro-2,3-dimethylbutane. Write mechanisms that explain how each product forms. Include all elementary steps (collisions), intermediates, formal electrical charges, and involved lone pair electrons.

8.3.3. Alkenes always yield Markovnikov products from HCl, HI, H_2SO_4, and H_2O addition. Puzzle: up to 1933, HBr addition sometimes yielded Markovnikov and sometimes anti-Markovnikov products, depending on reaction conditions.

   a. M. S. Kharasch and F. R. Mayo (1933) found that peroxide traces cause anti-Markovnikov H-Br orientation.
   1. Peroxides – often minor organic compound impurities, formed by slow reaction with O_2.

8.3.4. H-Br addition observations:
   a. Very small peroxide traces $\Rightarrow$ some Markovnikov and some anti-Markovnikov addition.
   b. UV light irradiation (known to homolyze HBr) and peroxide absence $\Rightarrow$ anti-Markovnikov orientation.
   c. Free radical inhibitor addition $\Rightarrow$ Markovnikov addition (section 4.15.1).

8.3.5. anti-Markovnikov H-Br addition mechanism.
   a. Initiation (section 4.15.2).
   b. Propagation: ($\Delta H^°_{rxn}$'s calculated from homolytic bond dissociation energies, section 4.6. in Chapter 1)

   1. $(CH_3)_3C-O\cdot + H-Br: \longrightarrow (CH_3)_3C-O-H + :Br\cdot$ $\Delta H^°_{rxn}$ (kJ/mol): -63

   2. $H_3C-CH=CH_2 + \cdot Br: \longrightarrow$
      - $H_3C-CHBr-CH_2\cdot$ (1°, slow)
      - $H_3C-CH(\cdot)-CH_2Br$ (2°) $\Delta H^°_{rxn}$ (kJ/mol): -12

   3. $H_3C-\overset{\cdot}{C}H-CH_2Br + H-Br: \longrightarrow H_3C-CH_2-CH_2Br + \cdot Br:$ (recycles in step (2)) $\Delta H^°_{rxn}$ (kJ/mol): -25

   c. Termination:

   4. $:Br\cdot + \cdot Br: + A \longrightarrow :Br:Br: + A^\star$

   5. $H_3C-\overset{\cdot}{C}H-CH_2Br + \cdot Br: \longrightarrow H_3C-CHBr-CH_2Br$

   1. A is an inert specie (such as Ar gas) that dissipates energy.

8.3.6. Electrophilic Br· regiospecifically attacks π-bond so that (when possible) the more stable carbon free radical forms.
   a. Factors stabilizing the free radical (unpaired electron delocalization by alkyl groups, section 4.13.35) also stabilize the transition state leading to the free radical $\Rightarrow$ lower $E_a \Rightarrow$ faster reaction. A 2° R· forms faster than a 1° R·.

b. Another factor (besides unpaired electron delocalization, a polar factor) is probably involved. Transition state structure central C atom has not only free radical character but also carbocation character. δ+ electrical charge delocalization provides additional stabilization.

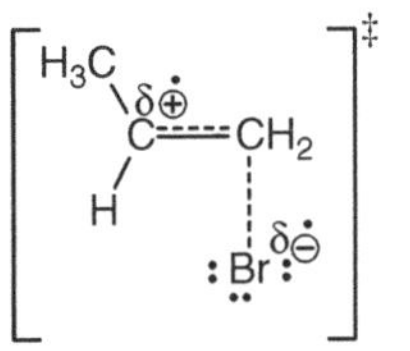

$$\left[\begin{array}{c} H_3C \\ \overset{\delta\oplus}{C} ===CH_2 \\ H \vdots \\ :Br:^{\delta\ominus} \end{array} \right]^{\ddagger}$$

c. A 1° vinyl C atom is less sterically hindered than a 2° vinyl C towards attacking Br· ⇒ same preference.

d. Factors a-c are all thought to operate simultaneously, except in extreme cases (very stable carbon radicals, very bulky attacking radicals (·CBr₃), or very electrophilic radicals (·CF₃)).

8.3.7. Steps (2) and (3) are *both* exothermic only for the H-Br reaction. An unfavorable endothermic step would be involved for H-Cl and H-I.

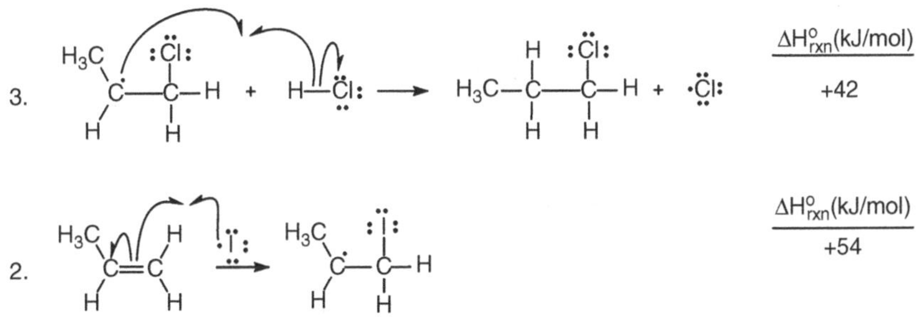

3. [mechanism with ΔH°_rxn (kJ/mol) = +42]

2. [mechanism with ΔH°_rxn (kJ/mol) = +54]

a. Problem 7. Write mechanisms consistent with the following reactions: Include all elementary steps (collisions), intermediates, formal electrical charges, and involved lone pair and unpaired electrons.

a. CH₂=CH–CH₂(CH₂)₄CH₃ + CHCl₃ —peroxide (trace)→ product

b. CH₂=CH–CH₂(CH₂)₄CH₃ + CBrCl₃ —peroxide (trace)→ product

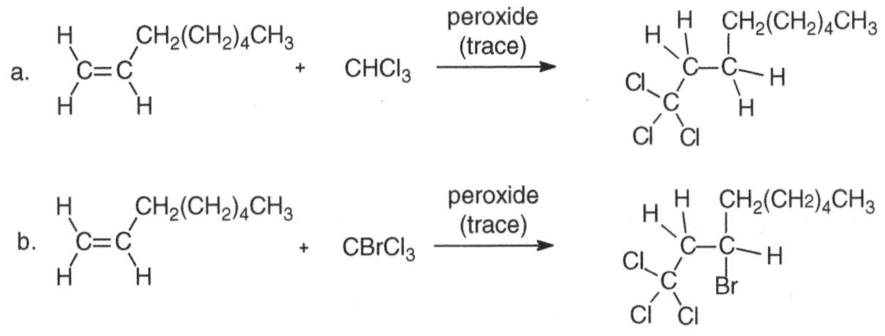

b. Problem 8. A Cl₂ / Cl₂C=CCl₂ solution is stable for long periods with no reaction at 25°C in the dark. When the mixture is irradiated with ultraviolet light, Cl₂ is rapidly consumed yielding hexachloroethane. Many product molecules form for each light photon absorbed. Reaction slows markedly when O₂ is bubbled through the solution. Write a mechanism that accounts for these facts, including the O₂ effect. Include all elementary steps (collisions), intermediates, formal electrical charges, and involved lone pair and unpaired electrons.

c. Problem 9. a. Equally consistent with the Kharasch / Mayo mechanism is the following:

1. $(H_3C)_3C-\ddot{O}-\ddot{O}-C(CH_3)_3 \longrightarrow 2\ (H_3C)_3C-\ddot{O}\cdot$

2. $(H_3C)_3C-\ddot{O}\cdot\ +\ :\ddot{Br}-H \longrightarrow (H_3C)_3C-\ddot{O}-\ddot{Br}:\ +\ \dot{H}$

3. $\dot{H}\ +\ C=C \longrightarrow -\overset{|}{\underset{H}{C}}-\overset{|}{\underset{|}{C}}-$

4. $-\overset{|}{\underset{H}{C}}-\overset{|}{\underset{|}{C}}-\ +\ :\ddot{Br}-H \longrightarrow -\overset{|}{\underset{H}{C}}-\overset{|}{\underset{:\ddot{Br}:}{C}}-\ +\ \dot{H}$ recycles

Assuming that the alkene is ethene and primary (1°) C atoms, bond homolysis energies are:

Bond	$\Delta H°_{rxn}$ (kJ/mol)	Bond	$\Delta H°_{rxn}$ (kJ/mol)
H-Br	366	C=C (π bond)	321
R-O-Br*	234	t-bu-O-O-t-bu**	159
1° C-H	423	1° C-Br	303

*Value from CRC Handbook for HO-Br. **Value from CRC Handbook.

Use Hess' Law to calculate $\Delta H°_{rxn}$(kJ/mol) for each step.

step 2 $\Delta H°_{rxn}$ (kJ/mol):
step 3 $\Delta H°_{rxn}$ (kJ/mol):
step 4 $\Delta H°_{rxn}$ (kJ/mol):

b. Suggest two reasons why this mechanism is not likely.

c. Electron spin resonance spectroscopy (esr) showed that the intermediate alkyl free radical is the same whether HBr or DBr is the reactant. How does this evidence confirm the Kharasch / Mayo mechanism?

8.4.1. Markovnikov alcohols are often made in the laboratory by acid catalyzed electrophilic water / alkene addition (EA).

8.4.2. Alkene hydration – reverse of dehydration.

    a. *Microscopic reversibility principle* – at equilibrium, a reverse reaction follows the same pathway (goes through the same transition state structures and intermediates) as the forward reaction. In other words: at equilibrium, a reverse reaction mechanism is the same as for the forward reaction except that steps occur in reverse order. Hydration – reverse dehydration. Example:

    b. Review question. How can reactant and product equilibrium concentrations be controlled? (See section 7.18.2d.)

    c. Problem 10. D_2O addition to 2-methylbut-2-ene (catalyzed by adding a trace of D_2SO_4) yielded the expected alcohol $(H_3C)_2C(OD)CHDCH_3$. When reaction was half complete, it was interrupted and unconsumed alkene was isolated. Mass spectrometric analysis showed the alkene contained almost no deuterium. What, specifically, does this show about the mechanism?

    d. Problem 11a. Either 2-methylbut-1-ene or 2-methylbut-2-ene hydration yields the same alcohol product. What is the alcohol structure? Use electron pushing arrows to write mechanism steps for these reactions. Include all elementary steps (collisions), intermediate products, formal electrical charges, and involved lone pair electrons.

e. **Problem 11b.** Each alkene was allowed to react *separately* with aqueous (dilute) H_2SO_4. When hydration was about half complete, reaction was interrupted and unconsumed alkene was recovered. In each case, only the original alkene was recovered, uncontaminated by the other isomer. Which is faster – carbocation attack by water (nucleophile), or carbocation deprotonation (return to alkene reactant)?

8.4.3. Rearranged alcohol products are consistent with carbocation intermediates (sections 6.15.1 and 7.10.11). Example:

8.4.4. Review: more stable carbocation ⇒ more stable transition state leading to its formation ⇒ lower E_a ⇒ faster carbocation formation (and overall reaction).

8.5.1. Hydration does not always give good alcohol product yields, especially when alkenes yield 1° or 2° carbocations.

8.5.2. Problems: insolubility, charring, rearrangements, and polymerizations.

8.5.3. Regioselective oxymercuration-demercuration gives high Markovnikov alcohol yields (> 90%), and some alcohols can be made that are impossible by acid catalyzed hydration.

    a. Reaction is done in a single flask (20 sec – 10 min), with H_2O/THF solvent. $NaBH_4$ reduction is done in basic solution (with OH^-) in about an hour without isolating intermediates. The oxymercuration mechanism is believed to be an EA variation:

CHAPTER 8   259

[Reaction mechanism scheme showing oxymercuration-demercuration of 2-methylpropene (isobutylene) with Hg(OAc)₂ in THF/H₂O followed by NaBH₄ reduction, giving the alcohol in 90% overall yield, plus acetic acid, H₃BO₃, Hg, Na⁺, and HSO₄⁻.]

b. <mark>Electrophile – Hg(OAc)⁺ ion forms a cyclic mercurinium ion</mark> (instead of a carbocation). OAc is an abbreviation for acetate (ethanoate) anion. *Caution:* mercury compounds are toxic (brain and kidney damage), and must be disposed properly. Mercury bonds to crucial enzyme thiol (-S-H) groups, inactivating them.
   1. Olah (1971) observed mercurinium ions spectroscopically.

c. <mark>Rearrangements usually do not occur ⇒ a cyclic mercurinium ion intermediate structure rather than a nonring carbocation.</mark>
   1. Example: 3,3-dimethylbutan-2-ol can be made by oxymercuration/demercuration, but not by acid catalyzed hydration because the carbocation intermediate would rearrange.

   [Reaction: 3,3-dimethyl-1-butene + Hg(OAc)₂ in H₂O/THF, then NaBH₄, then H₂SO₄(dil)/H₂O/25°C → 3,3-dimethylbutan-2-ol as the only product (no rearrangement).]

   —OAc = —O—C(=O)—CH₃ (acetate)

   2. Calculations show + charge mostly on Hg, very little on C.
   3. NaBH₄ reduction mechanism is not well understood – possibly involves free radicals.

4. **Oxymercuration / demercuration is a stereospecific *anti*-addition.** Example: only (Z)-products form.

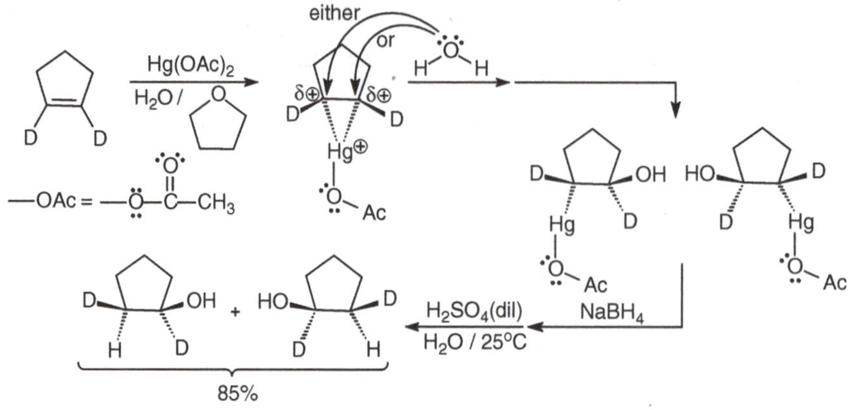

[Reaction scheme: cyclopentene-d2 + Hg(OAc)₂, H₂O/THF → mercurinium ion intermediate → water attacks either face → organomercury alcohol intermediates → NaBH₄, H₂SO₄(dil), H₂O/25°C → trans-diol products, 85%]

8.6.1. Alkoxymercuration – ether products form. Markovnikov orientation.

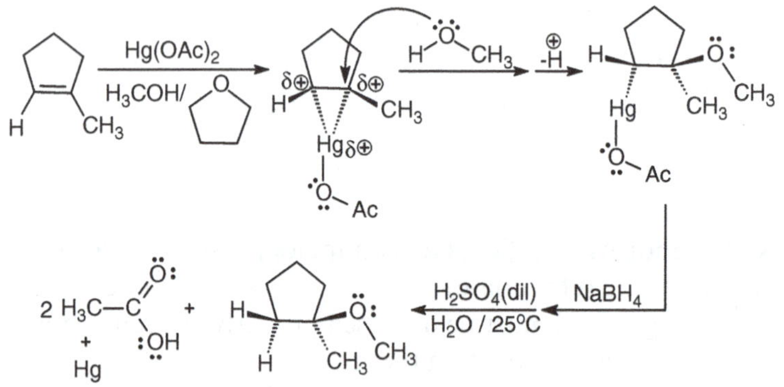

[Reaction scheme: methylcyclopentene + Hg(OAc)₂, H₃COH/THF → mercurinium ion → methanol attacks → organomercury methyl ether → NaBH₄, H₂SO₄(dil), H₂O/25°C → methyl ether product + 2 H₃C–C(=O)–OH + Hg]

a. **Problem 12.** What product forms when mercuric acetate reacts with propene in methanol solvent followed by reduction with NaBH₄?

8.7.1. We know only one anti-Markovnikov alkene addition (HBr with peroxides, section 8.3.3). Borane (BH₃) also adds mostly anti-Markovnikov, regioselectively, (H. C. Brown, Nobel Prize, 1979). The trialkylborane product can be efficiently oxidized to an alcohol.

a. Borane is a Lewis acid and electrophile (B has only six valence electrons). Diborane (and alkylboranes) are pyrophoric (green flame). Borane and diborane structures:

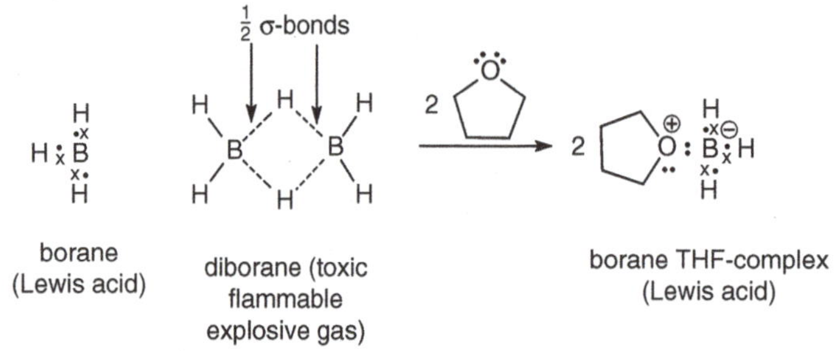

[Structures: borane BH₃ (Lewis acid); diborane B₂H₆ with ½ σ-bonds (toxic flammable explosive gas); + 2 THF → 2 borane THF-complex (Lewis acid)]

b. The solvent is a dry ether (often THF or $H_3COCH_2CH_2OCH_2CH_2OCH_3$ (diglyme)) which forms an acid/base complex that helps the borane dissolve.
c. H_2O_2/KOH oxidation is done without isolating the intermediate products. A chiral R-group configuration is retained during the oxidation.
d. Example:

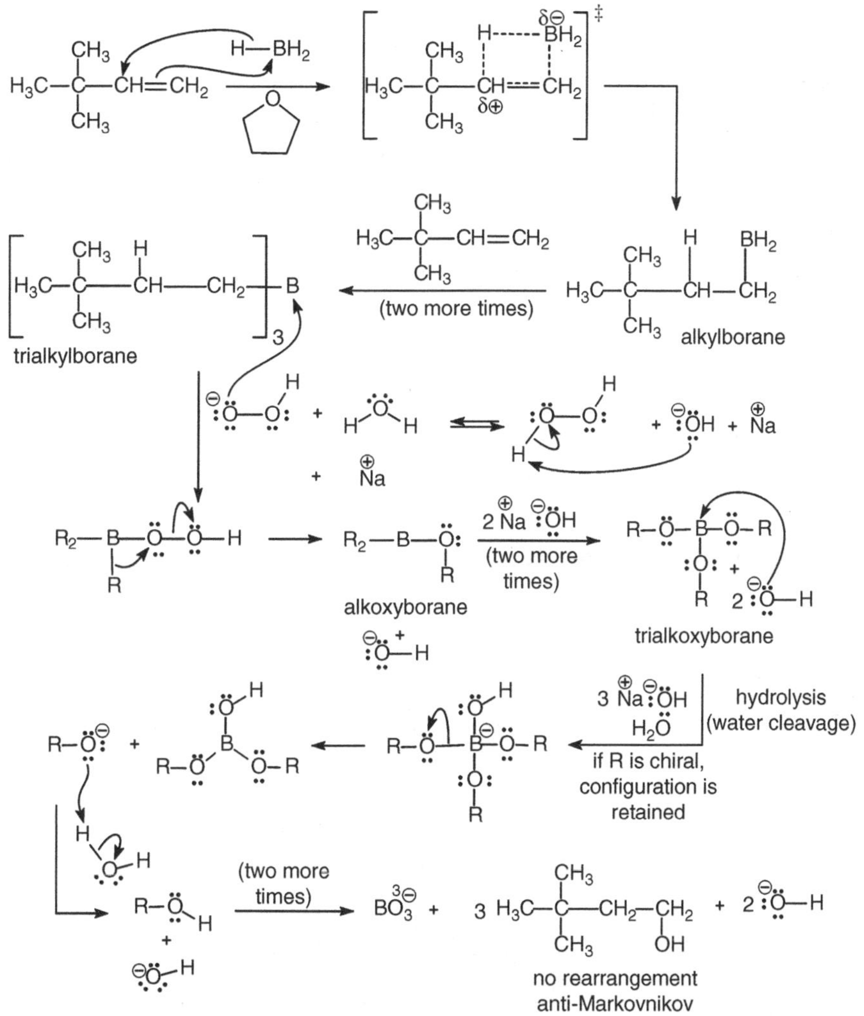

[reaction scheme showing hydroboration-oxidation of 3,3-dimethyl-1-butene with BH_3 in THF, forming alkylborane, then trialkylborane, followed by oxidation with H_2O_2/NaOH through alkoxyborane and trialkoxyborane intermediates, hydrolysis with water cleavage; if R is chiral, configuration is retained; final products: BO_3^{3-} + 3 $H_3C-C(CH_3)_2-CH_2-CH_2-OH$ + 2 $\overset{\ominus}{:}\ddot{O}-H$; no rearrangement, anti-Markovnikov]

e. Problem 13. Label Lewis acids, bases and spectator ions in the reactions below. Emphasize Lewis definitions by drawing electron pushing arrows.

1.  $\overset{\oplus}{Li}$ + $\overset{\ominus}{H:}$ + $H-\ddot{O}-H$ $\longrightarrow$ $\overset{\oplus}{Li}$ + $:\ddot{O}^{\ominus}-H$ + H—H

2. $(H_3CCH_2)_3B$ + $H-\overset{H}{\underset{H}{\ddot{N}}}-H$ $\longrightarrow$ $(H_3CCH_2)_3\overset{\ominus}{B}-\overset{H}{\underset{H}{\overset{\oplus}{N}}}-H$

3. $H-\overset{H}{\underset{H}{B}}$ + $H_3C-\overset{CH_3}{\underset{CH_3}{\ddot{N}}}-CH_3$ $\longrightarrow$ $H-\overset{H}{\underset{H}{\overset{\ominus}{B}}}-\overset{CH_3}{\underset{CH_3}{\overset{\oplus}{N}}}-CH_3$

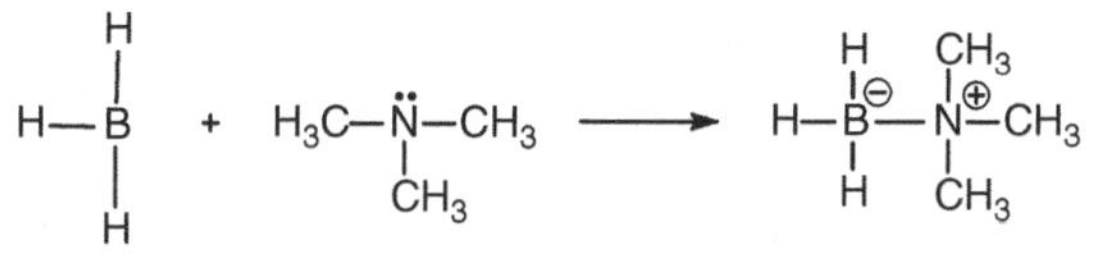

4. 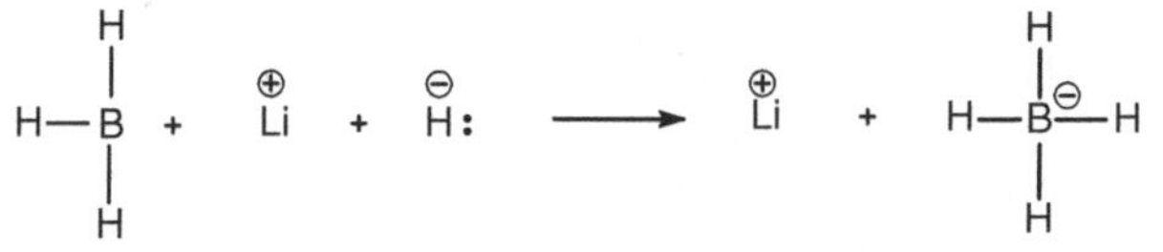 $H-\overset{H}{\underset{H}{B}}$ + $\overset{\oplus}{Li}$ + $\overset{\ominus}{H:}$ $\longrightarrow$ $\overset{\oplus}{Li}$ + $H-\overset{H}{\underset{H}{\overset{\ominus}{B}}}-H$

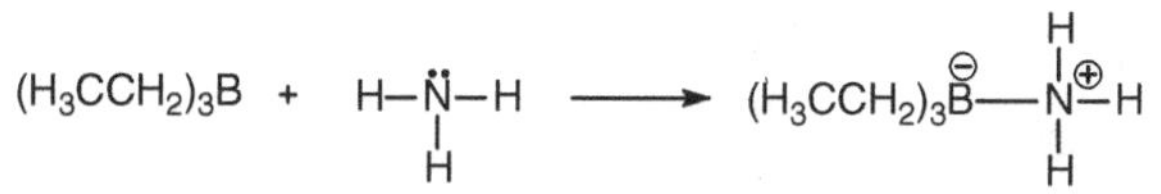

8.7.2. No free carbocation exists ⇒ no carbocation rearrangement. A C-B bond forms simultaneously as a B-H bond breaks. That same H bonds to δ+ C.

8.7.3. The transition state C-B bond is more fully formed than the C-H bond; a δ- exists on B and a δ+ on C. The δ+ charge is on the more substituted C atom (that can best delocalize it) ⇒ anti-Markovnikov orientation. The H atom adds to C as if it were an H:⁻ ion. Orientation is retained during H_2O_2 oxidation.

  a. BH$_3$ (or BH$_2$R or BHRR') adds to the less sterically crowded C atom ⇒ anti-Markovnikov orientation also.
  b. Some alcohols cannot be made from alkenes any other way.

$$H_3CCH_2CH_2CH_2CH=CH_2 \xrightarrow{H_3B:O(THF)} \xrightarrow{H_2O_2/KOH} H_3CCH_2CH_2CH_2\underset{H}{C}H-\underset{OH}{C}H_2 \quad 90\%$$

$$H_3CCH_2-\underset{\underset{1\%}{\uparrow}}{C}(CH_3)=\underset{\underset{99\%}{\uparrow}}{C}H_2 \qquad H_3C\underset{\underset{98\%}{\uparrow}}{C}H=\underset{\underset{2\%}{\uparrow}}{C}(CH_3)-CH_3$$

  c. *Syn*-addition – both boron and H add from the same face:

(cyclopentene-CH₃ + BH₃/THF dry → two syn-addition organoborane products (3:1 B substituted); then H₂O₂/KOH → two alcohol products, 86% overall)

  1. Stereospecific reaction – A (Z)-alkene isomer yields only the (R,R) and (S,S) pair, and no (R,S) or (S,R).

  d. Other substitutions (besides -OH) can be made from organoboranes. Heating with acetic acid replaces boron with an H atom (a reduction, section 8.14.4).

$$R-B\begin{pmatrix}\\\end{pmatrix} \xrightarrow[\text{heat}]{H_3CCOOH} R-H + H_3CCO-B\begin{pmatrix}\\\end{pmatrix}$$

e. **Problem 14.** Show which alkene(s) would be used to make each alcohol by either oxymercuration / demercuration or by hydroboration / oxidation. Fill in the table.

Alcohol	Oxymercuration / demercuration alkene	Hydroboration oxidation alkene
H₃C(CH₂)₃CH₂OH		
H₃CCH₂CH₂CHCH₃ OH		
H₃CCH₂CHCH₂CH₃ OH		
CH₃ H₃CCH₂CHCH₂OH		
CH₃ H₃CCH₂CCH₃ OH		

f. **Problem 15.** What products would be expected from 1,2-dimethylcyclopentene hydroboration/oxidation? Draw three dimensional (perspective) structures where needed.

1,2-dimethylcyclopentene + (H₃CCH₂)₂O : BH₃ / dry → H₂O₂, KOH / H₂O → ? + ?

(perspective drawings)

8.8.1. Alkenes add Cl₂ or Br₂ to form vicinal (1,2-) dihalides, an oxidation reaction (section 8.14.4). I₂ can be used, but vicinal diiodides easily decompose.

a. Although each X₂ X atom has eight electrons, X₂ is electrophilic. Examples:

H–Ö:⁻ + :Br–Br: ⟶ [H–Ö------Br------Br:]‡ ⟶ H–Ö–Br: + :Br:⁻

H₃N: + :Cl–Cl: ⟶ [H₃N------Cl------Cl:]‡ ⟶ H₃N⁺–Cl: + :Cl:⁻

8.8.2. Alkene halogenation is <u>not</u> a free radical reaction when done at 0 to 25°C in the dark (free radicals do not form).

H₃C\C=C/H + :Br–Br: —hexane, H₂CCl₂, HCCl₃ or CCl₄ (solvents)→ H₃C–C(Br)–C(Br)H–H
H/ \H   red/brown   colorless

a. An alkene will decolorize red/brown Br₂ within 1 minute at 25°C in the dark– a simple alkene test.

8.8.3. X$_2$ / alkene addition is an EA variation – a cyclic halonium ion forms after step 1 rather than a carbocation.

   a. X$^+$ equivalent (Lewis acid) adds to an alkene (Lewis base) in step 1 and forms a new σ-bond to each C atom.
   b. Alternative view – nucleophilic π-electrons attack one X atom and the other X atom simultaneously leaves (as an X$^-$ ion, a good leaving group). Alkene π-electrons polarize the X-X σ-bond as they approach. This step can be viewed as an S$_N$2 reaction at X (instead of C).

8.8.4. Evidence for Br$_2$ addition mechanism:
   a. More substituted alkene (with alkyl groups, section 7.8.6) ⇒ more reactive towards Br$_2$ ⇒ step 1 is an <u>electrophilic</u> attack.
   b. Other added nucleophiles compete with Br$^-$ in step 2 ⇒ a two step process and bromonium ion intermediate.

   1. Aqueous NaCl does not react with ethene unless Br$_2$ is added.

8.9.1. With aqueous Cl$_2$, Br$_2$, or I$_2$, chloro-, bromo-, or iodohydrins can be the major products.
   a. H$_2$O adds to whichever C is more electrophilic (has the larger δ+ charge), and also has the weaker C-X bond. Although ring opening is written below as an S$_N$2 reaction, it has some S$_N$1 character.

b. **Problem 16.** Ethene reacts with bromine (Br$_2$) in methanol solvent yielding not only 1,2-dibromoethane, but another product whose formula is C$_3$H$_7$OBr. *a.* Draw the product structure and a mechanism like the one above showing how it forms. *b.* What would be the analogous product with propene?

c. **Problem 17.** 3-Bromopropene reacts with Br$_2$ / H$_2$O yielding chiefly (80%) 2,3-dibromopropanol (1° alcohol). On the other hand, propene reacts under the same conditions yielding 1-bromopropan-2-ol (2° alcohol). Explain why the first reaction yields mostly 1° alcohol, while the latter yields mostly 2° alcohol.

8.9.2. Reactions (such as alkene halogenation) often follow specific three dimensional pathways. Stereochemical pathway terminology review:

a. *Regioselective reaction* (section 8.3.1a1): bonds could be made or broken in two (or more) different orientations, but one (or more) orientation(s) is (are) preferred over others. See examples in sections below.

   8.5.1 – oxymercuration
   8.7.1 – hydroboration

b. *Regiospecific reaction* (section 8.3.1a): same as regioselective except preference for one orientation over another is 100%. No minor products form. See examples in the sections below.

   8.3.1a – Markovnikov addition
   8.3.5 – anti-Markovnikov H-Br addition

8.9.3. Definitions below are not always used consistently.
  a. *Stereoselective reaction*: one (or a few of several possible) stereoisomer(s) is (are) formed or destroyed in considerable preference to other stereoisomers that might have been formed or destroyed. That is, a preference for one (or a few of several possible) diasteriomer(s) is (are) observed *on one side* of a reaction.
  b. *Stereospecific reaction*: one particular stereoisomer yields only one (or a few of several possible) stereoisomeric product(s). Another different stereoisomer yields (exclusively) a different (or a few different) diasteriomeric product(s). That is, preferences for particular stereoisomers are observed *on both sides* of a reaction.
   1. Stereospecific reactions are fairly common. See examples in the sections below.
      4.16.7a6a – carbene addition
      5.16.10 – free radical substitution
      6.12.2 – S_N2
      7.14.3 – E2
      8.7.4 – hydroboration
      8.10.2b (Chapt. 7) – hydrogenation
      8.9.1 – bromination)
  c. All stereospecific reactions must also be stereoselective, but the converse is not always true – reactions can be stereoselective but not stereospecific.

8.9.4. Stereochemical analysis provides more evidence for the X_2 addition mechanism.

8.9.5. Example: Br_2 / *cis*-but-2-ene addition creates two new chiral C atoms.
  a. Step 1 (rate determining) – either top or bottom face electrophilic attack (EA) forms a cyclic bromonium ion (first proposed by I. Roberts and G. E. Kimball, 1937).
  b. Step 2 (fast) – backside (S_N2-like) bromonium ion attack by Br^- inverts C.
   1. Example: Br_2 reacts stereospecifically with *cis*-diasteriomer yielding only (*R,R*) and (*S,S*) product enantiomers, but no (*R,S*) diasteriomer. *Trans*-but-2-ene + Br_2 reaction is also stereospecific, yielding only (*R,S*) diasteriomer and no (*R,R*) or (*S,S*).

c. Problem 18a. What is the relationship between the two bromonium ions formed by Br⁺ attachment to top and bottom *trans*-but-2-ene faces? In what relative proportions are they formed? *b.* Answer the *a.* questions for *cis*-but-2-ene. *c.* Answer the questions for *trans*-pent-2-ene and, *d. cis*-pent-2-ene.

d. Problem 19a. Predict the two bromine addition products for *trans*-pent-2-ene. Are the two bromide ion attack pathways equally likely? Why are optically inactive products obtained? *b.* Answer the *a.* questions for *cis*-pent-2-ene.

e. Example: alkyne π-bonds add Br_2 similarly as alkenes (section 9.9). Br_2 addition to propyne (which has no stereoisomers) yields more *trans* than *cis* product. Reaction is stereoselective, but not stereospecific.

H—C≡C—CH₃ propyne (no stereoisomers) → Br₂ (one mol) / hexane (solvent) 0°C / dark → (E)- or *trans*- more + (Z)- or *cis*- less

1. Unless a reactant and a product both have stereoisomers, a reaction cannot be stereospecific.

f. *Syn*-addition – both groups (Y and Z) add from the same molecular side (face) – can cause stereospecificity. Example: hydrogenation (section 8.10.2b in Chapter 7).

but no (R,S)

g. *Anti*-addition – two atoms add from opposite molecular faces – can cause stereospecificity. Example: alkene bromination – one Br atom adds from top face and other Br atom adds from bottom face.
   1. Cl_2 and Br_2 / alkene additions are nearly always *anti*.
   2. *Anti*-stereospecificity is consistent with a cyclic halonium ion intermediate.
   3. If both *syn*- and *anti*- addition occurred simultaneously (from an open carbocation), all three products, (R,R), (S,S) and (R,S), would be expected from either *cis*- or *trans*-reactant. If these three products formed in equal amounts, reaction would be neither stereospecific nor stereoselective. Even if *syn*- or *anti*-addition occurs exclusively, rotation about an open carbocation C-C bond would yield all three stereoisomers (contrary to facts).

4. Problem 20a. 2-Chlorobutane E2 dehydrohalogenation yields but-2-enes in the ratio 6 (*trans*) : 1 (*cis*). Suggest reason(s) for this moderate stereoselectivity. Assume that the higher yield product (*trans*) forms faster – that E_{act} is the controlling factor and reactant conformations interconvert rapidly at the reaction temperature. b. 3-Bromobutane E2 dehydrohalogenation yields a ratio 3 (*trans*) : 1 (*cis*). Why is the *trans* preference smaller for this reaction?

h. Problem 21a. Cl_2 / H_2O addition to *cis*-but-2-ene yields only *threo* chlorohydrin, while *trans*-but-2-ene yields only *erythro* chlorohydrin. Assuming these results are typical (they are), what is the halohydrin formation stereochemistry (*syn* or *anti*)? b. Use electron pushing arrows to write a mechanism for the *threo* product formation. Include all elementary steps (collisions), intermediates, formal electrical charges, and involved lone electron pairs. Do not show transitions state structures. c. Which step leads to product racemization?

8.9.6. Further halonium ion evidence.

a. Georg Olah (1967) – direct nuclear magnetic resonance (NMR, Chapter 13) spectroscopic observation in superacid.

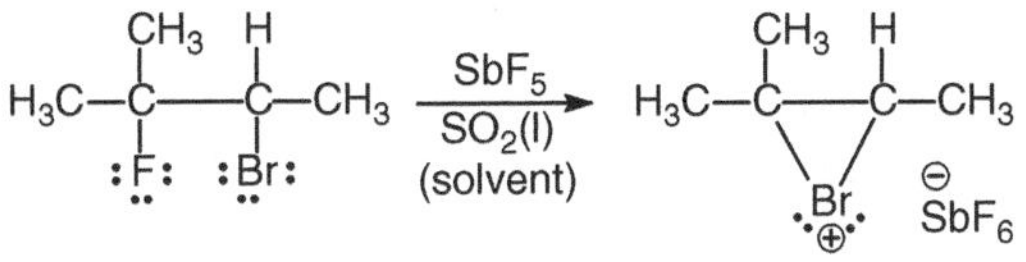

b. Problem 22. Olah observed halonium ions form by the above reaction when X = I, Br, and Cl, but a carbocation forms when X = F. Suggest a reason why F forms a carbocation.

8.9.7. Why is a halonium ion preferred over a nonring carbocation? Every atom except H obeys octet rule.

a. The advantage of having eight electrons around C outweighs the disadvantage of having a + charge on halogen.

8.9.8. *Anti*-addition exception – strongly resonance stabilized (benzylic) carbocation.

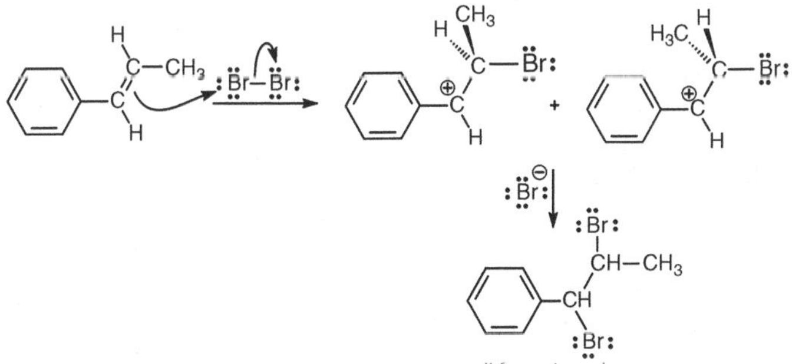

all four stereoisomers

8.10. Catalytic hydrogenation was covered in sections 7.3.8, 7.8.3, and 8.10.1 in Chapter 7.

8.11. Carbenes were covered in section 4.16.6.

8.12.1. Oxiranes (epoxides) – similar to halonium ions, but more stable – undergo similar reactions (*anti*-nucleophilic attack). Example:

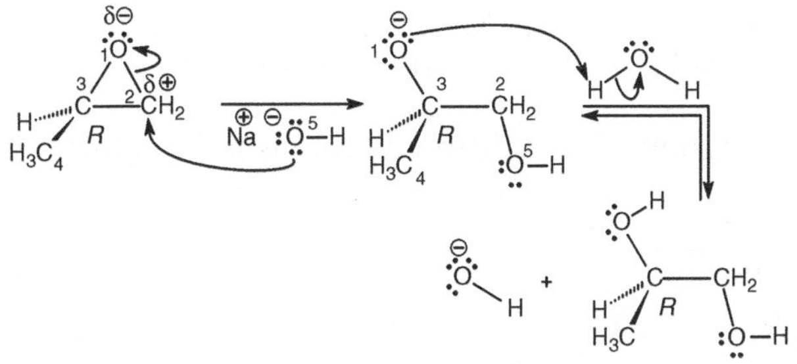

8.12.2. Oxirane – three membered ring ether, but unusually reactive due to angle strain – treated separately from ether functional group – synthesis substrate – biological defense product.

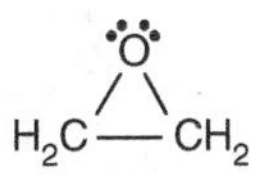

oxirane
epoxide
ethylene oxide

8.12.3. Oxirane preparations:
   a. 1,2-halohydrins.

   b. Peroxycarboxylic acid alkene oxidation (section 8.14.4).

   1. A peroxycarboxylic acid can be made by mixing a carboxylic acid (RCOOH) and hydrogen peroxide (H_2O_2), an oxidizing agent.

8.12.4. Alkene oxidization – concerted, stereospecific *syn*-addition mechanism. Example:

8.12.5. Oxiranes can be isolated from nonaqueous solvents, but the ring hydrolyzes (cleaves in water, section 2.16.5) in H$_2$O solvent, yielding a 1,2-diol.

8.13.1. Acid catalyzed oxirane ring opening (hydrolysis) is much faster than ordinary ether C-O cleavage. Inversion and ^{18}O isotope data suggest an S$_N$2-like mechanism (no open carbocation), but with considerable S$_N$1 character (transition state carbocation character).

a. H$_2$^{18}O nucleophile attacks the more sterically hindered 2° C atom with larger δ+ charge. Electrical attraction outweighs steric hindrance.

b. Ring opening is stereospecific (exclusive *anti*-OH group addition).

and no (R,S) ≡ (S,R)

8.13.2. Unlike ordinary ethers, oxiranes are cleaved by strong bases via an S_N2 mechanism. Steric factors control orientation (normal for S_N2 reactions).

8.13.3. In industry, oxirane is anionically polymerized (section 8.16) to polyethylene glycols (useful solvents) and carbowaxes (used in gas chromatography).

8.13.4. Alkene groups can be oxidized in the body to oxiranes (epoxides) → diols via epoxide hydrolase enzyme. Diols are eliminated in urine.

8.14.1. Epoxidation / hydrolysis is a mild oxidation that cleaves a C=C π-bond but not the σ-bond, yielding a 1,2-diol.

8.14.2. Oxidation – an organic molecule loses electrons (to oxidizing agent). Reduction – an organic molecule gains electrons (from reducing agent).

8.14.3. Inorganic redox examples:

$$Mg^0 \rightarrow Mg^{2+} + 2e^- \text{ oxidation}$$

$$e^- + Fe^{3+} \rightarrow Fe^{2+} \text{ reduction}$$

$$2e^- + Ca^{2+} \rightarrow Ca^0 \text{ reduction}$$

$$2\,Zn + O_2 = 2\,ZnO \text{ redox reaction}$$

(Zn is oxidized, oxygen is reduced. Electrons lost by Zn = electrons gained by O.)

8.14.4. Oxidation number assignment rules:
  a. H atoms are usually +1.
  b. O atoms are usually –2.
  c. –4 (methane) ≤ C atoms ≤ +4 (CO_2) – assign values so that all oxidation numbers in a molecule add to zero. Examples:

[Lewis structure equation: CH₄ + 2 O₂ → CO₂ + 2 H₂O with oxidation numbers labeled]

[Lewis structure: ethylene oxidized to ethylene glycol with oxidation numbers labeled]

1. C atom oxidation number increases from reactant to product. Each C atom formally loses electron(s).

d. Counting electrons can be tedious since an organic product often has a different number of atoms than reactant. Organic chemists use simplified rules:
   1. During an *oxidation reaction* – an organic molecule gains oxygen (or another electronegative element) atoms and/or loses hydrogen atoms.
   2. During a *reduction reaction* – an organic molecule loses oxygen (or another electronegative element) atoms and/or gains hydrogen atoms. Example:

   [Lewis structure: ethylene + H₂/Pt (reduce) → ethane, with oxidation numbers labeled]

   3. No net C oxidation state change occurs when a molecule gains or loses both H and O (or another electronegative atom) ⇒ hydration / dehydration are not redox reactions.

   [Lewis structure: ethylene + H₂SO₄(dil) / H₂O / heat ⇌ ethanol, with oxidation numbers labeled]

e. Redox reactions can be complicated. You are not responsible for mechanism details, but should learn reactants, reagents, and products.

8.14.5. Another mild oxidation: osmium tetroxide *syn*-hydroxylation.

[Reaction scheme: cyclopentene + OsO₄ → osmate ester (via *syn*-addition) → cis-1,2-cyclopentanediol (via H₂O₂ or R₃N⁺–O⁻); cis (only) 65%]

a. OsO₄ is very toxic, volatile, and expensive – used in catalytic (trace) amounts if H₂O₂ or R₃N⁺-O⁻ (a 3° amine oxide) is added. H₂O₂/H₂O hydrolyzes osmate ester and reoxidizes osmium to OsO₄.

b. Example:

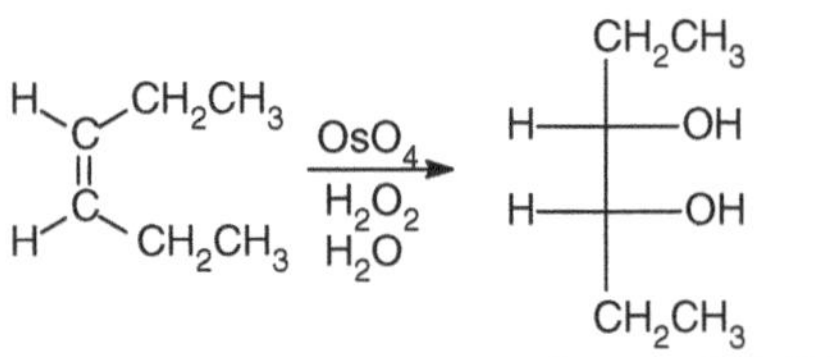

$$H_2C=C(CH_2CH_3)_2 \text{ (cis)} \xrightarrow[H_2O]{OsO_4, H_2O_2}$$

Fischer projection:
H—CH₂CH₃ top, H—OH, H—OH, CH₂CH₃ bottom
meso-hexane-3,4-diol

c. Cold, dilute aqueous KMnO₄/KOH is less toxic and cheaper, but yields are somewhat lower – also follows a *syn*-addition pathway.
   1. Example:

   $$3\ H_2C=CH_2 + 2\ KMnO_4 + 4\ H_2O \longrightarrow 3\ H_2C(OH)\text{—}CH_2(OH) + 2\ MnO_2(s) + 2\ KOH$$

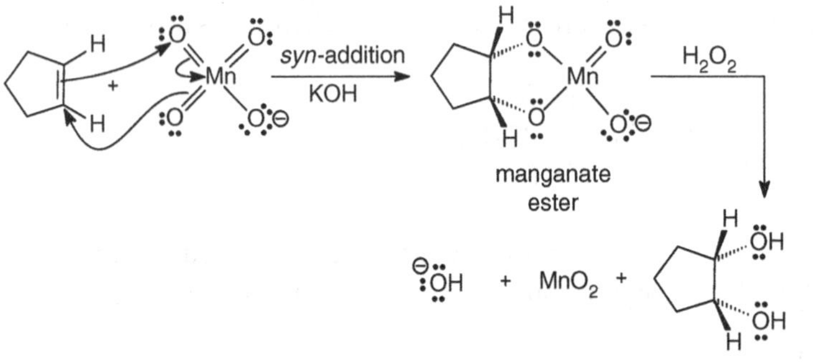

   [Mechanism diagram: cyclopentene + MnO₄⁻ → syn-addition/KOH → manganate ester → H₂O₂ → cis-cyclopentane-1,2-diol + ⁻OH + MnO₂]

   cis (only) 49%

   2. Overoxidation must be avoided (when KMnO₄ solution is too acidic, warm or concentrated).
   3. Baeyer alkene test – purple KMnO₄ color disappearance (and brown MnO₂ precipitate appearance) ⇒ positive test.
d. Oxidizing agent choice depends on costs, scale, yields, and convenience.
e. Problem 23. Mild *cis*-but-2-ene oxidation with KMnO₄(dil) / KOH / H₂O / 0°C yields butane-2,3-diol whose mp is 34°C. *trans*-But-2-ene oxidation under the same conditions also yields butane-2,3-diol, but with mp 19°C. Both diol products are optically inactive. The mp 19°C diol (I) was resolvable into two optically active fractions, but the mp 34°C diol (II) could not be resolved. *a*. Draw structures for diols I and II (3 dimensional perspective drawings, label chiral C configurations (*R* or *S*)). *b*. Assuming these results are typical (they are), what is the mild KMnO₄ oxidation stereochemistry? *c*. Peroxyacid alkene oxidation (aqueous) yields opposite results – *cis*-but-2-ene yields diol I and *trans*-but-2-ene yields diol II. What is the peroxyacid stereochemistry?

f. Problem 24. The following reactions were carried out and products separated by careful distillation, recrystallization, or chromatography. Tell how many fractions (different products) will be collected for each reaction. Draw each fraction's product(s) showing three dimensional perspectives. Tell whether each collected fraction would be optically active or inactive.

a. 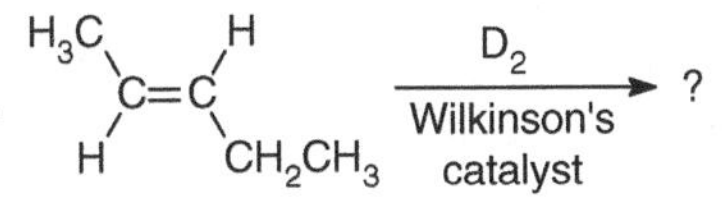
$$\underset{H}{\overset{H_3C}{>}}C=C\underset{CH_2CH_3}{\overset{H}{<}} \xrightarrow[\text{Wilkinson's catalyst}]{D_2} ?$$

$(C_5H_{10}D_2)$

b. $? \xrightarrow[\text{KOH / H}_2\text{O} \\ 0°C]{KMnO_4(dil)} ?$

(S)(Z)-pent-3-en-2-ol $(C_5H_{12}O_3)$

c. 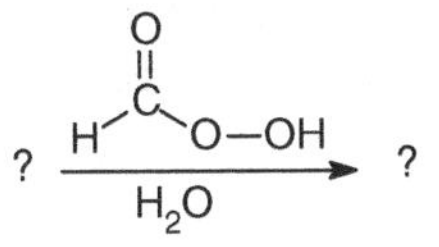 $? \xrightarrow[H_2O]{H-\overset{O}{\overset{\|}{C}}-O-OH} ?$

(S)(Z)-pent-3-en-2-ol $(C_5H_{12}O_3)$

d. $? \xrightarrow[\text{hexane (solvent)} \\ 0°C]{Br_2} ?$

(E)-4-methylhex-2-ene
(racemic) $(C_7H_{14}Br_2)$

e. $? \xrightarrow[\text{KOH / H}_2\text{O} \\ 0°C]{KMnO_4(dil)} ?$

(S)-HOCH₂CH(OH)CH=CH₂ $(C_4H_{10}O_4)$

f. $? \xrightarrow[Ni]{H_2} ?$

(R)-2-ethyl-3-methylpent-1-ene (C_8H_{18})

g. Problem 25. Show what unsaturated compound you would start with to make the following products. Draw reactant structures, reagents, and 3 dimensional (perspective) product structures.
   1. (R,R)(S,S)-2,3-dichloropentane (*erythro*)

   2. *meso*-3,4-hexanediol

   3. *meso*-3,4-hexanediol (from a different alkene)

   4. (R,S)(S,R)-3-bromobutan-2-ol (*threo*)

5. racemic (2,3-d_2)-butane (CH$_3$CHDCHDCH$_3$)

h. Problem 26. Starting from cyclopentanol, show how you could make stereochemically pure *cis*-cyclopentane-1,2-diol, and *trans*-cyclopentane-1,2-diol.

8.15.1. Strong alkene oxidation does not yield diols (addition), but cleaves a C=C π- and σ-bonds. Ozone (section 6.3.8) is a highly reactive oxidizing agent. Its formation is endothermic.

$$\tfrac{3}{2} O_2 + 142 \tfrac{kJ}{mol} \longrightarrow O_3$$

a. Ozone (O_3) is bubbled through an alkene solution in an inert solvent (CH$_2$Cl$_2$). Solvent is evaporated, leaving a viscous (explosive) ozonide, to which H$_2$O and Zn or H$_3$CSCH$_3$ (reducing agents) are added. Each C=C sp^2 C atom becomes a C=O. Reducing agent prevents H$_2$O$_2$ formation, which would oxidize product aldehydes further. Learn reactants, reagents, and products.

b. Ozonolysis is an analytical method. (Oxidize alkene, then analyze products to help determine structure.)

c. Ozone is a potent lung irritant – may attack unsaturated fatty acids in cell membranes – sometimes used in swimming pools instead of Cl$_2$.

d. Problem 27. Draw an alkene structure that would yield the following ozonolysis products upon treatment with 1. O_3, and 2. S(CH$_3$)$_2$:

1. 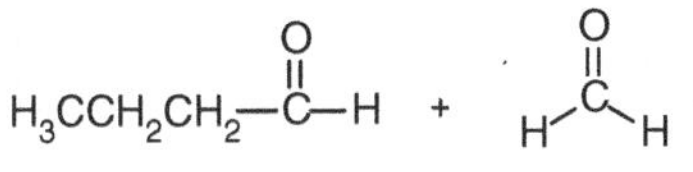 H₃CCH₂CH₂—C(=O)—H + H—C(=O)—H

2. 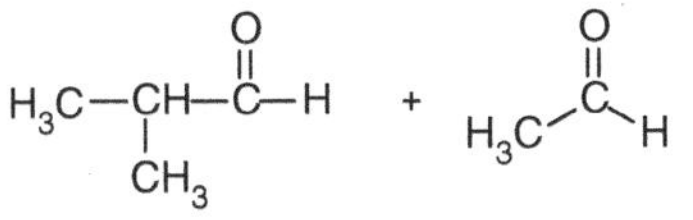 H₃C—CH(CH₃)—C(=O)—H + H₃C—C(=O)—H

3. 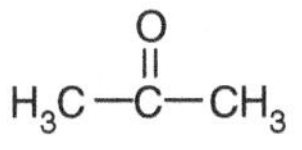 H₃C—C(=O)—CH₃   only

4. 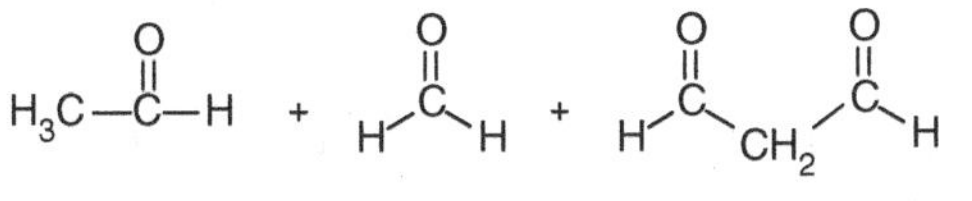 H₃C—C(=O)—H + H—C(=O)—H + H—C(=O)—CH₂—C(=O)—H

5. 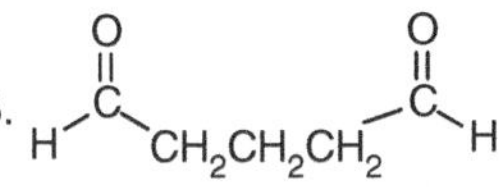 H—C(=O)—CH₂CH₂CH₂—C(=O)—H   only

- e. **Problem 28** Draw the ozonolysis (followed by (CH₃)₂S) products expected from:
  1. cyclopentene

  2. 1-methylcyclopentene

  3. 3-methylcyclopentene

  4. cyclohexa-1,3-diene

  5. cyclohexa-1,4-diene

- f. **Problem 29.** Both cyclohexene and octa-1,7-diene would yield hexa-1,6-dial upon ozonolysis. How else might you distinguish the two dienes?

g. **Problem 30** *a*. A C$_{10}$H$_{16}$ hydrocarbon absorbs one mol of H$_2$ upon hydrogenation. How many rings does it have? *b*. Ozonolysis yields cyclodecane-1,6-dione. What is the original hydrocarbon?

cyclodecane-1,6-dione

8.15.2. Warm concentrated KMnO$_4$ (or NaIO$_4$ with a catalytic amount of KMnO$_4$) is an even more vigorous oxidizing agent.

  a. KMnO$_4$ hydroxylates C=C, then IO$_4^-$ cleaves the σ-bond, and reoxidizes Mn^{4+} back to MnO$_4^-$. Each end of the C=C becomes a C=O group, but aldehydes oxidize further to carboxylic acids. Methanoic acid oxidizes to CO$_2$.

  b. Problem 31. Alkenes can be burned to CO$_2$ and H$_2$O with excess O$_2$.
   1. Write balanced combustion reactions for but-1-ene, *cis*-but-2-ene, and *trans*-but-2-ene, respectively. How many mols of CO$_2$ and H$_2$O are produced in each reaction?

   2. Which alkene corresponds to which measured heat of combustion, ΔH°_{rxn}? Fill in the table.

Alkene	ΔH^o_{rxn} (kJ/mol)
?	2712
?	2707
?	2719

3. Problem 32. Which pentene (pent-1-ene, *cis*-pent-2-ene, *trans*-pent-2-ene) corresponds to which measured heat of combustion, ΔH^o_{rxn}?

	ΔH^o_{rxn} (kJ/mol)
?	3365
?	3376
?	3369

8.16.1. Polymer (many parts) – a large molecule made of many equivalent building blocks.

8.16.2. Monomer (one part) – reactant from which a polymer is made.

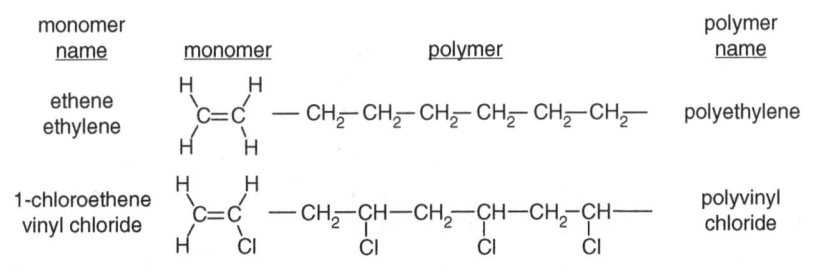

monomer name	monomer	polymer	polymer name
ethene ethylene	H₂C=CH₂	—CH₂—CH₂—CH₂—CH₂—CH₂—CH₂—	polyethylene
1-chloroethene vinyl chloride	H₂C=CHCl	—CH₂—CH(Cl)—CH₂—CH(Cl)—CH₂—CH(Cl)—	polyvinyl chloride

a. *Chain growth polymer* – growth occurs from the reactive (carbocation, free radical, or carbanion) chain end.
b. Alkenes undergo "addition" polymerization – monomers add into a long chain without any atom loss.

8.16.3. *Dimerization*– two molecules add.

a. Example (acid catalyzed): 2-methylbutene (isobutene, isobutylene) yields two isomeric alkenes: 2,4,4-trimethylpent-1-ene and 2,4,4-trimethylpent-2-ene, which consume one mol of H_2(g) to yield 2,2,4-trimethylpentane (isooctane), a gasoline "octane" enhancer.

1. Dimerization mechanism: EA (carbocation electrophile (Lewis acid)) – initial product is a larger carbocation – can either eliminate or react with another C=C forming an even larger carbocation.
   b. *Cationic* polymerization occurs when stable carbocation intermediates form, alkene concentration is high and with the right catalysts and temperatures.
   c. Polyisobutylene is a butyl rubber component.
   d. Problem 33. What would be the oxidation products expected when dimers 1 and 2 react with (1) ozone then $(H_3C)_2S$, or (2) $NaIO_4$ / $KMnO_4$?

8.16.4. Alkylation – joining an alkene to an alkane.
   a. Acid catalyzed isobutene alkylation with 2-methylpropane (isobutane) yields isooctane directly. Mechanism:

   1. Intramolecular (within the same molecule) 1,2-hydride shifts were seen in carbocation rearrangements (see section 6.15.1). Intermolecular (between two molecules) hydride ion transfer is a similar Lewis acid/base reaction.
   2. Alkylation is a chain reaction similar to those of sections 4.3.1 and 4.13.5, but involving carbocations instead of free radicals.
   3. Alkanes are extremely weak bases – a carbocation must be a strong (Lewis) acid to pull off a hydride ion (H:⁻).

8.16.5. Summary – carbocations (electrophiles and Lewis acids) can
   a. combine with a nucleophile (S_N1).
   b. rearrange to a more stable carbocation.
   c. eliminate (E1) to form an alkene.
   d. add to an alkene forming a larger carbocation.
   e. abstract a hydride ion from an alkane.

8.16.6. All carbocation reactions are driven by the octet rule – the need to share a fourth electron pair at the central C atom.

8.16.7. *Low density polyethylene* (mol. mass ≈ 20,000 g/mol) is made by heating ethene (200°C) under pressure (3000 atm) with O_2 or RO-OR' (initiators) in a free radical chain (addition) polymerization.
   a. Polyethylene – nonpolar, alkane-like (mp –100°C) – used in sandwich bags, films, dishes, etc.

b. When monomer runs out, a termination step occurs, either coupling (two chains together), or reaction with R-O·.

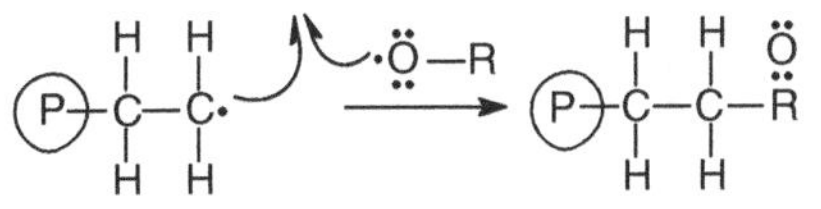

$$\text{P}-\text{CH}_2-\text{CH}_2\cdot + \cdot\text{Ö}-\text{R} \longrightarrow \text{P}-\text{CH}_2-\text{CH}_2-\ddot{\text{O}}-\text{R}$$

P = polymer

8.16.8. Polymer properties are modified if ethene has substituent groups, such as -CH$_3$, -Cl, -COOCH$_3$, -F, or phenyl.

a. Polypropylene (mp ~ 150°C) – dishwasher safe dishes, plastic parts, underwear. Addition always occurs so that the more stable secondary free radical intermediate forms.

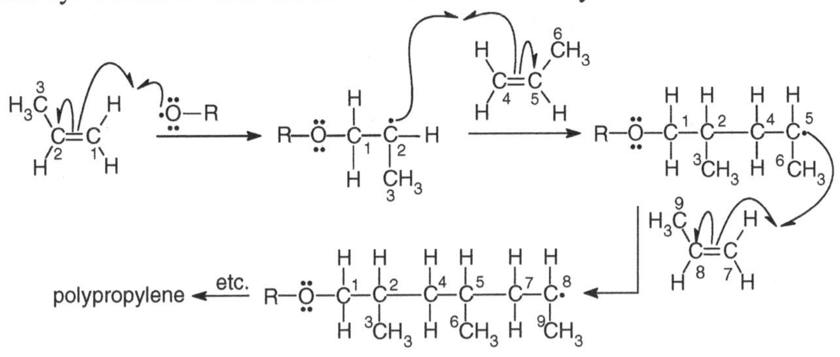

b. Polyvinyl chloride (sprinkler pipe, toys, plastic parts, floor coverings, dishes).

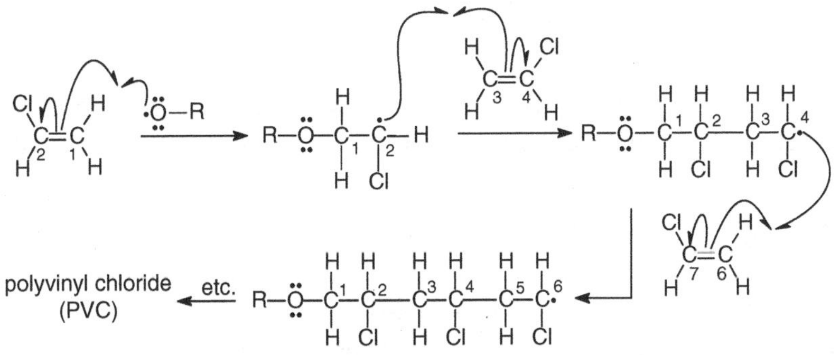

c. Saran (sandwich wraps, packaging, seat covers). A copolymer (two monomers, randomly ordered). Being phased out.

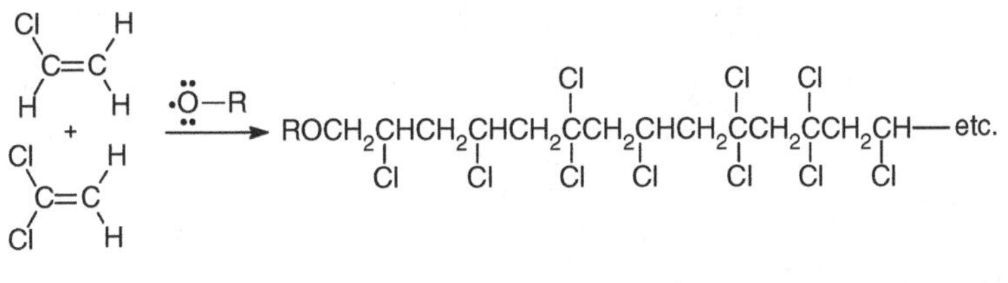

d. Polymethylmethacrylate (safety glass for auto windshields).

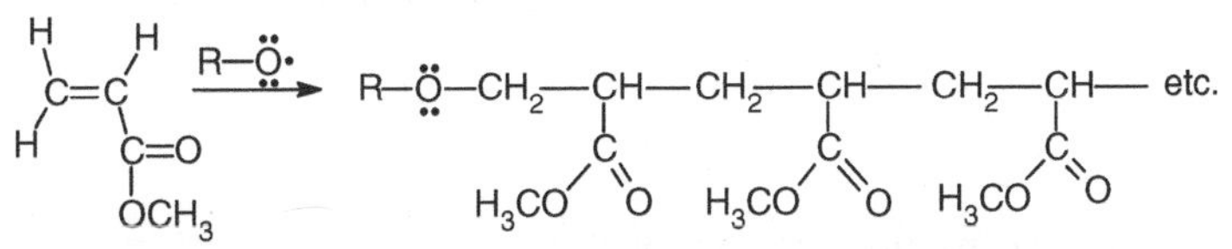

methyl acrylate

e. Teflon (polytetrafluoroethylene, PTFE – nonstick surfaces, wire insulation, plastic parts).

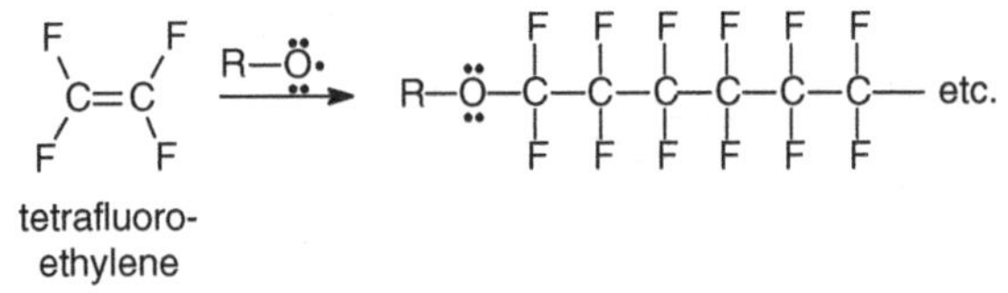

tetrafluoroethylene

f. Polystyrene (hamburger containers, insulated cups, egg cartons, transparent containers, lenses).

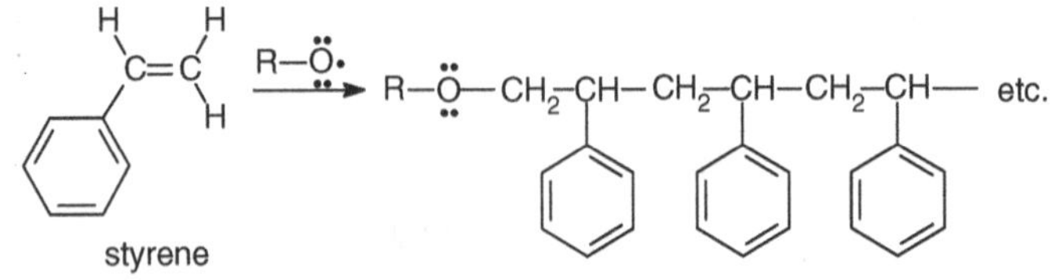

styrene

g. Orlon (polyacrylonitrile, fabrics) –

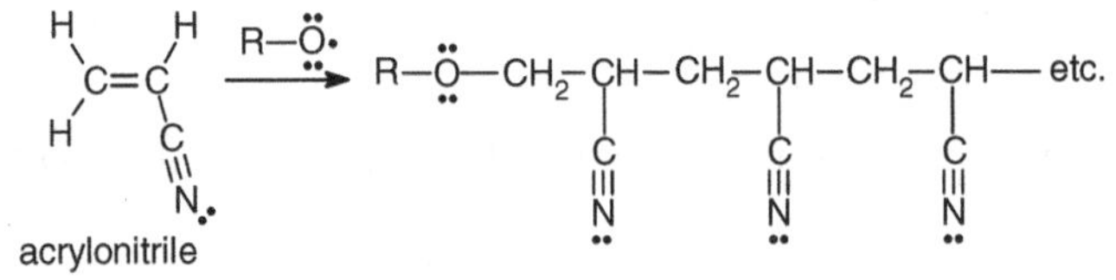

acrylonitrile

8.16.9. Methyl α-cyanoacrylate is *Super glue*. Base or metal oxide traces catalyze polymerization. A carbanion-like intermediate (enolate anion) is extra stable, due to resonance stabilization.

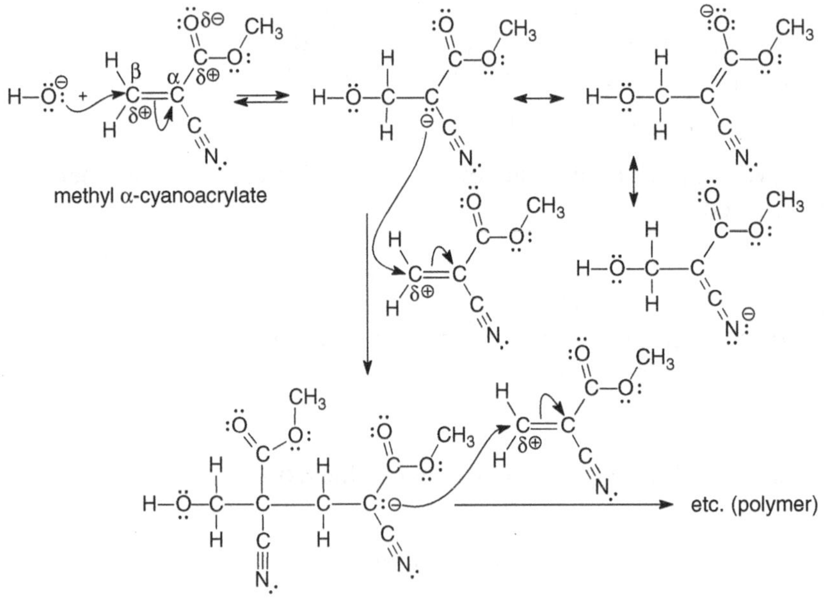

methyl α-cyanoacrylate

8.16.10. Small rings and bicyclics have some unusual properties.

8.16.11. Cyclopropane undergoes addition as if it were an alkene, but it is usually less reactive than propene.

$$\text{cyclopropane} \xrightarrow{\text{Ni / H}_2 \text{/ 80°C}} \text{H}_2\text{C}-\text{CH}_2-\text{CH}_2 \text{ (with H on each end)}$$

$$\xrightarrow{\text{Cl}_2, \text{ FeCl}_3 \text{ (catalyst)}} \text{H}_2\text{C}-\text{CH}_2-\text{CH}_2 \text{ (Cl on each end)}$$

$$\xrightarrow{\text{H}_2\text{SO}_4\text{(dil)}, \text{ H}_2\text{O}} \text{H}_2\text{C}-\text{CH}_2-\text{CH}_2 \text{ (H and OH ends)}$$

$$\text{cyclobutane} \xrightarrow{\text{H}_2 \text{/ catalyst, heat / pressure}} \text{H}_2\text{C}-\text{CH}_2-\text{CH}_2-\text{CH}_2 \text{ (H on each end)}$$

8.16.12. Cyclobutane hydrogenates, but is much less reactive than cyclopropane in other additions.

8.16.13. An electron rich region outside the ring is susceptible to electrophilic attack (the first step in ring opening).

$$\text{cyclopropane} + \overset{\delta+}{\text{H}}-\overset{..}{\underset{..}{\text{O}}}-\text{SO}_3\text{H} \longrightarrow \text{H}_2\overset{\oplus}{\text{C}}-\text{CH}_2-\text{CH}_2\text{H} \longrightarrow \text{products}$$

a. Problem 34. Draw structures and name the main organic products (if any) expected for the reactions:
   1. cyclopropane + Cl$_2$/FeCl$_3$(Lewis acid catalyst) →

   2. cyclopropane(excess) + Cl$_2$ (300°C) →

   3. cyclopentane + Cl$_2$/FeCl$_3$(Lewis acid catalyst) →

   4. cyclopentane(excess) + Cl$_2$(300°C) →

   5. cyclopentane + H$_2$SO$_4$(conc) →

   6. cyclopentene + Br$_2$/hexane(solvent)/0°C →

   7. cyclopentene(excess) + NBS / heat →

8. 1-methylcyclohexene + HCl(dry gas) →

9. 1-methylcyclohexene + Br$_2$/H$_2$O →

10. 1-methylcyclohexene + HBr/peroxides →

11. bromocyclohexane + KOH/CH$_3$CH$_2$OH/heat →

12. cyclopentene + KMnO$_4$(dil)/KOH/0°C →

13. cyclopentene + HCO$_2$OH/H$_2$O →

14. cyclopentene + C$_6$H$_5$CO$_2$OH/CH$_2$Cl$_2$(dry solvent) →

15. cyclopentene + KMnO$_4$(conc)/KOH/heat →

16. 3-bromocyclopentene + KOH/CH$_3$CH$_2$OH/heat →

17. cyclohexene + H$_2$SO$_4$(conc, trace amount) → (C$_{12}$H$_{20}$)?

18. cyclopentene + CHCl$_3$/*t*-BuOK →

19. cyclopentene + CH$_2$I$_2$/Zn(Cu) →

20. chlorocyclopentane + (CH$_3$CH$_2$)$_2$CuLi

21. 3-methylcyclopentene + 1. O$_3$ and 2. (CH$_3$)$_2$S →

22. 1-methylcyclohexene + 1. (BH$_3$)$_2$/THF(dry solvent) and 2. H$_2$O$_2$ / H$_2$O / NaOH(dil) →

23. 1-methylcyclohexene + 1. Hg(OAc)$_2$/H$_2$O and 2. NaBH$_4$ →

8.16.14. Characterization test – easy to carry out reaction that shows an observable change (color, gas evolution, precipitation, or solid dissolution).
  a. Alkene – dissolve in CH_2Cl_2 or hexane and add Br_2 ⇒ red/brown color disappears.
  b. Alkene – Baeyer test – add $KMnO_4$/KOH (cold, dilute) ⇒ purple color disappears, brown precipitate, MnO_2, appears.
  c. Alkene – dissolves in cold H_2SO_4(conc) – unlike alkanes, alkenes do not form a separate layer.

8.16.15. Characterization tests do not prove alkene presence. Other functional groups (cyclopropane, for example) can react with these reagents. Cyclopropane usually reacts slower than alkenes.

8.16.16. Simple alkyl halides and alkenes are good synthetic precursors – often cheap and commercially available. Successful synthesis suggestions:
  a. Work from a target molecule backwards (retrosynthetic analysis) to simpler reactants (with four or fewer C atoms unless instructed otherwise).
  b. Include adequate detail (often on reaction arrows) – reactants, reagents, conditions (temperatures, concentrations, pressures, solvents, etc.) and products. Do not show mechanism steps unless required.
  c. Reactions usually occur at specific functional groups.
  d. Example: synthesize the compound below from any alkenes:

  e. Potential problems:
    1. Top two routes – oxirane ring may open.
    2. Middle route – involves S_N2 reaction at a 2° C atom ⇒ low yield due to competing E2.
    3. Middle route – -O⁻ ion may remove -OH proton (react as a base rather than a nucleophile), since S_N2 is slower for a 2° C atom.
  f. Bottom route – avoids problems. Peroxyacid does not oxidize the -OH group in the last step.

8.16.17. Be able to assemble reactions into multi-step syntheses. Flash cards may be helpful.

8.17.1 Olefin metathesis (Greek, *change position*), Y. Chauvin, R. Grubbs, R. Schrock, Nobel Prize, 2005 – interchanges two *alkylidene* (=CHR) units – R may contain many functional groups.

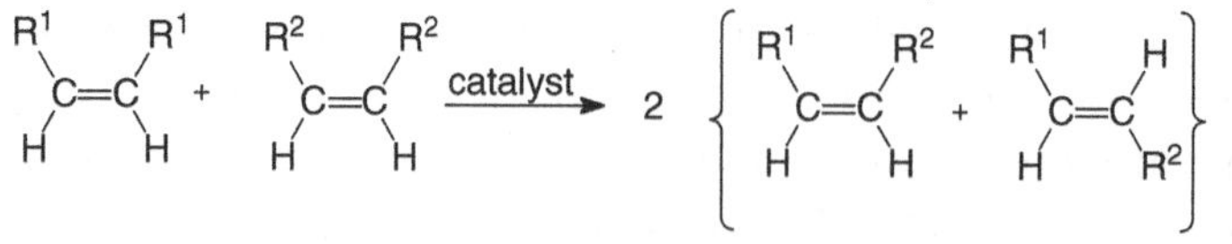

286 SEQUENTIAL ORGANIC CHEMISTRY I

a. Example: Phillips triolefin process (1950's):

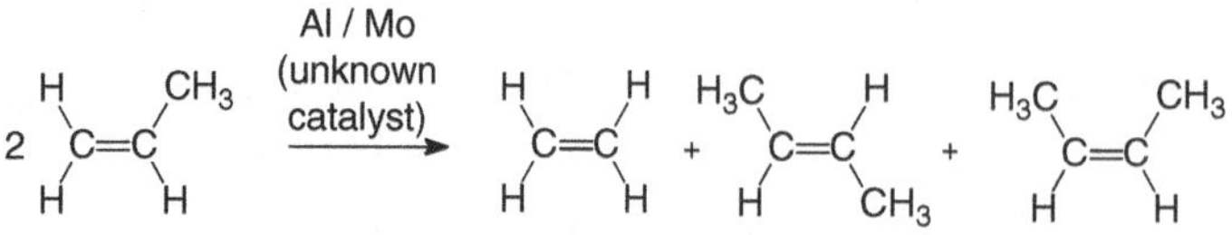

b. Catalysts:

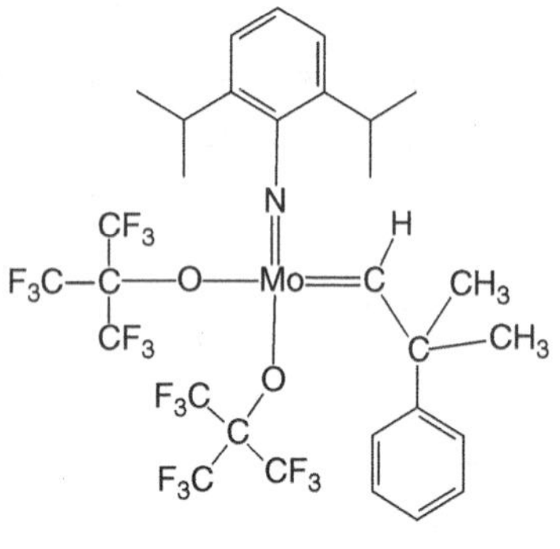

a Schrock catalyst
(decomposes
in air)

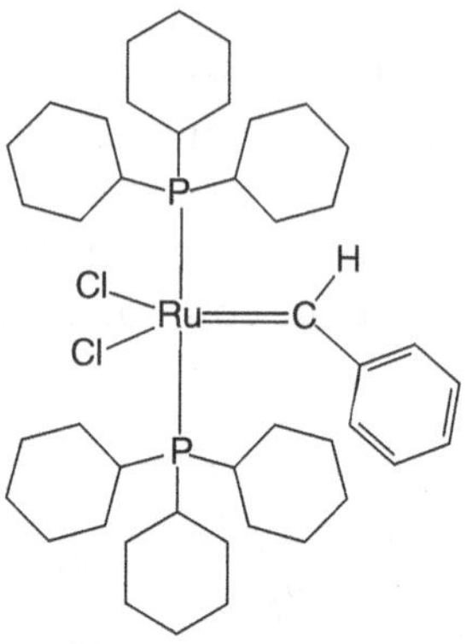

a Grubbs catalyst

8.17.2. Reversible Schrock and Grubbs reactions:
  a. Examples:

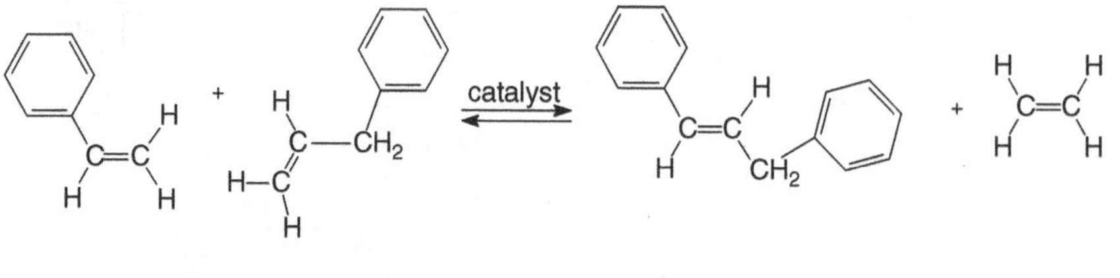

  1. Ethene gas loss drives equilbrium to completion.
  b. Example:

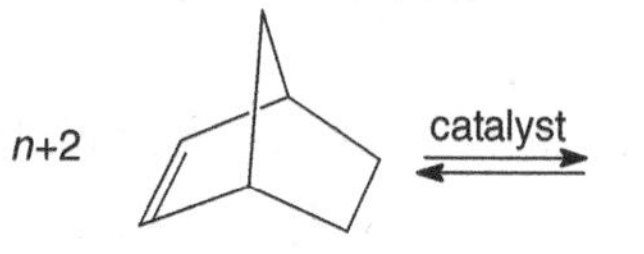

  norbornene polynorbornene + heat

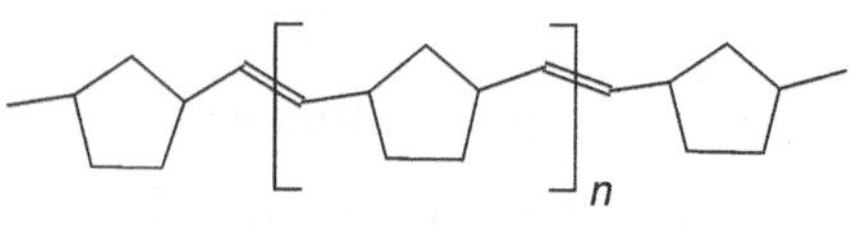

  1. Ring strain energy release drives equilibrium to completion.

8.17.3. Mechanism (Chauvin)

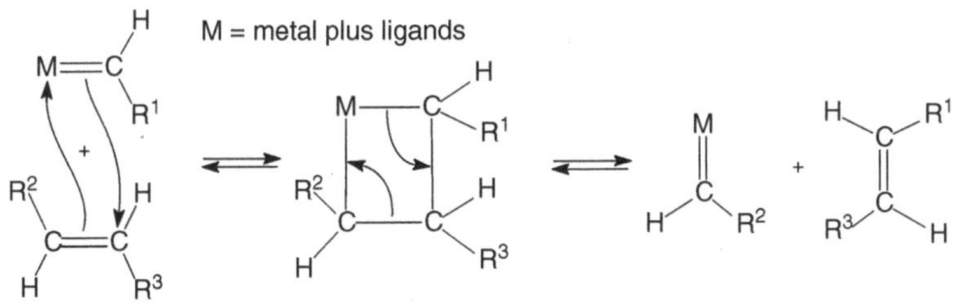

8.17.4. Problem 35. Predict products that would form when 2-methylpropene reacts with reagents shown. In some cases, no reaction may occur.

a. H_2/Ni

b. Cl_2 / CH_2Cl_2(solvent) / 0°C / dark

c. Br_2 / hexane(solvent) / 0°C / dark

d. I_2 / hexane(solvent) / 0°C / dark

e. HBr(dry gas) / exclude O_2

f. HBr(dry gas) / $(H_3C)_3C$-O-O-$C(CH_3)_3$ (trace) / ultraviolet light

g. HI(dry gas)

h. HI(dry gas) / $(H_3C)_3C$-O-O-$C(CH_3)_3$ (trace) / ultraviolet light

i. H_2SO_4(dil) / H_2O

j. Br_2 / H_2O (excess, solvent)

k. Br_2 / NaCl(excess) / H_2O (solvent)

l. H_2SO_4(conc, trace) → (C_8H_{16} – draw structure)?

m. 2-methylpropane + HF

n. KMnO$_4$(dil) / KOH / H$_2$O (solvent) / 0°C.

o. KMnO$_4$(conc) / KOH / H$_2$O (solvent) / 100°C

p. Peroxymethanoic acid (HCO$_2$OH) / H$_2$O

q. 1. O$_3$, 2. (H$_3$C)$_2$S

r. 1. Hg(OAc)$_2$ / H$_2$O 2. NaBH$_4$

s. 1. (BH$_3$)$_2$ 2. H$_2$O$_2$ / H$_2$O / NaOH(dil)

8.17.5. Problem 36. Describe simple chemical tests that would distinguish between:
   a. 2-methylbutane and 2-methylbutene

   b. hex-2-ene and 2-bromo-2-methylpropane

   c. 2-chloropentane and *n*-heptane

   d. 3-bromopropene and hex-1-ene

8.17.6. Problem 37. Propose structures for **A** and **B**.

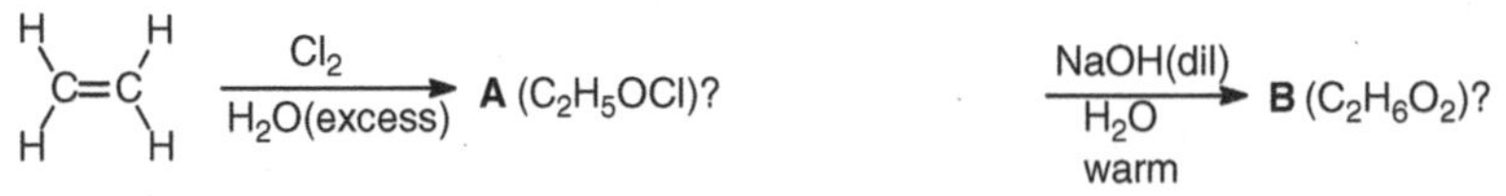

H$_2$C=CH$_2$ →[Cl$_2$ / H$_2$O(excess)] **A** (C$_2$H$_5$OCl)?   →[NaOH(dil) / H$_2$O / warm] **B** (C$_2$H$_6$O$_2$)?

8.17.7. Problem 38. A hydrocarbon I adds one mol of H$_2$ (Pt catalyst) yielding *n*-hexane. When I is vigorously oxidized with KMnO$_4$ (conc) / KOH / H$_2$O / heat, a single carboxylic acid containing 3 C atoms is isolated. Draw a structure for hydrocarbon I. Write reactions to justify your answer.

8.17.8. Problem 39. Show how you would make each product starting from an alkene.
   a. 2-methylpropan-2-ol

   b. 2-iodopropane

   c. 1-bromo-2-methylpropane

   d. 1-chloro-2-methylbutan-2-ol

   e. 2-methylpentane

   f. 2,3-dimethylbutane-2,3-diol

CHAPTER 9

9.1.1. Alkynes – hydrocarbons containing a C≡C triple bond.

    a. Ethyne (acetylene) is more unsaturated than ethene (can absorb two mols of H_2 over Pt catalyst).

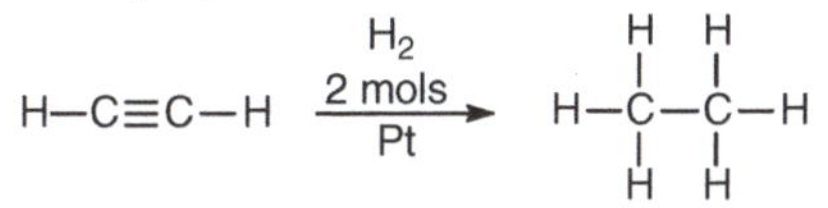

$$H-C\equiv C-H \xrightarrow[Pt]{H_2 \text{ 2 mols}} H-\underset{H}{\overset{H}{C}}-\underset{H}{\overset{H}{C}}-H$$

    b. Non-ring alkynes (two elements of unsaturation) have a lower C : H ratio (formulas C_nH_{2n-2}, $n > 2$) than non-ring alkenes.

    c. Alkynes undergo similar electrophilic addition and oxidation reactions as alkenes.

9.1.2. Alkynes are occasionally found in nature and drugs.

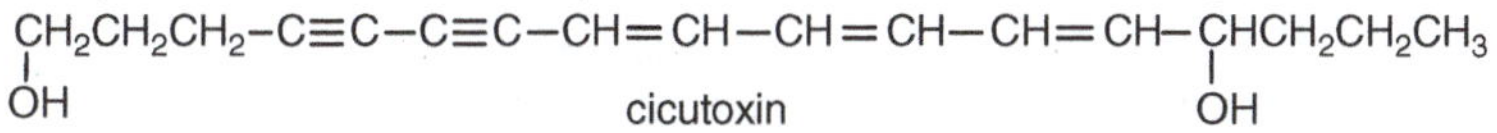

cicutoxin
from water hemlock
(toxic to predators)

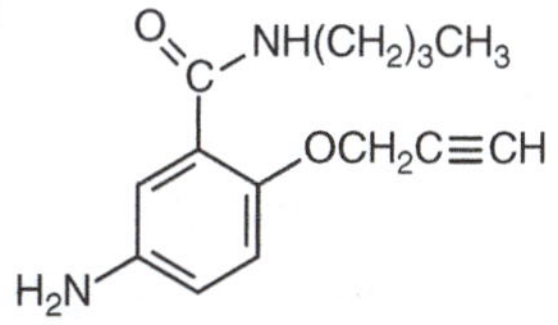

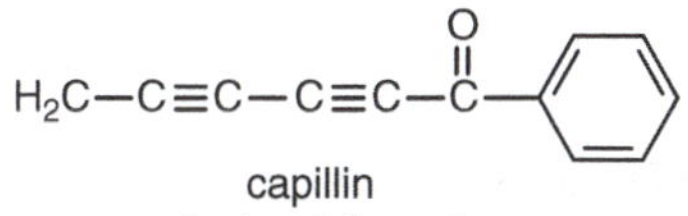

capillin
plant anti-fungal

parsalmide
(an analgesic)

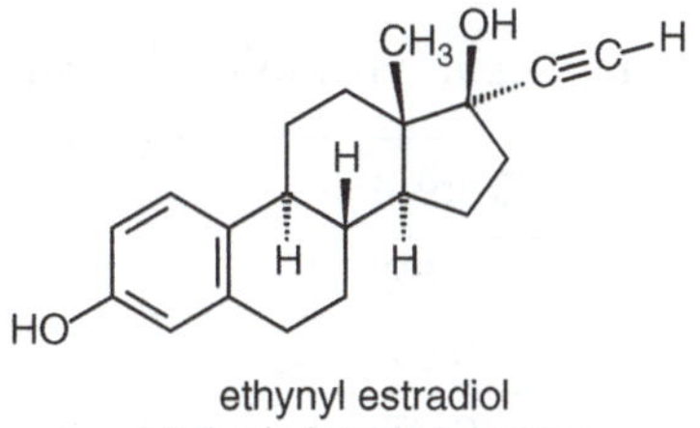

ethynyl estradiol
synthetic female hormone
(birth control pills)

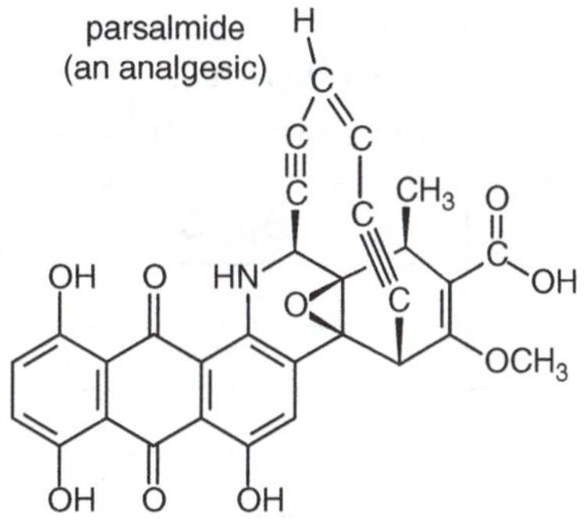

Dynemicin A
(antibacterial,
antitumor? agent)

9.2.1. Alkyne nomenclature.

    a. See functional group priorities (section 2.17.4).

291

b. Find longest continuous chain containing C≡C and name as alkane (assuming no higher priority functional groups are present). Change "ane" ending to "yne."
   1. Number triple bond position as small as possible.
   2. If a higher priority functional group is present and C≡C is part of parent chain, change "yne" suffix to "yn." Add higher priority group suffix.
c. Examples:

H—C≡C—H
ethyne
acetylene

H—C≡C—CH₃
propyne
methylacetylene

H—C≡C—CH₂CH₃
1-butyne
but-1-yne
ethylacetylene

H₃C—C≡C—CH₃
2-butyne
but-2-yne
dimethylacetylene

H₃C—C≡C—CH(CH₃)—CH₃ (with CH₃ branch)
4-methyl-2-pentyne
4-methylpent-2-yne
isopropylmethylacetylene

H₃C—CH(CH₃)—C≡C—CH₂—CH(Br)—CH₃
6-bromo-2-methyl-3-heptyne
6-bromo-2-methylhept-3-yne

H₃C—CH(OH)—C≡C—H
3-butyn-2-ol
but-3-yn-2-ol

H₂C=C(CH₃)—C≡C—CH₃
2-methyl-1-penten-3-yne
2-methylpent-1-en-3-yne

H₃C—C≡C—CH(OCH₃)—CH₂CH₃
4-methoxy-2-hexyne
4-methoxyhex-2-yne

Ph—C≡C—H
1-phenylethyne
ethynylbenzene?
phenylacetylene

Ph—C≡C—Ph
1,2-diphenylethyne
diphenylacetylene

H₃C—CH(CH₃)—C≡C—CH(CH₃)—CH₃
2,5-dimethyl-3-hexyne
2,5-dimethylhex-3-yne

H—C≡C—CH₂—OH
2-propyn-1-ol
prop-2-yn-1-ol
hydroxymethylacetylene
propargyl alcohol

9.3.1. Alkyne properties.
a. Nonpolar – about like alkanes and alkenes.
b. Water insoluble, but soluble in most organic solvents (hydrocarbons, ethers, di- and trichloromethanes, and alcohols).
c. Lower densities than water (1.00 g/mL).
d. Low boiling points about like alkanes and alkenes having similar molecular mass.

Alkyne	Bp °C	Density
ethyne	-84	0.62
propyne	-23	0.67
but-1-yne	8	0.67
but-2-yne	27	0.69
pent-1-yne	40	0.70
pent-2-yne	55	0.71
hex-1-yne	71	0.72
oct-1-yne	125	0.75
non-1-yne	151	0.76
dec-1-yne	174	0.77

e. Smell – mildly offensive to stinky.

9.4.1. Welders burn ethyne (acetylene) with oxygen – heats metal hotter (flame 2800°C) than ethane or ethene. Key factor – heat per mol of products (right column).

	ΔH°_{rxn} kJ	ΔH°_{rxn} $\frac{kJ}{mol\ products}$
$H_3C-CH_3 + \frac{7}{2} O_2 \longrightarrow \underbrace{2\ CO_2 + 3\ H_2O}_{5\ mols}$	-1561	-312
$H_2C=CH_2 + 3\ O_2 \longrightarrow \underbrace{2\ CO_2 + 2\ H_2O}_{4\ mols}$	-1410	-352
$HC{\equiv}CH + \frac{5}{2} O_2 \longrightarrow \underbrace{2\ CO_2 + 1\ H_2O}_{3\ mols}$	-1326	-442

a. Acetylene is thermodynamically unstable and tends to explode.

	ΔH°_{rxn} $\frac{kJ}{mol\ ethyne}$
$H-C{\equiv}C-H \longrightarrow 2\ C + H_2$	-234
$2\ C + H_2 \xrightarrow{\frac{5}{2} O_2} 2\ CO_2 + H_2O$	-1090

  1. It is stored in steel cylinders with ground firebrick / acetone which help stabilize it.

9.4.2. Industrial acetylene preparations.

  a. From methane (natural gas). At high T, positive ΔS°_{rxn} causes ΔG°_{rxn} to be negative ($-T\Delta S^{\circ}_{rxn}$ becomes larger negative than positive ΔH°_{rxn}).

$$2\ CH_4 \xrightarrow[0.01\ sec]{1500°C} H-C{\equiv}C-H + 3\ H_2$$

  b. From coke (roasted coal):

$$3\ C + CaO \xrightarrow[2500°C]{electric\ furnace} \underset{calcium\ carbide}{Ca^{2+}[:C{\equiv}C:]^{2-}} + CO$$

  c. The latter is the "miner's lamp" reaction.

9.5.1. Lewis dot, LCAO, and MO ethyne structures were shown in sections 1.17.2–1.17.3.

  a. Problem 1. CO_2 is a linear molecule (bond angle 180°), while H_2O is bent (bond angle 104.5°). How are C and O hybridized? Draw a linear combination of atomic orbitals (LCAO) picture for CO_2. Label atomic hybrid orbitals ($1s$, sp, sp^2, sp^3, or n) and molecular orbitals (MO's) as σ or π.

b. **Problem 2.** Draw an LCAO picture for the carbide dianion (C_2^-) in the miner's lamp reaction. Label orbitals as in previous problem.

c. **Problem 3.** Draw the seven C_6H_{10} alkyne isomer structures and assign IUPAC names.

9.1.3. While ≡C-H homolysis is more difficult, heterolysis is easier for ethyne than for ethene and ethane.

H—C≡C—H →(homolysis) H—C≡C• + H•

H—C≡C—H →(heterolysis) H—C≡C:⁻ + H⁺

9.6.1. An H atom can be acidic when bonded to an electronegative atom (such as N, O, S, and X). (See section 2.9.1.)

9.6.2. Effective carbon electronegativity depends on its hybridization:

$$\text{most } sp > sp^2 > sp^3 \text{ least}$$

a. Example: *terminal* alkynes (with a ≡C-H bond) are acidic enough to react with the very strong base amide ion ($^-NH_2$).

b. Review: ammonia often reacts as a weak base. It can also react as an acid:

weaker base   weaker acid   stronger acid   stronger base

weaker acid   weaker base   stronger base   stronger acid
$K_a = 10^{-35}$ (amide ion)

c. Lithium or sodium amide can be made by a metal / ammonia redox reaction.

2 H—N(H)(H) + 2 Na →(FeCl₃ catalyst) 2 [H—N:—H]⁻ + 2 Na⁺ + H—H

sodium amide

d. Ethyne is a stronger acid than ammonia – converted to conjugate base by reaction with amide anion. Internal alkynes do not react.

H—C≡C—H + H₂O ⇌ H—C≡C:⁻ + H₃O⁺

weaker acid   weaker base   stronger base   stronger acid
$K_a = 10^{-25}$ (acetylide anion)

H—C≡C—H + ⁻NH₂ + Na⁺ ⟶ H—C≡C:⁻ + NH₃ + Na⁺

stronger acid   stronger base   weaker base   weaker acid
                (amide ion)   (acetylide anion)

H₃C—C≡C—CH₃ + ⁻NH₂ + Na⁺ ⇌ no reaction
no acidic H atoms

e. Sodium (or lithium) acetylide can also be made by a redox reaction with the metal.

H—C≡C—H⁺¹ + 2 Na⁰ ⟶ 2 H—C≡C:⁻ + 2 Na⁺¹ + H—H⁰

sodium acetylide (alkynide)

f. What will happen if sodium acetylide (itself a strong base) is added to water?

H—C≡C:⁻ Na⁺ + H₂O ⇌ H—C≡C—H + ⁻OH + Na⁺

stronger base   stronger acid   weaker acid   weaker base

9.6.3. Acidity and basicity series (section 2.6.5) – ethyne and conjugate base acetylide anion included.

stronger acid

H₂SO₄, HI, HNO₃ > HBr > HCl > H₃O⁺ > H₃C—C(=O)—OH > NH₄⁺ > H₂CO₃ > HCN

weaker acid

H—C≡N > H₂O > ROH > R—C≡C—H > NH₃ > H₂ > CH₄

weaker base

⁻O—SO₃H, I⁻, NO₃⁻ < Br⁻ < Cl⁻ < H₂O < H₃C—C(=O)—O⁻ < NH₃ < HCO₃⁻

stronger base

< ⁻C≡N: < ⁻OH < ⁻OR < ⁻C≡C—R < ⁻NH₂ < H:⁻ < ⁻CH₃

a. ROH represents an alcohol.

9.6.4. Terminal alkynes are acidic enough to neutralize Grignard or alkyllithium compounds.

H₃C—C≡C—H + Br—Mg—CH₂CH₃ ⟶ H₃C—C≡C—Mg—Br
$\delta^{\ominus}$ $\delta^{2\oplus}$ $\delta^{\ominus}$ weaker base

stronger acid   stronger base   + CH₃CH₃

weaker acid

9.6.5. Ethynide (acetylide) anion lone electron pair occupies an *sp* orbital (~½ s and ~½ p character).
  a. More *s* character ⇒ closer electrons approach nucleus and stronger attraction – more stable. Tightly held electrons ⇒ less available to donate (to a Lewis acid) ⇒ weaker base.

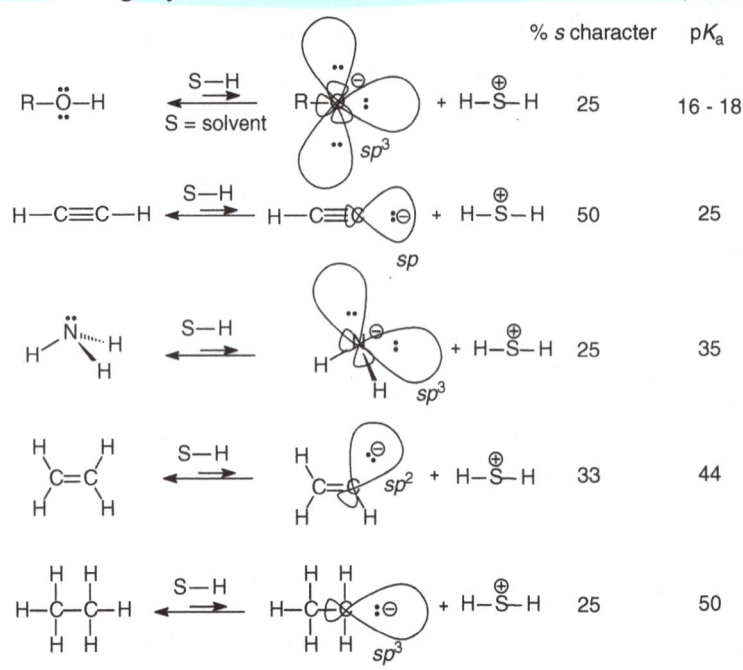

	% s character	pK_a
R–Ö–H ⇌ (S=solvent) R–Ö:⊖ (sp^3) + H–S–H⊕	25	16 - 18
H–C≡C–H ⇌ H–C≡C:⊖ (*sp*) + H–S–H⊕	50	25
H–N̈H–H ⇌ H–N̈:⊖ (sp^3) + H–S–H⊕	25	35
H₂C=CH₂ ⇌ H₂C=CH:⊖ (sp^2) + H–S–H⊕	33	44
H₃C–CH₂–H ⇌ H₃C–CH₂:⊖ (sp^3) + H–S–H⊕	25	50

9.7.1. Carbon-carbon bond formation is a major synthesis goal.
  a. Most carbon-carbon bond forming synthetic reactions are ionic.
  1. Intermediately electronegative (2.5) C can have either a + (or δ+) or - (or δ-) charge. A C-A bond is polarized (δ+ on C) when C is bonded to an electronegative (electron withdrawing) atom (A = O, N, or X).

  $H_3C^{\delta+}–C^{\delta+}(=O^{\delta-})H$ $H_3C^{\delta+}–Cl^{\delta-}$ $H_3C^{\delta+}–C≡N^{\delta-}$

  2. An electropositive (usually metal) atom bonded to C results in δ- on C ⇒ bond polarity is reversed in organometallics (sections 3.5.7-3.5.11).

  $H_3CCH_2CH_2CH_2^{\delta-}–Li^{\delta+}$ $H_3CCH_2CH_2CH_2^{\delta-}–Mg^{\delta 2+}–Br^{\delta-}$

9.7.2. C-C bond synthesis – a - (or δ-) C nucleophile usually attacks a + (or δ+) C electrophile.
  a. Review – synthetic C-C bond forming processes:
  1. Grignard synthesis (with aldehydes and ketones to make alcohols, section 3.5.8d, not fully discussed yet).

  2. Corey-House reaction (section 3.5.11).

3. Cyanide ion S$_N$2 reaction (section 6.9.1).

4. Alkynide ion S$_N$2 reaction.

b. Corey-House is not as synthetically useful as the others because alkane products have no functional groups that can be further modified.

9.7.3. Laboratory alkyne preparations.

9.7.4. Alkynide ion reaction with methyl or 1° R-X's.

Liquid ammonia is used as a solvent in this reaction. See Solomons, T. W. G., Fryhle, C. B., Snyder, S. A. "Organic Chemistry," 11 ed. p 316.

a. E2 competes with S$_N$2 for 2° R-X's. E2 dominates for 3° R-X's.

9.7.5. Alkynide reaction with carbonyl (C=O) compounds.

a. Electronegative oxygen atom polarizes the carbonyl group (C=O) σ- and π-bonds (section 18.2.1).

b. δ+ C is electrophilic (attacked by nucleophiles).

[Reaction scheme: H₃C-CH(CH₃)-C≡C-H → (1. NaNH₂, 2. Ph-CHO, 3. H₂SO₄ (dil)/H₂O/25°C (neutralize pH)) → H₃C-CH(CH₃)-C≡C-CH(OH)Ph]

Ph = phenyl = —C₆H₅

[Reaction scheme: cyclohexanone → (1. Na⁺ :C≡C-H⁻, 2. H₂SO₄ (dil)/H₂O/25°C (neutralize pH)) → 1-ethynylcyclohexan-1-ol]

9.8.1. 1,1- (geminal) or 1,2- (vicinal) dihalide dehydrohalogenation.

  a. Vicinal dihalides can be made by X_2 addition to alkenes.

[Reaction scheme: propene + Br₂/hexane cold/dark → 1,2-dibromopropane → KOH/H₃CCH₂OH heat → propenyl bromide (vinyl halide) + :Br⁻ + H₂O → Na⁺:NH₂⁻/NH₃(l) −33°C → H₂SO₄(dil)/H₂O/25°C (neutralize pH) → :NH₃ + :Br⁻ + H₃C-C≡C-H]

  b. Reaction (E2) can be stopped after one step, yielding vinyl halides.
  c. Drastic conditions are needed for the second, vinyl halide dehydrohalogenation – (1) molten (fused) KOH / 200°C, (2) KOH / ROH / 200°C sealed tube, or (3) 1. NaNH₂ / 150°C, 2. H₂O.

[Reaction scheme: H₃CCH₂-CHBr-CHBr-CH₃ → KOH (fused) (excess) 200°C → H₂SO₄(dil)/H₂O/25°C (neutralize pH) → H₃C-C≡C-CH₂CH₃ + 2:Br⁻ + 2 H₂O, 45%]

[Reaction scheme: H₃C-CH₂-CH₂-CH₂-CHCl₂ → NaNH₂ 150°C → H₂SO₄(dil)/H₂O/25°C (neutralize pH) → H₃C-CH₂-CH₂-C≡C-H, 55%]

  d. A conjugated (two double bonds separated by a single bond) diene is usually preferred over an alkyne when possible. Reaction conditions (temperatures, concentrations, etc.) affect relative yields. Stability order is:

  more stable conj. diene > alkyne > allene less stable

[Reaction scheme: cyclohexene + Br₂/hexane cold/dilute → 1,2-dibromocyclohexane → K⁺ :O-H⁻/H₃CCH₂OH heat → 3-bromocyclohexene + :Br⁻ + H₂O → K⁺ :O-H⁻/H₃CCH₂OH heat → 1,3-cyclohexadiene + H₂O + :Br⁻]

e. Peter Stang (U. of Utah) has solvolyzed vinyl triflates:

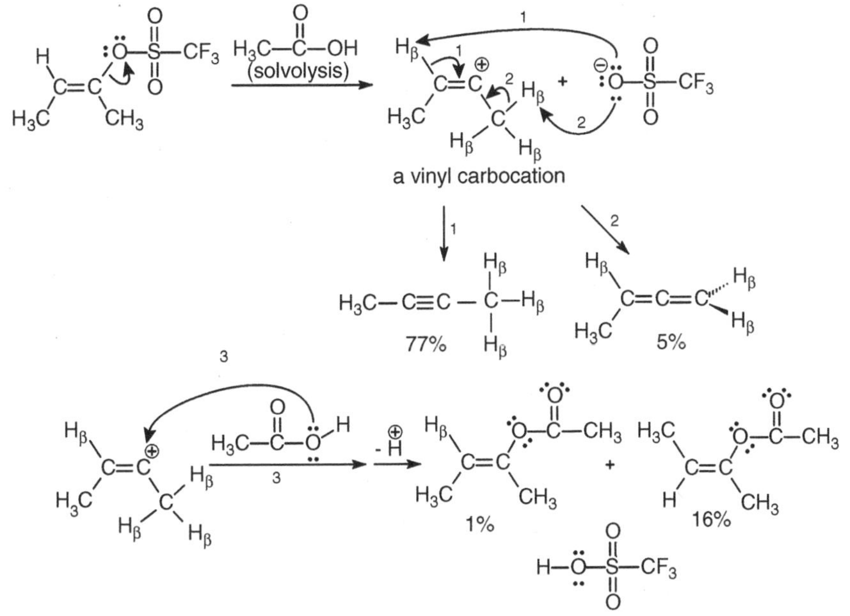

[Reaction scheme showing vinyl triflate solvolysis with acetic acid forming a vinyl carbocation intermediate, which partitions into three pathways: pathway 1 giving H₃C—C≡C—C(Hβ)₃ (77%), pathway 2 giving an allene-type product (5%), and pathway 3 giving enol acetate products (1% and 16%) plus triflic acid.]

1. *Triflate* (short for trifluoromethane sulfonate), $F_3CSO_3^-$, is a very fast leaving group, needed to form a highly unstable (highly reactive) *vinyl* carbocation intermediate. Vinyl carbocations are less stable (relative to the triflate from which they form) than $^+CH_3$ (relative to H_3C-X from which it forms).

f. Problem 4. Show how you could synthesize prop-1-yne from:
   1. 1,2-dibromopropane

   2. prop-1-ene

   3. 2-bromopropane

   4. propan-1-ol

   5. 1,2-dichloropropane

   6. ethyne

9.9.1. Alkynes are less stable (and often more reactive) than alkenes, due to π-electron repulsions.
  a. Bond dissociation (homolysis) energies: (the second π-bond is weaker than the first.) See section 1.17.4f1.

Bond	Total energy (kJ/mol)	Type of bond	Approximate energy (kJ/mol)
C–C	377	σ-bond	377
C=C	728	π-bond	321
C≡C	965	second π-bond	268

9.9.2. Like alkenes, alkynes undergo EA reactions, but they can consume two mols of reagents. Reactions can often be stopped at the alkene stage.

9.9.3. Alkyne reactions:

9.9.4. Catalytic hydrogen addition (reduction).

$$H-C \equiv C-CH_2CH_3 \xrightarrow{2H_2 / Pt} H-CH_2-CH_2-CH_2CH_3 \quad 100\%$$

$$H_3C-C \equiv C-CH_3 \xrightarrow{2H_2 / Pt} H_3C-CH_2-CH_2-CH_3 \quad 100\%$$

  a. Can be stopped at alkene with partially deactivated (poisoned) Lindlar's catalyst. *Syn*-stereoselective (section 8.10.2a (Chapter 7) and 8.9.3b) reaction yields *cis* (Z) products. Both H atoms add from the same face on Pd surface.

  Lindlar's catalyst: BaSO₄ (powdered) coated with Pd / quinoline (poison)

  $H_3C-C \equiv C-CH_2CH_2CH_3 \xrightarrow[\text{HOCH}_3 \text{ (solvent)}]{H_2, \text{ Lindlar's catalyst}}$ (Z)-alkene with H₃C and CH₂CH₂CH₃ on same side — *cis*-only, no *trans*

  $H-C \equiv C-CH_2CH_2CH_2CH_3 \xrightarrow[\text{HOCH}_3 \text{ (solvent)}]{H_2, \text{ Lindlar's catalyst}}$ (Z)-alkene with H and CH₂CH₂CH₂CH₃ on same side

  1. Ni₂B (nickel boride) is a Lindlar's catalyst alternative that is easier to make and often gives better yields.

2. **Problem 5.** Most alkene preparations yield predominantly the more stable *trans* isomer (where possible). How might you convert a 75% *trans* / 25% *cis*-pent-2-ene mixture into pure *cis*-pent-2-ene (moderate yield)?

b. Chemical (Birch) reduction – disubstituted alkynes yield mostly *trans* alkenes with Na or Li in NH_3 (*l*, bp = –33°C, dry ice cooled). Mechanism – *anti*-stereoselective due to electron cloud and alkyl group (steric) repulsions.

$NH_3(l) + \overset{\cdot}{Na} \longrightarrow NH_3 \cdot e^{\ominus} + Na^{\oplus}$
(solvated electron)
deep blue solution

trans-mostly, less *cis*

c. Stereochemically pure alkenes are valuable starting materials for stereospecific syntheses.

9.9.5. Halogen addition. Alkynes are much less reactive than alkenes despite their relative instability and reactivity.

a. Often a mixture of tetrahalide, dihalides (*cis* and *trans*), and unreacted alkyne is obtained with one mol of Br_2.

b. A three membered ring halonium ion from an alkyne has sp^2 hybridized C atoms (ideal bond angles 120°), and larger ring strain than a halonium ion from an alkene (with sp^3 hybridized C atoms).

$$H_3C-C \equiv C-H \xrightarrow[\text{cold / dark}]{\text{Br}_2 \text{ (one mol)}} \underset{\substack{sp^2 \text{ C atoms} \\ \text{(very strained ring)}}}{\overset{\oplus \ddot{\text{Br}}:}{\underset{H_3C \quad H}{C=C}}} + :\ddot{\text{Br}}:^{\ominus} \longrightarrow \text{products}$$

$$\underset{H}{\overset{H_3C}{>}}C=C\underset{H}{\overset{H}{<}} \xrightarrow[\text{cold / dark}]{\text{Br}_2 \text{ (one mol)}} \underset{\substack{sp^3 \text{ C atoms} \\ \text{(not as much ring strain)}}}{\overset{H_3C \ \oplus\ddot{\text{Br}}: \ H}{\underset{H \quad\quad H}{C-C}}} + :\ddot{\text{Br}}:^{\ominus} \longrightarrow \text{products}$$

1. Reactivity diagram. Alkyne reaction has higher E_a.

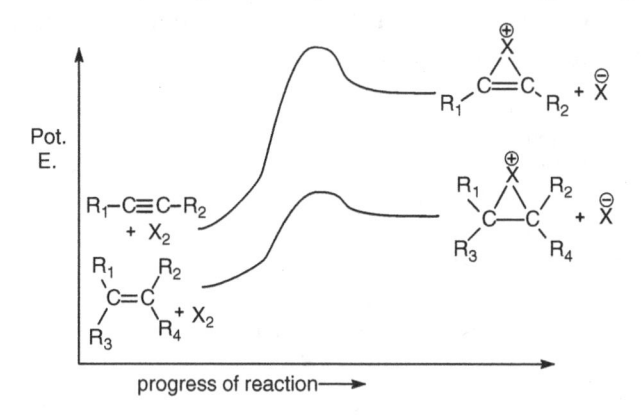

c. **Problem 6.**
  1. Write reactions for the two stage bromine addition to but-2-yne that yields ultimately 2,2,3,3-tetrabromobutane.

  2. Realizing that bromine atoms are electronegative and pull the π-electrons toward themselves, would 2,3-dibromobut-2-ene product from the first reaction be more or less reactive towards Br_2 than but-2-yne? Why?

  3. To maximize the 2,3-dibromobut-2-ene product yield would you (1) bubble but-2-yne into a Br_2(excess) / hexane solution, *or* (2) add Br_2 / hexane solution dropwise to a but-2-yne solution?

9.9.6. HX electrophilic addition (EA) to an alkyne is about as fast as HX / alkene EA.
  a. This seems inconsistent with a high energy vinyl carbocation intermediate.

[Mechanism: H–C≡C–CH₂CH₂CH₃ + H–Br (no peroxides, Markovnikov EA) → 2° vinyl carbocation (very unstable) + :Br:⁻ → vinyl bromide H₂C=C(Br)–CH₂CH₂CH₃]

[Mechanism: H₃C–C≡C–CH₂CH₃ + H–Br (one mol, no peroxides, Markovnikov EA) → mixture of (Z) plus (E) vinyl bromides]

  b. Reactivity is governed by intermediate carbocation stability *relative to the reactant from which it forms.*
  c. Vinyl carbocation is highly unstable relative to a vinyl halide. An alkyne is less stable initially than a vinyl halide. E_a is not too large ⇒ EA occurs at nearly the same rate as alkenes.
    1. Reaction diagram.

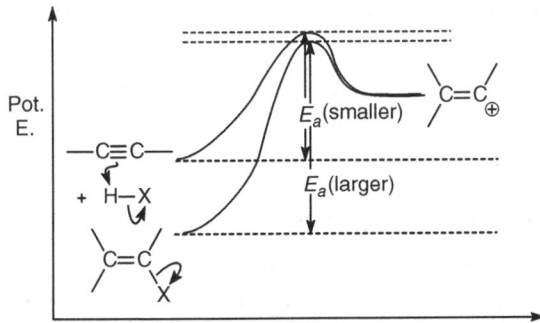

[Potential energy diagram showing —C≡C— + H–X with E_a(smaller) to vinyl cation, compared to alkene with E_a(larger)]

  d. H–Br / peroxides addition occurs anti-Markovnikov similarly as for alkenes (section 8.3.3).

[H–C≡C–CH₂CH₂CH₃ + H–Br (one mol, R–O–O–R) → cis vinyl bromide with Br and propyl on same side]

9.9.7. Alkyne hydration (Markovnikov) with H₂SO₄(dil)/H₂O/HgSO₄. Initial product is an unstable enol. Hg^{2+} ion catalyzes reaction.

[Mechanism showing H–C≡C–CH₂CH₃ + Hg²⁺SO₄²⁻ / H₂SO₄(aq) (Markovnikov) → mercurinium-like intermediate → addition of H₂O → vinyl alcohol (enol), usually unstable]

vinyl alcohol
(enol)
usually unstable

a. Tautomers – different structures (isomers) in rapid equilibrium with each other. Hydrogen tautomerism is the most common type.
b. Acid catalyzed keto-enol tautomerization mechanism – equilibrium usually lies far towards keto form.

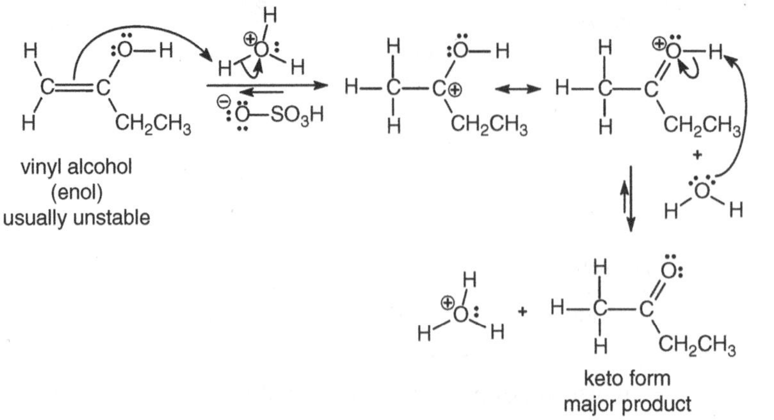

vinyl alcohol
(enol)
usually unstable

keto form
major product

c. C-H and C=O bonds (together) are stronger than C=C and O-H bonds (together). H is less acidic when bonded to C than O. Equilibrium lies toward the weaker acid/base pair.
d. Problem 7. Propyne hydration (H_2SO_4(dil) / H_2O / $HgSO_4$) yields propan-2-one (acetone) rather than propanal. Is the electrophilic addition step orientation Markovnikov or anti-Markovnikov? Why?

9.9.8. Like alkenes, alkynes react with borane (in an ether solvent, section 8.7.1).
a. Both π-bonds react even when only ⅓ mol of borane is added. Adding two mols of 2-methylbut-2-ene to borane yields the less reactive di(secondary-isoamyl)borane or disiamylborane, that still has one H to add to an alkyne. Disiamylborane is a sterically hindered, less reactive form of borane.

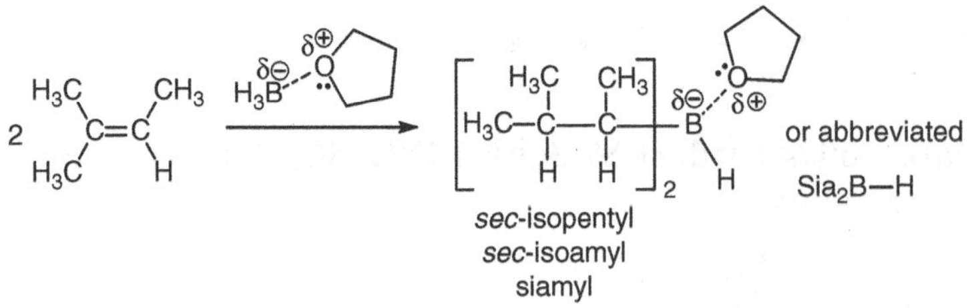

sec-isopentyl
sec-isoamyl
siamyl

or abbreviated
Sia_2B-H

b. Sia$_2$B-H adds anti-Markovnikov to alkynes. Oxidation yields the enol which tautomerizes to the keto form.

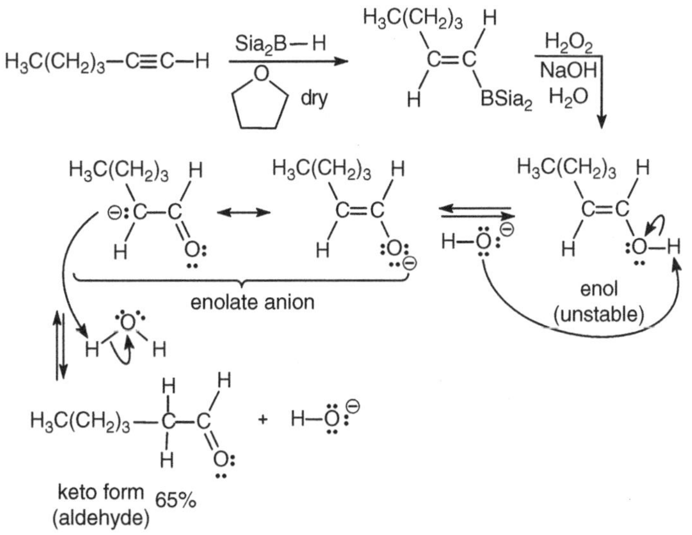

1. Hydroxide ion catalyzes keto-enol tautomerization.

c. Problem 8. Show how you could make the following from ethyne. You may use any inorganic reagents that you need. More than one synthetic step may be needed in some cases. Show all reaction conditions.

1. ethene

2. ethane

3. 1,2-dibromoethane

4. 1-chloroethene

5. 1,2-dichloroethane

6. ethanal

7. but-1-yne

8. but-2-yne

9. *cis*-but-2-ene

10. *trans*-but-2-ene

11. pent-1-yne

12. pent-2-yne

13. hex-3-yne

d. Problem 9. Show how you could synthesize each of the following from but-2-yne using any other needed organic or inorganic reagents.
   1. *cis*-but-2-ene

   2. *trans*-but-2-ene

   3. *meso*-2,3-dibromobutane

   4. (*R,R*)(*S,S*)-3-chlorobutan-2-ol (*threo*, racemic)

   5. *meso*-butane-2,3-diol

   6. racemic butane-2,3-diol

   7. butan-2-one

9.10.1. Mild alkyne oxidation (near neutral pH) is similar to mild alkene oxidation (section 8.14.5). Dehydration steps at the end lead to a 1,2-dicarbonyl. (Reaction actually has more steps than shown.)

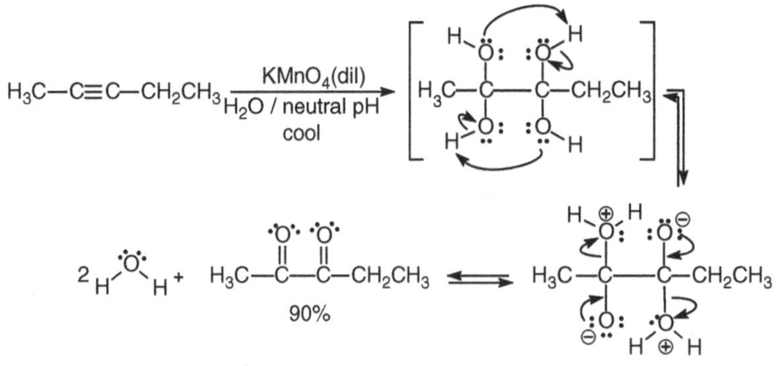

9.10.2. Severe basic oxidation cleaves both π-bonds and the σ-bond. Products are carboxylic acids after pH neutralization.

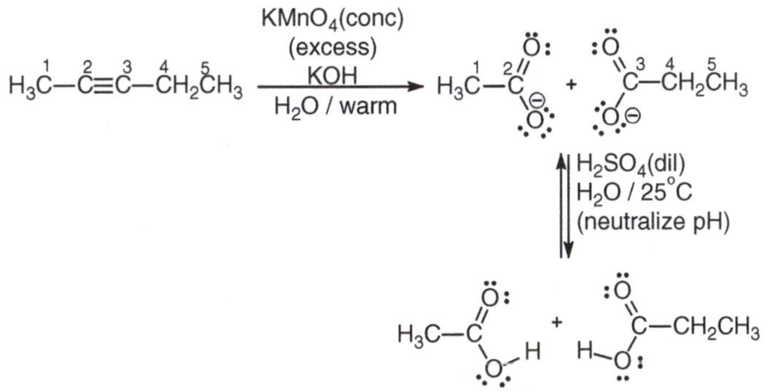

a. Terminal alkyne oxidation yields one mol of CO_2.

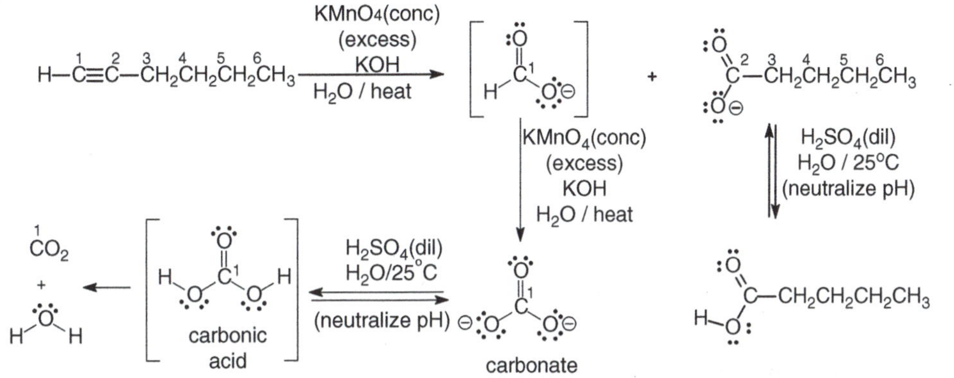

9.10.3. Alkyne oxidation with excess ozone (section 8.15.1) cleaves both π-bonds and the σ-bond.

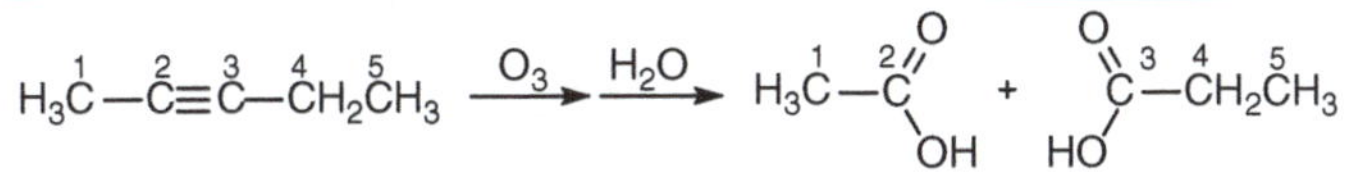

a. Acid product identification helps identify the original alkyne.
b. Problem 10. What ozonolysis (with 1. O_3 2. H_2O) products would be obtained from:
  1. pent-1-yne

  2. pent-2-yne

  3. 3-methylbut-1-yne

  4. penta-1,3-diene

  5. penta-1,4-diene

9.10.4. Synthesis example: make *cis*-2-methylhex-4-en-3-ol from ethyne and other organics with four or fewer C atoms.

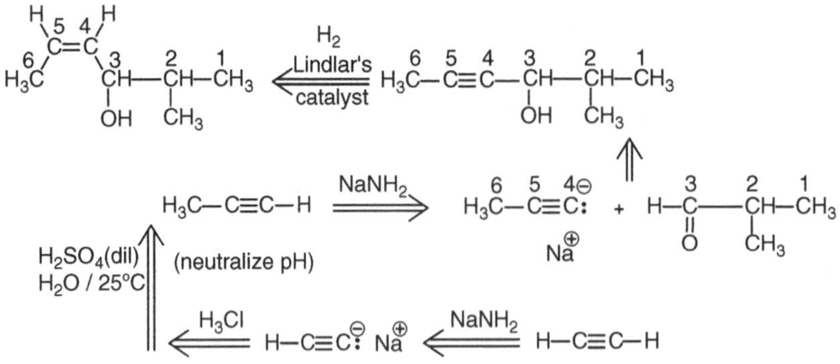

$$H_3C-C\equiv C-H \xrightarrow{NaNH_2} H_3C-C\equiv C:^{\ominus} Na^{\oplus} + H-\underset{\underset{CH_3}{|}}{\overset{\overset{O}{\|}}{C}}-CH-CH_3$$

$$\Downarrow$$

$$H_3C-C\equiv C-CH(OH)-CH(CH_3)-CH_3 \xrightarrow[\text{catalyst}]{H_2 \text{ Lindlar's}} \underset{H_3C}{\overset{H}{\underset{|}{C}}}=\underset{H}{\overset{}{C}}-CH(OH)-CH(CH_3)-CH_3$$

$$H_2SO_4(\text{dil}) \Big\uparrow \text{ (neutralize pH)}$$
$$H_2O / 25°C$$

$$\xleftarrow{H_3Cl} H-C\equiv C:^{\ominus} Na^{\oplus} \xleftarrow{NaNH_2} H-C\equiv C-H$$

9.10.5. Problem 10. Give product structures and names expected when but-1-yne reacts with:

1. 1 mol H_2 / Ni

2. 2 mols H_2 / Ni

3. 1 mol Br_2 / hexane (solvent)

4. 2 mols Br_2 / hexane (solvent)

5. 1 mol HCl(dry gas)

6. 2 mols HCl(dry gas)

7. H_2SO_4(dil) / H_2O / $HgSO_4$

8. $LiNH_2$

9. Product from 8 + CH_3CH_2Br

10. Product from 8 + $(CH_3)_3CCl$

11. CH_3CH_2MgBr

12. Product from 11 + H$_2$O

13. 1. O$_3$ 2. H$_2$O

14. KMnO$_4$(hot,conc) / KOH / H$_2$O

9.10.6. **Problem 11.** Describe chemical tests (reactions) that would allow you to distinguish between:
  a. pent-2-yne and *n*-pentane

  b. penta-1,3-diene and *n*-pentane

9.10.7. **Problem 12.** *Muscalure* is the common housefly sex pheromone. Draw structures for synthetic intermediates A and B, and muscalure.

n-C$_{13}$H$_{27}$C≡C—H + Li—CH$_2$CH$_2$CH$_2$CH$_3$ ⟶ A?   + CH$_3$CH$_2$CH$_2$CH$_3$

A + *n*-C$_8$H$_{17}$Br ⟶ B? (C$_{23}$H$_{44}$)

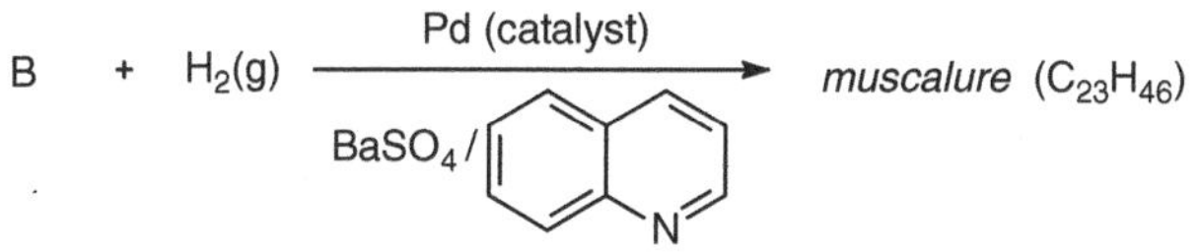

B + H$_2$(g) $\xrightarrow[\text{BaSO}_4 / \text{quinoline}]{\text{Pd (catalyst)}}$ muscalure (C$_{23}$H$_{46}$)

CHAPTER 10

10.1.1. Alcohols – from Arabic *al-kuhl* "the powder" or later "the essence (of wine)," have a hydroxyl (−OH) functional group bonded to an sp^3-hybridized C atom. IUPAC revised rules (1997) – put number just before parent suffix.

$$sp^3$$

H₃C—Ö:—H   H₃CCH₂—Ö:—H   H₃CCH₂CH₂—Ö:—H   H₃C—CH—CH₃ (with :Ö—H)

methanol   ethanol   n-propanol   2-propanol
methyl alcohol   ethyl alcohol   n-propyl alcohol   propan-2-ol
(wood alcohol)   (grain alcohol)   isopropanol
                                                      isopropyl alcohol
                                                      (rubbing alcohol)

H₃CCH₂CH₂CH₂—Ö:—H

n-butanol
n-butyl alcohol

10.1.2. Simple alcohols are cheap and commercially available. Useful as:
   a. synthetic reactants
   b. solvents
   c. fuels
   d. antiseptics
   e. beverages
   d. cosmetics
   f. drugs
   g. anti-freezes

10.2.1. Alcohol molecular structure – similar to water, except one H atom is replaced by an alkyl group.
   a. H-O-H bond angle < sp^3 ideal (109.5°). Lone pair electron clouds are larger than those of σ-bonds ⇒ lone pair electron repulsions squeeze angle down.
   b. Methyl — hydrogen steric interaction widens methanol C-O-H bond angle.

0.96 Å, 104.5° (water); 1.4 Å, 0.96 Å, 108.9° (methanol)

311

10.2.2. Methyl, 1°, 2°, 3°, benzylic, or allylic alcohol class is determined by R-groups bonded to the *carbinol* C atom (−OH bonded). Same class alcohols share distinctive properties.

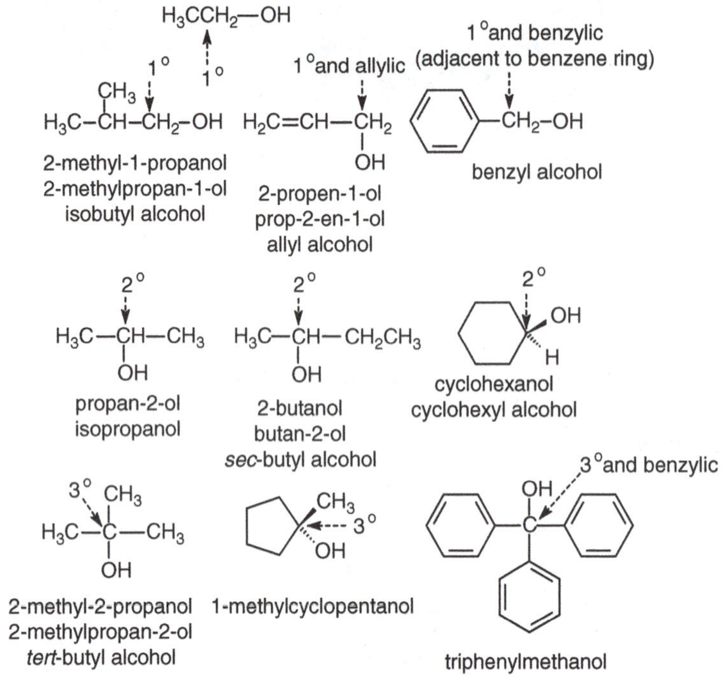

[Structures shown:
- 2-methyl-1-propanol / 2-methylpropan-1-ol / isobutyl alcohol (1°)
- 2-propen-1-ol / prop-2-en-1-ol / allyl alcohol (1° and allylic), H₃CCH₂—OH (1°)
- benzyl alcohol (1° and benzylic, adjacent to benzene ring)
- propan-2-ol / isopropanol (2°)
- 2-butanol / butan-2-ol / sec-butyl alcohol (2°)
- cyclohexanol / cyclohexyl alcohol (2°)
- 2-methyl-2-propanol / 2-methylpropan-2-ol / tert-butyl alcohol (3°)
- 1-methylcyclopentanol (3°)
- triphenylmethanol (3° and benzylic)]

a. **Enols** (vinyl alcohols), with −OH bonded to an *sp²* C, have some alcohol properties, but are enough different to be treated as a separate family. **A corresponding keto form is usually more stable than an enol form.**

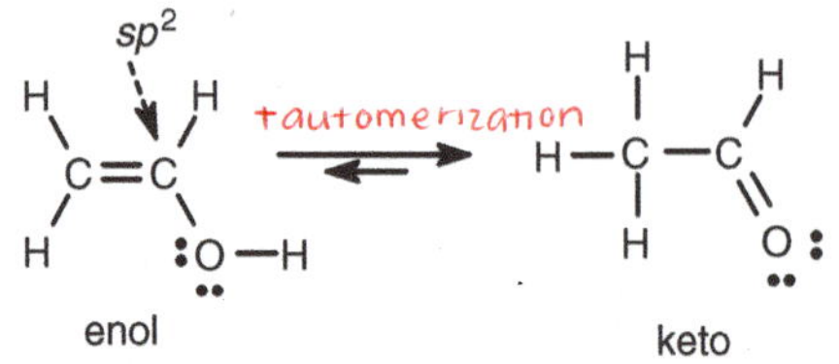

[Diagram: enol (H₂C=CH—OH with sp² label) ⇌ tautomerization ⇌ keto (H₃C—CHO)]

b. Phenols, Ar−OH, also have −OH bonded to an *sp²* C. They are enough different from alcohols to be treated as a separate family (Chapter 17).
1. Ar is a general symbol for phenyl (or other) aromatic (aryl) ring.

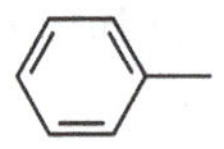

[Structure: simplest aryl group (phenyl)]

2. Phenol is an IUPAC parent name – naming rules are in section 17.15.2. A few examples:

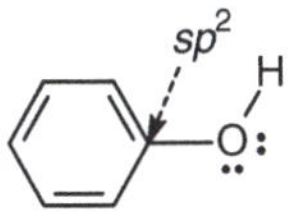

phenol

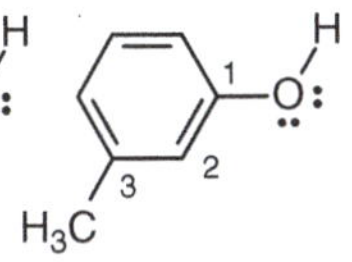

3-methylphenol
meta-methylphenol
meta-cresol

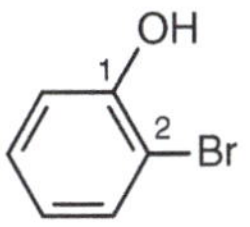

2-bromophenol
ortho-bromophenol

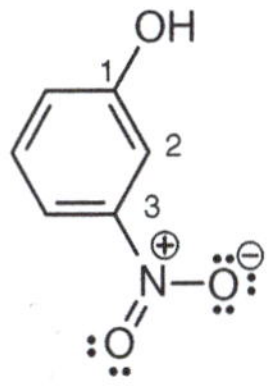

3-nitrophenol
meta-nitrophenol

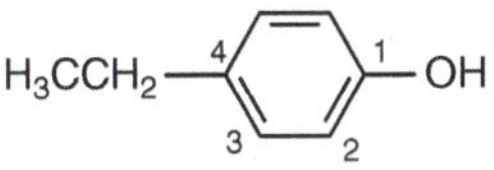

4-ethylphenol
para-ethylphenol

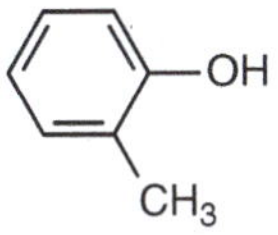

2-methylphenol
ortho-methylphenol
ortho-cresol

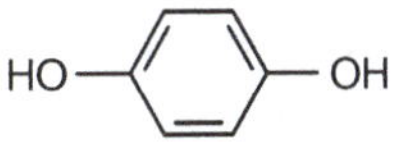

1,4-benzenediol
benzene-1,4-diol
hydroquinone
(a diphenol)

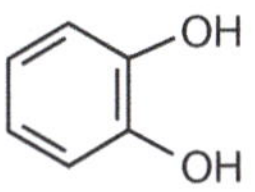

1,2-benzenediol
benzene-1,2-diol
catechol

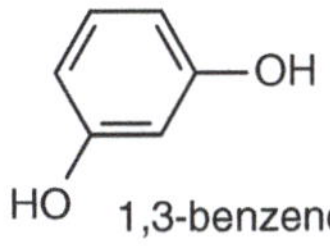

1,3-benzenediol
benzene-1,3-diol
meta-resorcinol

10.3.1. IUPAC alcohol naming rules:
  a. Find and name (parent alkane name) longest carbon chain.
  b. Drop last "e" and add "ol", parent alcohol suffix.
  c. If a higher priority functional group is present (section 2.17.4), –OH group is named as a branch (substituent) – hydroxy.
  d. Number –OH position as a prefix (as small as possible).
     1. –OH number takes priority over lower priority substituents.
     2. Examples:

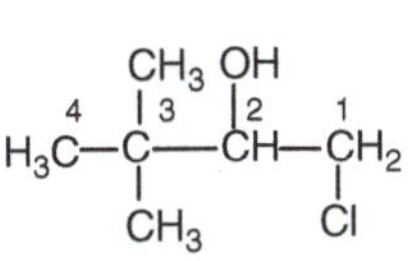

1-chloro-3,3-dimethyl-2-butanol
1-chloro-3,3-dimethyl-butan-2-ol

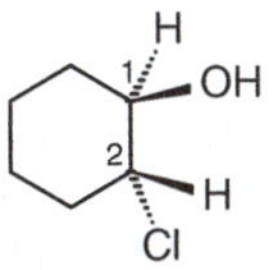

(1R,2R)-2-chlorocyclohexanol
trans-2-chlorocyclohexanol

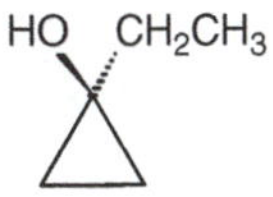

1-ethylcyclopropanol

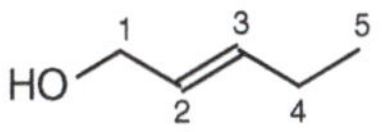

trans-pent-2-en-1-ol
(1997 revised name)

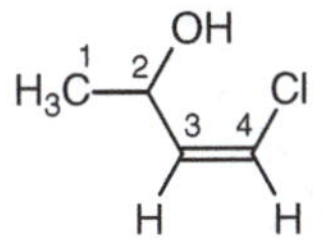

(Z)-4-chlorobut-3-en-2-ol

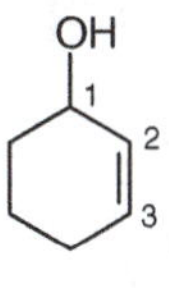

cyclohex-2-en-1-ol

4-hydroxy-2-pentanone
4-hydroxypentan-2-one

4-hydroxy-4-methyl-hexanoic acid

2-hydroxymethyl-cyclohexanone

***trans*-3-(2-hydroxyethyl)-cyclopentanol**

e. More than one −OH group – do not drop "e" but add parent suffixes diol, triol, etc. Common name glycol means vicinal 1,2-diol. Common names are based on alkene from which glycol is made (confusing).

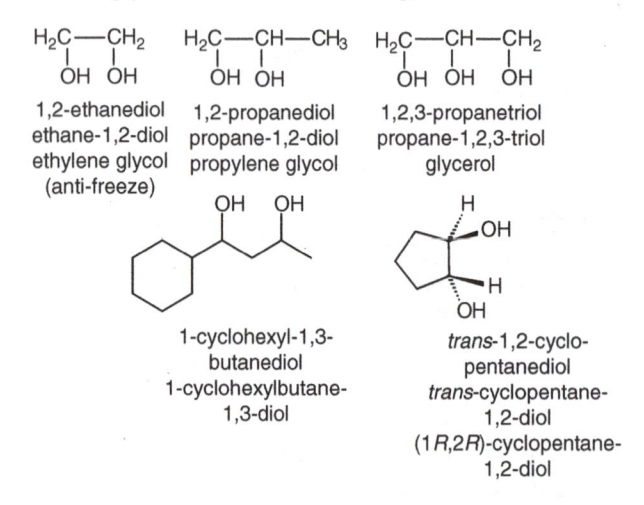

1,2-ethanediol
ethane-1,2-diol
ethylene glycol
(anti-freeze)

1,2-propanediol
propane-1,2-diol
propylene glycol

1,2,3-propanetriol
propane-1,2,3-triol
glycerol

1-cyclohexyl-1,3-butanediol
1-cyclohexylbutane-1,3-diol

***trans*-1,2-cyclo-pentanediol**
trans-cyclopentane-1,2-diol
(1*R*,2*R*)-cyclopentane-1,2-diol

f. Problem 1. Ignoring stereoisomers, draw the eight possible isomeric C$_5$H$_{11}$OH alcohols. Write IUPAC names, and label as 1°, 2°, or 3°.

hydrogen bonding ↑ alcohol's bp, mp, density, etc!

10.4.1. Alcohol properties – a cross between nonpolar alkane and polar water. Hydrogen bonding raises ethanol's bp 103°C above that of dimethyl ether (methoxymethane). Dipole-dipole attractions raise dimethyl ether's bp 17°C higher than propane's.

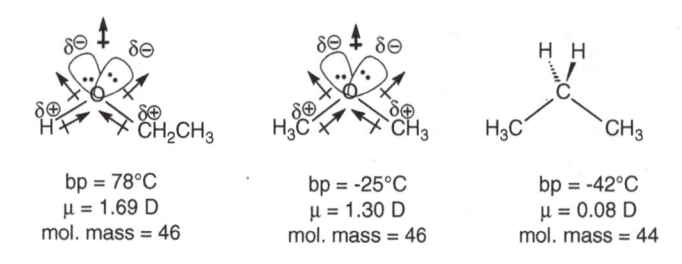

bp = 78°C
μ = 1.69 D
mol. mass = 46

bp = -25°C
μ = 1.30 D
mol. mass = 46

bp = -42°C
μ = 0.08 D
mol. mass = 44

a. Intermolecular R—OH H-bonds are possible, but not for R-O-R':

b. **Problem 2.** Refer to the common functional groups table (notes section 2.17.4). Which functional groups can form hydrogen bonds to themselves? To water molecules?

c. **Problem 3.** For many 2-substituted ethanols (GCH$_2$CH$_2$OH, where G = −OH, −NH$_2$, −F, −Cl, −Br, −OCH$_3$, −NHCH$_3$, -or −NO$_2$), the *gauche* conformation is more stable than *anti*. Draw a structure of this conformation, and suggest a reason for this behavior.

10.4.2. Alcohols have higher bp's, mp's, densities, and solubilities than similar molecular mass alkanes.

Compound	mol. mass	bp °C	Density (g/mL)	Solubility (g/100 g H$_2$O)
H$_3$COH	32.0	64.5	0.793	infinite
H$_3$CCH$_3$	30.1	−88.5	gas	insoluble
H$_3$CCH$_2$CH$_2$OH	60.1	97	0.80	infinite
H$_3$CCH$_2$CH$_2$CH$_3$	58.1	0	gas	insoluble

(Continued)

Compound	mol. mass	bp °C	Density (g/mL)	Solubility (g/100 g H$_2$O)
H$_3$CCHOHCH$_3$	60.1	82	0.79	infinite
(H$_3$C)$_3$CH	58.1	–12	gas	insoluble
H$_3$C(CH$_2$)$_3$OH	74.1	118	0.810	7.9
H$_3$C(CH$_2$)$_3$CH$_3$	72.2	36	0.626	insoluble
(H$_3$C)$_3$COH	74.1	83	0.789	infinite
H$_3$CCH$_2$CH(OH)CH$_3$	74.1	99.5	0.806	12.5
(H$_3$C)$_2$CHCH$_2$OH	74.1	108	0.802	10.0
C$_6$H$_5$—OH	94.1	182	1.071	9.3

10.4.3. Trends:

a. Higher molecular mass ⇒ higher mp, bp, density, and less H$_2$O soluble.
b. More isomeric branching ⇒ lower bp, density and more H$_2$O soluble. (Some exceptions exist.)
c. Fewer than four C atoms ⇒ infinitely H$_2$O soluble (miscible). More than five C atoms ⇒ limited H$_2$O solubility.
  1. Nonpolar alkyl groups are *hydrophobic* (water hating).
  2. Polar, H-bonding –OH groups are *hydrophilic* (water loving).
d. Larger R-groups make molecules more alkane-like (nonpolar).

e. Polyols have even higher bp's and solubilities, (ethane-1,2-diol, bp 197°C).
  1. Diols up to 7 C atoms are H$_2$O soluble.
f. Low boiling point alcohols have strong medicinal or fruity odors.
g. Like alkanes, most alkyl halides, alkenes, and alkynes, alcohols are flammable.
h. Problem 4. Although ethoxyethane (bp 35°C) (CH$_3$CH$_2$OCH$_2$CH$_3$) has a much lower bp than butan-1-ol (bp 118°C), the two compounds have nearly the same water solubility (8 g / 100 g H$_2$O at 25°C). Explain why by drawing hydrogen bonds between molecules.

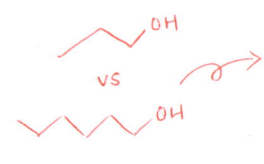

i. Problem 5. The disaccharide sucrose (table sugar), $C_{12}H_{22}O_{11}$, is a big molecule, extremely water soluble. How might its water solubility be explained?

sucrose can form many H-bonds to H_2O molecules.

j. Problem 6. Without referring to literature values, put the following compounds in order of increasing boiling point – lowest on the left and highest on the right.
hexan-3-ol, *n*-hexane, 2-methylpentan-2-ol, *n*-octanol, *n*-hexanol

10.5.1. Industrial alcohol sources. Methanol was made by distilling wood chips at high temperature excluding air.
  a. Newer method: methanol is made from *synthesis gas* (CO and H_2 mixture).

$$:C\equiv O: \;+\; 2\,H_2 \xrightarrow[\text{ZnO - }Cr_2O_3\text{(catalyst)}]{300\text{ - }400°C\,/\,200\text{ - }300\text{ atm}} H_3COH$$

  b. Synthesis gas is made from methane, alkanes, or coal (sources of C and H_2).

$$3\,C \;+\; 4\,H_2O \xrightarrow{\text{high temperature}} CO_2 \;+\; 2\,CO \;+\; 4\,H_2$$

  c. Many people went blind or died during Prohibition (1919-1933) from methanol ingestion (fatal dose, about 100 mL). CH_3OH is a not too toxic solvent (compared to alkyl halides) and fuel (Indy cars 1965–2006, then ethanol until 2013, and until now, E-85 (85% ethanol/15% gasoline). For every 1.0 g of gasoline burned, 1.7 g of methanol must be burned. Alcohol fires are easier to put out than gasoline fires, but burn with a nearly colorless flame.

10.5.2. Ethanol toxicity – hypnotic (sleep producer), liver damage (*cirrhosis* – scarring, reduced metabolic function, fatal dose about 200 mL ≈ 7/8 cup).

10.5.3. Ethanol sources:
  a. Sugar fermentation – probably world's oldest organic reaction – Babylonia (and others) about 6000 B.C. Molasses came from sugar cane or beets.

$$C_6H_{12}O_6 \xrightarrow{\text{yeast enzymes}} 2\,H_3CCH_2OH \;+\; 2\,CO_2$$
glucose ethanol

  1. Glucose conversion to ethanol is about 95%, but reaches a maximum 12-15% aqueous concentration.
  2. Distillation concentrates ethanol to 40–50% (80–100 proof) in hard liquors.

  b. Also from starch (grains – corn, wheat, rye, or barley).
    1. Cook or soak grain, add malt (sprouted barley) which converts starch into simple sugars (glucose).
  c. Lesser amounts of fusel oil (isopentyl-, *n*-propyl-, and isobutyl alcohols, and 2-methyl-1-butanol) are found when fermenting plant materials.
  d. Ethanol forms a *binary* (two component) *azeotrope* (bp 78.15°C, 95% ethanol / 5% H$_2$O) with a lower bp than either pure ethanol (78.3°) or water (100°C).
    1. The last 5% of water cannot be separated by fractional distillation, but can be removed by distilling a *ternary* (three component) azeotrope (made by adding benzene, toluene, or cyclohexane). H$_2$O can also be absorbed by adding a drying agent, such as CaO or CaSO$_4$, then Mg metal.

10.5.4. From WW II until about 20 years ago, ~90% of industrial ethanol was made by ethene hydration (from catalytic reforming). Government subsidized fermentation is common now, mostly from corn.

$$H_2C=CH_2 \xrightarrow[300°C\ /\ catalyst]{H_2O,\ 100-300\ atm} H_3C-CH_2-OH$$

  a. Catalysts: P$_2$O$_5$, WO$_2$, or specially treated clays.
  b. Drinkable ethanol is heavily taxed and regulated (by the Federal Bureau of Alcohol, Tobacco, and Firearms), is expensive, and has a state tax stamp on it.
  c. Denatured alcohol – fuel or solvent grade ethanol is not taxed and regulated and is much cheaper. It contains small amounts of an impurity (methanol, methyl isobutyl ketone, aviation gasoline) to prevent people from drinking it.
  d. Cars are equipped to run on gasohol (90% gasoline/10% ethanol) without modification. Running pure ethanol or 85% ethanol/ 15% gasoline (E-85) requires corrosion resistant fuel lines and carburetor parts. Fuel/air mixture must be richer and mileage is lower than for gasoline.

10.5.5. Isopropanol is made by propene hydration. It doesn't dry or absorb through skin as much as ethanol (or methanol), but is about as toxic as methanol.

$$H_3C-CH=CH_2 \xrightarrow[\substack{300°C\ /\ catalyst \\ (Mark.\ addition)}]{H_2O,\ 100-300\ atm} H_3C-\underset{OH}{\underset{|}{CH}}-CH_3$$

10.5.6. Petroleum – a nonrenewable fuel source. Burning increases atmospheric CO$_2$ (greenhouse gas).

10.5.7. Carbohydrates (C$_n$H$_{2n}$O$_n$) are renewable (from photosynthesis – plant chemical energy storage).

$$CO_2 + H_2O + energy \rightarrow carbohydrates + O_2$$

10.5.8. Man and animals burn carbohydrates, utilizing stored energy:

$$\text{carbohydrates} + O_2 = H_2O + CO_2 + \text{energy}$$

10.5.9. Burning fermentation alcohol (in an engine) produces no net atmospheric CO_2 increase, but currently about as much ethanol is burned as is required to produce it.

10.6.1. Like water, alcohols are weak acids or weak bases (sections 2.4 and 9.6.5).

a. [reaction of H₃CCH₂OH + H₂O ⇌ H₃CCH₂O⁻ + H₃O⁺]

$$K_a = \frac{[H_3CCH_2O^-][H_3O^+]}{[H_3CCH_2OH]} = 1.3 \times 10^{-16}$$

b. Strong acids dissociate nearly completely in alcohol solvents.

R—OH + H₂SO₄ ⇌ R—OH₂⁺ + HSO₄⁻

stronger base   stronger acid   weaker acid   weaker base

$$K_{eq} = \frac{[ROH_2^+][HSO_4^-]}{[ROH][H_2SO_4]} = \text{large}$$

R—OH + ⁻NH₂ + Na⁺ ⇌ R—O⁻ + H—NH₂ + Na⁺

stronger acid stronger base   weaker base weaker acid

$$K_{eq} = \frac{[RO^-][NH_3]}{[ROH][NH_2^-]} = \text{large}$$

Acid	K_a	pK_a
HCl	$1 \times 10^{+7}$	−7
H_3CCOOH	1.6×10^{-5}	4.8
C₆H₅—OH	1.0×10^{-10}	10.0
H_3COH	3.2×10^{-16}	15.5
H_2O	1.8×10^{-16}	15.74
H_3CCH_2OH	1.3×10^{-16}	15.9
$ClCH_2CH_2OH$	5.0×10^{-15}	14.3

Acid	K_a	pK_a
Cl_3CCH_2OH	6.3×10^{-13}	12.2
$(H_3C)_2CHOH$	3.2×10^{-17}	16.5
$(H_3C)_3COH$	1.0×10^{-18}	18.0
cyclohexanol (with H, OH)	1.0×10^{-18}	18.0

c. Be able to put above acids in increasing acidity order.
d. Alkoxides can be made by an alcohol + active metal redox reaction. (Water reacts similarly.) Unreactive 2° and 3° alcohols react faster with potassium.

$$2\,H-\overset{+1}{\overset{H}{O}}\!:^{-2} + 2\,\overset{0}{Na} \longrightarrow 2\,H-\overset{+1}{\overset{H}{O}}\!:^{\ominus}_{-2} + \overset{0}{H_2} + 2\,\overset{\oplus\,+1}{Na}$$

$$2\,H_3CCH_2-\overset{+1}{\overset{H}{O}}\!:^{-2} + 2\,\overset{0}{Na} \longrightarrow 2\,H_3CCH_2-\overset{\ominus}{O}\!:^{-2} + \overset{0}{H_2} + 2\,\overset{\oplus\,+1}{Na}$$

$$2\,H_3C-\underset{CH_3}{\overset{CH_3}{\underset{|}{\overset{|}{C}}}}-\overset{+1}{\overset{H}{O}}\!:^{-2} + 2\,\overset{0}{K} \longrightarrow 2\,H_3C-\underset{CH_3}{\overset{CH_3}{\underset{|}{\overset{|}{C}}}}-\overset{\ominus}{O}\!:^{-2} + \overset{0}{H_2} + 2\,\overset{\oplus\,+1}{K}$$

1. **Problem 7.** 2-Methylpropan-2-ol reacted with Na metal (one mol). When all the Na was consumed, bromoethane (one mol) was added and the mixture was warmed. After workup, a product (formula $C_6H_{14}O$) was obtained. What is it? Use electron pushing arrows to write a mechanism for its formation. Include all elementary steps, intermediates, formal electrical charges, and involved lone pair electrons.

e. R–OH's that are unreactive with Na, K, or $NaNH_2$ will react with sodium hydride (NaH).

$$R-\overset{..}{\underset{..}{O}}-H + \overset{\ominus}{H} + \overset{\oplus}{Na} \rightleftharpoons R-\overset{..}{\underset{..}{O}}\!:^{\ominus} + \overset{\oplus}{Na} + H-H$$

stronger acid   stronger base   weaker base   weaker acid

10.6.2. Alkoxide ions, R-O⁻, are usually stronger bases than hydroxide ion, H-O⁻.

a. Review: what is the strongest base that can be dissolved in H₂O? (H-O⁻ ion.) Review: which way does equilibrium lie?

$$R\text{-}\overset{\ominus}{\underset{..}{O}}\text{:} + H\text{-}\overset{..}{\underset{..}{O}}\text{-}H + \overset{\oplus}{Na} \rightleftharpoons R\text{-}\overset{..}{\underset{..}{O}}\text{:} + \overset{H}{\underset{H}{\overset{..}{O}^{\ominus}}} + \overset{\oplus}{Na}$$

stronger base   stronger acid   weaker acid   weaker base

R is <u>neither</u> H <u>nor</u> CH₃

10.6.3. What is the strongest base that can be dissolved in any given solvent? (That solvent's conjugate base.)

$$\overset{\oplus}{Na} + \text{:}\overset{\ominus}{NH_2} + H_3C\text{-}\overset{..}{\underset{..}{O}}\text{:}\text{-}H \rightleftharpoons H_3C\text{-}\overset{\ominus}{\underset{..}{O}}\text{:} + H\text{-}\overset{..}{N}H_2 + \overset{\oplus}{Na}$$

stronger base   stronger acid   weaker base   weaker acid
              (solvent)

10.6.4. Sections 9.6.3 and 10.6.1b show alcohol and alkoxide ion acidity and basicity relative to other functional groups.

a. More substituted carbinol C atom ⇒ weaker acid and stronger conjugate base (in alcohol solution).

— more methyl groups = weaker acid

stronger acid   weaker acid

$$H_3C\text{-}OH > H_3CCH_2\text{-}OH > \underset{CH_3}{\overset{CH_3}{HC}}\text{-}OH > \underset{CH_3}{\overset{CH_3}{H_3C\text{-}C}}\text{-}OH$$

weaker base   stronger base

$$H_3C\text{-}\overset{\ominus}{\underset{..}{O}}\text{:} < H_3CCH_2\text{-}\overset{\ominus}{\underset{..}{O}}\text{:} < \underset{CH_3}{\overset{CH_3}{HC}}\text{-}\overset{\ominus}{\underset{..}{O}}\text{:} < \underset{CH_3}{\overset{CH_3}{H_3C\text{-}C}}\text{-}\overset{\ominus}{\underset{..}{O}}\text{:}$$

— more methyl groups w/ oxygen anion = stronger base

b. Aqueous alcohol acidities were once explained (erroneously) by alkyl group inductive electron release in the conjugate base (alkoxide ion). Larger R-group ⇒ less stable alkoxide ion, and less acidic conjugate acid alcohol.

Already has a full
- charge. O does not
want additional
electron density.

c. Gas phase methyl, 1°, 2°, and 3° alcohol acidities and basicities are reversed (methanol is weakest acid).

$$\text{weaker acid} \quad\quad\quad \text{gas phase} \quad\quad\quad\quad \text{stronger acid}$$

$$H_3C-OH \; < \; H_3CCH_2-OH \; < \; \underset{CH_3}{\overset{CH_3}{HC}}-OH \; > \; \underset{CH_3}{\overset{CH_3}{H_3C-C}}-OH$$

$$\text{stronger base} \quad\quad\quad\quad\quad\quad\quad\quad \text{weaker base}$$

$$H_3C-\ddot{O}:^{\ominus} \; > \; H_3CCH_2-\ddot{O}:^{\ominus} \; > \; \underset{CH_3}{\overset{CH_3}{HC}}-\ddot{O}:^{\ominus} \; > \; \underset{CH_3}{\overset{CH_3}{H_3C-C}}-\ddot{O}:^{\ominus}$$

d. Liquid phase alcohol acidity is now thought to be a solvation effect – a bulky alkoxide ion is harder for water (or alcohol) to solvate ⇒ less stable ⇒ lower equilibrium concentration ⇒ less acidic alcohol (conjugate acid).

close solvent approach – stabilizes anion – reduces basicity

hindered solvation – destabilizes anion – increases basicity

e. Enhanced 2-chloroethanol acidity is due to the inductive effect.

anion is stabilized by inductive delocalization of – charge

f. Phenol (section 17.15) is 10^8 times more acidic than cyclohexanol due to the resonance effect. Hydroxide ion almost completely neutralizes phenol.

[mechanism diagram showing phenol + water equilibrium with resonance-stabilized phenoxide anion (major), labeled "phenol weaker acid", "weaker base", "major", "hybrid – charge delocalized – anion stabilized stronger base", "stronger acid"; and below, phenol + OH⁻ + Na⁺ → phenoxide + H₂O + Na⁺, labeled "stronger acid stronger base" → "weaker base weaker acid"]

g. Alkoxide anions are usually strong bases and nucleophiles (section 11.14).
h. Problem 8. Which alcohol in each pair is more acidic? Explain why.
   1. propan-2-ol or 1,1,1,3,3,3-hexafluoropropan-2-ol
      1,1,1,3,3,3-hexafluoropropan-2-ol is a stronger acid

   2. propan-1-ol or propan-1,2,3-triol
      Propan-1,2,3-triol bc it contains more -OH's than propan-1-ol

i. Problem 9. Which alcohol in previous problem is the stronger nucleophile?
   Nucleophilicity is usually inverse of acidity, so propan-2-ol & propan-1-ol are stronger nucleophiles.

j. Problem 10.
   a. First order kinetics are observed when 3-bromo-2,2-dimethylbutane is heated with $CH_3CH_2O^-\ Na^+$(dil) in CH_3CH_2OH (solvent) or with CH_3CH_2OH alone. Elimination product alkenes I and II are formed. Use electron pushing arrows to draw a mechanism that explains their formation. Include all elementary steps, intermediates, formal electrical charges, and involved lone pair electrons.

   [structures of alkenes I, II, III shown: I = 2,3-dimethyl-2-butene; II = 2,3-dimethyl-1-butene; III = 3,3-dimethyl-1-butene]

b. Second order kinetics are observed when the 3-bromo-2,2-dimethylbutane is heated with $CH_3CH_2O^- Na^+$(conc) and elimination product alkene III forms exclusively. By what mechanism does alkene III form? Why are no I and II formed?

c. What substitution product(s) would you expect to compete in reaction *a*? Reaction *b*? Explain.

10.7.1. Alcohols were made before. Review examples:
  a. Alkyl halide substitution – S_N2 (section 6.7) or S_N1 (section 6.13).

  b. Alkene hydration (section 8.4). Markovnikov addition.

  c. Oxymercuration - demercuration (section 8.5). Markovnikov addition.

  d. Hydroboration - oxidation (section 8.7). Anti-Markovnikov addition.

e. Mild alkene oxidation (*syn*-hydroxylation, section 8.14.5) to 1,2-diols.

f. Mild alkene oxidation (*trans*-hydroxylation, section 8.13.1).

10.8.1. Victor Grignard's (Nobel Prize, 1912) organomagnesium reagents (sections 3.5.7 and 9.7.2).

a. fast RI > RBr > RCl >> RF slow
b. RF is very unreactive and rarely used. R can be methyl, 1°, 2°, 3°, benzylic, allylic, aryl or vinyl.
c. Simplified formation:

d. Actual structure contains two, three, or four R-Mg-X units associated with several ether (solvent) molecules. The equilibrium below also exists.

$$2\ R{-}MgX \rightleftharpoons R_2Mg + MgX_2$$

1. C-Mg bond is about 34% ionic / 66% covalent. With δ- on C, Grignard reagents are very strong bases (and nucleophiles).
   e. Despite their ionic character, Grignard reagents are soluble in fairly nonpolar ether solvents. Unless an ether solvent is used, the Grignard reagent coats the Mg metal surface, and reaction stops.

   f. Coupling (section 3.5.9) is usually slow at 35°C. Example:

   $$2\ H_3C-Mg-Br \xrightarrow{dry\ ether} H_3C-CH_3 + byproducts$$

10.8.2. Organolithiums are made similarly, except hydrocarbon solvents can be used.

$$H_3CCH_2CH_2CH_2-Br \xrightarrow[2\ Li]{hexane\ (solvent)} H_3CCH_2CH_2CH_2-Li + LiBr(s)$$

$$H_2C=CHCl \xrightarrow[2\ Li]{pentane\ (solvent)} H_2C=CHLi + LiCl(s)$$

$$C_6H_5-Br \xrightarrow{dry\ ether\ (solvent)} C_6H_5-Li + LiBr$$

10.8.3. Organolithiums react similarly to Grignard reagents. They are somewhat more reactive (43% ionic, 57% covalent), but couple less. R-Li slowly reacts with ether:

$$R-Li + H-CH_2-CH_2-OCH_2CH_3 \rightarrow R-H + H_2C=CH_2 + Li^+\ ^-OCH_2CH_3$$

10.9.1. Aldehyde and ketone C=O reaction (basic nucleophilic addition mechanism, section 18.12).

R', R = H, alkyl, or aryl

a. Acid/water addition (HCl(dil), H$_2$SO$_4$(dil), or NH$_4$Cl(dil)) liberates R–OH from the magnesium salt (an acid/base reaction) and dissolves the mixed Mg(OH)X salt after the nucleophilic addition (Grignard) step. NH$_4$Cl(aq) (weak acid) minimizes product alcohol dehydration.

b. Examples:

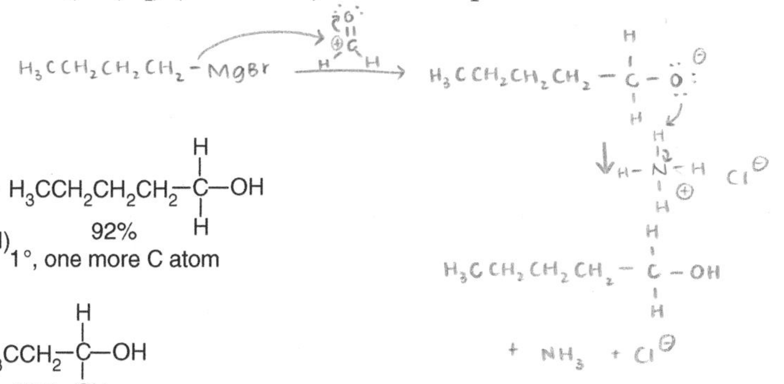

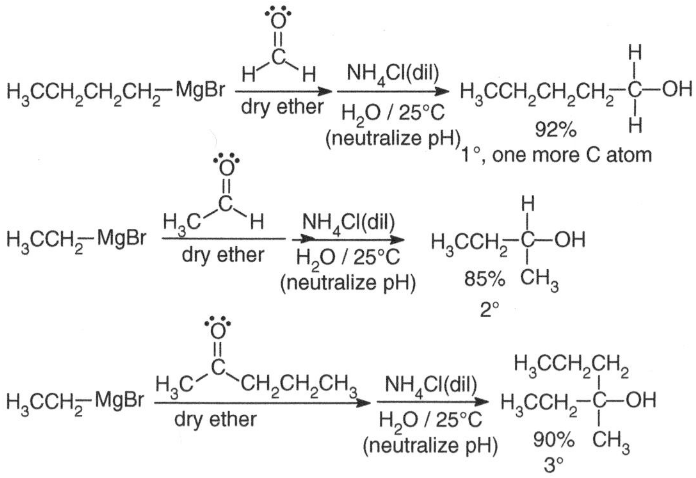

H$_3$CCH$_2$CH$_2$CH$_2$–MgBr $\xrightarrow[\text{dry ether}]{}$ H–C(=O)–H $\xrightarrow[\text{H}_2\text{O / 25°C (neutralize pH)}]{\text{NH}_4\text{Cl(dil)}}$ H$_3$CCH$_2$CH$_2$CH$_2$–CH(H)–OH   92%   1°, one more C atom

H$_3$CCH$_2$–MgBr $\xrightarrow[\text{dry ether}]{}$ H$_3$C–C(=O)–H $\xrightarrow[\text{H}_2\text{O / 25°C (neutralize pH)}]{\text{NH}_4\text{Cl(dil)}}$ H$_3$CCH$_2$–CH(CH$_3$)–OH   85%   2°

H$_3$CCH$_2$–MgBr $\xrightarrow[\text{dry ether}]{}$ H$_3$C–C(=O)–CH$_2$CH$_2$CH$_3$ $\xrightarrow[\text{H}_2\text{O / 25°C (neutralize pH)}]{\text{NH}_4\text{Cl(dil)}}$ H$_3$CCH$_2$–C(H$_3$CCH$_2$CH$_2$)(CH$_3$)–OH   90%   3°

H$_3$CCH$_2$CH$_2$CH$_2$–MgBr $\xrightarrow{\text{H–C(=O)–H}}$ H$_3$CCH$_2$CH$_2$CH$_2$–C(H)(H)–O:$^\ominus$

↓ H–N(H)(H)–H$^\oplus$ Cl$^\ominus$

H$_3$CCH$_2$CH$_2$CH$_2$–C(H)(H)–OH + NH$_3$ + Cl$^\ominus$

10.9.2. A new C-C bond joins two smaller molecules into a larger molecule. C-C bond forming methods review (section 9.7.2):

a. Coupling – Corey-House reaction (making alkanes, section 3.5.11).
b. Alkene polymerization (section 8.16).
c. Acetylide or cyanide ion reaction with alkyl halides, aldehydes, and ketones (section 9.7.4).

10.9.3. Summary – Grignard reaction products:

C=O group type	Product
methanal	1° alcohol, (one C atom longer chain)
other aldehydes	2° alcohols
ketones	3° alcohols

a. More than one synthesis can yield a desired product. The one where fragments are smallest is usually best.

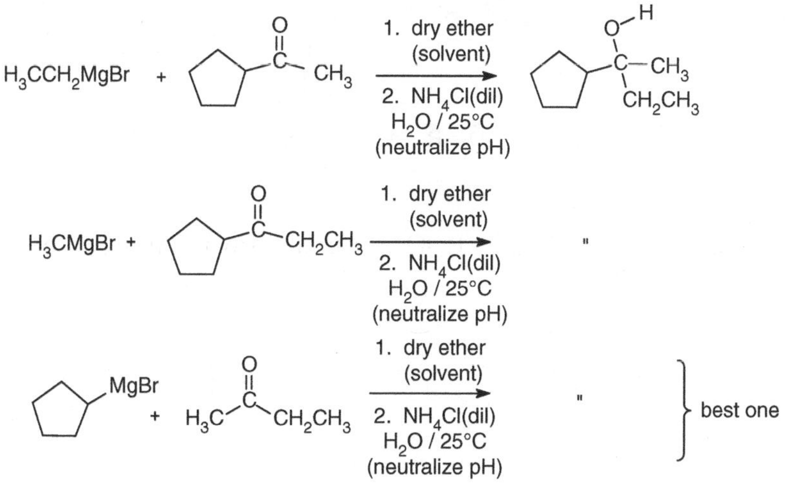

H$_3$CCH$_2$MgBr + cyclopentyl–C(=O)–CH$_3$ $\xrightarrow[\text{2. NH}_4\text{Cl(dil) H}_2\text{O / 25°C (neutralize pH)}]{\text{1. dry ether (solvent)}}$ cyclopentyl–C(OH)(CH$_3$)(CH$_2$CH$_3$)

H$_3$CMgBr + cyclopentyl–C(=O)–CH$_2$CH$_3$ $\xrightarrow[\text{2. NH}_4\text{Cl(dil) H}_2\text{O / 25°C (neutralize pH)}]{\text{1. dry ether (solvent)}}$ "

cyclopentyl–MgBr + H$_3$C–C(=O)–CH$_2$CH$_3$ $\xrightarrow[\text{2. NH}_4\text{Cl(dil) H}_2\text{O / 25°C (neutralize pH)}]{\text{1. dry ether (solvent)}}$ "   } best one

10.9.4. More examples:

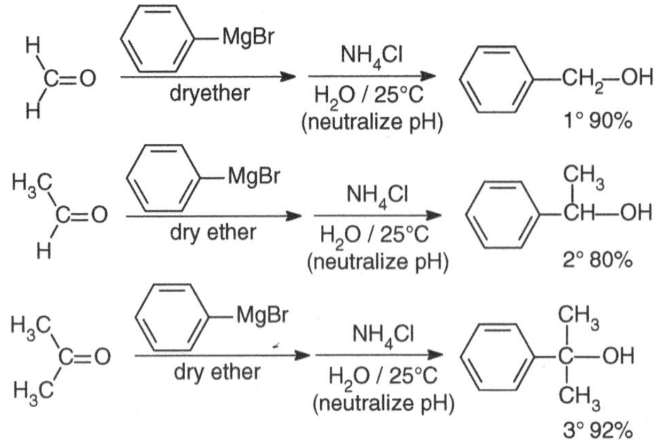

[Reaction 1: H₂C=O + PhMgBr / dry ether, then NH₄Cl / H₂O / 25°C (neutralize pH) → Ph–CH₂–OH, 1° 90%]

[Reaction 2: CH₃CHO + PhMgBr / dry ether, then NH₄Cl / H₂O / 25°C (neutralize pH) → Ph–CH(CH₃)–OH, 2° 80%]

[Reaction 3: (CH₃)₂C=O + PhMgBr / dry ether, then NH₄Cl / H₂O / 25°C (neutralize pH) → Ph–C(CH₃)₂–OH, 3° 92%]

10.9.5. Acid halide or ester C=O Grignard reaction (better with esters).

  a. Acid halides and esters are carboxylic acid *derivatives* where –X or –OR replace the –OH group (Chapter 21).

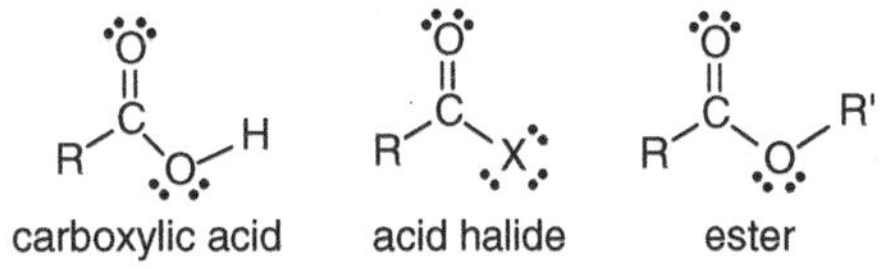

  [Structures shown: carboxylic acid (R–C(=O)–O–H), acid halide (R–C(=O)–X), ester (R–C(=O)–O–R')]

  b. Two mols of RMgX are needed per mol of acid halide or ester. Mechanism: basic nucleophilic acyl substitution (section 21.5). Example (ester case):

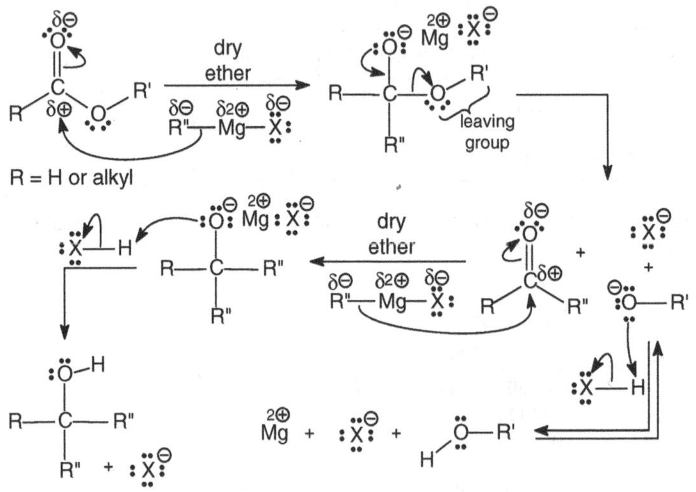

  [Mechanism diagram showing: R–C(=O)–O–R' attacked by R''–Mg–X in dry ether, forming tetrahedral intermediate with Mg²⁺ X⁻, then R'–O⁻ leaving group departs giving ketone R–C(=O)–R'', which is attacked by second R''–Mg–X to give tetrahedral alkoxide R–C(O⁻)(R'')(R''), then protonation by X–H gives R–C(OH)(R'')(R'') + X⁻, plus Mg²⁺ + X⁻ + H–O–R']

  R = H or alkyl

  c. A strong base and slow leaving group, R'–O⁻ leaves anyway because second step is exothermic (section 21.5.7).

d. Example:

e. Example:

10.9.6. Oxirane (ethylene oxide) reaction – lengthens chain by two C atoms, and yields a 1° alcohol.

10.10.1. Grignard reagents react with H₂O yielding alkanes (section 3.5.8), an indirect alkyl halide reduction. Deuterium can be added.

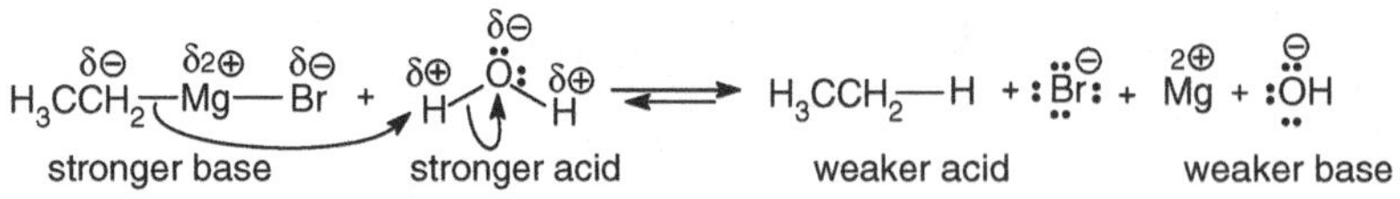

a. $K_a = 10^{-50}$ for ethane (a very weak acid).
b. Example:

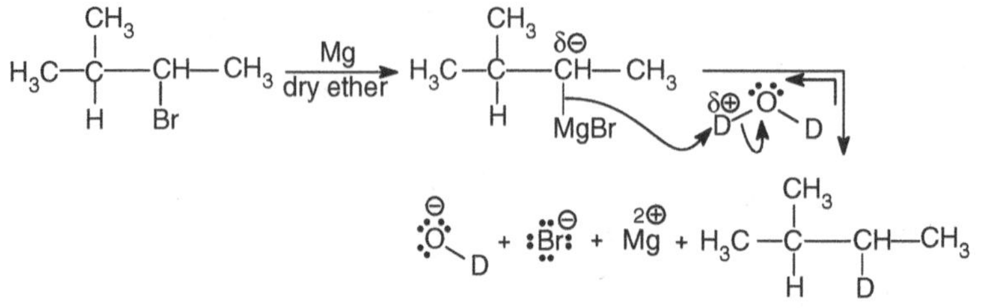

10.10.2. Any functional group as or more acidic (electrophilic) than a terminal alkyne will neutralize a Grignard reagent.

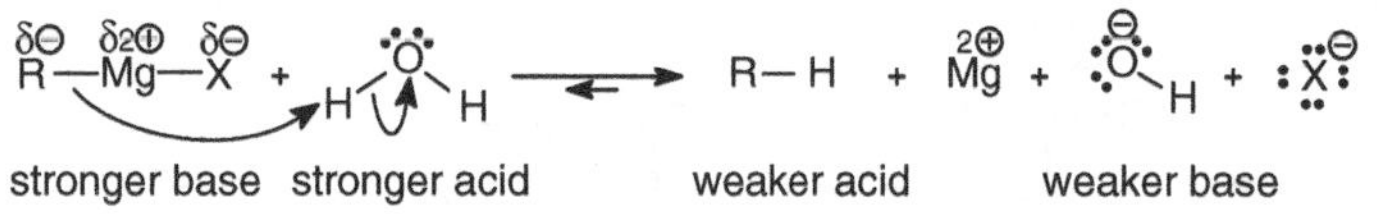

a. A Grignard reaction is often done under a dry N₂ or Ar atmosphere.

10.10.3. Some substances and groups must be excluded during Grignard reagent formation. Neutralize Grignard reagents: H₂O, R−OH, R−NHR', RCOOH. Attack Grignard reagents: O₂, CO₂, R−NO₂, RCOOR', R−COR', R−CHO, R−C≡N, and RR'C=N-R".

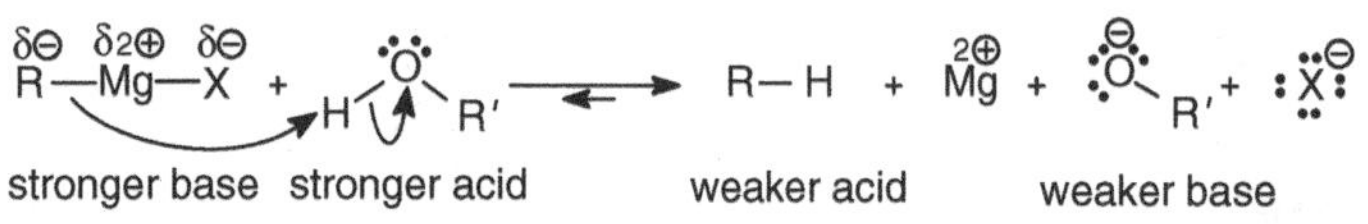

a. Inert substances that can be present: alkanes, alkenes (usually), ROR (ethers), or Ar-Cl.

10.10.4. *Protecting groups* (sections 13.10 and 18.11.20) can sometimes circumvent group exclusions above.

10.11.1. Hydride ion (H:⁻, a strong base and nucleophile) reduces a C=O group to an alcohol after pH neutralization (section 18.21).

a. A B-H or Al-H bond in NaBH₄ or LiAlH₄ is partly covalent, making B-H or Al-H weaker bases (than NaH) but stronger nucleophiles. Al-H bonds are more ionic than B-H bonds because B is more electronegative than Al. LiAlH₄ is much more reactive than NaBH₄.

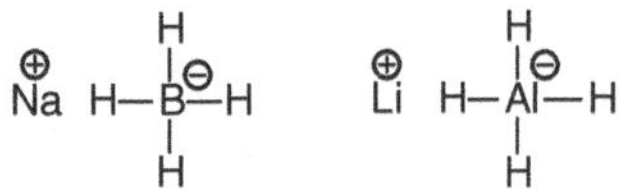

10.11.2. NaBH₄ only reduces aldehydes and ketones, and is often easier to use.

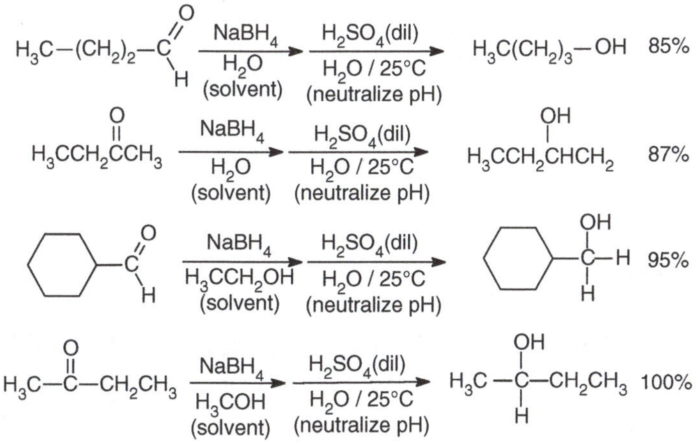

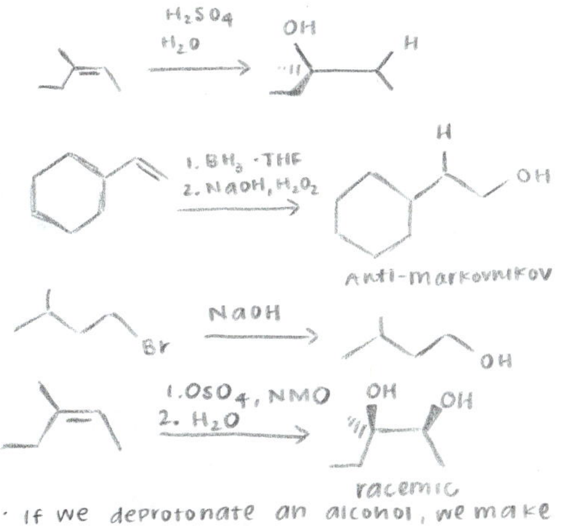

$$H_3C-(CH_2)_2-CHO \xrightarrow[\text{H}_2\text{O (solvent)}]{\text{NaBH}_4} \xrightarrow[\text{H}_2\text{O / 25°C (neutralize pH)}]{\text{H}_2\text{SO}_4\text{(dil)}} H_3C(CH_2)_3-OH \quad 85\%$$

$$H_3CCH_2COCH_3 \xrightarrow[\text{H}_2\text{O (solvent)}]{\text{NaBH}_4} \xrightarrow[\text{H}_2\text{O / 25°C (neutralize pH)}]{\text{H}_2\text{SO}_4\text{(dil)}} H_3CCH_2CH(OH)CH_2 \quad 87\%$$

cyclohexyl-CHO $\xrightarrow[\text{H}_3\text{CCH}_2\text{OH (solvent)}]{\text{NaBH}_4} \xrightarrow[\text{H}_2\text{O / 25°C (neutralize pH)}]{\text{H}_2\text{SO}_4\text{(dil)}}$ cyclohexyl-CH(OH)-H 95%

$$H_3C-CO-CH_2CH_3 \xrightarrow[\text{H}_3\text{COH (solvent)}]{\text{NaBH}_4} \xrightarrow[\text{H}_2\text{O / 25°C (neutralize pH)}]{\text{H}_2\text{SO}_4\text{(dil)}} H_3C-CH(OH)-CH_2CH_3 \quad 100\%$$

10.11.3. LiAlH₄ reduces all C=O functional groups. Examples:

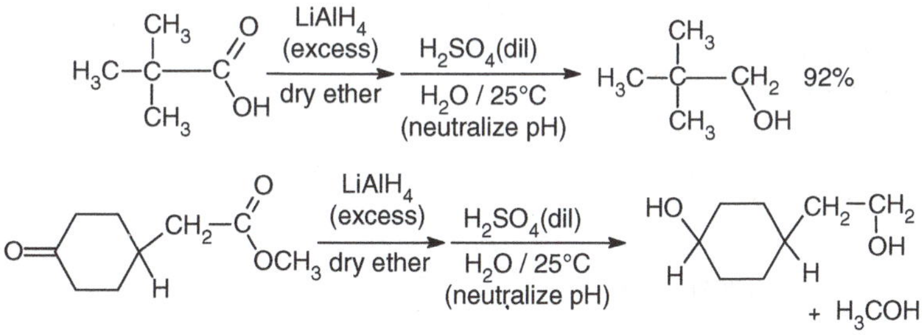

(CH₃)₃C-COOH $\xrightarrow[\text{dry ether}]{\text{LiAlH}_4 \text{ (excess)}} \xrightarrow[\text{H}_2\text{O / 25°C (neutralize pH)}]{\text{H}_2\text{SO}_4\text{(dil)}}$ (CH₃)₃C-CH₂OH 92%

4-oxo-cyclohexyl-CH₂-COOCH₃ $\xrightarrow[\text{dry ether}]{\text{LiAlH}_4 \text{ (excess)}} \xrightarrow[\text{H}_2\text{O / 25°C (neutralize pH)}]{\text{H}_2\text{SO}_4\text{(dil)}}$ 4-HO-cyclohexyl-CH₂-CH₂OH + H₃COH

a. NaBH₄ reacts slowly with water, but LiAlH₄ reacts violently. H₂ can catch fire in air.

$$\text{LiAlH}_4 + 4 \text{H}_2\text{O} \longrightarrow 4 \text{H}_2(g) + \text{LiOH} + \text{Al(OH)}_3 + \text{heat}$$

10.11.4. Aldehydes and ketones can be reduced to alcohols catalytically.
   a. A common catalyst is a reactive Ni powder (Raney Ni) containing absorbed H₂. It is made by treating Ni / Al alloy with conc. NaOH solution.
   b. A C=C is easier to reduce catalytically than a C=O. NaBH₄ and LiAlH₄ usually do not reduce C=C.

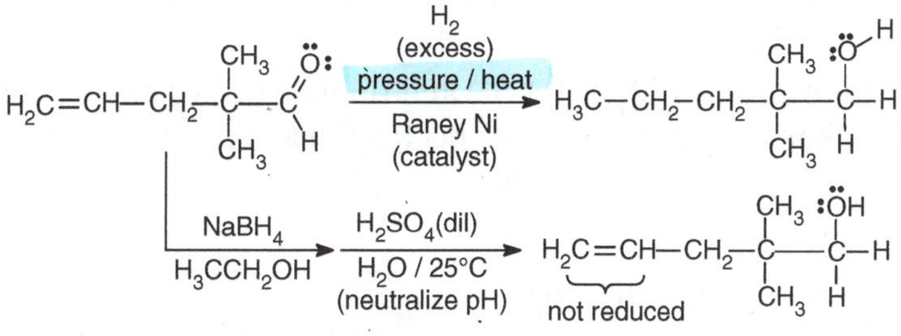

H₂C=CH-CH₂-C(CH₃)₂-CHO $\xrightarrow[\text{Raney Ni (catalyst)}]{\text{H}_2 \text{ (excess) pressure / heat}}$ H₃C-CH₂-CH₂-C(CH₃)₂-CH₂OH

H₂C=CH-CH₂-C(CH₃)₂-CHO $\xrightarrow[\text{H}_3\text{CCH}_2\text{OH}]{\text{NaBH}_4} \xrightarrow[\text{H}_2\text{O / 25°C (neutralize pH)}]{\text{H}_2\text{SO}_4\text{(dil)}}$ H₂C=CH-CH₂-C(CH₃)₂-CH₂OH (C=C not reduced)

CHAPTER 10   331

c. LiAlH₄ easily reduces (unreactive) esters. High temperature and H₂ pressure are needed to reduce esters catalytically. CuO · CuCr₂O₄ is an even more active catalyst than Raney Ni.

$$R-\underset{\underset{O}{\|}}{C}-OR' \xrightarrow[\substack{175°C \\ 5000\ psi}]{\substack{H_2 \\ CuOCuCr_2O_4 \\ (active\ catalyst)}} R-CH_2-OH + R'-OH$$

or LiAlH₄ / dry ether, then H₂SO₄ (dil) / H₂O / 25°C (neutralize pH)

10.11.5. Reactant availability limits alcohol preparation reactions. Alkenes, alkyl halides, aldehydes, ketones, carboxylic acids, acid halides, and esters are commonly made from alcohols (rather than the reverse).

10.12.1. Thiols – sulfur analogues of alcohols.

10.12.2. IUPAC parent name: *thiol*; branch name *mercapto-* ("mercury capturing agent").

   a. Name as alkane, but do not drop "e." Add "thiol."

   H₃C—SH   H₃CCH₂CH₂CH₂—SH   HS—CH₂CH₂—OH
   methanethiol   butane-1-thiol   2-mercaptoethan-1-ol
   methyl mercaptan   n-butyl mercaptan

10.12.3. These two compounds are in skunk oil.

   H₃C—CH(CH₃)—CH₂—CH₂—SH   H₃CCH=CHCH₂—SH
   3-methylbutanethiol   but-2-ene-1-thiol

10.12.4. Thiols are about as acidic as phenols. NaOH neutralizes them.

H₃CCH₂—S—H + H—O—H ⇌ H₃CCH₂—S:⁻ + H—O⁺H—H $pK_a = 10.5$
weaker acid   weaker base   stronger base   stronger acid

H₃CCH₂—S—H + :O⁻—H + Na⁺ → H₃CCH₂—S:⁻ + H—O—H
stronger acid   stronger base   weaker base   weaker acid

   a. Thiolate anions are more stable than alkoxide anions, despite O electronegativity > S electronegativity. A – charge on S is delocalized more than on O because S is larger – size factor dominates.

10.12.5. S_N2 reactions can yield thiols. Hydrosulfide (H-S⁻) and thiolate (R-S⁻) anions are weaker bases than hydroxide or alkoxide, but are stronger nucleophiles. S is more polarizable than O, and is less solvated.

H₃CCH₂—Br + Na⁺ :S—H⁻ $\xrightarrow{H_2O\ /\ warm}$ H₃CCH₂—S—H + Na⁺ + :Br:⁻

10.12.6. Mild thiol oxidation yields disulfides (important in biochemistry). Mild reduction reforms thiols.

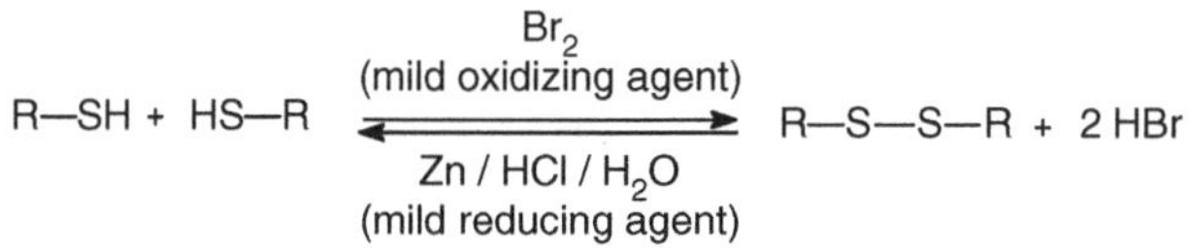

$$R\text{—}SH + HS\text{—}R \underset{\substack{Zn/HCl/H_2O \\ \text{(mild reducing agent)}}}{\overset{\substack{Br_2 \\ \text{(mild oxidizing agent)}}}{\rightleftharpoons}} R\text{—}S\text{—}S\text{—}R + 2\,HBr$$

10.12.7. KMnO$_4$, HNO$_3$, or NaOCl (bleach) (strongly) oxidize thiols to sulfonic acids. Example:

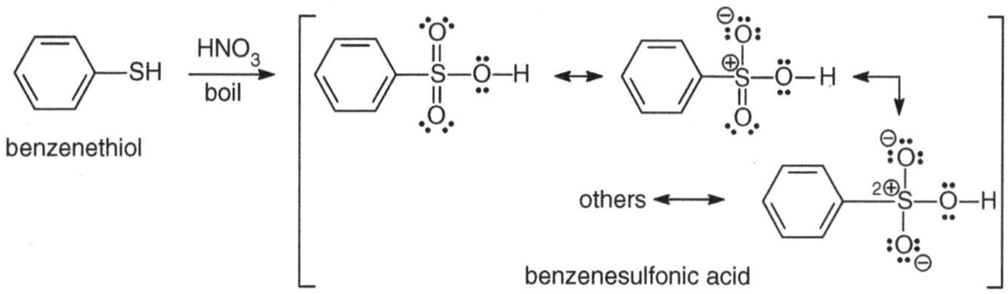

Ph—SH $\xrightarrow{\text{HNO}_3,\ \text{boil}}$ benzenesulfonic acid (resonance structures shown)

benzenethiol → benzenesulfonic acid

CHAPTER 11

11.1.1. Alcohols can be converted into many other functional groups. Examples:

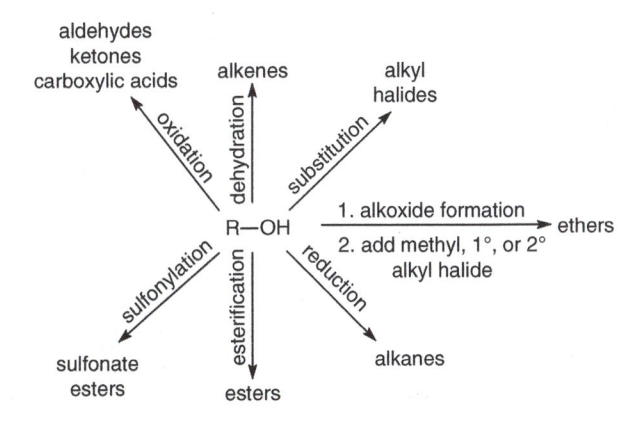

- oxidation → aldehydes, ketones, carboxylic acids
- dehydration → alkenes
- substitution → alkyl halides
- 1. alkoxide formation; 2. add methyl, 1°, or 2° alkyl halide → ethers
- reduction → alkanes
- esterification → esters
- sulfonylation → sulfonate esters

(center: R—OH)

11.1.2. Like alkenes (section 8.14.4), alcohols can be oxidized.

11.1.3. C can be assigned oxidation numbers atoms in an organic molecule.

335

336 SEQUENTIAL ORGANIC CHEMISTRY I

11.2.1. Oxidation reaction table. Reagents do not usually oxidize alcohols all the way to CO_2. [O] represents an oxidizing agent in left column.

Oxidizing Agent	Reactions and Products
Basic: $KMnO_4$/ KOH (hot conc) NaOCl (warm)	methyl $H_3C-OH \xrightarrow{[O]} H-C(=O)-O^\ominus \xrightarrow{H_2SO_4(dil)}_{H_2O/25°C} H-C(=O)-OH \xrightarrow{[O]} CO_2 + H_2O$ 1° $R-CH_2-OH \xrightarrow{[O]} R-C(=O)-O^\ominus \xrightarrow{H_2SO_4(dil)}_{H_2O/25°C} R-C(=O)-OH$ 2° $RR'CH-OH \xrightarrow{[O]} R-C(=O)-R'$ 3° $RR'R''C-OH \xrightarrow{[O]}$ no reaction
Acidic:* HNO_3 H_2CrO_4, CrO_3/ H_2SO_4(aq) $Na_2Cr_2O_7$/ H_2SO_4(aq)	methyl $H_3C-OH \xrightarrow{[O]} H-C(=O)-H \xrightarrow{H_2SO_4(dil)}_{H_2O/25°C} H-C(=O)-OH \xrightarrow{[O]} CO_2 + H_2O$ (if distilled) 1° $R-CH_2-OH \xrightarrow{[O]} R-C(=O)-H \xrightarrow{H_2SO_4(dil)}_{H_2O/25°C} R-C(=O)-OH$ (if distilled) 2° $RR'CH-OH \xrightarrow{[O]} R-C(=O)-R'$ 3° $RR'R''C-OH \xrightarrow{dehydration}$ alkenes $\xrightarrow{[O]}$ oxidation products (not usually useful)
Near neutral $CrO_3 \cdot 2$ N⏜ Collins reagent $Cl^\ominus$ $CrO_3 \cdot H-N^\oplus$⏜ pyridinium chlorochromate (PCC) in CH_2Cl_2 (solvent)	methyl $H_3C-OH \xrightarrow{[O]} H-C(=O)-H$ 1° $R-CH_2-OH \xrightarrow{[O]} R-C(=O)-H$ 2° $RR'CH-OH \xrightarrow{[O]} R-C(=O)-R'$ 3° $RR'R''C-OH \xrightarrow[\text{slightly acidic}]{dehydration}$ alkenes ⟶ oxidation products (not usually useful) $\xrightarrow{\text{slightly basic}}$ no reaction

* $H_3C-C(=O)-CH_3$ or $H_3C-C(=O)-OH$ are common solvents. Dilute chromic acid in acetone is the Jones' reagent, milder than usual. $KMnO_4$ and HNO_3 require carefully controlled conditions or over-oxidation may occur. NaOCl is milder, often works with acid sensitive functional groups (alkenes) present.

a. Examples:

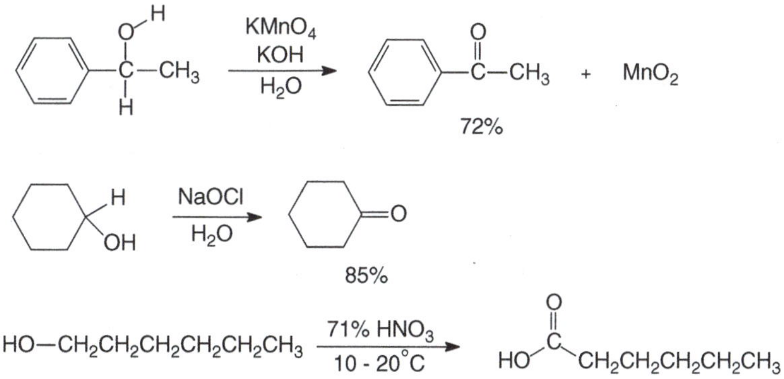

PhCH(OH)CH₃ —KMnO₄, KOH, H₂O→ PhC(O)CH₃ + MnO₂ (72%)

Cyclohexanol —NaOCl, H₂O→ Cyclohexanone (85%)

HO—CH₂CH₂CH₂CH₂CH₂CH₃ —71% HNO₃, 10-20 °C→ HOOC—CH₂CH₂CH₂CH₂CH₃

b. Chromic acid (yellow) forms when $Na_2Cr_2O_7$ dissolves in H_2SO_4 / H_2O, and when CrO_3 dissolves in H_2O. Oxidizing agent is either chromic acid or hydrogen chromate anion.

$Na_2Cr_2O_7 + H_2\ddot{O}: + 2\ H_2SO_4 \longrightarrow 2\ HO-Cr(=O)_2-OH + 2\ Na^{\oplus} + 2\ HSO_4^{\ominus}$
(chromic acid)

$CrO_3 + H_2\ddot{O}: \longrightarrow HO-Cr(=O)_2-OH \xrightleftharpoons[]{H_2\ddot{O}:} H_3O^{\oplus} + {}^{\ominus}\ddot{O}-Cr(=O)_2-OH$
(hydrogen chromate)

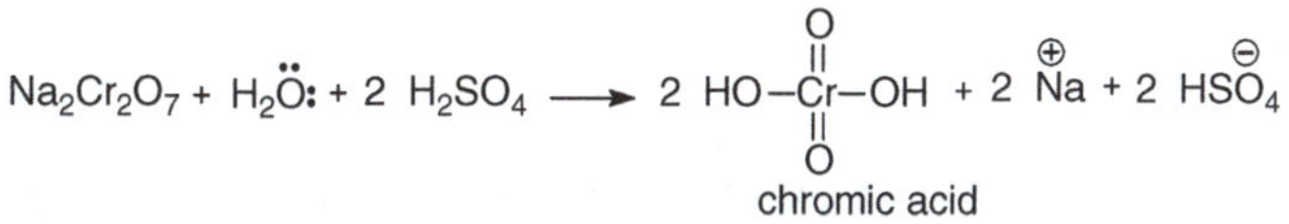

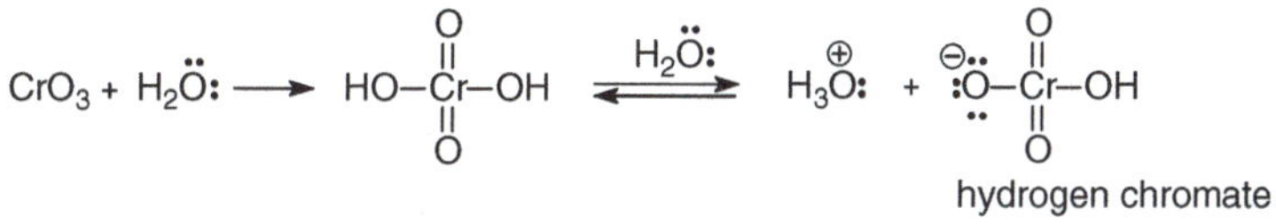

11.2.2. Possible, partial oxidation mechanisms with (a) hypochlorous acid and (b) hydrogen chromate anion:

(handwritten notes:)
1° alcohol → aldehyde → carboxylic acid
2° alcohol → ketone
3° alcohol → no rxn

R—CH₂—OH —PCC or PDC→ R—CHO

(a) [mechanism with hypochlorous acid, producing R'C(=O)R, Cl⁻, and H₃O⁺]

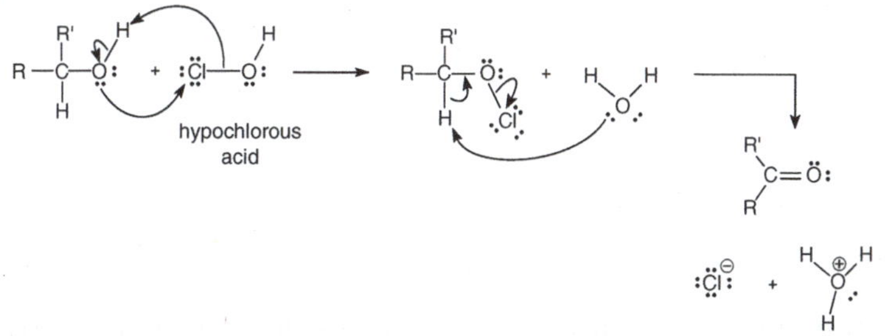

(b) [mechanism with hydrogen chromate anion showing formation of chromate ester from 2-propanol, then elimination to give acetone, H₃O⁺, and reduced Cr species]

chromate ester

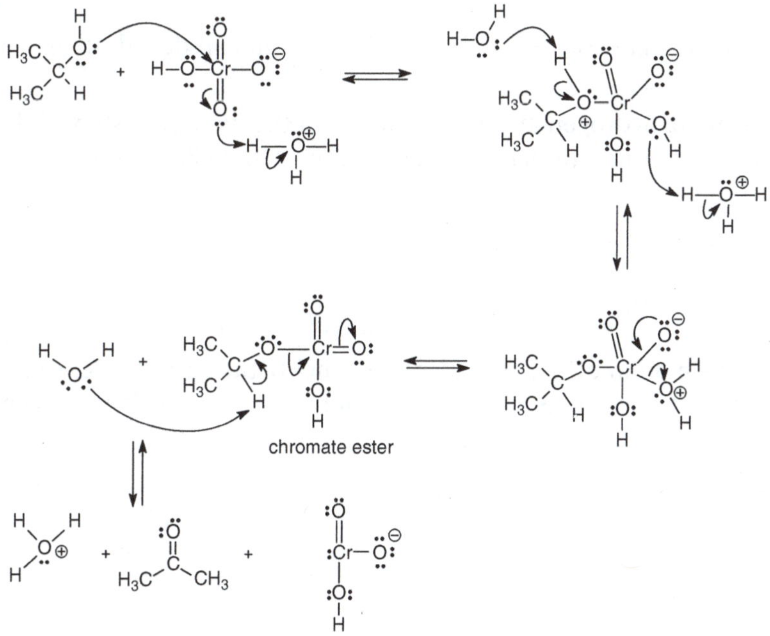

a. $HCrO_3^-$ chromium oxidation number is +4, but Cr^{+3} (green) is the actual byproduct. $HCrO_3^-$ oxidizes more alcohol, so the actual process is more complex than shown above.

b. *Problem 1.* Thousands of motorists in Great Britain were (politely) stopped by police and asked to blow into a "breathalyser," a glass tube containing silica gel (SiO_2) impregnated with specific chemicals, leading into a plastic bag. If, for more than half the length of the tube, the original yellow color turned green, the motorist often looked unhappy and turned red. What chemicals were impregnated on the silica gel? What caused it to turn green, and why did the motorist turn red?

11.2.3. Selective oxidizing agent – oxidizes one functional group but not another. Many such reagents exist.

11.3.1. $CrO_3 \cdot 2\,N\bigcirc$ (Collins reagent) is slightly basic; $CrO_3 \cdot H-N^{\oplus}\bigcirc\;Cl^{\ominus}$ (pyridinium chlorochromate, or PCC) is slightly acidic. Oxidation stops at the aldehyde and C=C's are not oxidized in dry H_2CCl_2 (solvent).

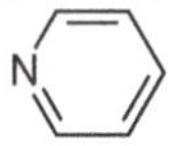

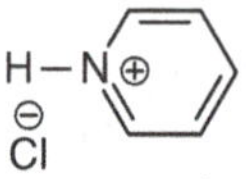

a. Aldehyde is in equilibrium with hydrate form (section 18.14) when H_2O is present. Hydrate easily oxidizes to carboxylic acid.

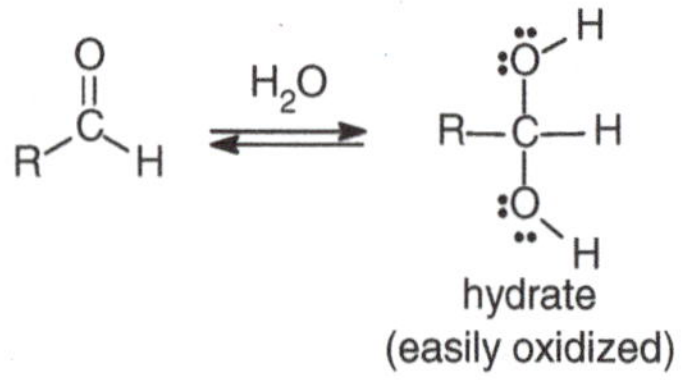

hydrate
(easily oxidized)

a. Spent (reduced) chromium reagents are suspect carcinogens – must be disposed as toxic waste (expensive).

11.3.2. A 3° alcohol does not normally oxidize in basic solution, but can dehydrate in acidic solution and product C=C can oxidize.

a. A 3° alcohol has no α-hydrogen atom to eliminate in the oxidation step, and does not usually yield useful products (aldehydes, ketones, or carboxylic acids).

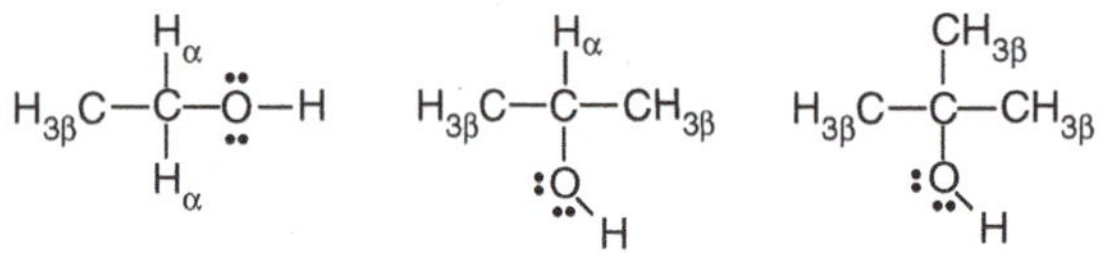

b. *Chromic acid test* – methanol, 1°, or 2° alcohols quickly turn a yellow chromic acid solution blue/green (Cr^{+3}), but 3° alcohols do not (react very slowly with heating).

11.3.3. Other oxidations: (1) dehydrogenation (industrial) with Cu, CuO, or Cu-Zn (catalyst), (2) Swern, and (3) Dess-Martin periodinane (DMP) oxidations.

(1) $H_3C-\underset{H}{\overset{O-H}{C}}-CH_2CH_3 \xrightarrow[400°C]{Cu-Zn} H_3C-\overset{O}{C}-CH_2CH_3 + H_2$

(2) cyclopentanol + $H_3C-S(=O)-CH_3$ / $Cl-C(=O)-C(=O)-Cl$, $(H_3CCH_2)_3N$ / H_2CCl_2 (solvent), -60°C → cyclopentanone (90%) + $CO_2(g)$ + $CO(g)$ + 2HCl + $H_3C-S-CH_3$

$H_3C(CH_2)_8-\underset{H}{\overset{OH}{C}}-H \xrightarrow{\quad "\quad} H_3C(CH_2)_8-\overset{O}{C}-H + "$ (85%)

(3) $R-\underset{R'}{\overset{OH}{C}}-H$ + DMP → $R-\overset{O}{C}-R'$ + [reduced iodinane byproduct] + 2 $H_3C-C(=O)-OH$

R' = H or R

a. Best reagents:

To oxidize	Product	Chromium reagent	Chromium-free reagent
2° alcohol	ketone	chromic acid, Collins', PCC	NaOCl, Swern, DMP
1° alcohol	aldehyde	Collins', PCC	Swern, DMP
1° alcohol	carboxylic acid	chromic acid	NaOCl

b. 2,2,6,6-Tetramethylpiperidinyl-1-oxy (TEMPO) is a relatively stable free radical oxidation catalyst. Addition speeds up HOCl (and other) oxidations.

11.4.1. The body oxidizes (detoxifies) ethanol – catalyzed by liver enzyme alcohol dehydrogenase (ADH). Oxidizing agent is nicotinamide adenine dinucleotide (NAD$^+$), derived from nicotinic acid (niacin or vitamin B$_3$).

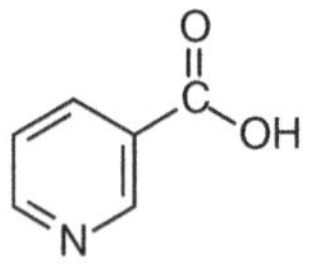

nicotinic acid (niacin)

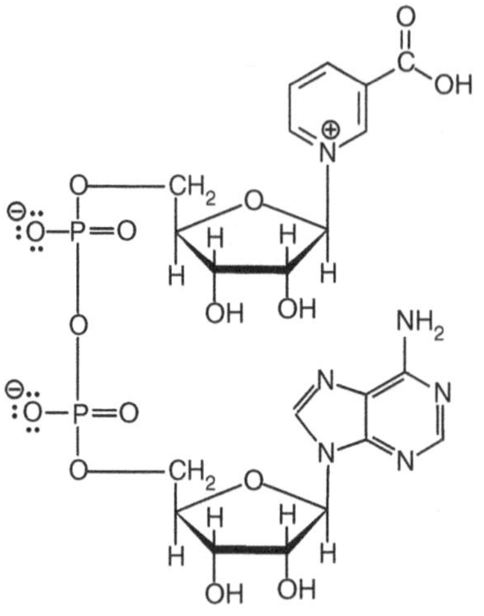

NAD$^\oplus$

a. Ethanal oxidizes further via aldehyde dehydrogenase enzyme (ALDH) to nontoxic metabolite ethanoic acid.

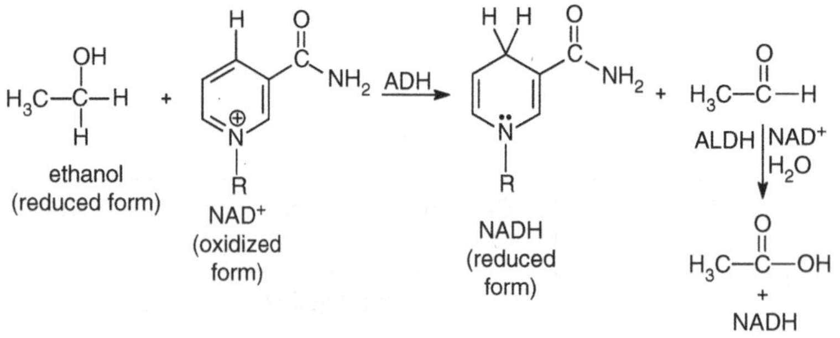

b. Other alcohols oxidize to more toxic products.

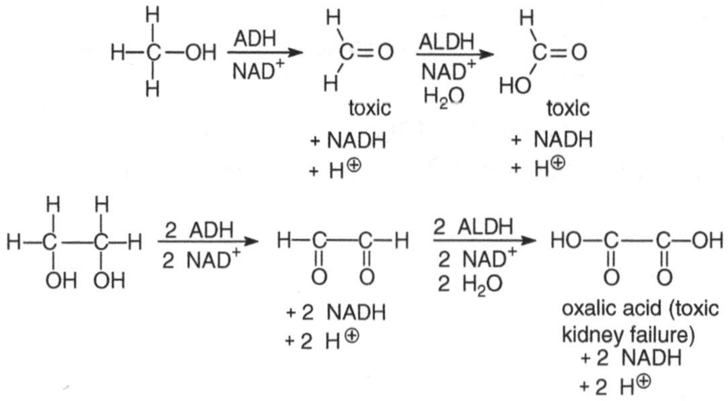

oxalic acid (toxic kidney failure)
+ 2 NADH
+ 2 H$^\oplus$

1. Excess ethanol treatment ties up ADH enzyme and allows methanol or ethylene glycol to pass through unoxidized.

11.5.1. O polarizes both C-O and O-H alcohol bonds.

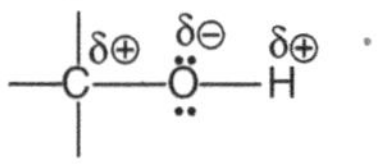

11.5.2. R-OH as a nucleophile – O lone pairs are weakly basic and nucleophilic.
   a. Example: 2-bromo-2-methylpropane methanolysis. The carbocation is a strong electrophile.

   b. An alcohol becomes a much stronger nucleophile when converted to alkoxide anion (sections 10.6.1 and 11.14.1), which will react with a weak electrophile (an alkyl halide).

   c. O-H bond breaks and C-O bond remains intact in neutral (becomes acidic) or basic nucleophilic reactions.

11.5.3. R-OH is a very weak electrophile (δ+ on C) and HO⁻ is a slow (strongly basic) leaving group. Reaction with acid converts –OH into a better leaving group ⁺-OH₂ (oxonium ion), and C becomes much more electrophilic.

   a. O-H bond remains intact and C-O bond breaks when R-OH is an electrophile.
   b. Only weak nucleophiles exist in acidic solution.

11.5.4. Like oxonium ions, R-OH sulfonate esters are more electrophilic. Alcohol C-O is not broken (chiral carbinol C configuration is retained) during sulfonate ester formation. Substitution is then possible (section 6.11.1b3) since sulfonates are good leaving groups. Inversion occurs during S_N2.

a. *p*-Toluenesulfonyl chloride preparation:

[Scheme: toluene (from catalytic reforming) + SO₃ / H₂SO₄(conc) → para-toluenesulfonic acid; then PCl₃ or SOCl₂ → para-toluenesulfonyl chloride + HCl + SO₂]

b. Sulfonates are often faster leaving groups than chlorides or bromides – weak bases due to - charge delocalization via resonance. "Triflate" (section 9.8.1e1) is a very fast leaving group.

[Resonance structures of benzenesulfonate anion]

c. Strongly basic nucleophiles can be added to alkyl sulfonates, but not to oxonium ions (section 11.5.3) since the latter solutions are acidic.

d. Problem 2. Suppose you made 2-tosylbutane from alcohol whose [α] = +6.9°·dm⁻¹·g⁻¹·mL. Reacting this substrate with NaOH(dil) / H₂O / warm yielded alcohol whose [α] = -6.9°·dm⁻¹·g⁻¹·mL. Although you don't know absolute configurations (*R* or *S*), what is the overall reaction stereochemistry?

11.6.1. H can replace (reduce) an -OH group two ways:

a. Dehydration (section 7.18.1) and hydrogenation (section 8.10.1 in Chapter 7).

[Scheme: cyclopentanol → H₂SO₄/heat → cyclopentene → H₂/Pt → cyclopentane]

b. Methyl, 1° and 2° alcohol tosylates are reduced by LiAlH₄.

[Scheme: cyclohexanol + TsCl / pyridine → cyclohexyl tosylate + pyridinium chloride; then LiAlH₄ (dry ether), H₂SO₄(dil) / H₂O / 25°C (neutralize pH) → *p*-toluenesulfonic acid + cyclohexane, 75%]

11.7.1. −X replaced −OH via the tosylate (section 11.5.4). R-OH reaction with H-X can accomplish −X substitution directly.

    a. H-X reactivity is in the same order as acidity.

$$\text{stronger H-I} > \text{H-Br} > \text{H-Cl} > \text{H-F weaker}$$

        1. Usually H-I gives poor yields, and H-F is unreactive.

    b. R-OH reactivity:

$$\text{fast allyl, benzyl} > 3° > 2° > 1° > \text{methyl slow}$$

    c. Methanol and 1° alcohols follow S_N2, but usually 2° and always 3° alcohols follow S_N1. E2 and E1 compete, diminishing yields. X^- is nucleophilic even in acidic solution.

11.7.2. Methanol, 1°, and 2° R-OH's react slowly with H-Cl. Yields are often poor.

    a. Cl^- is a weaker nucleophile than Br^- because it is effectively larger (solvated) and less polarizable (harder).

    b. An -OH group complexes more strongly with $ZnCl_2$ than with H^+, which speeds up reaction. $Zn[(OH)Cl_2]^-$ is a better leaving group than H_2O.

1. *Lucas test* – distinguishes 1°, 2°, and 3° alcohols). 1° alcohols react with ZnCl$_2$/HCl(conc) slowly (> 6 min), 2° alcohols faster (1-5 min), and 3° very fast (< 1 min). Alkyl chlorides are almost insoluble in the Lucas solution and form a separate layer above the aqueous layer. 1°, 2° and 3° alcohols are distinguished by how fast the alkyl halide layer forms.

c. Rearranged products are often found. (Exception: most 1° alcohols. Why?).

> 96%

11.7.3. Mechanism summary:
   a. R-OH (weak Lewis base) initially reacts with HX to form an oxonium ion. Slow leaving group -OH converts into a fast leaving group $^+$-OH$_2$.
   b. Rearranged products are consistent with carbocation intermediates. Reactivity order parallels carbocation stabilities (section 11.7.1b).

11.7.4. Two factors affecting S$_N$1 vs. S$_N$2 competition:
   a. Steric hindrance towards nucleophilic attack (slows down S$_N$2 vs. S$_N$1).
   b. Positive charge dispersal in the carbocation intermediate (speeds up S$_N$1 vs. S$_N$2).
      1. Example: 2,2-dimethylpropan-1-ol is 1°, but steric hindrance slows S$_N$2. Main product is rearranged, although a distinct 1° carbocation probably does not exist.

      2. Problem 3. 1-Chloro-2,2-dimethylpropane cannot be prepared from the alcohol 2,2-dimethylpropan-1-ol because (neopentyl) carbocation intermediate rearranges fast and completely. How might you make 1-chloro-2,2-dimethylpropane in good yield (*hint*: from an alkane substrate)?

3. Example: 2-chloropropan-2-ol – slow reaction since intermediate carbocation (and associated transition state) is (are) inductively destabilized:

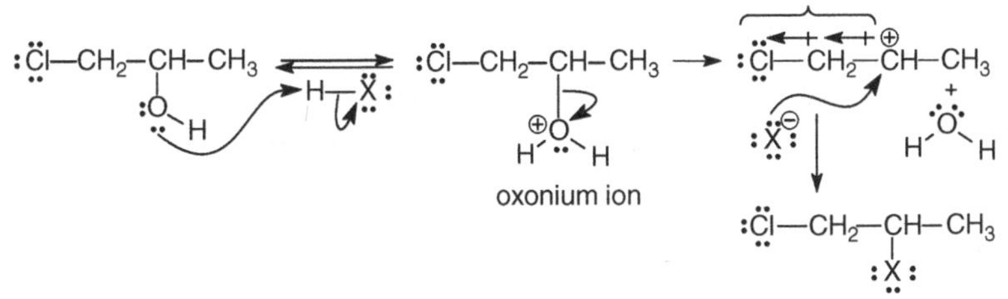

[mechanism diagram showing Cl–CH₂–CH(OH)–CH₃ reacting with HX to form oxonium ion intermediate, then carbocation, then Cl–CH₂–CH(X)–CH₃]

oxonium ion

4. Problem 4. Arrange each alcohol set in increasing reactivity order towards H-Br(dry gas), slowest on the left and fastest on the right.
   a. butan-2-ol, 2-methylpropan-1-ol, 2-methylpropan-2-ol
      strongest   *weakest*

   b. pentan-3-ol, 2-fluoropentan-3-ol, 2,2-difluoropentan-3-ol, 1-fluoropentan-3-ol

   c. As many of the eight isomeric $C_5H_{11}OH$ alcohols as possible. (It will be necessary to group some of them by class.)

5. Problem 5. Reaction of either pure pentan-2-ol or pure pentan-3-ol with HCl(dry gas) yields a 2-chloropentane / 3-chloropentane product mixture.
   a. Use electron pushing arrows to write a mechanism showing how pentan-2-ol is converted into 3-chloropentane. Include all elementary steps (collisions), intermediates, formal electrical charges, and involved lone pair electrons.

   b. Write a similar mechanism showing how pentan-3-ol is converted into 2-chloropentane.

6. Problem 6. (S. Winstein and H. J. Lucas, 1939).
   a. *Threo*-(2R,3S)(2S,3R)-3-bromobutan-2-ol is converted into racemic 2,3-dibromobutane, and *erythro*-(2R,3R)(2S,3S)-3-bromobutan-2-ol is converted into *meso*-2,3-dibromobutane upon treatment with HBr(conc). What is the reaction stereochemistry (configuration inversion or retention)?

   b. Optically active *threo*-(2S,3S)-3-bromobutan-2-ol yields *racemic* 2,3-dibromobutane when treated with HBr(conc). Write a mechanism that would explain this stereochemistry. *Hint:* the mechanism involves a "neighboring group effect."

11.7.5. Only nucleophiles that are as weak or weaker bases than R-OH can exist in acidic solution (or they would be neutralized).

    a. Weak nucleophiles ⇒ slower S_N2.

11.8.1. More recent (commercially available) PCl_3, PCl_5, PBr_3, $SOCl_2$ and $SOBr_2$ are often better reagents to replace methyl, 1° and 2° -OH groups with halogen. PI_3 is unstable, but is made *in situ* (in the reaction mixture) with P and I_2. These reagents do not work well with 3° R-OH's.

$$2P + 3I_2 \longrightarrow 2PI_3$$

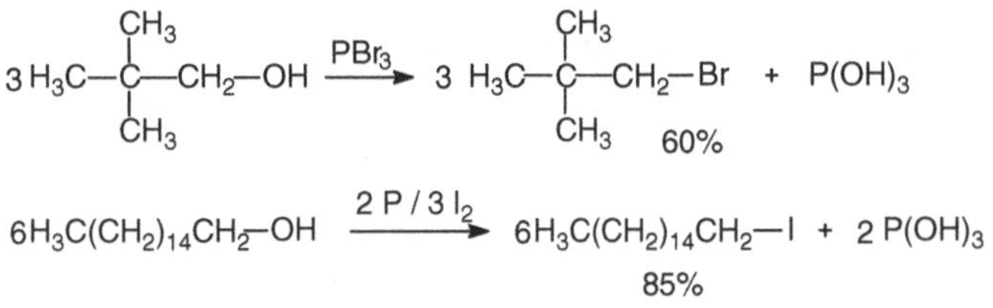

$$3\,H_3C-C(CH_3)_2-CH_2-OH \xrightarrow{PBr_3} 3\,H_3C-C(CH_3)_2-CH_2-Br + P(OH)_3$$
$$60\%$$

$$6\,H_3C(CH_2)_{14}CH_2-OH \xrightarrow{2P/3I_2} 6\,H_3C(CH_2)_{14}CH_2-I + 2\,P(OH)_3$$
$$85\%$$

    a. R-Cl yields are better with $SOCl_2$ than PCl_3 or PCl_5, especially with 3° R-OH's.

    b. Usually no carbocations form with these reagents ⇒ rearrangements are uncommon (especially when < 0°C), and yields are excellent. A 3° amine (pyridine or *N,N*-diethylethanamine, for example) is often added to neutralize the HX byproduct. Chiral R-OH's usually invert. Mechanism:

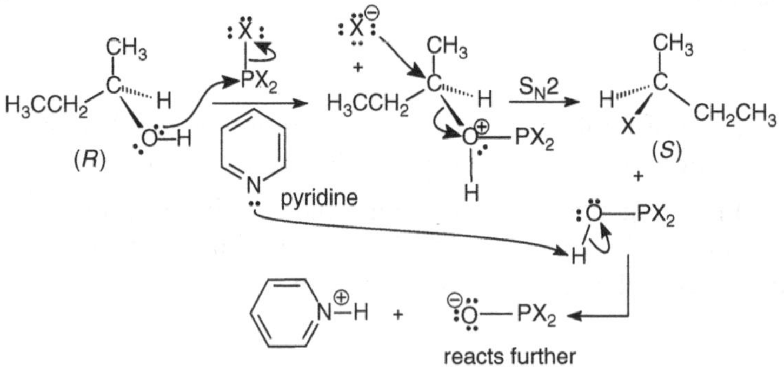

        1. Steric hindrance slows S_N2 for 3° alcohols, which explains their poor yields. Side reactions occur if a carbocation forms (via S_N1).

11.9.1. Configuration is retained when thionyl chloride ($SOCl_2$) converts a chiral R-OH to R-Cl. However, inversion *does* occur if a 3° amine is added. Mechanism example (without 3° amine):

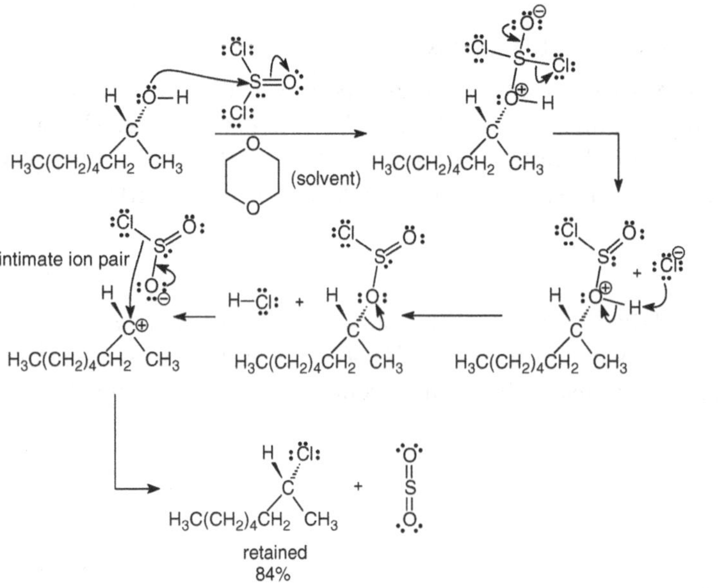

11.9.2. Preferred reagents summary:

Alcohol class	Chloride	Bromide	Iodide
1°	SOCl$_2$	PBr$_3$ or HBr*	P / I$_2$
2°	SOCl$_2$	PBr$_3$	P / I$_2$*
3°	HCl	HBr	HI*

*works only for selected cases.

11.10.1. 2° and 3° alcohols often dehydrate to alkenes in good yield (section 7.18), via an E1 mechanism (after -OH group protonation). Rearrangements are common.

bp 161°C

bp 83°C (80%)
(distilled out of reaction mixture)

a. Full dehydration reaction diagram (see section 7.18):

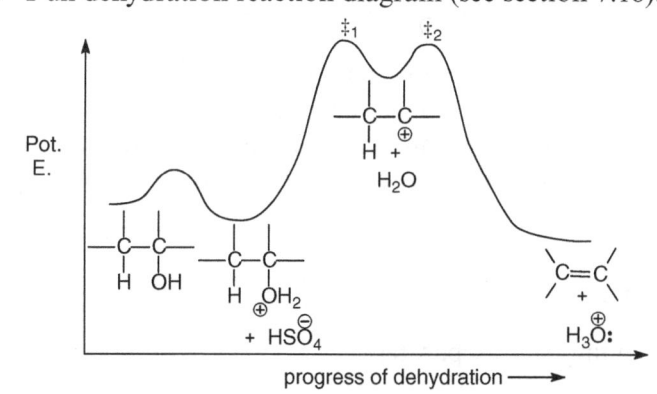

11.10.2. 1° alcohol yields are often lower due to rearrangements and multiple products.

70% Zaitsev
(mixture of cis and trans)

30% Hofmann

a. **Problem 7.** Optically active butan-2-ol retains its optical activity indefinitely in basic aqueous solution, but is rapidly converted into optically inactive (racemic) butan-2-ol in dilute sulfuric acid solution. Write an acid catalyzed racemization mechanism. Show electron pushing arrows, intermediates, formal charges, and involved lone pair electrons.

b. **Problem 8.** Heating 2,2-dimethylpropan-1-ol with H_2SO_4(conc) slowly converts it into two alkene products (formula C_5H_{10}) in an 85:15 ratio. Draw the alkene structures and use electron pushing arrows to draw a mechanism for their formation. Show all collisions (elementary steps), intermediates, formal electrical charges, and involved lone pair electrons. Which one is the major product? Explain why.

11.10.3. Bimolecular condensation (dehydration) to ethers will be covered in section 13.7.

11.10.4. Problem solving – determining and writing mechanisms.

  a. Are electrophiles, nucleophiles, or free radicals involved?

11.10.5. Suggestions / common mistakes:

  a. Free radicals are unlikely unless initiators, light, and/or high temperatures (~300°C) are involved.
  b. Strong electrophiles cannot coexist with strong nucleophiles.
     1. Acidic solutions contain strong electrophiles and weakly basic nucleophiles.
     2. Basic solutions contain strong nucleophiles and weakly acidic electrophiles.
  c. Draw all bonds to a reaction center – otherwise it is easy to forget an H-atom.
  d. Draw each collision (elementary step) separately.
  e. Curved arrows point the direction electron(s) move. A double-headed arrow means two electrons, a half-headed arrow means one electron.
  f. Keep track of formal electrical charges and involved lone electron pairs.

11.10.6. Example: reaction below yields a cyclized minor product. How does it form?

a. Try to correlate product C atoms with reactant C atoms.

b. Do electrophilic groups exist? If not can any groups be made electrophilic by reaction with acid? Yes, the alkene and alcohol.

1. Since product has no oxygen, this does not appear to lead anywhere.
2. Alcohol dehydration ⇒ a different carbocation could form.

c. Once an electrophile forms, what nucleophile might react with it – intramolecularly or intermolecularly? π-Bond must react intramolecularly in this case.
d. Once nucleophile attacks, how does resulting intermediate convert to product? Often a subsequent (acid/base) step is needed – a β-elimination in this case.

11.11.1. Pinacol rearrangement mechanism is similar to the example, but will not be covered.

11.11.2. Periodic acid oxidizes 1,2-diols, a useful ozonolysis alternative (section 8.15) – used extensively in sugar analysis (Chapter 23).

11.12.1. Fischer esterification – alcohols and carboxylic acids yield carboxylic esters. Alcohol sulfonate esters were made in section 11.5.4.

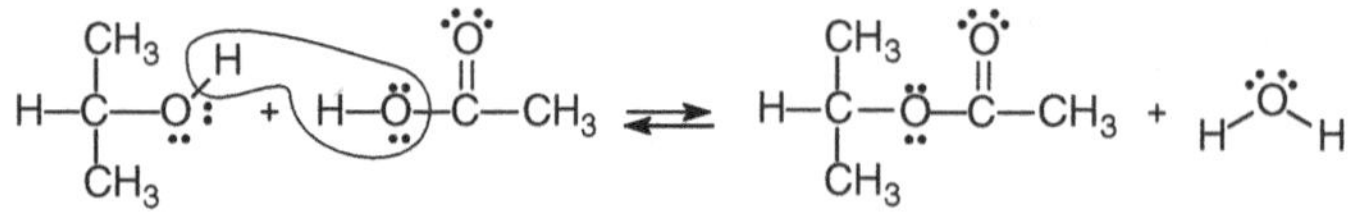

ester
isopropyl ethanoate
isopropyl acetate

  a. Water removal (as an azeotrope, by adding a dehydrating agent, or a large R-OH or RCOOH excess) drives equilibrium to completion.
  1. Alcohol groups in bad tasting drugs are often converted into esters to mask the taste.

11.12.2. Better ester yields are obtained when R-OH reacts with an acid chloride (more exothermic).

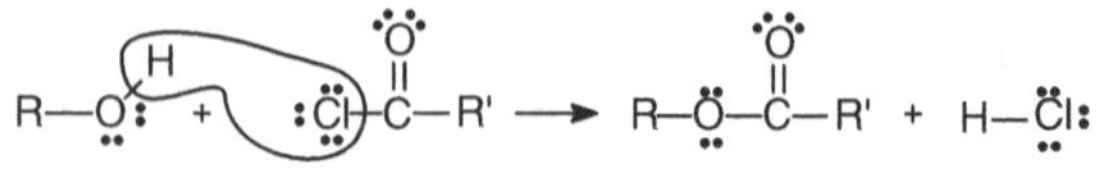

11.12.3. Mechanism is in sections 20.10, 21.5.2, and 21.10.5.

11.13.1. Alcohol reaction with acids yields inorganic esters.

  a. The nitrate ester N atom has oxidation number +5, while that of C is lower. N can oxidize C intramolecularly, yielding a vigorous explosion (stronger than black gunpowder). Products are gases (N_2, CO_2, H_2O, etc.).

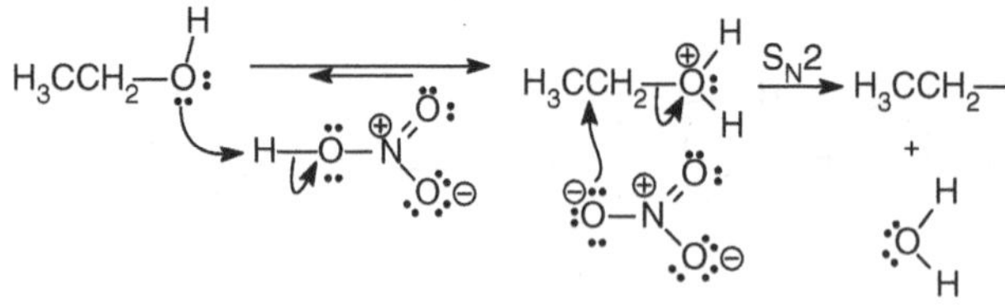

  nitrate ester

  b. Alfred Nobel (1866) added diatomaceous earth to nitroglycerin, stabilizing the latter enough to mold into sticks (dynamite).

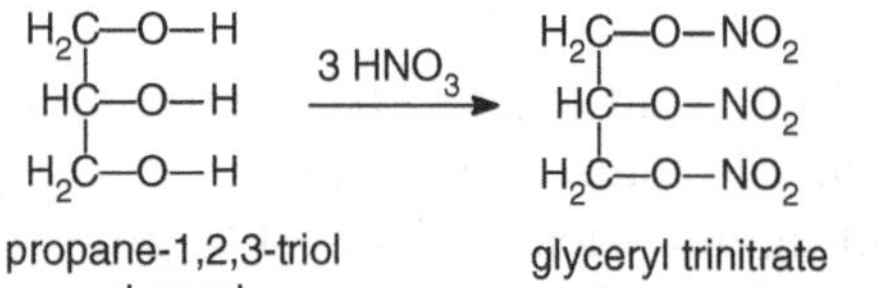

  propane-1,2,3-triol
  glycerol
  glycerine

  glyceryl trinitrate
  nitroglycerin
  (a misnomer, actually a trinitrate ester)

  c. Other common high explosives:

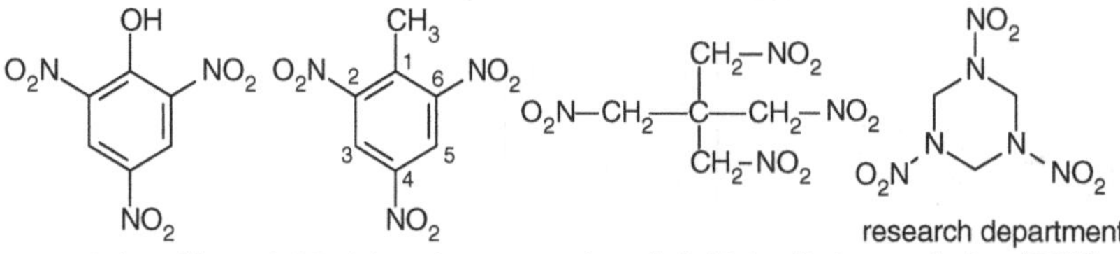

  picric acid   2,4,6-trinitrotoluene   pentaerythritol tetranitrate   research department explosive (RDX)
                 (TNT)   (PETN)

  d. In small amounts, nitroglycerin relieves angina (chest pain).

11.13.2. Adding concentrated (anhydrous) sulfuric acid (via Markovnikov EA at 0°C) to alkenes yields toxic alkyl hydrogen sulfate esters, or full sulfate esters (as *deliquescent* solids – rapidly absorb atmospheric water vapor and dissolve themselves.)

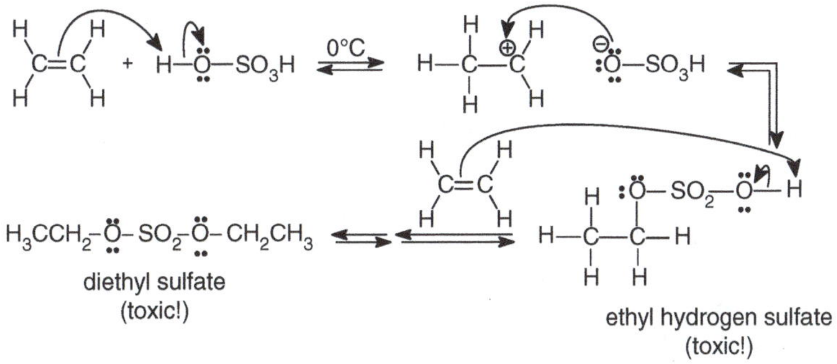

diethyl sulfate (toxic!)

ethyl hydrogen sulfate (toxic!)

a. Methyl and 1° hydrogen sulfates and sulfates are powerful "alkylating agents," i.e., they react with nucleophiles via S_N2 because they are fast leaving groups (similar to sulfonates).

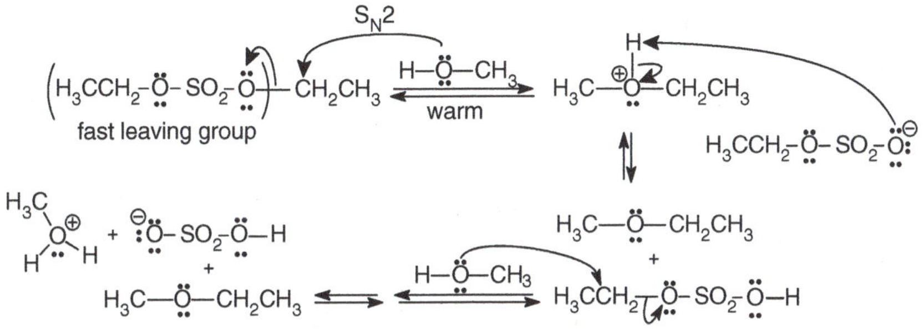

b. Example:

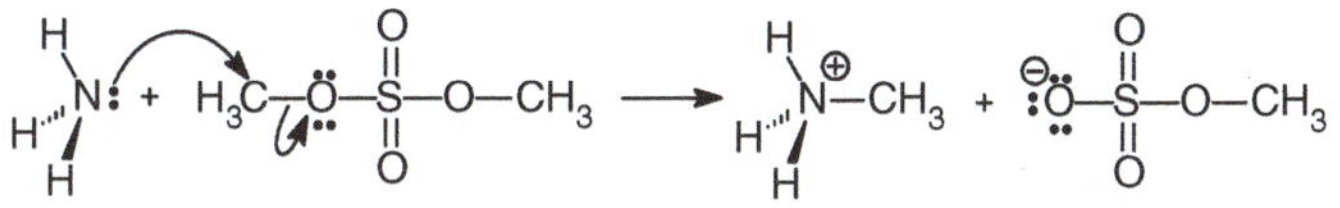

c. Hydrogen sulfates and sulfates can be *hydrolyzed* (cleaved with boiling water) to make alcohols. Yields are better than direct hydration (section 8.4 and 10.7.1b) for alkenes that would form 1° and 2° carbocations (section 8.5.1).
  1. Example:

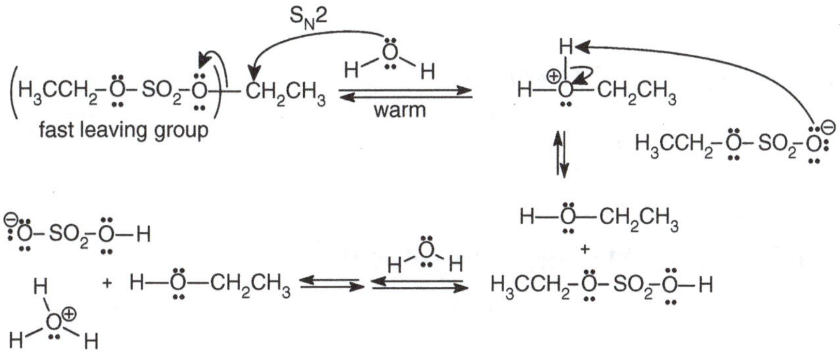

2. Example:

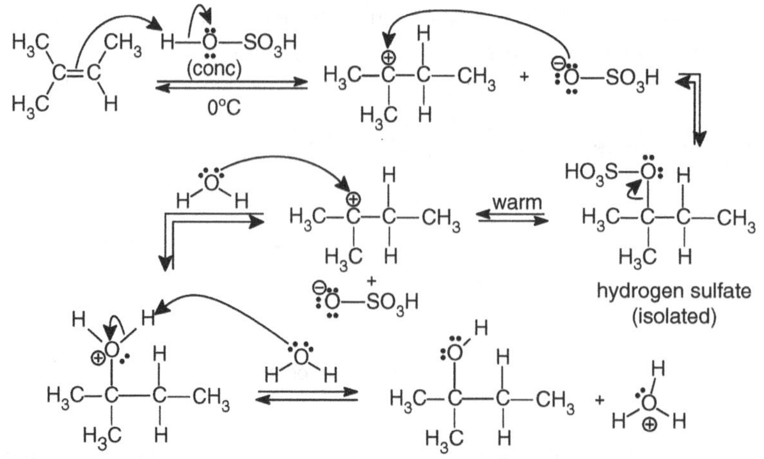

d. "Washing" an alkane or alkyl halide with cold H_2SO_4(conc) removes alkene impurities as hydrogen sulfates (which are soluble in H_2SO_4).

e. Problem 9. Draw the chief product that forms when propan-2-ol reacts with the reagents below. Indicate if no reaction occurs.

1. H_2SO_4(cold, conc)

2. $KMnO_4$(dil) / KOH / H_2O / 0°C

3. CrO_3 / H_2SO_4(conc)

4. HBr(conc) / H_2O

5. Br_2 / hexane (solvent) / 0°C / dark

6. product from 4 + Mg / ether (dry solvent)

7. P + I₂

8. Na

9. H₂ / Ni

10. H₃C-Mg-Br / ether (dry solvent)

11. NaOH(dil) / H₂O

12. tosyl chloride / KOH(trace, catalyst) / warm / exclude H₂O

f. Problem 10. Describe simple chemical tests (that could be done in test tubes) that would distinguish:
   1. *n*-butanol and *n*-octane

   2. *n*-butanol and *n*-bromopentane

   3. 2-methylbutan-2-ol and pentan-1-ol

   4. *n*-butanol and 2-bromoethanol

11.13.3. Alcohols form mono-, di-, and tri-esters with H_3PO_4. Phosphate esters are less reactive than sulfonates. They are important in biochemistry, for example, in the DNA structure. Hydrolysis is exothermic, but slow at pH 7. Enzymes catalyze hydrolysis.

    a. Cancer, diabetes, and obesity involve phosphate ester formation abnormalities.

11.14.1. Williamson ether synthesis (sections 10.6.1d1 and 13.5) involves strongly nucleophilic alkoxide anions.

    a. Methyl and 1° R-X's work better than 2°, as usual for S_N2. 3° R-X's yield only E2 elimination products.

11.14.2. Multi-step synthesis problem solving – competent students can accomplish this. Retrosynthetic analysis (working backwards) may help.

    a. Example: make 3-ethylpentane-2,3-diol from alcohols containing four or fewer C atoms and any inorganic reagents that you need.

    b. Example: convert cyclohexanol into 1-bromo-2-methylcyclohexane using alcohols with four or fewer C atoms and any inorganic reagents that you need.

11.14.3. Problem 11. Show how you would transform *n*-butanol into the following products. You may use any inorganic reagents that you need.
  a. bromobutane

  b. iodobutane

  c. butane hydrogen sulfate

  d. sodium *n*-butoxide

  e. pentanenitrile

  f. *n*-butanal

  g. *n*-butanoic acid

  h. *n*-butane

  i. *n*-butane-1-*d* (CH$_3$CH$_2$CH$_2$CH$_2$D)

  j. *n*-octane

11.14.3. Problem 12. Starting from four (or fewer) C atom alcohols, outline possible syntheses for the following products. Use any necessary reagents and solvents.
  a. 2-chloropropane

b. 2-tosylethane

c. potassium 2-methylprop-2-oxide

d. 2-methylpropane

e. butane-2-*d* (CH$_3$CH$_2$CHDCH$_3$)

f. 3-methylhexane

g. 2-methylpropanoic acid

h. butan-2-one

11.14.4. Problem 13. Show how you would synthesize each of the following from cyclohexanol.
  a. cyclohexene

  b. cyclohexane

  c. *trans*-1,2-dibromocyclohexane

  d. *cis*-cyclohexane-1,2-diol

  e. *trans*-cyclohexane-1,2-diol

f. hexane-1,6-dial

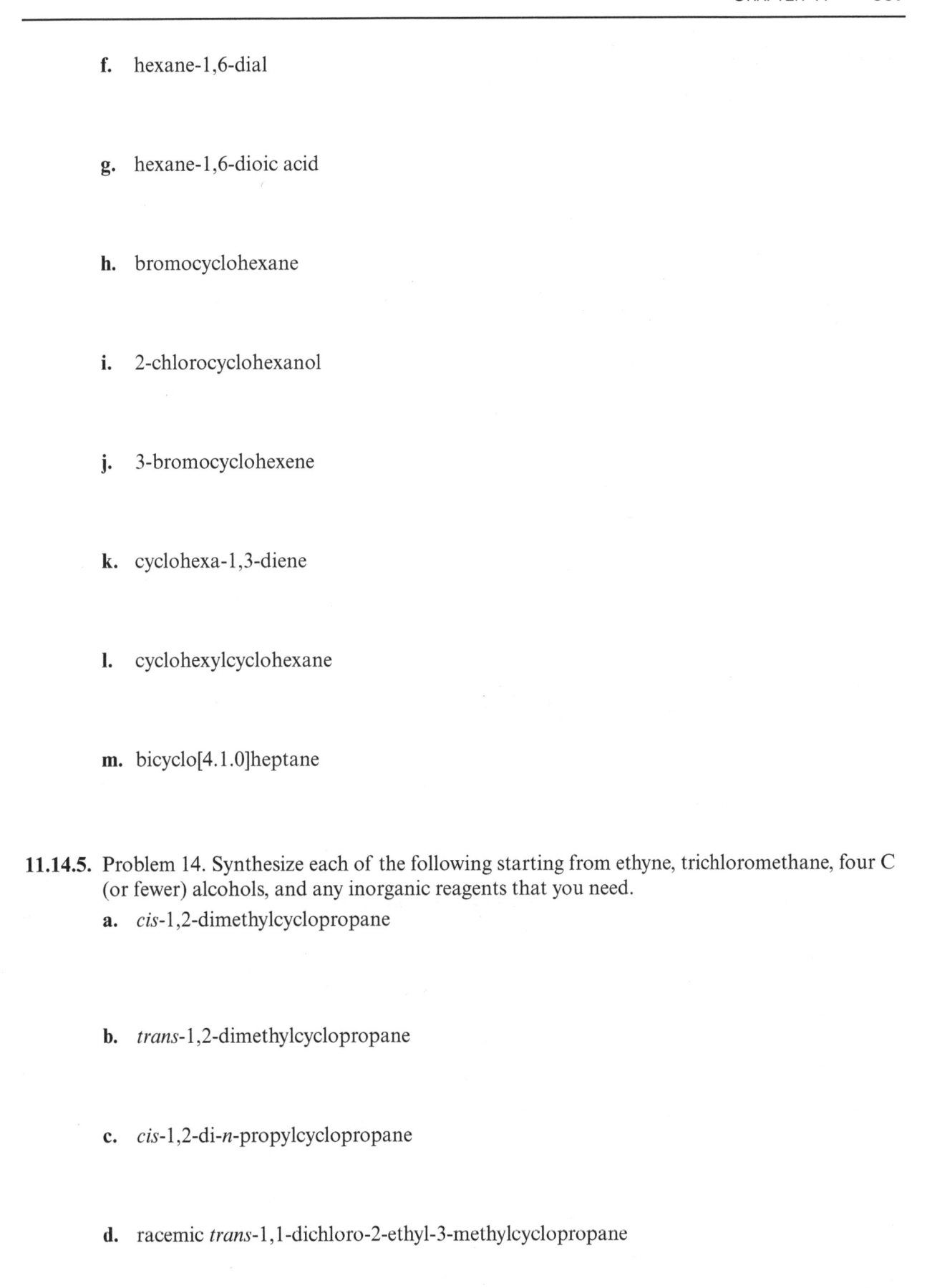

g. hexane-1,6-dioic acid

h. bromocyclohexane

i. 2-chlorocyclohexanol

j. 3-bromocyclohexene

k. cyclohexa-1,3-diene

l. cyclohexylcyclohexane

m. bicyclo[4.1.0]heptane

11.14.5. Problem 14. Synthesize each of the following starting from ethyne, trichloromethane, four C (or fewer) alcohols, and any inorganic reagents that you need.

a. *cis*-1,2-dimethylcyclopropane

b. *trans*-1,2-dimethylcyclopropane

c. *cis*-1,2-di-*n*-propylcyclopropane

d. racemic *trans*-1,1-dichloro-2-ethyl-3-methylcyclopropane

11.14.6. Problem 15. What simple chemical tests would allow you to distinguish:

    a. cyclopropane from propane

    b. cyclopropane and propene

    c. 1,2-dimethylcyclopropane and cyclopentane

    d. cyclobutane and but-1-ene

    e. cyclopentane and pent-1-ene

    f. cyclopentane and cyclopentene

    g. cyclohexanol and *n*-butylcyclohexane

    h. 1,2-dimethylcyclopentene and cyclopentanol

    i. cyclohexane, cyclohexene, cyclohexanol, bromocyclohexane

CHAPTER 12

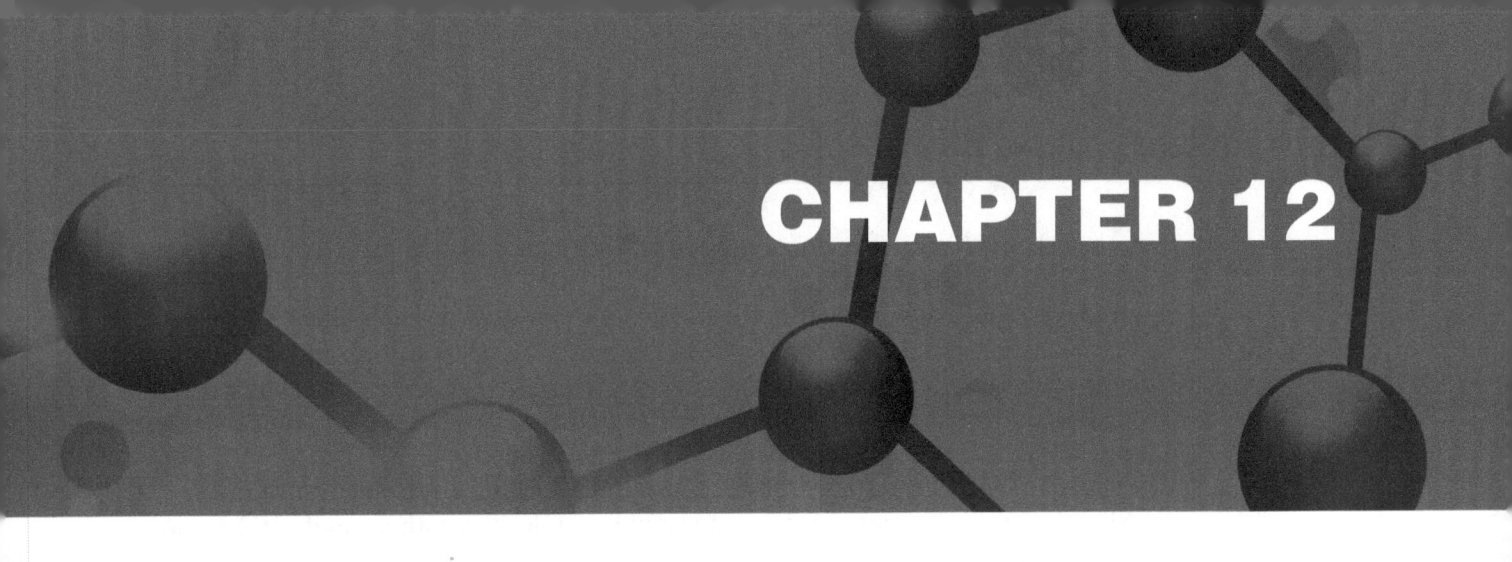

12.1.1. Organic compound identification – a basic necessity. Separation and purification (section 2.1.10) are usually required first.

    a. Full characterization is required before application development.

12.1.2. Unknown compound isolation and characterization. Strategy:

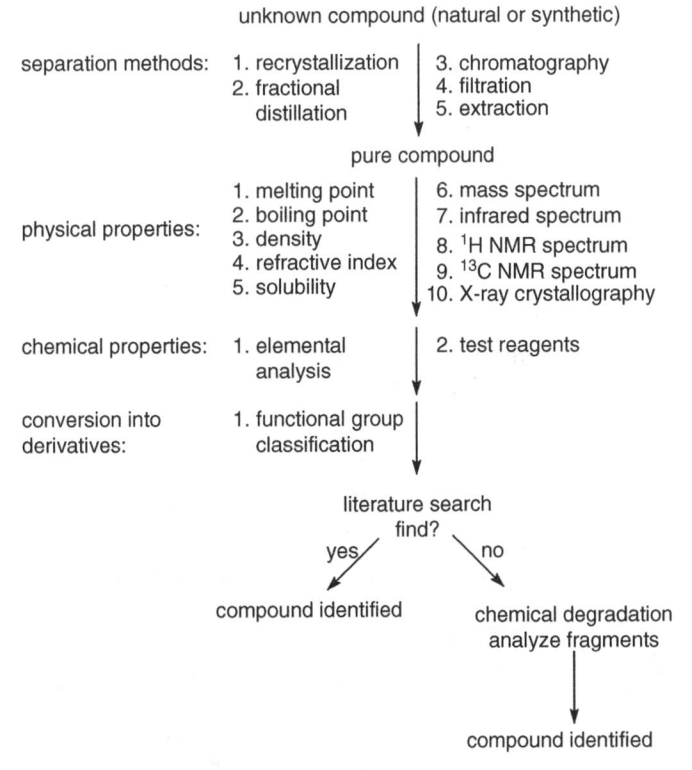

```
                    unknown compound (natural or synthetic)
                                    |
separation methods:  1. recrystallization    3. chromatography
                     2. fractional           4. filtration
                        distillation         5. extraction
                                    |
                              pure compound
                                    |
                     1. melting point        6. mass spectrum
                     2. boiling point        7. infrared spectrum
physical properties: 3. density              8. ¹H NMR spectrum
                     4. refractive index     9. ¹³C NMR spectrum
                     5. solubility          10. X-ray crystallography
                                    |
chemical properties: 1. elemental            2. test reagents
                        analysis
                                    |
conversion into      1. functional group
derivatives:            classification
                                    |
                            literature search
                                 find?
                          yes  /        \  no
                              /          \
                    compound identified    chemical degradation
                                           analyze fragments
                                                  |
                                          compound identified
```

    a. Steps are not always followed in order.

    b. Physical and chemical properties alone are usually insufficient to identify small amounts of complex new compounds.

    c. Spectroscopy – light / matter interaction study – yields much structural evidence, requires a very small sample, and usually does not destroy it.

12.1.3. Four common spectroscopic techniques are often used in order:

    a. Mass spectrometry – bombard molecules with high energy electrons to ionize and fragment. Ion observation reveals molecular mass and sometimes molecular formula.

Fragments help identify functional groups present. Samples are destroyed, but very small amounts (1×10^{-6} g) are needed.

    b. Infrared (IR) spectroscopy – observes atomic vibrations which helps elucidate functional groups. Sample is not destroyed. Very small samples can be used.

    c. Nuclear magnetic resonance (NMR) – observes hydrogen (or some other) nuclear environment which helps determine molecular structure and functional groups. Sample is not destroyed, but larger samples (~1 mg) are often needed.

    d. Ultraviolet (UV) spectroscopy – observes molecular orbital electronic transitions, revealing electronic structure. It is most useful for molecules with π-bonds. It does not destroy the sample.

12.1.4. Absorption spectroscopy – (infrared, NMR, and UV) involves interactions between light and matter.

12.2.1. Light – a form of pure energy (see section 5.4.2).

    a. Light waves have a wavelength (λ). Frequency (ν) is inversely proportional to λ:

$$\nu = \frac{c}{\lambda}$$

cycle = one unique wave unit, no part of which is repeated.
ν = (nu) cycles/sec (Hertz)
c = light speed = 2.998×10^{10} cm/sec
λ = (lambda) wavelength (cm)

    b. Photon – smallest possible light particle.

        1. Planck's equation – gives a single light photon energy having wavelength λ.

$$E_{photon} = h\nu = \frac{hc}{\lambda}$$

h = Planck's constant = 6.626×10^{-34} J · sec.

    c. Photon energy decreases with increasing λ and with decreasing ν.

	wavelength (m)		energy (kJ/mol)	molecular effect
higher frequency Shorter wavelength ↑	10^{-10}	γ-ray	10^7	
	10^{-8}	X-ray	10^5	ionization
Pot. E.		vacuum UV	10^3	electronic transitions
		near UV		
	10^{-6}	visible	10^2	
	10^{-4}	infrared	10	molecular vibrations
	10^{-2}	microwave	10^{-2}	rotational motion
lower frequency longer wavelength	$10^0 = 1$	radio	10^{-4}	
	10^2	long wave	10^{-6}	nuclear spin transitions

12.2.2. A photon's energy can either be absorbed or transmitted (reflected) when it collides with a molecule.

    a. Absorption – a molecule traps a photon (frequency ν), causing a transition to an excited energy state. The photon disappears in the process.

    b. Spectrum – a light intensity (number of photons passing through a sample) graph (y-axis) *vs.* light frequency (x-axis).

1. A molecule can absorb a photon (whose $E_{photon} = h\nu$) when $E_{photon} = \Delta E$, where ΔE is an energy gap between two electronic, vibrational, or rotational quantum levels. Light intensity at frequency ν passing through the sample diminishes.

12.3.1. Infrared region – from $\lambda \approx 1 \times 10^{-2} - 8 \times 10^{-5}$ cm. Common infrared spectrometers cover the range $\lambda \approx 2.5 - 25 \times 10^{-4}$ cm, or $2.5 - 25\ \mu$ (microns), where $1\ \mu = 1 \times 10^{-4}$ cm $= 1 \times 10^{-6}$ m, corresponding to energies of $4.6 - 46$ kJ/mol.

   a. Modern spectrometers often use inverse wavelength, $1/\lambda$ (cm^{-1}), since it is directly proportional to frequency. $1/\lambda$ is the number of complete waves in 1.00 cm (called the wavenumber).
   b. Example: How many wavenumbers correspond to $11\ \mu$ wavelength light?

$$\frac{1}{11\ \mu} \times \frac{10000\ \mu}{cm} = \frac{910}{cm} = 910\, cm^{-1}$$

12.3.2. Infrared region E_{photon}'s are about the same order of magnitude as ΔE's between atomic vibrational levels in molecules.

   a. Vibrational ΔE's are small compared to gaps between molecular electronic states (which correspond to ultraviolet photons).

12.4.1. Two bonded atoms vibrate as if held together by a spring, except vibrational frequencies are quantized. Only certain frequencies are permitted.

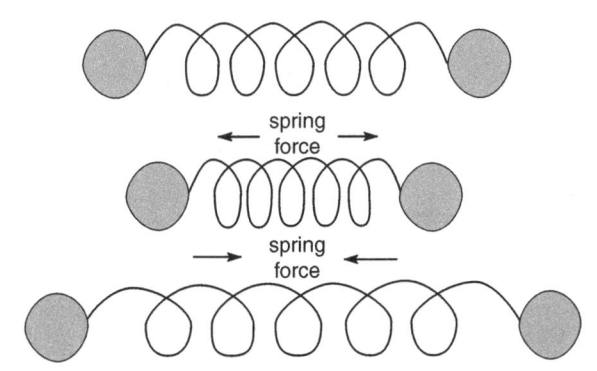

12.4.2. Vibrational frequency depends on particle masses and on stiffness (bond strength).
   a. Larger masses $\Rightarrow$ lower frequency
   b. Stronger bond $\Rightarrow$ higher frequency
   c. Examples:

Bond	Bond energy kJ/mol	Approximate stretching wavenumber (cm^{-1})
C-H	420	3000
C-D	420	2100
C-C	350	1200
C=C	611	1660
C≡C	840	2200
C-N	305	1200

(Continued)

Bond	Bond energy kJ/mol	Approximate stretching wavenumber (cm^{-1})
C=N	615	1650
C≡N	891	2200
C-O	360	1100
C=O	745	1700

1. A C-H bond vibrates at higher frequency than a C-D bond because D is more massive than H.
2. Bond vibrational frequency is in the order (A represents C, N or O):

   higher C≡A > C=A > C–A lower

   the same order as bond strength.

12.4.3. Normal vibrational modes – mathematically predicted simple (fundamental) molecular vibrations. Quantum mechanics predict that a nonlinear molecule has $3n-6$ *allowed* normal modes, where n is the number of atoms in a molecule. $3n-5$ normal modes exist for linear molecules.

   a. For H$_2$O, $3(3)-6 = 3$ normal modes exist, shown below.

   symmetric stretching   asymmetric stretching

   bending (scissoring)

   b. -CH$_2$- group vibrational modes.

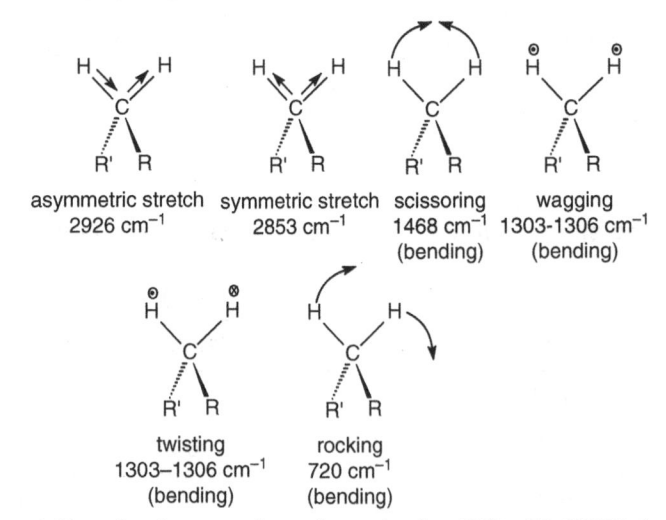

   asymmetric stretch symmetric stretch scissoring   wagging
   2926 cm^{-1} 2853 cm^{-1} 1468 cm^{-1} 1303-1306 cm^{-1}
                                           (bending)   (bending)

   twisting   rocking
   1303–1306 cm^{-1} 720 cm^{-1}
   (bending)   (bending)

   c. $3(5) - 6 = 9$ normal modes exist for CH$_4$. H$_3$CCH$_3$ has $3(8) - 6 = 18$ normal modes.
      1. Even simple molecules have many normal vibrational modes – each compound has its own distinct set. An infrared spectrum is like a fingerprint – unique to each compound.

12.5.1. Bond dipole moment magnitude oscillates as atoms move if two bonded atoms have different electronegativites.

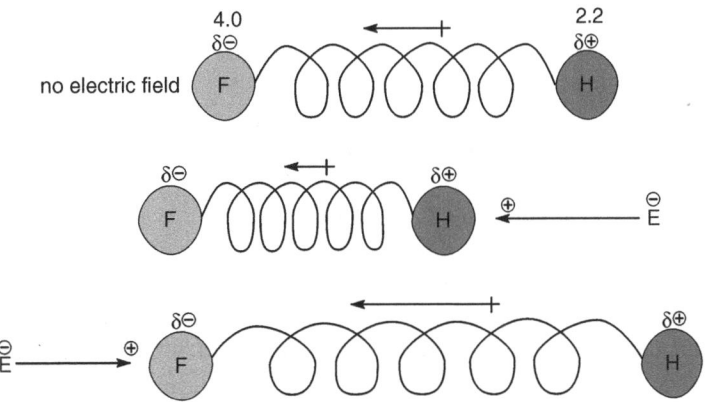

a. Resonance condition – when an infrared photon electric field oscillates (pushes and pulls) at the same frequency as a bond vibration. A light wave has an oscillating electric field.

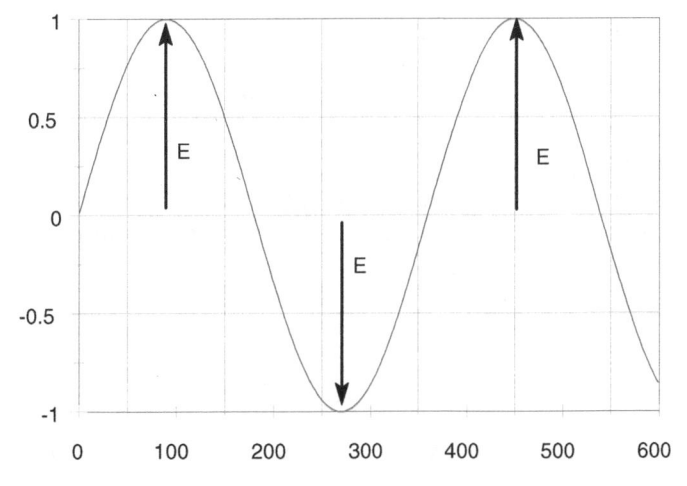

b. A photon can be absorbed if an oscillating dipole and a resonance condition exist. A resonant bond jumps to a higher vibrational energy state. The infrared photon disappears. Its energy is incorporated by the bond (makes the bond vibrate more vigorously).

  1. Symmetric and asymmetric stretching vibrations in CO_2.

  symmetric bond stretching
  no net dipole moment change
  infrared inactive

  asymmetric bond stretching
  net oscillating dipole moment change
  infrared active

2. Vibrational transition energy schematic:

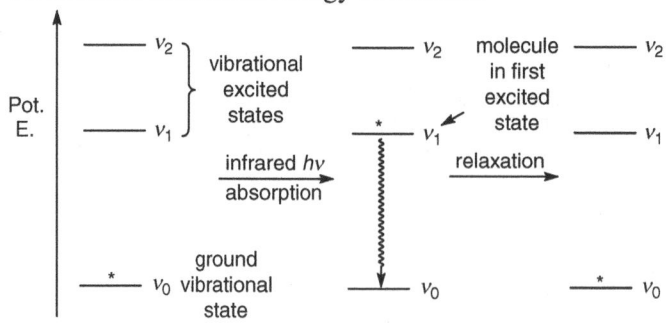

  c. Slowly scanning frequencies produces a spectrum. Infrared light emerging from a sample at an absorbed frequency is dimmer than that at a transmitted (unabsorbed) frequency. Intensity is detected by a phototube (similar to a camera light meter). Phototube electrical output is amplified and plotted. Absorptions appear as peaks on a graph.
   1. Greater oscillating bond dipole moment amplitude ⇒ greater absorption probability and stronger peak.
   2. Many weakly absorbing molecular vibrations can result in strong absorption overall (for example, sp^3 C-H stretch, 3000–2800 cm^{-1}).
   3. An infrared inactive vibration (and its associated *forbidden transitions*) may nevertheless yield weak peaks due to thermal molecular distortions. Collisions, rotations, and nearby vibrations result in instantaneous dipoles which can interact with infrared light and absorb.
  d. Infrared light intensity is plotted on the vertical axis as either: (1) % transmission (%T) (common), or (2) absorbance (A) (rare for IR, more common in UV spectroscopy).

$$\%T = \frac{I_{observed}}{I_{initial}} \times 100$$

$$A = \log_{10}\frac{I_{initial}}{I_{observed}} = \log_{10}\frac{1}{T}$$

  where $I_{initial}$ is the light intensity before it enters a sample, and $I_{observed}$ is the light intensity after it has passed through a sample.
   1. The logarithmic absorbance scale better retains weak peak sensitivity.
  e. Characteristic absorption frequencies usually reveal functional groups. A given functional group bond vibration is often independent of other molecular vibrations.

12.6.1. Continuous scan (dispersive) infrared spectrometer schematic:

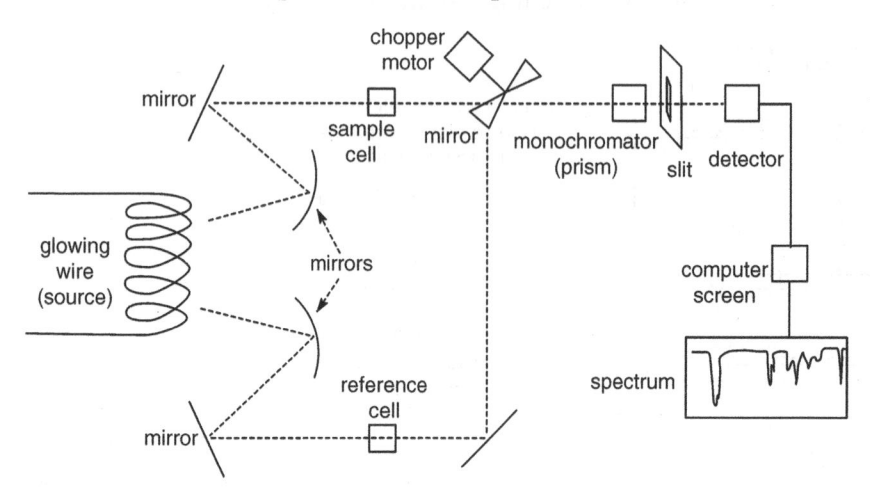

  a. Glowing wire – generates all infrared wavelengths.

b. Infrared light passes through a sample cell and a reference cell. Often a thin liquid film is sandwiched between infrared transparent (NaCl, KBr, or AgCl, etc.) plates.
   1. Solids can be run by mixing with powdered KBr and pressing into a clear window. Or they may be dissolved in a solvent, CCl_4, CH_2Cl_2, CS_2, or mineral oil (saturated hydrocarbons).
c. Mirrors are attached to a chopper motor fan blades. Sample path light enters the monochromator half of the time. Reference path light enters the other half of the time. Sample and reference light intensity *difference* produces an electric detector signal.
d. The monochromator allows only one frequency at a time to enter the detector.
e. Amplified detector signal generates peaks.
   1. $100\%T$ baseline is at the top of the page. Peaks are negative.

12.6.2. Dispersive instruments are now obsolete. They had expensive prisms or gratings and intense light sources. They had to be aligned and calibrated regularly, and took 2 – 10 min to run a spectrum. But they are conceptually simpler than modern instruments.

12.6.3. Modern Fourier transform infrared (FT-IR) instrument – based on Michelson interferometer (first built in 1881):

a. When monochromatic infrared light (having only one wavelength λ, from a He-Ne laser) encounters the beam splitter, half of the light reflects towards a fixed mirror, and half passes through toward a mirror that moves (over a few mm distance) at a constant speed (s). One beam splitter type is polished KBr whose surface is lightly coated with Ge. When the moving mirror is at point 0, the mirror path lengths are the same, and the two waves recombine in phase at the beamsplitter reconstituting the original light. Half of this light reflects downward (toward the sample) and the other half reflects back toward the source. Whenever the mirror path lengths are different (when the moving mirror is not at point 0), destructive interference occurs. The output intensity is zero when the mirror path lengths differ by $¼\lambda$, (when the reflected waves are 180° out of phase). Intensity is a function of the distance the mirror moves, or since it moves at a constant speed, intensity is also a function of time ("time domain"). Only one intensity maximum (at point 0) exists in the range $-¼\lambda$ to $+¼\lambda$.

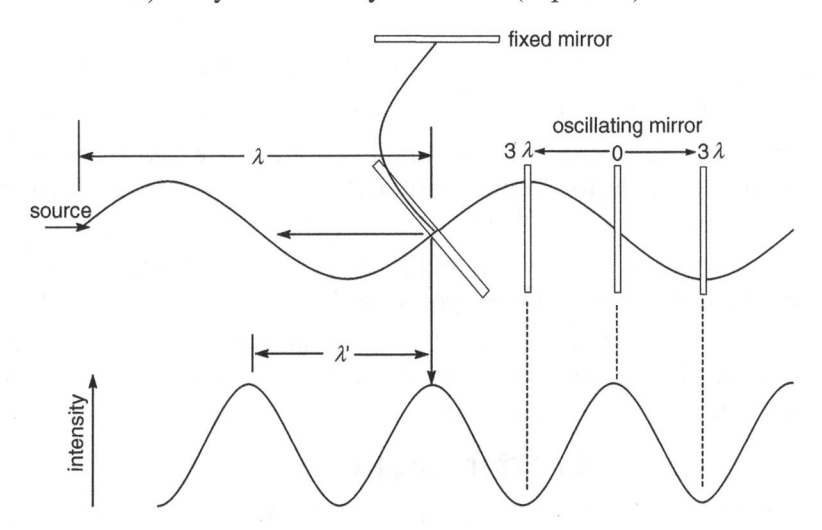

b. Mathematically, the output wave (interferogram) is described by:

$$I(x) = 0.5\ I(\bar{v}) \cos(2\pi\bar{v}x) = 0.5\ I(\bar{v}) \cos(2\pi\bar{v}st) \qquad \text{where } \bar{v} = \frac{1}{\lambda}$$

   1. $\bar{v}$ is the laser light wavenumber, $I(x)$ is the interferogram intensity as a function of the mirror position (x), $I(\bar{v})$ is the original laser light intensity as a function of wavenumber, s is the mirror speed, and t is the time. $I(\bar{v})$ is a constant for a given

wavenumber ($\bar{v}$) – true for monochromatic laser light, and $x = st$. The factor 0.5 appears because only half of the original light goes to the detector.

2. Notice that output frequency ($2\pi\bar{v}s$) involves moving mirror speed (s). If s is chosen correctly, output frequency can be reduced into the audio range (20–20000 Hz). In a spectrometer, s is small enough that λ' is at least 8 orders of magnitude larger than λ (where c is the speed of light):

$$\frac{c}{2\pi\bar{v}s} = \frac{\lambda'}{2\pi} = \frac{1}{2\pi\bar{v}'}$$

3. Example: λ for the He-Ne laser is 6.33×10^{-7} m. The corresponding frequency is:

$$v = \frac{c}{\lambda} = \frac{\frac{2.998 \times 10^8 \text{ m}}{\text{sec}}}{6.33 \times 10^{-7} \text{ m}} = \frac{4.74 \times 10^{14}}{\text{sec}} = 4.74 \times 10^{14} \text{ Hz}$$

If $\lambda' = 2.998 \times 10^4$ m, then the interferogram frequency v' would be 10000 Hz (in the audio range):

$$v' = \frac{c}{\lambda'} = \frac{\frac{2.998 \times 10^8 \text{ m}}{\text{sec}}}{2.998 \times 10^4 \text{ m}} = \frac{10000}{\text{sec}} = 10000 \text{ Hz}$$

In order to achieve $v' = 10000$ Hz with the 4.74×10^{14} Hz laser light, the mirror would have to move at a speed (s):

$$\frac{\lambda}{2\pi} = \frac{c}{\frac{2\pi s}{\lambda}} = \frac{c\lambda}{2\pi s}$$

and solving for s:

$$s = \frac{c\lambda}{\lambda'} = \frac{\frac{2.998 \times 10^8 \text{ m}}{\text{sec}} \times 6.33 \times 10^{-7} m}{2.998 \times 10^4 \text{ m}} = \frac{6.33 \times 10^{-3} \text{ m}}{\text{sec}} \times \frac{100 \text{ cm}}{\text{m}} = \frac{0.633 \text{ cm}}{\text{sec}}$$

4. Detectors exist that can accurately follow intensity and wave frequency around 10000 Hz, but even modern detectors cannot respond fast enough to follow high frequency infrared waves such as 4.74×10^{14} Hz.

12.6.4. When the light contains more than one wavelength, the interferometer effectively adds all the waves to each other, yielding a single interferogram wave.

a. Example: three wavelengths (λ, 2λ, and 3λ) having equal amplitude (intensity) are added into an interferogram. The interferometer yields the sum intensity curve as the mirror moves from -8λ to $+8\lambda$ at speed s over time t.

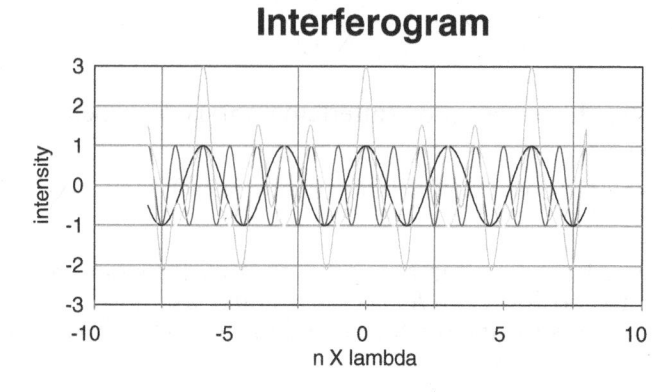

Interferogram

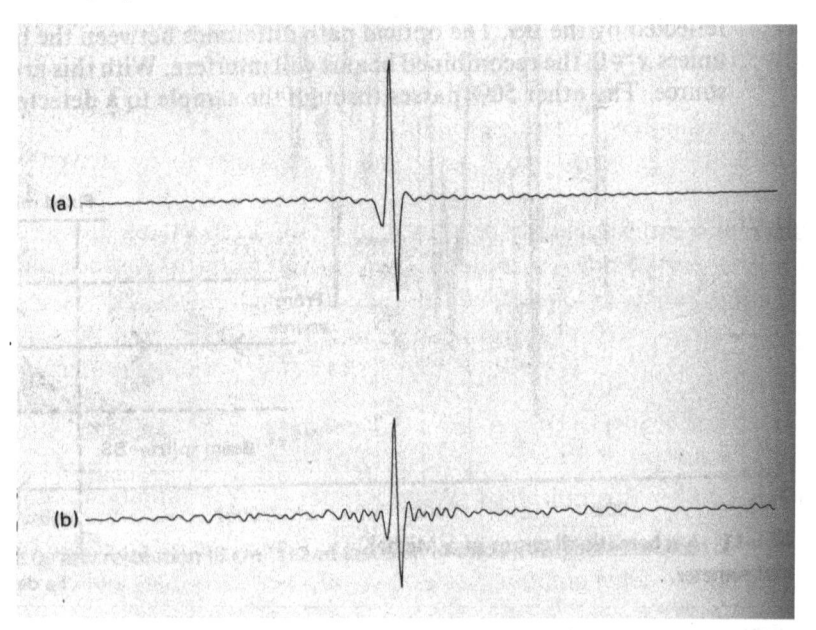

12.6.5. Output from a hot wire source contains a full wavelength spectrum between $2.5 < \lambda < 25$ μm. This spectrum's interferogram again has maximum intensity when the moving mirror passes through point 0.

* interferograms taken from Lambert, J. B., Shurvell, H. F., Lightner, D., Cooks, R. G., "Introduction to Organic Spectroscopy," Macmillan Publishing Co., New York, (1987), p 146.

   a. This "interferogram" that contains all IR frequencies, (a) above, passes through a sample into the detector, (b) above. The sample in this case is polystyrene film. The sample absorbs some frequencies, so the emerging interferogram is different from the original.
   b. An interferogram (intensity *vs.* time graph) looks nothing like a conventional infrared spectrum and is not very useful itself, but it contains all conventional spectrum information. Interferograms (original and sample) in digital (a series of points) form are Fourier transformed. This process interprets the observed intensity *vs.* time graph in terms all its constituent wavelengths, i.e., it "decodes" the interferogram by creating a "frequency domain." It determines all frequencies and intensities that must be present to generate the observed interferogram. The original spectrum (before it passed through the sample) is subtracted point by point from the sample spectrum. When intensity differences (or %*T*'s) are plotted *versus* frequency, a conventional infrared spectrum is revealed.

12.6.6. Some FT-IR advantages:
   a. Spectra can be run in < 2 sec.
   b. It is more sensitive. Less energy is needed, since all frequencies pass through the sample at once.
   c. Several scans can be run and added together (time averaging) to improve the signal-to-noise ratio. Smaller samples can be run this way.
   d. Wavelength is calibrated against that of a precisely known He-Ne laser, improving accuracy. Data collection can be triggered at accurate distance (or time) intervals, for example, whenever the laser interferogram crosses through a zero intensity point.
   e. Modern instruments often have an attenuated total reflectance (ATR) platform. Liquid and solid sample spectra can be run on a ZnSe or diamond surface, which greatly improves intensity reproducibility.

12.7.1. Infrared spectra can be very complex. The table below helps interpret simple functional group absorptions.
   a. Wavenumbers listed are for negative peak minima.
   b. Somewhat arbitrary descriptions: strong, medium, weak, shoulder, and broad are indicated as (s), (m), (w), (sh), and (br), respectively.
   c. Common functional groups can be identified from fundamental region peaks (4000 – 1400 cm^{-1}), which can usually be assigned unambiguously.
   d. Fingerprint region (1400 – 200 cm^{-1}) peak assignments must usually be viewed with skepticism.
   e. An unknown compound cannot usually be identified by its IR spectrum alone (unless the fingerprint matches that of a known sample). Other information (mass spectrum, NMR spectra, and/or chemical analyses) is needed.

12.7.2 Familiarize yourself with alkane, alkyl halide, alkene, alkyne, and alcohol family infrared absorptions. Other functional group absorptions will be added as we come to them.
   a. Commonly used infrared absorptions:

Vibration	Structural feature	Wavenumber range (cm^{-1})	Intensity shape
sp^3 -C-C- stretch	alkyl groups	800–1200	variable
sp^3 -C-H stretch	alkyl groups	2800–3000	strong
-C(=O)-H stretch	aldehydes	2700–2780 / 2820–2900	weak, 2820–2900 often obscured by alkane -C-H stretch
sp^3 -C-H bend	alkyl groups	1430–1470	medium
–H$_2$C-H bend	methyl groups	1370–1385	medium (if -C(CH$_3$)$_{2\ or\ 3}$ is present it is often two peaks)
sp^2 -C=C-stretch	alkenes	1640–1680	variable
sp^2 -C=C-H stretch	alkenes	3020–3080	medium
sp^2 C≡C-H stretch	aryl (phenyl) groups	3000–3100	medium
sp^2 -C=C-H out of plane bends	alkenes	880–900	RR'C=CH$_2$ (terminal)
		790–840	RR'C=CR"H (trisubstituted)
		910–920 / 990–1000	RHC=CH$_2$ (monosubstituted), two bands
		965–975	*trans*-disubstituted

Vibration	Structural feature	Wavenumber range (cm^{-1})	Intensity shape
		675–730	$\underset{R}{\overset{H}{C}}=\underset{R'}{\overset{H}{C}}$ (*cis*-disubstituted) variable
sp -C≡C- stretch	alkynes	2100–2260	variable
sp -C≡<u>C</u>-<u>H</u> stretch	alkynes	3300	medium
sp^2 —C=C— stretch (ring vibrations)	aryl (phenyl) groups	1500, 1600	variable, two bands; each peak often double (four peaks total), but two or three peaks often weak
sp^2 —C=<u>C</u>–<u>H</u> out of plane bends	branched aryl (phenyl) groups	690–710 730–770	two bands; monosubstituted benzene
		735–770	1,2-disubstituted benzene (*ortho*)
		690–710 750–810	two bands; 1,3-disubstituted benzene (*meta*)
		810–840	1,4-disubstituted benzene (*para*)
sp^2 —C=<u>C</u>–<u>H</u> bending overtones and combinations	branched aryl (phenyl) groups	1667–2000	weak, two to five peaks
sp^3 -C-X stretch, for X = Cl, Br, or I	alkyl halide	200–845	medium/strong
-C-O stretch	alcohols, ethers, carboxylic acids and esters	1050	strong, 1° sp^3 C-OH
		1100	strong, 2° sp^3 C-OH
		1150	strong, 3° sp^3 C-OH
		1140–1230	strong, sp^2 Ar-OH
		1000–1300	two bands for esters and unsymmetrical ethers
-O-H stretch	alcohols or phenols	3200–3600	strong, broad, 3610–3640 if not H-bonded

Vibration	Structural feature	Wavenumber range (cm^{-1})	Intensity shape
	carboxylic acids	2500–3300	strong, broad
-OH bend	carboxylic acids	920, 1400	broad, weak
sp^2 -C=O stretch	aldehydes	1675–1725	strong, below 1700 if conjugated
	ketones	1675–1725	strong, 1770 if part of a strained ring, below 1700 if conjugated
	carboxylic acids	1680–1725	strong, somewhat broad, below 1700 if conjugated
	esters	1715–1740	strong, 1770 if RCOOAr or RCOOC=C.
	acid chlorides	1750–1810	strong
	anhydrides	1740–1840	strong, usually two peaks
	amides	1640–1690	strong
R-N-H stretch	amines, amides	3050–3550	medium, broad
-C-N stretch	amines, amides	1030–1230	weak, 3° amine often 2 peaks
sp^2 Ar-N stretch	aryl amines or aryl amides	1180–1360	strong, two peaks
-HN-H bend	amines, amides	690–710 1560–1650	medium, broad medium
-RN-H bend	2° amines, 2° amides	1530–1570	medium
sp -C≡N stretch	nitriles	2200–2300	variable
sp^2 —N=O stretch	nitro	1515–1560 1345–1385	medium, two peaks

12.7.3. Alkane sp^3 C-H stretching is about 2800–3000 cm^{-1}, C-C stretching vibration is about 1200 cm^{-1}, sp^3 C-H bending is 1430–1470 cm^{-1}, and methyl C-H bending is about 1370–1390 cm^{-1}. Example spectra below are from "The Spectral Database for Organic Compounds" (Japan, National Institute of Advanced Industrial Science and Technology). An easy way to access it is to Google on "spectral database" and click on the link: "AIST: Spectral Database for Organic Compounds."

CHAPTER 12   371

a. Example: *n*-octane. C-H stretching and bending are observable, but C-C stretching is absent. This spectrum alone would not distinguish *n*-octane from most other alkanes.

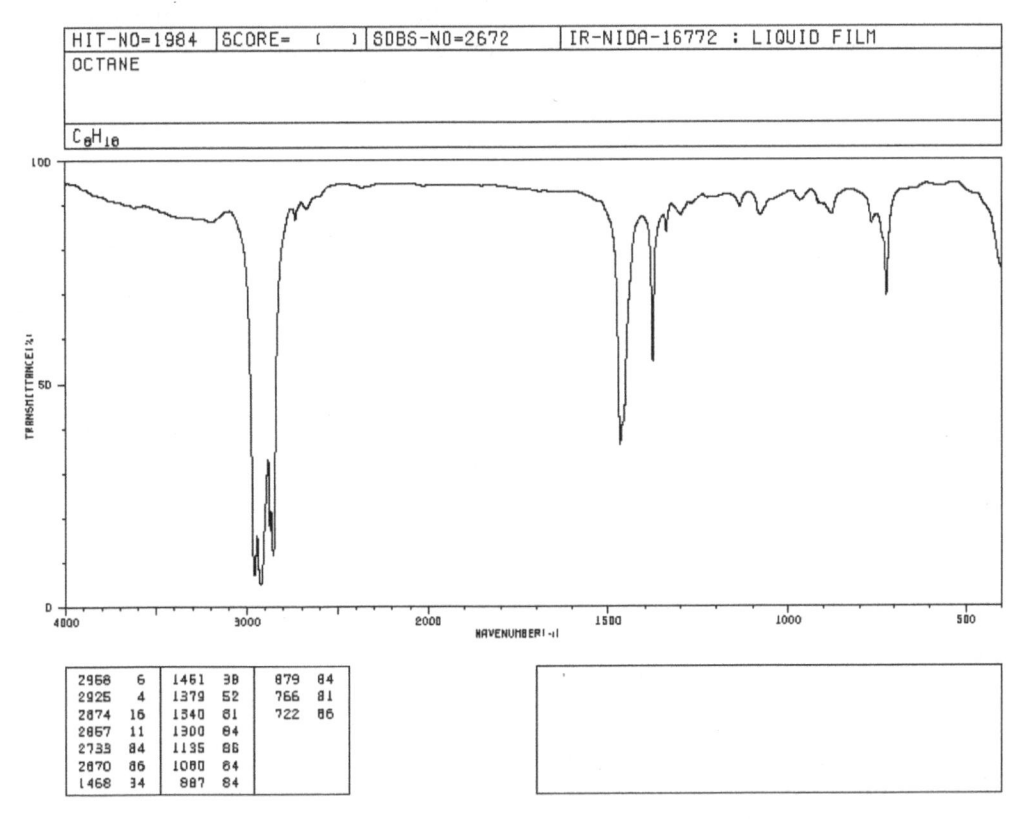

b. Example: 2-methylheptane. Note: *dimethyl split* at 1367 and 1379 cm^{-1}, a C-H bending that is characteristic of -CH(CH$_3$)$_2$ or -C(CH$_3$)$_3$.

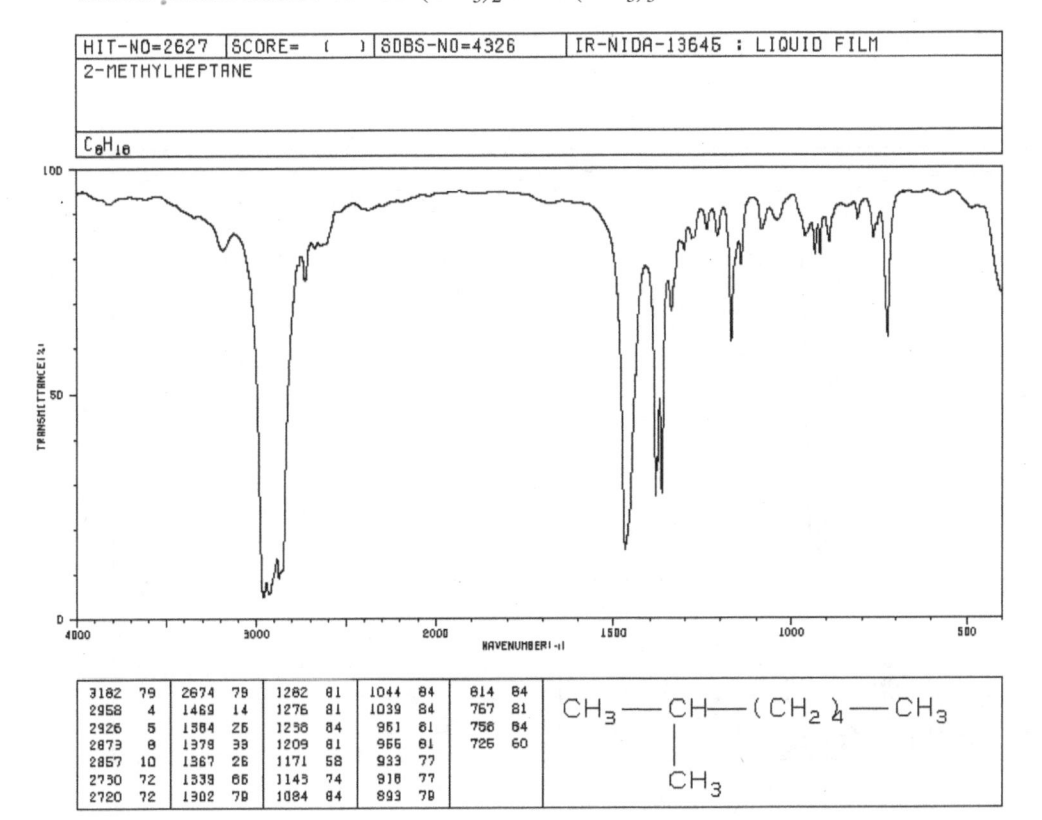

12.7.4. Ethene, mono-, di-, and trisubstituted alkenes have sp^2 C-H stretching at 3000–3100 cm^{-1}, moderate, (higher wavenumbers than sp^3 C-H stretching due to the stronger sp^2–$1s$ bond). Isolated C=C stretching is about 1640–1680 cm^{-1} (absent if C=C is symmetrical, otherwise weak to moderate). Out of plane sp^2 C-H bending vibrations at 675 – 1000 cm^{-1} help elucidate substitution pattern (*cis-*, *trans-*, *terminal*, etc.).

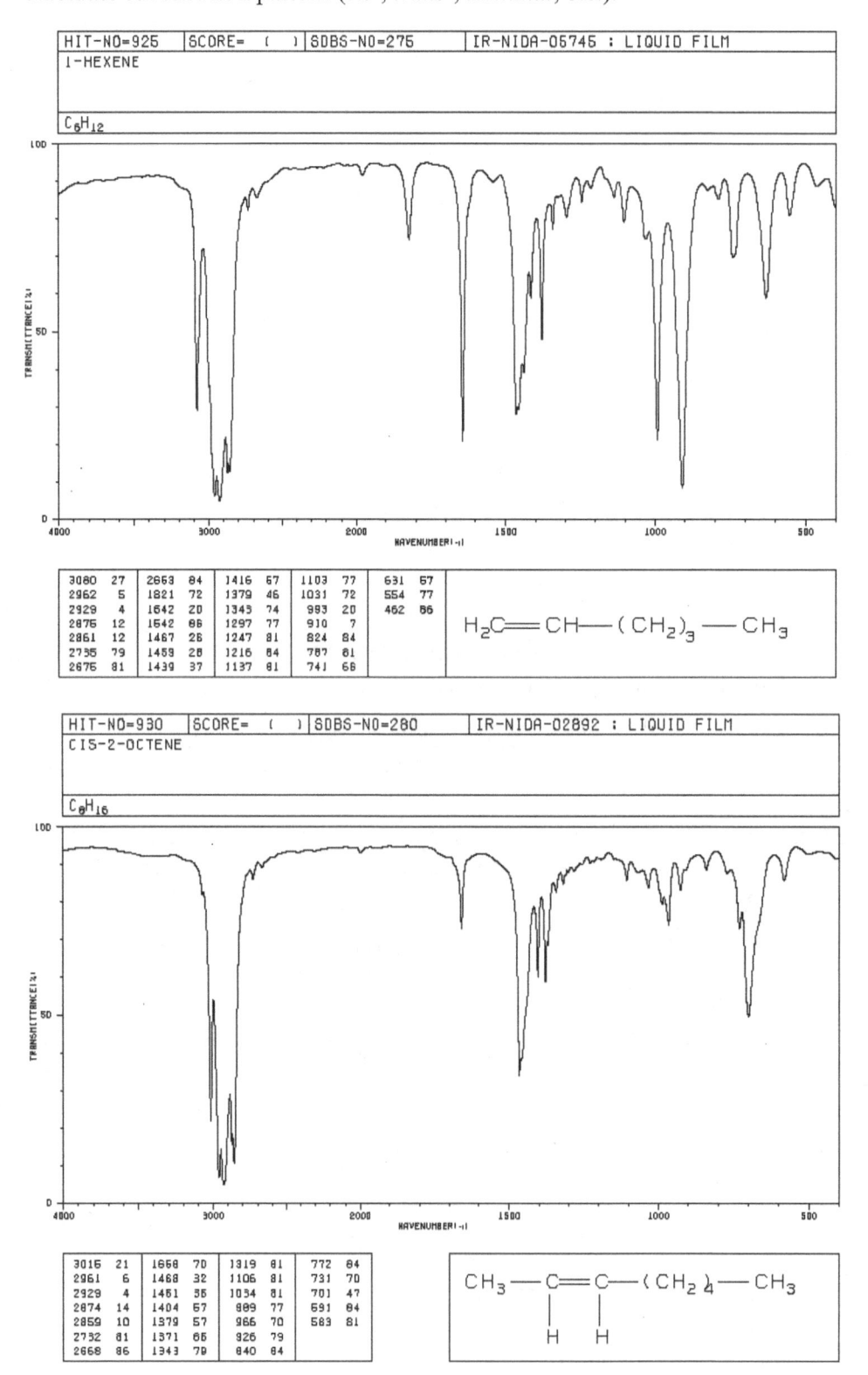

HIT-NO=925 SCORE= ( ) SDBS-NO=275 IR-NIDA-05745 : LIQUID FILM
1-HEXENE
C$_6$H$_{12}$

3080	27	2663	84	1416	67	1103	77	631	67
2962	5	1821	72	1379	46	1031	72	554	77
2929	4	1642	20	1343	74	993	20	462	86
2875	12	1642	86	1297	77	910	7		
2861	12	1467	26	1247	81	824	84		
2735	79	1459	28	1216	84	787	81		
2675	81	1439	37	1137	81	741	66		

H$_2$C=CH—(CH$_2$)$_3$—CH$_3$

HIT-NO=930 SCORE= ( ) SDBS-NO=280 IR-NIDA-02892 : LIQUID FILM
CIS-2-OCTENE
C$_8$H$_{16}$

3015	21	1658	70	1319	81	772	84
2961	6	1468	32	1106	81	731	70
2929	4	1461	35	1034	81	701	47
2874	14	1404	57	989	77	691	84
2859	10	1379	57	966	70	583	81
2732	81	1371	65	926	79		
2668	86	1343	79	840	84		

CH$_3$—C=C—(CH$_2$)$_4$—CH$_3$
         | |
         H   H

a. Conjugated alkenes (with two double bonds separated by a single bond) have less π-electron density between double bonded C atoms. Bonds are not as strong (stiff), and at least some absorption is at lower wavenumbers (1620–1640 cm^{-1}).

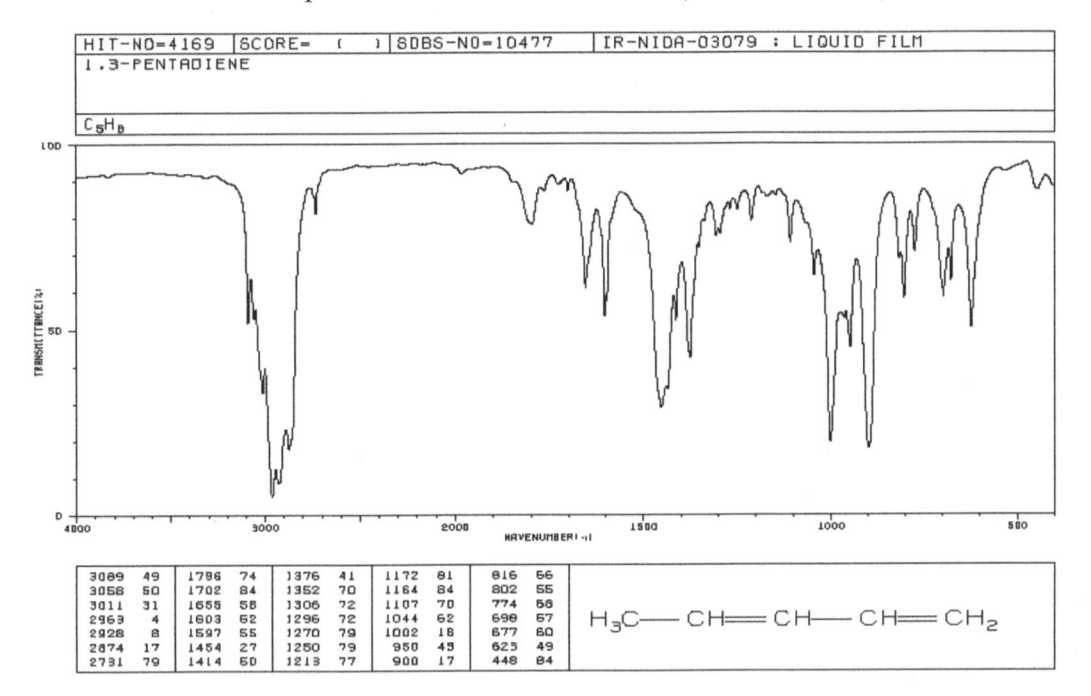

| HIT-NO=4169 | SCORE= ( ) | SDBS-NO=10477 | IR-NIDA-03079 : LIQUID FILM |

1,3-PENTADIENE

C_5H_8

3089	49	1796	74	1376	41	1172	81	816	56
3058	50	1702	84	1352	70	1164	84	802	55
3011	31	1655	58	1306	72	1107	70	774	56
2963	4	1603	62	1296	72	1044	62	698	57
2928	8	1597	55	1270	79	1002	18	677	60
2874	17	1454	27	1250	79	950	45	625	49
2731	79	1414	60	1213	77	900	17	448	84

H_3C—CH=CH—CH=CH_2

b. π-Electron delocalization is even greater in benzene rings. The 1½ bond C≡≡C stretches are often four peaks, two near 1600 and two near 1500 cm^{-1}. Often one of each pair is weak.

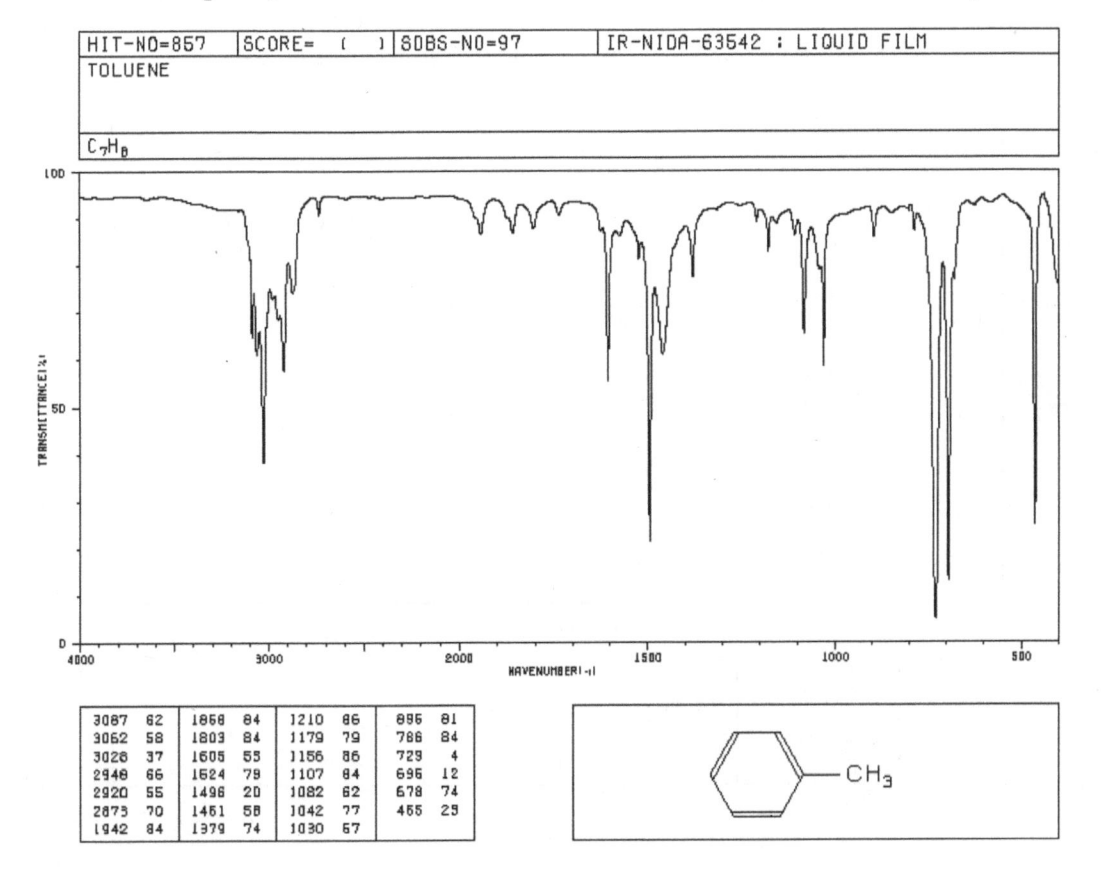

| HIT-NO=857 | SCORE= ( ) | SDBS-NO=97 | IR-NIDA-63542 : LIQUID FILM |

TOLUENE

C_7H_8

3087	62	1868	84	1210	86	895	81
3062	58	1803	84	1179	79	786	84
3028	37	1605	53	1156	86	729	4
2948	66	1524	79	1107	84	695	12
2920	55	1496	20	1082	62	678	74
2875	70	1461	58	1042	77	455	29
1942	84	1379	74	1030	67		

(benzene ring)—CH_3

12.7.5. Alkyne *sp* C-H stretching is at 3300 cm^{-1}, even higher than *sp^2* C-H stretching due to *sp*-C-H bond greater *s*-character. C≡C stretching is absent if symmetrical. Unsymmetrical alkyne C≡C bonding is stronger (stiffer) than in ethene, and absorbs at 2100–2200 cm^{-1} (weak to moderate).

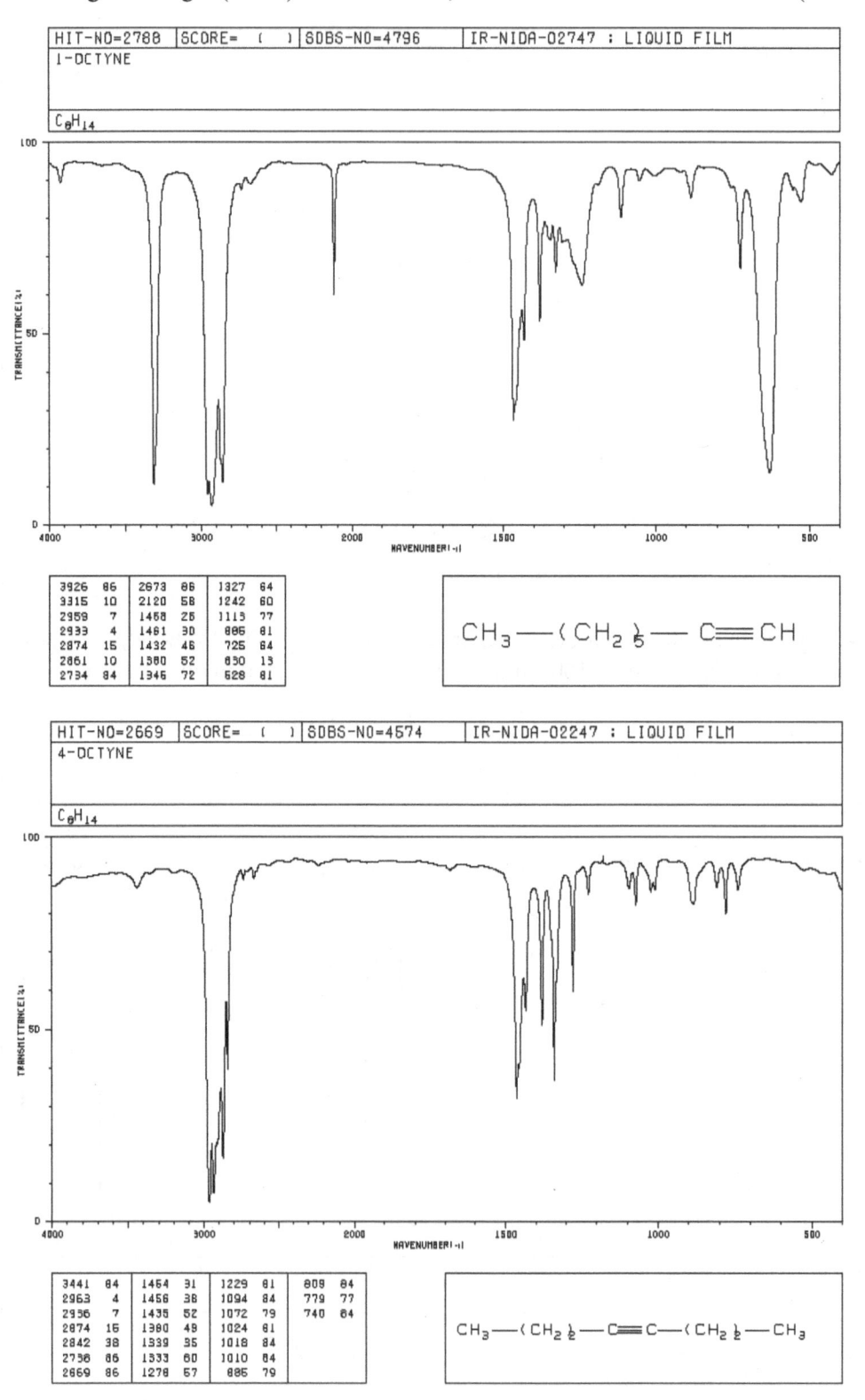

12.7.6. 1-Chloropropane has the usual sp^3 C-H stretchings and bendings. C-Cl stretching is at about 200–845 cm^{-1}, but hard to identify.

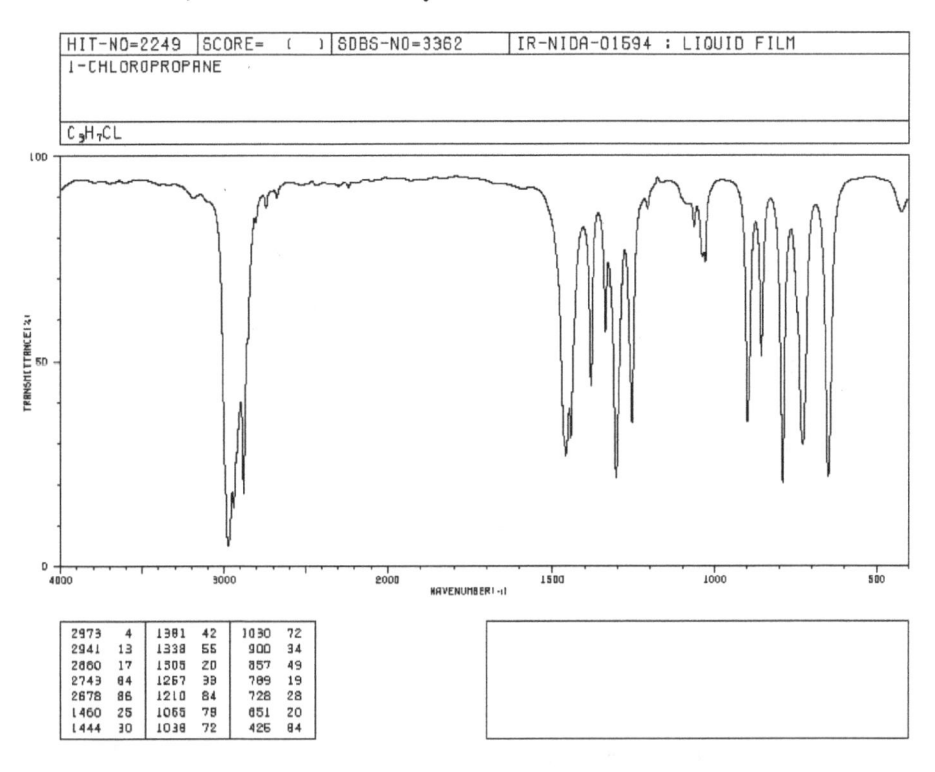

12.8.1. Alcohol O-H stretching is 3200–3600 cm^{-1} (strong, broad due to hydrogen bonding). A 1° C-O stretch is often near 1050 cm^{-1}, 2° C-O stretch near 1100 cm^{-1}, and 3° C-O stretch near 1150 cm^{-1}. Phenol or aryl or vinyl ether sp^2 C-O stretch is at 1200–1280 cm^{-1}.

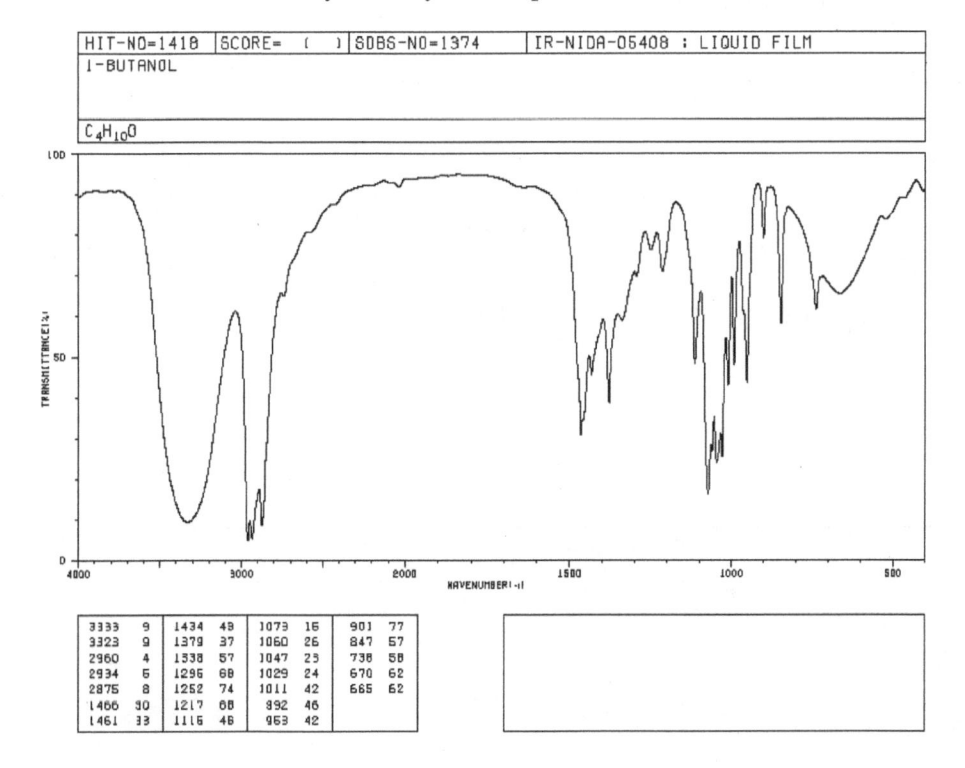

a. Ethers have no –O-H stretching, but do have C-O stretching.

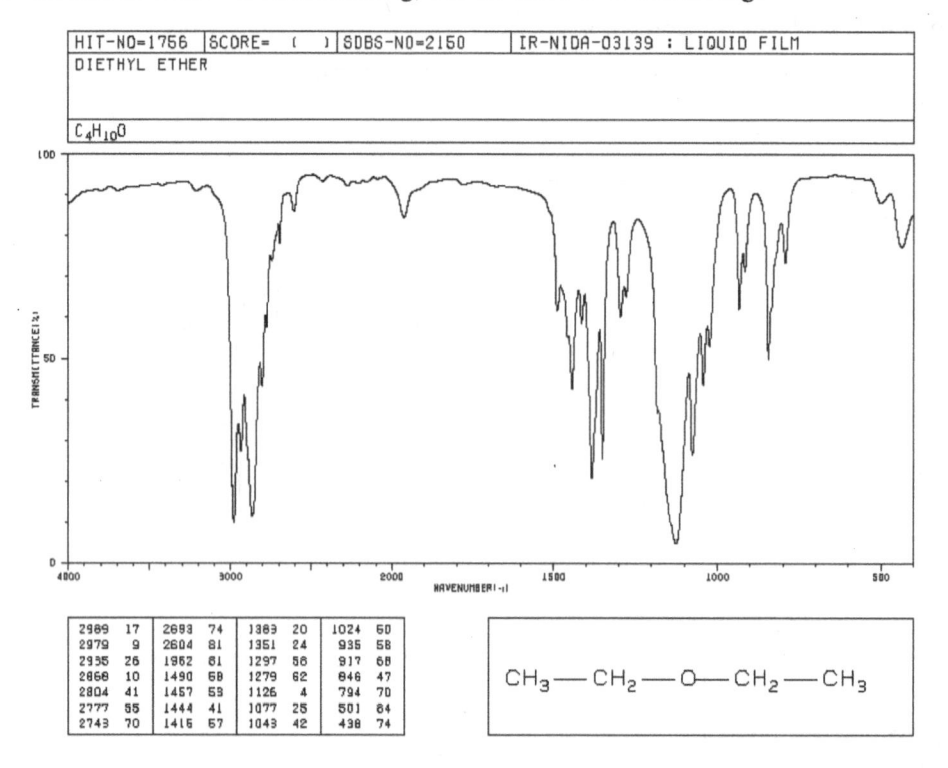

12.8.2. Amine N-H stretching is about 3300–3500 cm^{-1} (moderate, broad due to hydrogen bonding, often with one or two spikes for 2° or 1° amines, respectively). C-N stretching is in the 1030–1230 cm^{-1} region. 3° amines have no N-H stretch.

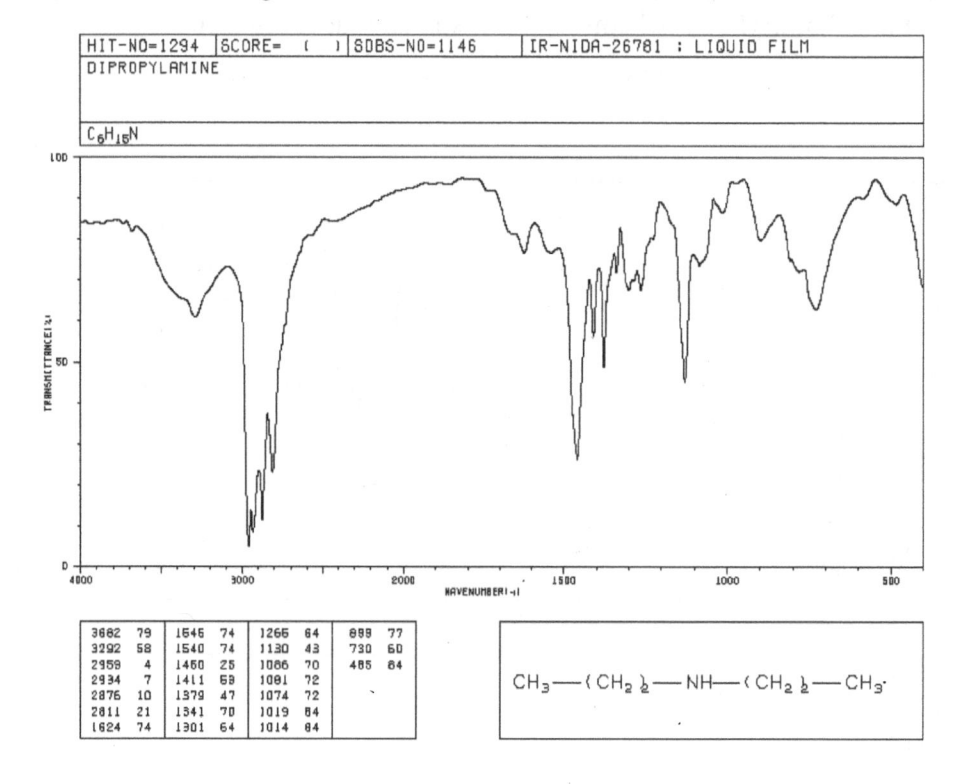

12.9.1. A C=O bond is stronger than C=C, and absorbs about 1710 cm^{-1} in aldehydes, ketones, and carboxylic acids (strong).

    a. Aldehydes have two C-H stretching bands at 2700–2780 and 2820–2900 cm^{-1}. The latter is partly obscured by sp^3 C-H stretching, but the former is diagnostic for aldehydes.

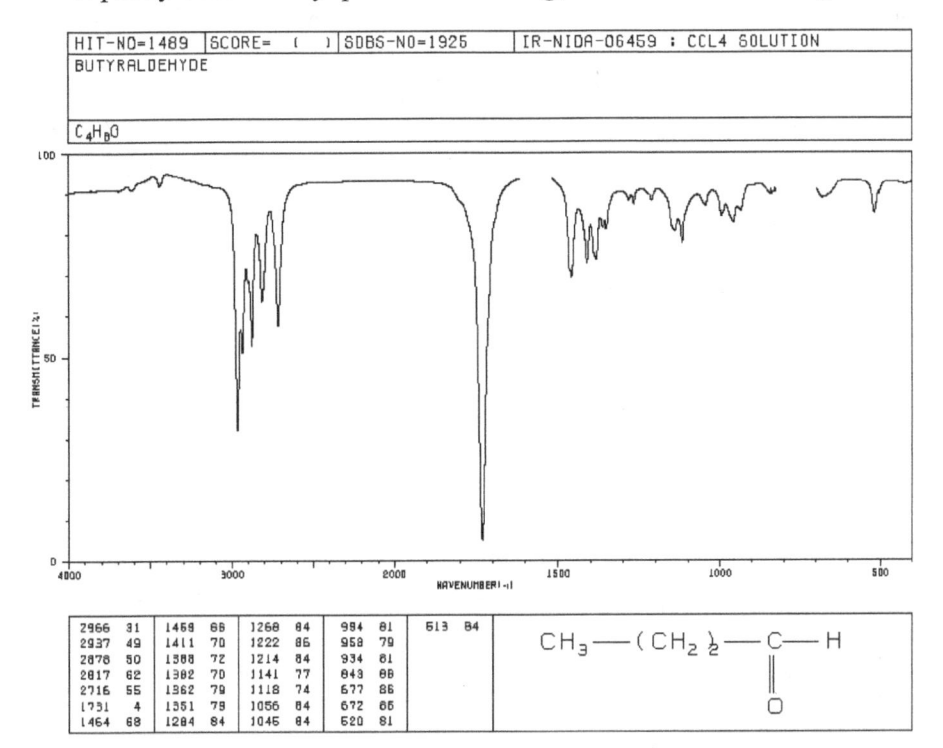

    b. The 2700 – 2780 cm^{-1} peak is not observed in ketones.

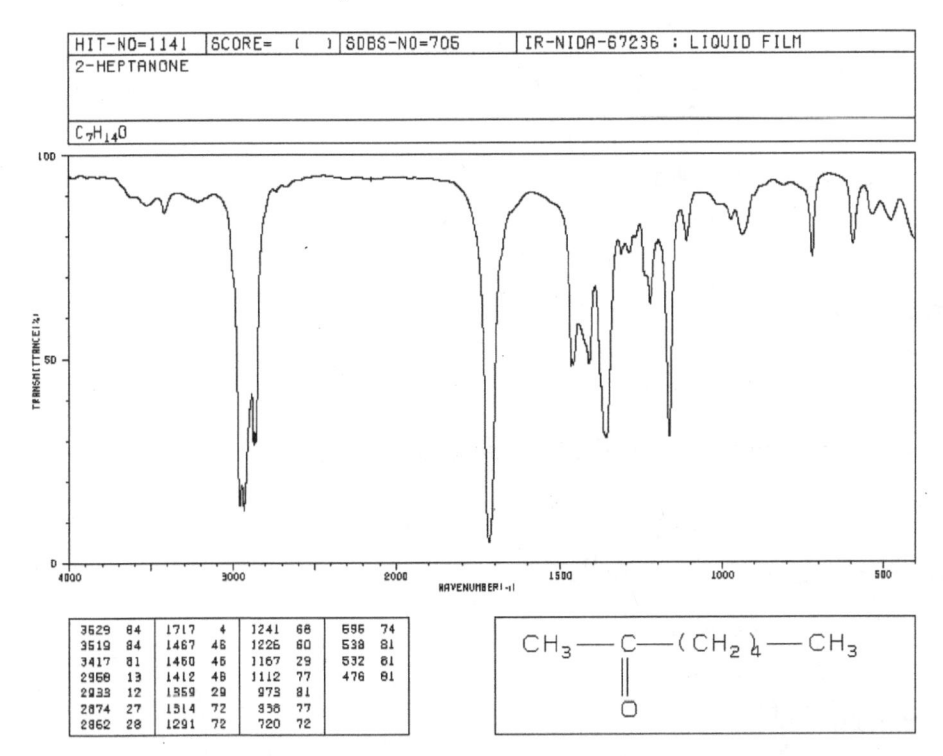

12.9.2. Carboxylic acids have O-H stretching (2500 –3500 cm^{-1}, strong, very broad) that partially obscures C-H stretching. H-bonding also usually broadens the C=O stretching. O-H bending bands are at 1400 and 920 cm^{-1}.

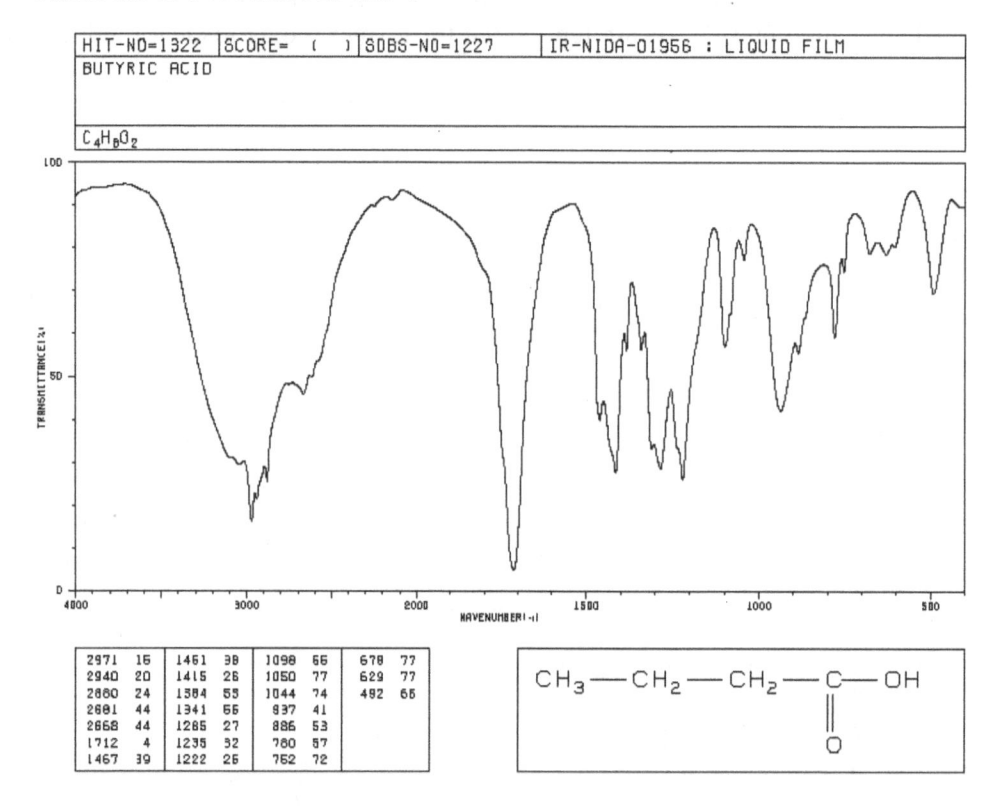

12.9.3. Conjugation with an adjacent C=C weakens a C=O bond, so absorption comes at about 1680–1690 cm^{-1}.

12.9.4. Similar resonance in amides lowers the C=O absorption wavenumber.

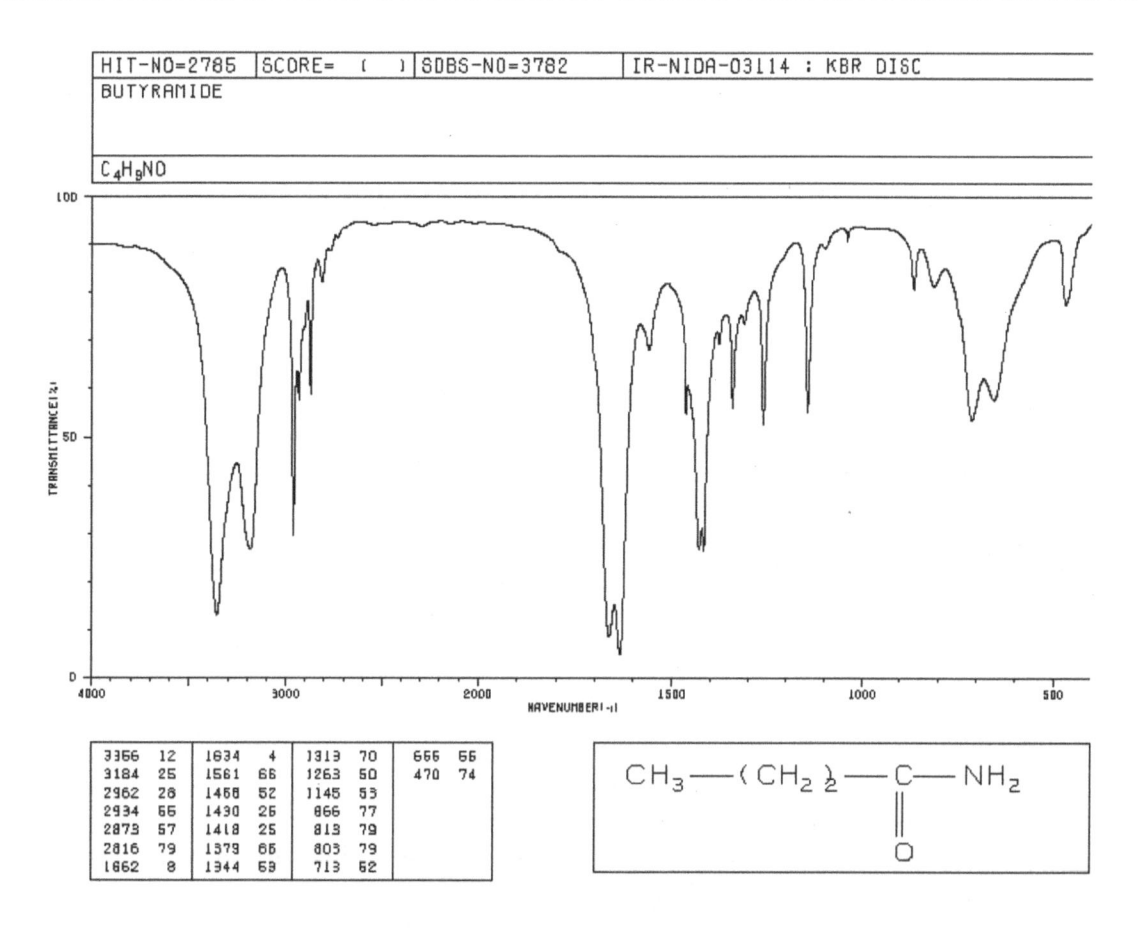

| HIT-NO=2785 | SCORE= ( ) | SDBS-NO=3782 | IR-NIDA-03114 : KBR DISC |
| BUTYRAMIDE |
| C$_4$H$_9$NO |

3356	12	1634	4	1313	70	656	66
3184	25	1561	66	1263	50	470	74
2962	28	1466	52	1145	53		
2934	56	1430	26	866	77		
2873	57	1418	25	813	79		
2816	79	1379	66	803	79		
1662	8	1344	59	713	62		

CH$_3$—(CH$_2$)$_2$—C(=O)—NH$_2$

- **a.** Amide C=O stretch might be confused with alkene C=C stretch. Other absorptions (N-H stretch) help clarify the situation. The N-H absorption wavenumber can be used to estimate H-bonding extent.

12.9.5. Strained ring carbonyl, or ester, anhydride, or acid halide C=O stretch come above 1710 cm^{-1}.

H$_3$C(CH$_2$)$_6$–C(=O)–O–CH$_2$CH$_3$ ← 1738 cm^{-1}

cyclobutanone ← 1785 cm^{-1}

H$_3$CCH$_2$–C(=O)–Cl ← 1792 cm^{-1}
propanoyl chloride

H$_3$C–C(=O)–O–C(=O)–CH$_3$
ethanoic anhydride
acetic anhydride
two peaks, 1766 and 1827 cm^{-1}

- **a.** Ring strain forces electron density into the C=O π-bond, increasing its strength.

12.10.1. C=N stretching occurs about 1660 cm^{-1}, similar to C=C except it is usually stronger.

12.10.2. Nitriles C≡N stretching occurs about 2200–2300 cm^{-1}, just higher than C≡C stretching (2100–2200 cm^{-1}).

380 SEQUENTIAL ORGANIC CHEMISTRY I

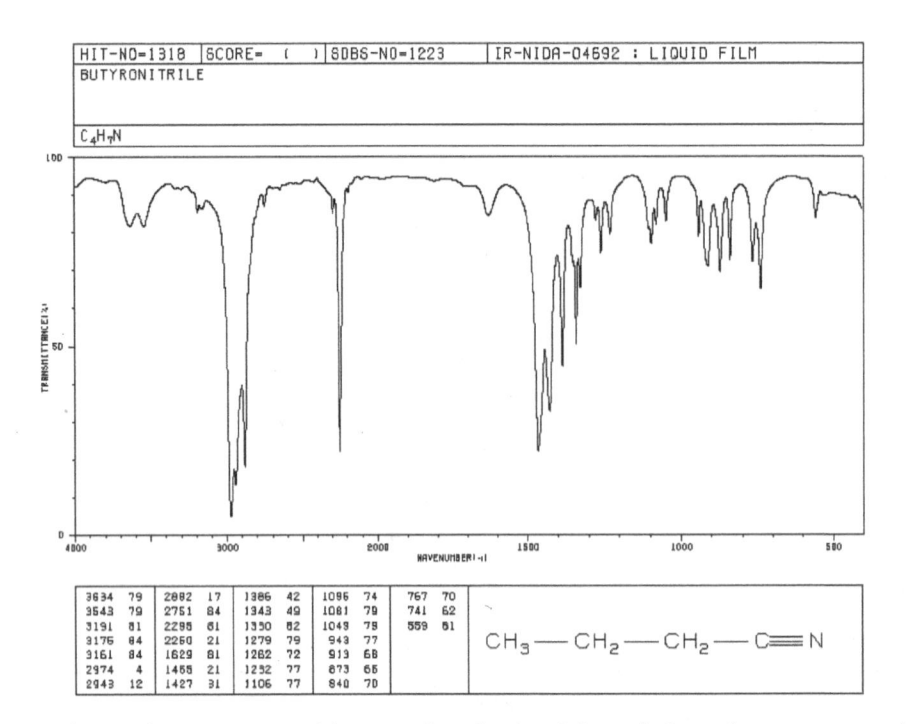

| HIT-NO=1318 | SCORE= ( ) | SDBS-NO=1223 | IR-NIDA-04692 : LIQUID FILM |

BUTYRONITRILE

C_4H_7N

3634	79	2882	17	1386	42	1096	74	767	70
3543	79	2751	84	1343	49	1081	79	741	62
3191	81	2295	61	1330	82	1049	79	559	81
3175	84	2250	21	1279	79	943	77		
3161	84	1629	81	1262	72	913	68		
2974	4	1455	21	1232	77	873	65		
2943	12	1427	31	1106	77	840	70		

$CH_3-CH_2-CH_2-C\equiv N$

12.11.1. Information summary (that can be obtained from infrared spectroscopy):

    a. Evidence for which functional groups are present.

    b. Evidence for functional group absences.

    c. Can prove compound identity if a spectrum matches a known compound spectrum.

12.12.1. Correct spectral interpretation requires practice.

12.12.2. Problem 1. A C_6H_{12} hydrocarbon has infrared absorptions at: 2920 (s), 2840 (s), 1450 (s) cm^{-1}. No absorptions appear above 2920, and none below 1450 until about 1250 cm^{-1}. Draw a structure consistent with these obervations.

12.12.3. Problem 2. The infrared spectrum is shown below for a $C_4H_{10}O$ compound. Draw a consistent structure. Justify your answer by identifying four peaks (for example, as sp^3 C-H stretch, etc.).

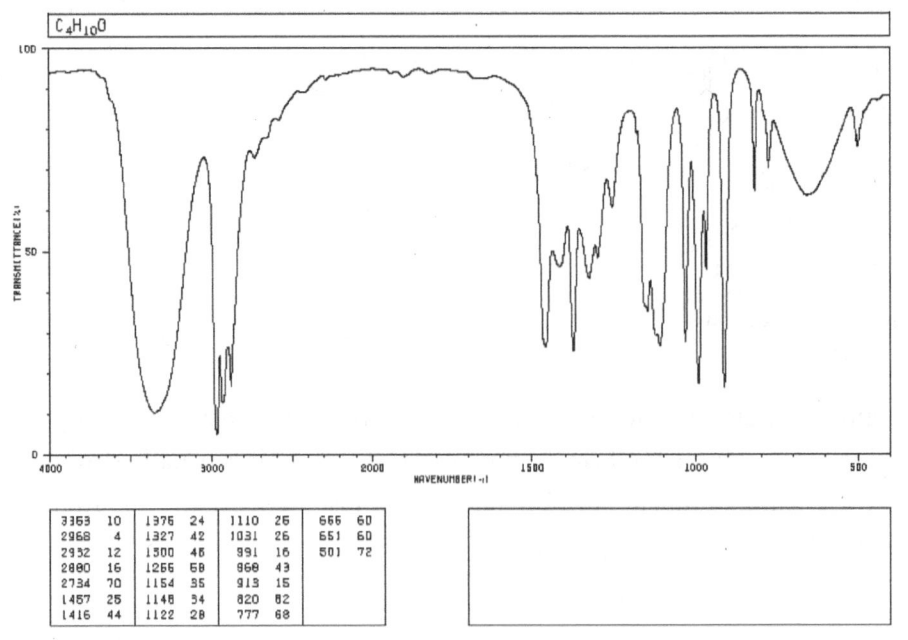

$C_4H_{10}O$

3353	10	1376	24	1110	26	666	60
2968	4	1327	42	1031	26	651	60
2932	12	1300	45	991	16	501	72
2880	16	1266	58	968	43		
2734	70	1154	35	913	15		
1457	25	1148	34	820	82		
1416	44	1122	28	777	68		

12.12.4. Problem 3. An infrared spectrum is shown for a C_5H_{10} hydrocarbon. Draw a likely structure. Justify your answer by identifying four peaks (as sp^3 C-H stretch, etc.).

```
C5H10
```

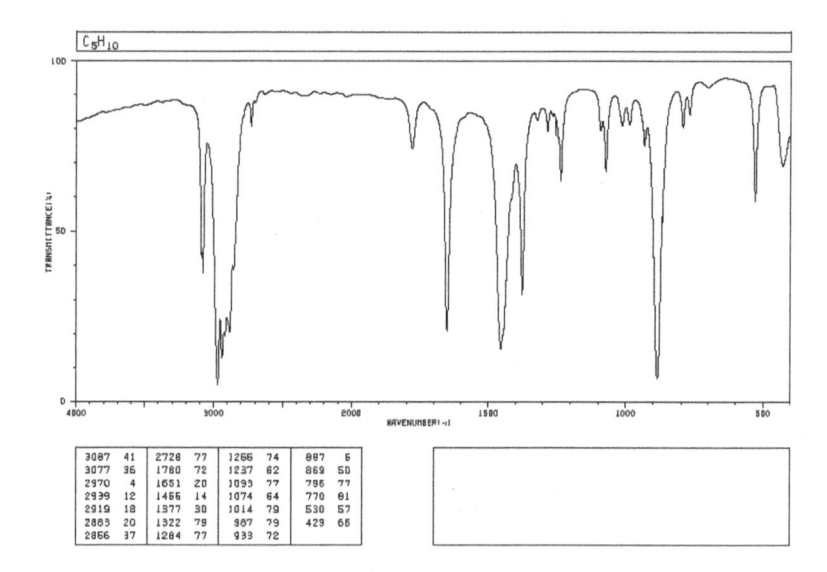

3087	41	2728	77	1266	74	897	6
3077	36	1780	72	1237	62	869	50
2970	4	1651	20	1093	77	796	77
2939	12	1456	14	1074	64	770	81
2919	18	1377	30	1014	79	530	57
2865	20	1322	78	987	79	429	66
2856	37	1284	77	933	72		

12.12.5. Problem 4. Upon hydrogenation (H_2 / Pt), compound A (C_4H_8O) is converted into B ($C_4H_{10}O$). A and B infrared spectra are shown below. Draw structures that are consistent with these data. Justify your answers by identifying four peaks in each spectrum (as sp^3 C-H stretch, etc.)

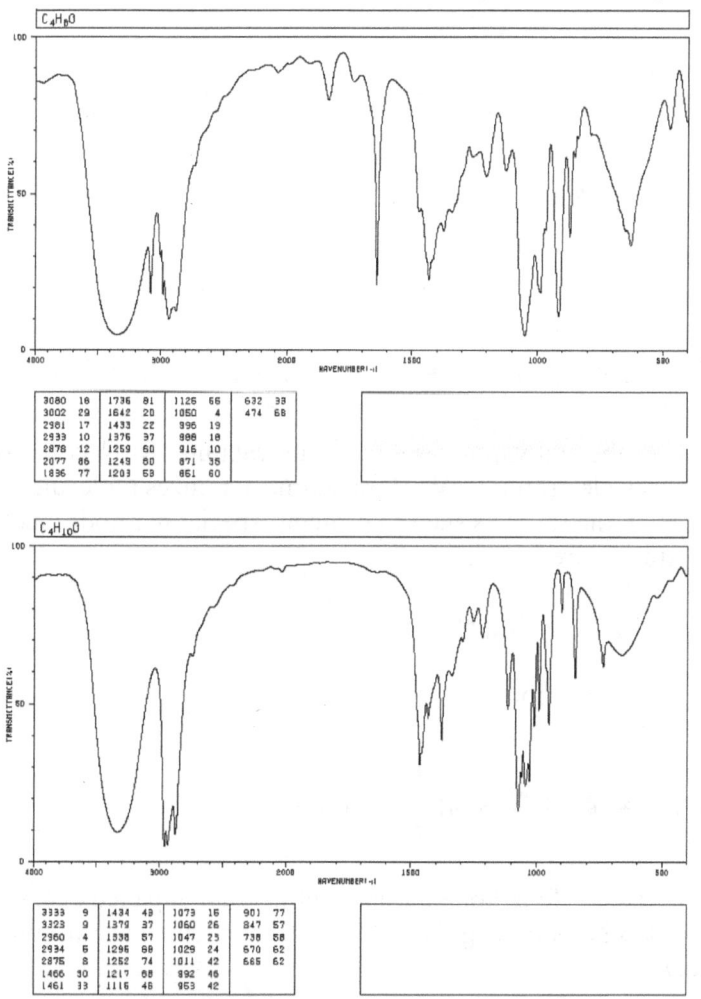

```
C4H8O
```

3080	16	1736	81	1126	66	632	39
3002	29	1642	20	1050	4	474	68
2981	17	1433	22	996	19		
2933	10	1376	37	988	18		
2878	12	1259	60	915	10		
2077	86	1249	60	871	35		
1636	77	1203	59	851	60		

```
C4H10O
```

3333	9	1434	49	1073	16	901	77
3323	9	1379	37	1060	26	847	57
2960	4	1338	57	1047	23	738	58
2934	6	1296	68	1029	24	670	62
2875	8	1252	74	1011	42	565	62
1466	50	1217	68	992	48		
1461	33	1116	48	953	42		

12.12.6. Problem 5. The *cis*-cyclopentane-1,2-diol infrared spectrum has an O-H stretching absorption at lower than usual wavenumber for alcohols. It does not disappear even in very dilute solution. *Trans*-cyclopentane-1,2-diol shows no such absorption. Suggest a possible explanation.

12.13.1. Mass spectrometry (MS) is now the most common method to find an unknown compound's molecular mass and formula, often the first step in identification.

    a. Other (older) methods: elemental analysis, freezing point depression, osmotic pressure. These can be tedious and require large, pure samples.

    b. MS does not involve light absorption or emission (section 12.1.3a).

    c. MS can determine protein molecular masses. Electrospray ionization – dissolve nonvolatile (high molecular mass) molecules in a solvent, then spray a hot droplet stream into evacuated source chamber.

12.13.2. Mass spectrometer tube schematic diagram*:

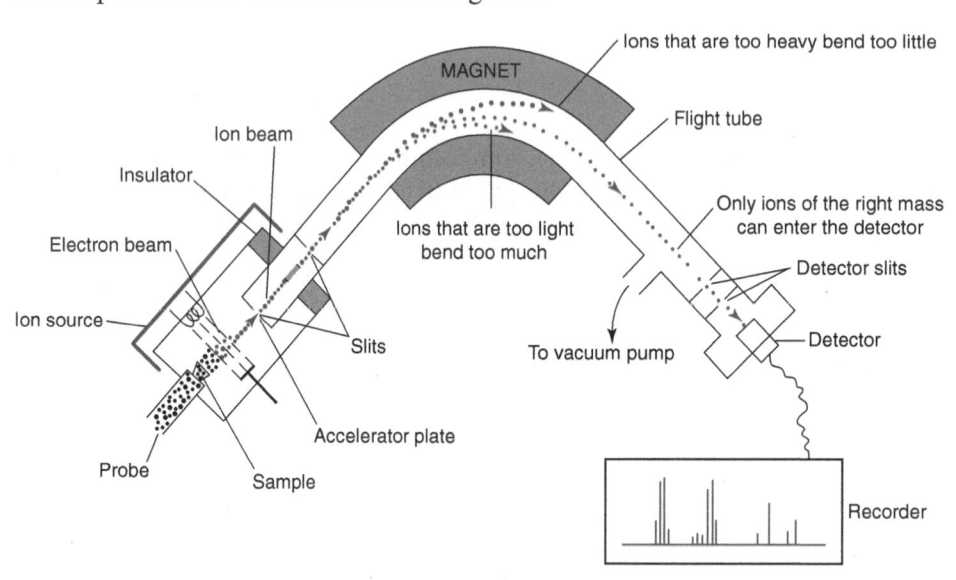

*taken from Wade, Leroy G., Simek, JAN W., ORGANIC CHEMISTRY, 9th Ed., © 2017. Reprinted by permission of Pearson Education, Inc., New York, New York.

    a. Sample is inserted via the probe and vaporizes under high vacuum. Most commonly, a collision with a high energy electron (70 eV = 6740 kJ/mol), ionizes molecules. For high molecular mass, nonvolatile samples such as proteins, special methods (such as electrospray ionization) are required.

$$M + e^- \rightarrow [M^{+\cdot}] + 2\,e^-$$

       1. The product $M^{+\cdot}$ is a radical cation.

       2. Example: methane.

       3. $M^{+\cdot}$ is an unusual carbocation – four atoms around central C, an unpaired electron, and one half C-H bond. Unless N (or other group V element) is present, it has an even mass to electrical charge ratio, *m/z*.

12.13.3. Electron impact also causes some molecular ions to fragment, yielding mostly odd *m/z* fragments unless a group V element is present.

a. Covalent bonds have homolysis energies around 420 kJ/mol.) The electron collision has more than enough energy to break bonds.

$$\begin{matrix} H & H \\ | & | \\ H-C-C-H \\ | & | \\ H & H \end{matrix} + e^{\ominus} \longrightarrow \begin{matrix} H & H \\ | & | \\ H-C-C^{\oplus} \\ | & | \\ H & H \end{matrix} + \dot{H} + 2\,e^{\ominus}$$

$$+ \; H-\underset{\underset{H}{|}}{\overset{\overset{H}{|}}{C}}{}^{\oplus} \; + \; \cdot\underset{\underset{H}{|}}{\overset{\overset{H}{|}}{C}}-H$$

+ other radicals and ions

12.13.4. Accelerator plates rapidly accelerate positive ions through a slit down the *flight tube*, usually before they undergo collision with another electron. Higher ions (+2 and +3) are rare beyond the accelerator plates.

a. Uncharged particles are not accelerated and are removed by the vacuum pump.

12.13.5. How much each ion trajectory is bent by the magnetic field depends on its mass to charge ratio (m/z), but z is normally +1. Lighter ions are easier to *magnetically deflect* than heavier ones since they have less momentum.

a. One ion after another can be focused on the detector by holding the accelerator plate voltage constant and scanning through a range of magnetic field strengths.
   1. Alternatively, magnetic field can be held constant and a range of accelerator plate voltages can be scanned.

b. Ion impact on the detector causes an electric current flow in the detector circuit, which is amplified and displayed as a peak (on a computer screen). Peak height (current intensity) is proportional to the number of ions flowing through the slit to the detector.

12.13.6. A mass spectrum is a bar graph (or table):

a. Horizontal axis: ion mass (atomic mass units (Daltons)) to charge ratio (m/z).

b. Vertical axis: relative peak intensity (electric current). The most intense (*base*) peak is arbitrarily assigned 100% intensity, and all other peak intensities are relative to it.

12.13.7. Example: methanol*. *Data taken from SDBS (section 12.7.3).

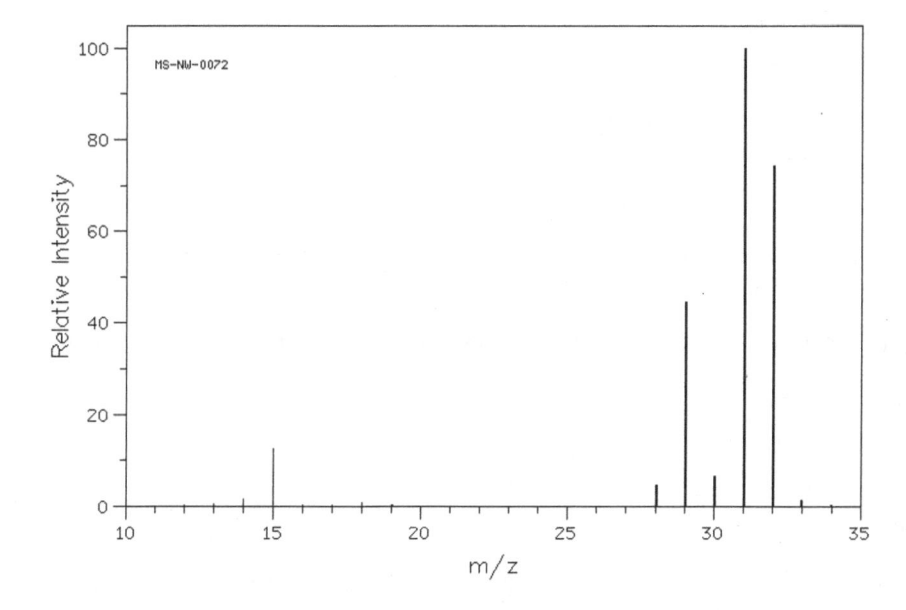

384 SEQUENTIAL ORGANIC CHEMISTRY I

m/z	relative intensity	m/z	relative intensity
14.0	1	30.0	6
15.0	12	31.0	100
28.0	4	32.0	74
29.0	44	33.0	1

a. Usually some molecular ions (or parent ions) survive long enough before fragmenting to detect. Fragments have lower mass than the original molecule, so the rightmost peak (except for heavy isotope peaks, see below) corresponds to the original molecule minus one electron. Since electron mass is negligible, this peak's *m/z* is the molecular mass.
   1. *m/z* = 32.0 is the molecular ion peak for methanol.
   2. If molecular ions fragment before detection, electron impact energy can be reduced (to 20 – 25 eV), or other ionization methods used that allow M$\overset{\oplus}{\cdot}$ detection.
b. What ion is responsible for the base peak (at *m/z* = 31.0)?

   1. Many *m/z* = 31.0 ions survive to hit the detector before fragmenting, since they are resonance stabilized.
c. What ion is responsible for the peak at *m/z* = 15.0?

d. The small peak at *m/z* = 33.0 is caused by a few ions that contain heavy isotopes.

   1. 98.89% of all C atoms are ^{12}C atoms, and 1.11% are ^{13}C atoms on Earth. 99.985% of all H atoms are ^{1}H atoms, and 0.015% are ^{2}H atoms.
   2. Ions containing ^{13}C isotope are the main contributors to the *m/z* = 33.0 peak.

12.13.8. Every isotope must be considered in MS.

12.13.9. Gas Chromatography Mass Spectrometry (GCMS) – volatile compound mixtures are separated by capillary gas chromatography (GC). Each substance passes successively through a transfer line directly into a mass spectrometer as it emerges off the column. A mass spectrum is obtained for each component as it elutes.

12.13.10. Infrared data, GC retention time and mass spectra comparison to those of authentic samples is sufficient to identify most common drugs.

12.13.11. GC-MS instruments usually have quadruple mass filters that separate ions by mass. An ion beam is accelerated initially parallel to four rods that have varying voltages on them. A particular mass ion is induced to follow a specific spiral pathway. Only when the voltage sequence is correct will the ion hit the target. Quadrupole detectors are very fast – can scan a broad mass range in under one sec.

12.14.1. High resolution mass spectrometry (HRMS): extra electrostatic or magnetic focusing stages determine particle masses within 1 part in ~10^6. These instruments are expensive.

a. Exact atomic masses have been determined:

Atom	Low resolution mass (amu)	High resolution mass (amu)
^1H	1.0	1.007825
^{12}C	12.0	12.000000
^{16}O	16.0	15.994914
^{14}N	14.0	14.003050

1. ^{12}C is the amu scale standard, whose mass = 12.000000 amu by definition. All other atomic masses are relative to this standard.

12.14.2. A chemical formula can be deduced from precise (exact) mass differences. Example:

Molecule	Low resolution mass (amu)	High resolution mass (amu)
CH_4N_2	44.0	1 C × 12.000000 = 12.000000 4 H × 1.007825 = 4.031300 2 N × 14.003050 = 28.006100 44.037400
C_3H_8	44.0	3 C × 12.000000 = 36.000000 8 H × 1.007825 = 8.062600 44.062600
CO_2	44.0	1 C × 12.000000 = 12.000000 2 O × 15.994914 = 31.989828 43.989828
C_2H_4O	44.0	2 C × 12.000000 = 24.000000 4 H × 1.007825 = 4.031300 1 O × 15.994914 = 15.994914 44.026214

12.14.3. Chemical formulas can often be deduced from isotope peaks if an HRMS instrument is not available.

- **a.** Peaks at higher *m/z* than the molecular ion (M$^{+\bullet}$) are due to heavy isotopes: one mass unit higher than M$^{+\bullet}$ is called M+1, two mass units higher is M+2, etc. Isotopes contribute intensity according to their natural abundances:

Element	Isotope	Contribution to M$^{+\bullet}$	Isotope	Contribution to M + 1	Isotope	Contribution to M + 2
hydrogen	^{1}H	99.985%	^{2}H	0.015%		
carbon	^{12}C	98.89%	^{13}C	1.11%		
nitrogen	^{14}N	99.6%	^{15}N	0.4%		
oxygen	^{16}O	99.8%			^{18}O	0.2%
sulfur	^{32}S	95.0%	^{33}S	0.8%	^{34}S	4.2%
chlorine	^{35}Cl	75.5%			^{37}Cl	24.5%
bromine	^{79}Br	50.5%			^{81}Br	49.5%
iodine	^{127}I	100.0%				

12.14.4. Example: H$_3$CCl is actually a mixture of several isotopically different molecules. Some possibilities:

[Structures shown: ^{1}H$_3$–^{12}C–^{35}Cl (major M$^{+\bullet}$); ^{1}H$_2$(^{1}H)–^{13}C–^{35}Cl (M+1); ^{1}H$_2$(^{2}H)–^{12}C–^{35}Cl (M+1); ^{1}H$_3$–^{12}C–^{37}Cl (M+2); ^{1}H$_2$(^{2}H)–^{13}C–^{35}Cl (M+2); (^{2}H)^{1}H$_2$–^{12}C–^{35}Cl (M+2); ^{1}H$_2$(^{1}H)–^{13}C–^{37}Cl (M+3); ^{1}H$_2$(^{2}H)–^{12}C–^{37}Cl (M+3); ^{1}H$_2$(^{2}H)–^{13}C–^{37}Cl (M+4); (^{2}H)^{1}H$_2$–^{12}C–^{37}Cl (M+4)]

- **a.** Most of these molecules are rare, and make small intensity contributions. The most important heavy atom contributors are ^{13}C and ^{37}Cl (since their natural abundances are higher than ^{2}H).
- **b.** Molecular ion region – major contributor to each peak is shown:

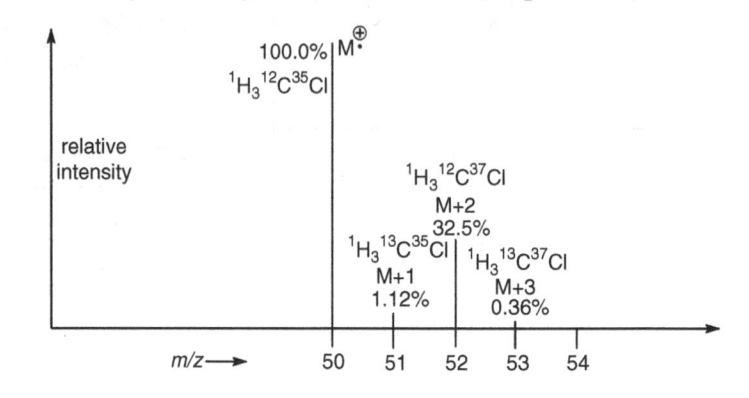

[Mass spectrum: M$^{+\bullet}$ at m/z 50, ^{1}H$_3^{12}$C^{35}Cl, 100.0%; M+1 at m/z 51, ^{1}H$_3^{13}$C^{35}Cl, 1.12%; M+2 at m/z 52, ^{1}H$_3^{12}$C^{37}Cl, 32.5%; M+3 at m/z 53, ^{1}H$_3^{13}$C^{37}Cl, 0.36%]

c. A compound that has an M+2 peak with about ⅓ the M$^{\oplus\cdot}$ intensity most likely contains chlorine.
  1. Example: 2-chloropropane:

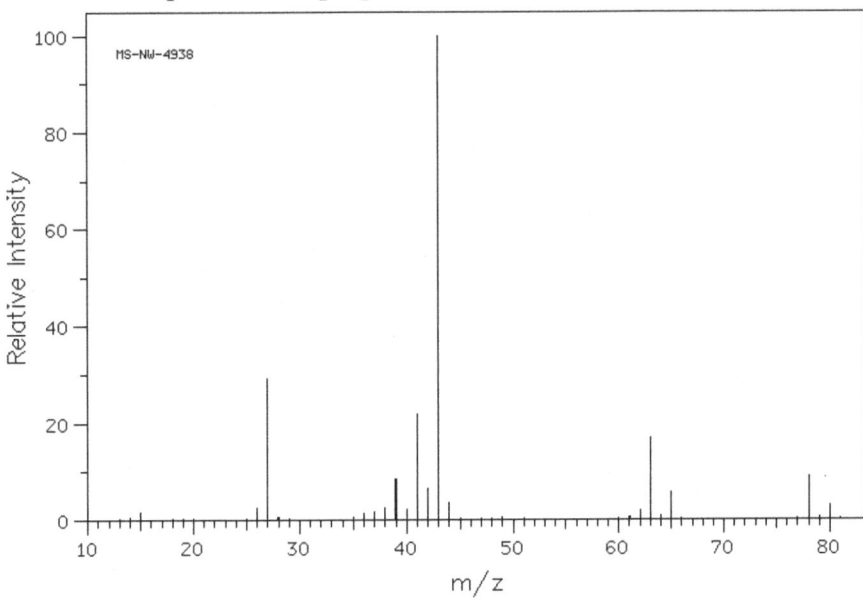

m/z	rel. I	m/z	rel. I	m/z	rel. I
15.0	1	39.0	8	62.0	1
26.0	2	40.0	2	63.0	17
27.0	29	41.0	21	65.0	5
36.0	1	42.0	6	78.0	9
37.0	1	43.0	100	80.0	3
38.0	2	44.0	6		

12.14.5. Relative isotope peak intensities can be predicted via the *binomial expansion.*

  a. Example: ethane. The chance that a C atom is a ^{12}C isotope is 98.89%, and the chance that a C atom is a ^{13}C isotope is 1.11%. Let *a* be the ^{12}C natural abundance, let *b* be the ^{13}C natural abundance, and the exponent is the number of C atoms in the molecule (2).

  The M$^{\oplus\cdot}$, M+1, and M+2 intensities are given by each binomial expansion term:

  $$(a + b)^2 = a^2 + 2ab + b^2$$

  $$= (98.89)^2 + 2(98.89)(1.11) + (1.11)^2$$

  $$= 9781 + 218 + 1.2$$

$$\% \text{ rel. M}^{\oplus} \text{ intensity} = \frac{9781}{9781 + 218 + 1.2} \times 100 = 97.81\%$$

$$\% \text{ relative M+1 intensity} = \frac{218}{9781 + 218 + 1.2} \times 100 = 2.18\%$$

$$\% \text{ relative M+2 intensity} = \frac{1.2}{9781 + 218 + 1.2} \times 100 = 0.012\%$$

1. Molecular ion region for ethane.

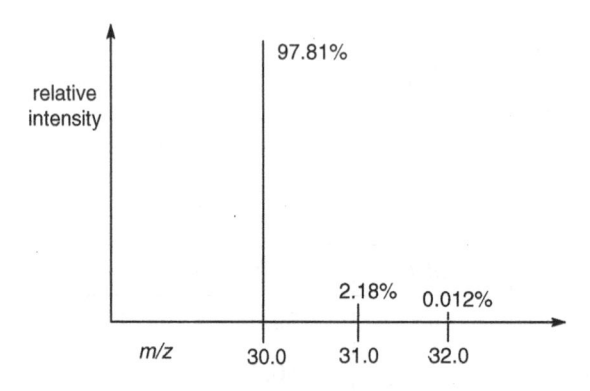

2. M+1 peak intensity relative to M$^{\oplus}$ reveals how many C atoms are present in the formula:

$$\frac{100.0}{97.81} = \frac{x}{2.18} \quad \text{solve for} \quad x = 2.23$$

(The M+1 peak is 2.23% as intense as the M$^{\oplus}$ peak.) Dividing the M+1 intensity by the ^{13}C natural abundance yields the number of C atoms in the structure:

$$2.23\% \times \frac{\text{C atom}}{1.11\%} = 2.01 \text{ C atom}$$

round to integer, 2 C atoms in ethane

b. Example: propane. Let a and b be isotopic abundances again, and the exponent is 3 since propane has three C atoms. M$^{\oplus}$, M+1, M+2, and M+3 intensities are given by each binomial expansion term:

$$(a + b)^3 = a^3 + 3a^2b + 3ab^2 + b^3$$

$$= (98.89)^3 + 3(98.89)^2(1.11) + 3(98.89)(1.11)^2 + (1.11)^3$$

$$= 9.67 \times 10^5 + 3.26 \times 10^4 + 3.66 \times 10^2 + 1.37$$

% relative M$^{\oplus}$ intensity = $\dfrac{9.67 \times 10^5}{9.67 \times 10^5 + 3.26 \times 10^4 + 3.66 \times 10^2 + 1.37} \times 100$ = 96.7 %

% relative M+1 intensity = $\dfrac{3.26 \times 10^4}{9.67 \times 10^5 + 3.26 \times 10^4 + 3.66 \times 10^2 + 1.37} \times 100$ = 3.26 %

% relative M+2 intensity = $\dfrac{3.66 \times 10^2}{9.67 \times 10^5 + 3.26 \times 10^4 + 3.66 \times 10^2 + 1.37} \times 100$ = 0.0366 %

% relative M+3 intensity = $\dfrac{1.37}{9.67 \times 10^5 + 3.26 \times 10^4 + 3.66 \times 10^2 + 1.37} \times 100$ = 0.000137 %

1. Calculate M+1 peak intensity relative to M$^{\oplus}$.

$$\dfrac{100.0}{96.7} = \dfrac{x}{3.26} \quad \text{solve for } x \quad x = 3.37\%$$

(The M+1 peak is 3.37% as intense as the M$^{\oplus}$ peak.) Dividing the M+1 intensity by the ^{13}C natural abundance yields the number of C atoms in the structure:

$$3.37\% \times \dfrac{\text{C atom}}{1.11\%} = 3.04 \text{ C atom}$$

round to integer, 3 C atoms in propane

c. Another example: for a certain compound M$^{\oplus}$ appears at 130.0 (rel. I = 100.0%), and M+1 is at 131.0 (rel. I = 7.87%), and a small M+2 peak is detectable (rel. I = 0.4%). How many C atoms does the compound contain?

$$7.87\% \times \dfrac{\text{C atom}}{1.11\%} = 7.09 \text{ C atom}$$

Round to integer, 7 C atoms. A likely formula $C_7H_{14}O_2$.

d. One more example: for a certain compound M$^{\oplus}$ appears at 130.0 (rel. I = 78.6%), and M+1 is at 131.0 (rel. I = 8.62%) and the M+2 peak is not detectable. How many C atoms does the compound contain?

$$\dfrac{100.0\%}{78.6\%} = \dfrac{x}{8.62\%} \quad \text{solve for } x = 11.0\%$$

$$11.0\% \times \dfrac{\text{C atom}}{1.11\%} = 9.91 \text{ C atom}$$

Round to integer, 10 C atoms. A likely formula is $C_{10}H_{10}$.

12.14.6. Intensity data from low resolution instruments can contain substantial experimental errors.
    a. Other hints:
      1. If M+2 is 98.0% as intense as M$^{\oplus}$, the compound most likely contains bromine. Example: 1-bromopropane.

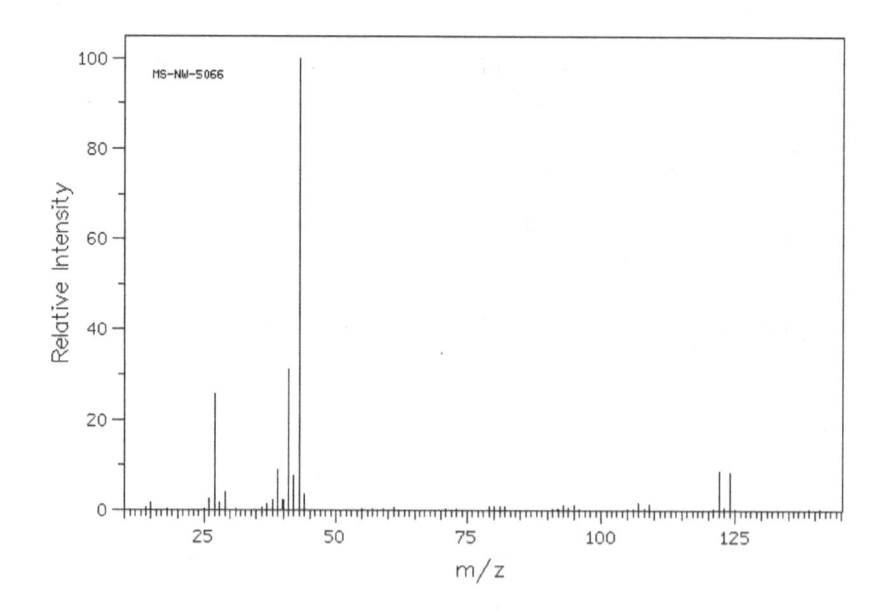

m/z	rel. I	m/z	rel. I	m/z	rel. I
15.0	1	39.0	9	95.0	1
26.0	2	40.0	2	107.0	1
27.0	25	41.0	31	109.0	1
28.0	1	42.0	7	122.0	8
29.0	4	43.0	100	124.0	8
37.0	1	44.0	3		
38.0	2	93.0	1		

2. A peak at 127 is consistent with I$^+$ cation. This, or a difference of 127 between M$^{\oplus}$ and another peak, is consistent with iodine in the compound. Example: iodoethane.

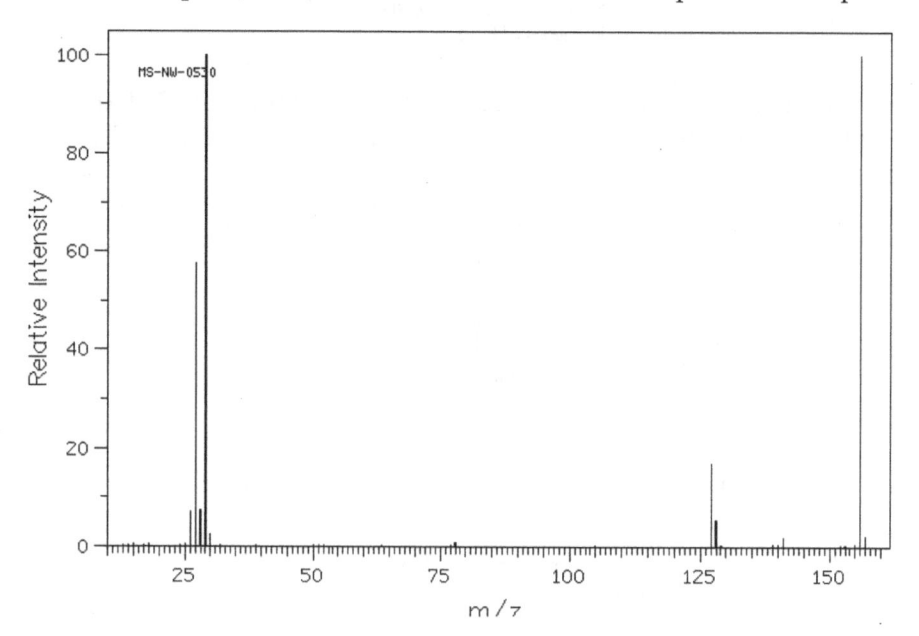

m/z	rel. I	m/z	rel. I	m/z	rel. I
26.0	7	30.0	2	156.0	100
27.0	57	127.0	16	157.0	2
28.0	7	128.0	5		
29.0	99	155.0	small		

3. If M$^{\oplus\cdot}$ is at an odd *m/z* and even fragments, then the compound likely contains a nitrogen atom. Example: ethanamine.

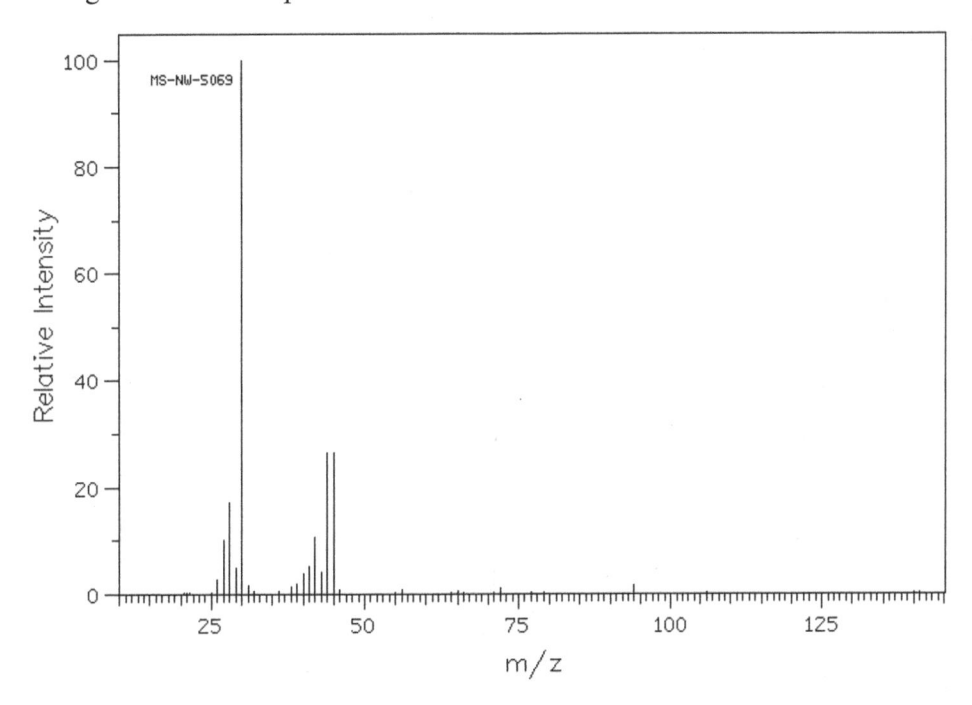

m/z	rel. I	m/z	rel. I	m/z	rel. I
26.0	2	31.0	1	42.0	10
27.0	10	38.0	1	43.0	4
28.0	17	39.0	1	44.0	26
29.0	4	40.0	3	45.0	26
30.0	100	41.0	5		

4. A compound with an M+2 peak that is 0.2 % as intense as M$^{\oplus\cdot}$ likely contains oxygen. (This is so small that it is often difficult to measure accurately).
5. If the M+2 peak is around 4.2% as intense as M$^{\oplus\cdot}$, sulfur is likely.

12.15.1. The electron that is *least* tightly held is the one that is lost from a molecule in the initial collision with a high energy electron.

   a. Ease of electron loss from a molecule: easy lone pair > C=C > C-C > C-H hard

b. Examples:

$$H_3C-\overset{\overset{\cdot\cdot}{\overset{\cdot\cdot}{O}}}{\underset{\|}{C}}-CH_3 + e^{\ominus} \xrightarrow{70\ eV} H_3C-\overset{\overset{\cdot\cdot}{\overset{\oplus}{O}}}{\underset{\|}{C}}-CH_3 \longleftrightarrow H_3C-\overset{\overset{\cdot\cdot}{\overset{\cdot\cdot}{O}\cdot}}{\underset{\oplus}{C}}-CH_3 + 2e^{\ominus}$$

$$R-\underset{H}{\overset{\cdot\cdot}{O}}: + e^{\ominus} \xrightarrow{70\ eV} R-\underset{H}{\overset{\cdot\cdot\oplus}{O}} + 2e^{\ominus}$$

$$\underset{R'}{\overset{R}{>}}C=C\underset{H}{\overset{H}{<}} + e^{\ominus} \xrightarrow{70\ eV} \underset{R'}{\overset{R}{>}}\overset{\oplus}{C}-\overset{\cdot}{C}\underset{H}{\overset{H}{<}} + 2e^{\ominus}$$

12.15.2. Ionic fragment analysis helps reconstruct the original molecule.

a. More stable fragments persist longer (in the flight tube) and yield more intense peaks. Example: *n*-hexane.

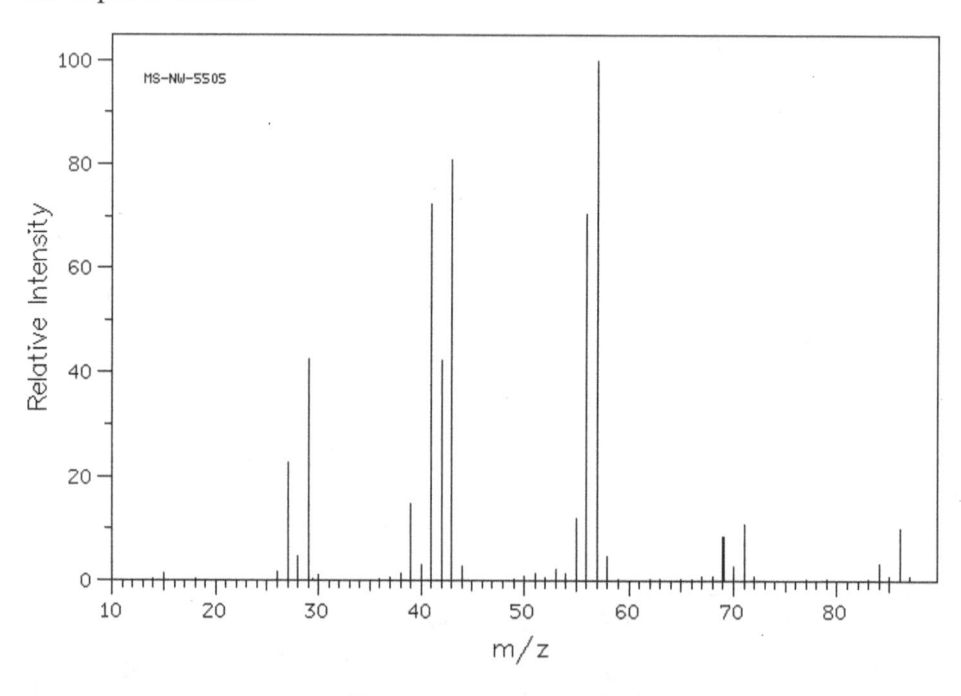

m/z	rel. I	m/z	rel. I	m/z	rel. I
15.0	1	41.0	72	57.0	100
26.0	1	42.0	42	58.0	4
27.0	22	43.0	80	69.0	8
28.0	4	44.0	2	70.0	2
29.0	42	51.0	1	71.0	11
30.0	1	53.0	2	84.0	3
38.0	1	54.0	1	86.0	10
39.0	14	55.0	12		
40.0	3	56.0	70		

b. The following processes account for some of the more intense fragments:

$$H_3CCH_2CH_2-\overset{\overset{H}{|}}{\underset{\underset{H}{|}}{C}} : \overset{\overset{H}{|}}{\underset{\underset{H}{|}}{C}}-CH_3 \xrightarrow[70\ eV]{e^{\ominus}} H_3CCH_2CH_2-\overset{\overset{H}{|}}{\underset{\underset{H}{|}}{\overset{1/2\ \oplus}{C}}} \cdot \overset{\overset{H}{|}}{\underset{\underset{H}{|}}{\overset{1/2\ \oplus}{C}}}-CH_3 + 2\ e^{\ominus}$$

n-hexane   M⁺•

Path 1:

$$H_3CCH_2\overset{\oplus}{CH}-\overset{\overset{H}{|}}{\underset{\underset{?}{|}}{C}}-H \longleftarrow H_3CCH_2CH-\overset{\overset{H}{|}}{\underset{\underset{H}{|}}{C}} \oplus + \cdot\overset{\overset{H}{|}}{\underset{\underset{H}{|}}{C}}-CH_3$$

m/z = 57.0   not detected

Path 2:

$$H_3CCH_2CH_2-\overset{\overset{H}{|}}{\underset{\underset{H}{|}}{C}}\cdot + \overset{\oplus}{\overset{\overset{H}{|}}{\underset{\underset{H}{|}}{C}}}-CH_3$$

not detected   *m/z* = 29.0

$$H_3CCH_2-\overset{\overset{H}{|}}{\underset{\underset{H}{|}}{C}} : \overset{\overset{H}{|}}{\underset{\underset{H}{|}}{C}}-CH_2CH_3 \xrightarrow[70\ eV]{e^{\ominus}} H_3CCH_2-\overset{\overset{H}{|}}{\underset{\underset{H}{|}}{\overset{1/2\ \oplus}{C}}} \cdot \overset{\overset{H}{|}}{\underset{\underset{H}{|}}{\overset{1/2\ \oplus}{C}}}-CH_2CH_3 + 2\ e^{\ominus}$$

n-hexane   M⁺•

$$?\ H_3C\overset{\oplus}{CH}-\overset{\overset{H}{|}}{\underset{\underset{H}{|}}{C}}-H \longleftarrow H_3C\overset{\overset{H}{|}}{\underset{\underset{H}{|}}{CH}}-\overset{\overset{H}{|}}{\underset{\underset{H}{|}}{C}} \oplus + \cdot\overset{\overset{H}{|}}{\underset{\underset{H}{|}}{C}}-CH_2CH_3$$

m/z = 43.0   not detected

c. The following process accounts for the fragment at *m/z* = 71.0

$$H_3C(CH_2)_3-\overset{\overset{H}{|}}{\underset{\underset{H}{|}}{C}} : \overset{\overset{H}{|}}{\underset{\underset{H}{|}}{C}}-H \xrightarrow[70\ eV]{e^{\ominus}} H_3C(CH_2)_3-\overset{\overset{H}{|}}{\underset{\underset{H}{|}}{\overset{1/2\ \oplus}{C}}} \cdot \overset{\overset{H}{|}}{\underset{\underset{H}{|}}{\overset{1/2\ \oplus}{C}}}-H + 2\ e^{\ominus}$$

n-hexane   M⁺•

$$H_3C(CH_2)_2\overset{\oplus}{CH}-\overset{\overset{H}{|}}{\underset{\underset{H\ ?}{|}}{C}}-H \longleftarrow H_3C(CH_2)_2\overset{\overset{H}{|}}{\underset{\underset{H}{|}}{CH}}-\overset{\overset{H}{|}}{\underset{\underset{H}{|}}{C}} \oplus + \cdot\overset{\overset{H}{|}}{\underset{\underset{H}{|}}{C}}-H$$

m/z = 71.0   *m/z* = 71.0   not detected

1. This peak is less intense presumably because a methyl free radical is less stable than a 1° free radical (formed in b above).
   d. The following process is very unfavorable (peak at m/z = 15.0 is small).

$$H_3C(CH_2)_3-CH_2-CH_3 \xrightarrow{-e^-, 70\,eV} [H_3C(CH_2)_3-CH_2\cdots CH_3]^{+\cdot} \rightarrow H_3C(CH_2)_3-CH_2\cdot + {}^{+}CH_3$$

n-hexane → M⁺· → not detected + m/z = 15.0

   1. Carbocation stability is more important than free radical stability in these fragmentations.
   e. Problem 6. *a.* Fragments corresponding to $C_4H_9^+$, $C_3H_5^+$, $C_2H_5^+$, and $C_2H_3^+$ are present in the 2,2-dimethylpropane mass spectrum. Draw likely structures for these ions. *b.* Write a balanced equation for $C_4H_9^+$ formation from the molecular ion ($C_5H_{12}^{+\cdot}$).

12.15.3. Another example: 2-methylpentane.

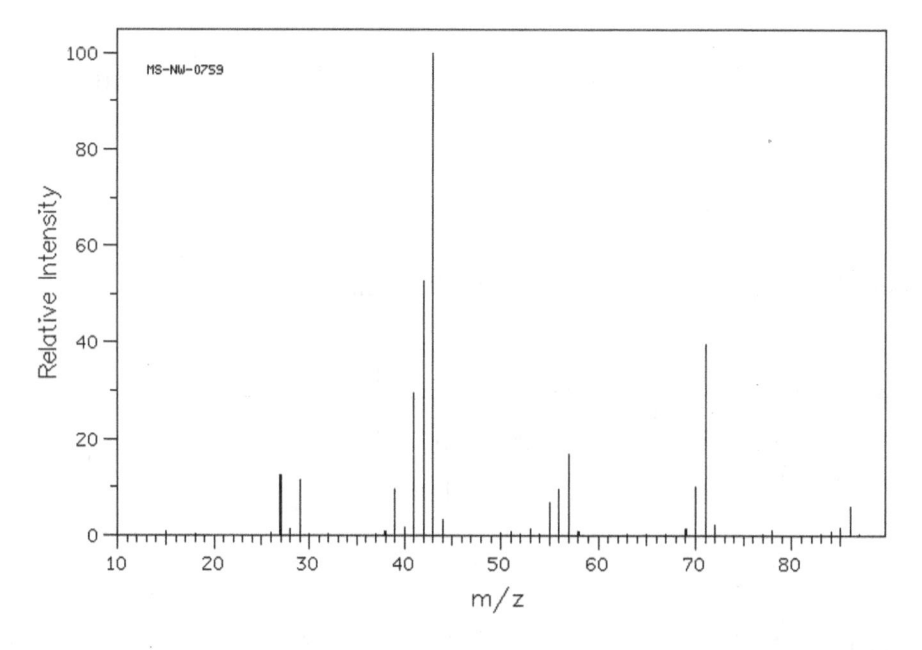

m/z	rel. I	m/z	rel. I	m/z	rel. I
27.0	12	43.0	100	70.0	10
28.0	1	44.0	3	71.0	39
29.0	11	53.0	1	72.0	2
39.0	9	55.0	6	78.0	1
40.0	1	56.0	9	85.0	1
41.0	29	57.0	17	86.0	6
42.0	52	69.0			

[Reaction scheme 1:]

H₃CCH₂CH₂–CH(CH₃)–CH₃ →(e⁻, 70 eV)→ [H₃CCH₂CH₂···CH(CH₃)–CH₃]⁺• (M⁺•, with ½⊕ charges) → H₃CCH₂CH₂• + ⁺CH(CH₃)–CH₃   m/z = 43.0

or

[Reaction scheme 2:]

H₃CCH₂CH₂–CH(CH₃):CH₃ →(e⁻, 70 eV)→ [H₃CCH₂CH₂–CH(CH₃)···CH₃]⁺• (M⁺•, with ½⊕ charges) → H₃CCH₂CH₂–CH⁺(CH₃) + •CH₃   m/z = 71.0

12.13.12. Another example – 2,4-dimethylpentane.

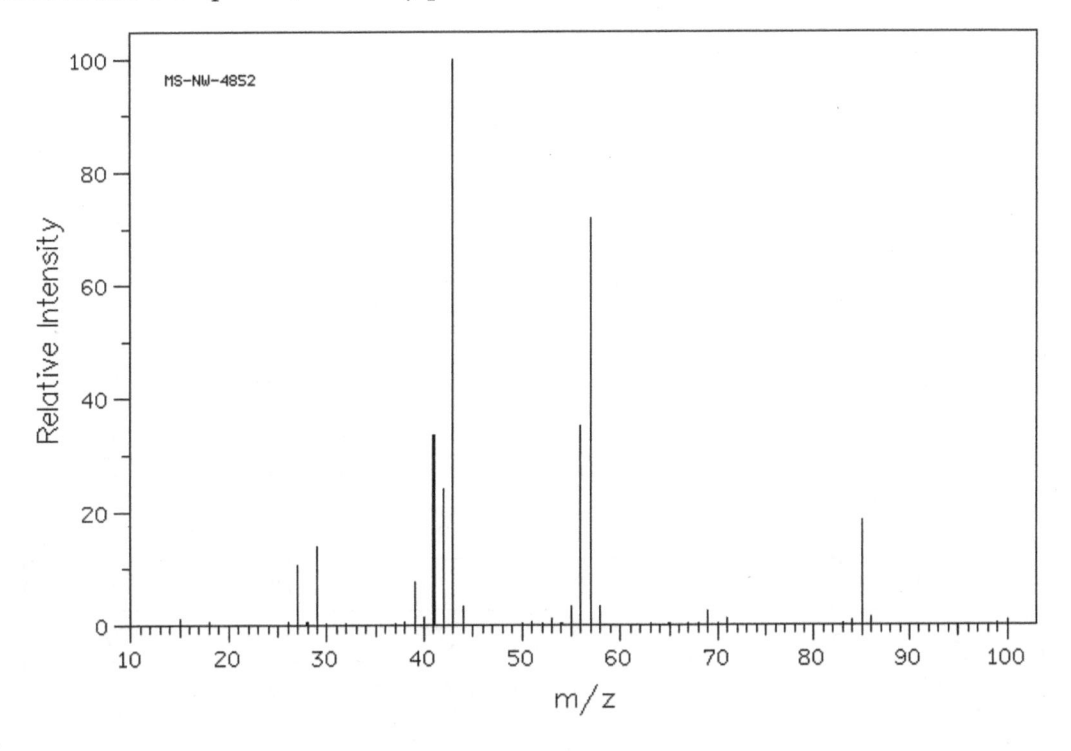

m/z	rel. I	m/z	rel. I	m/z	rel. I
15.0	1	43.0	100	69.0	2
27.0	10	44.0	3	71.0	1
29.0	14	53.0	1	85.0	18
39.0	7	55.0	3	86.0	1
40.0	1	56.0	35	100.0	small
41.0	33	57.0	71		
42.0	23	58.0	3		

a. Problem 7: suggest structures for fragments at m/z = 43.0, 57.0, and 85.0.

12.15.4. Example: hex-2-ene.

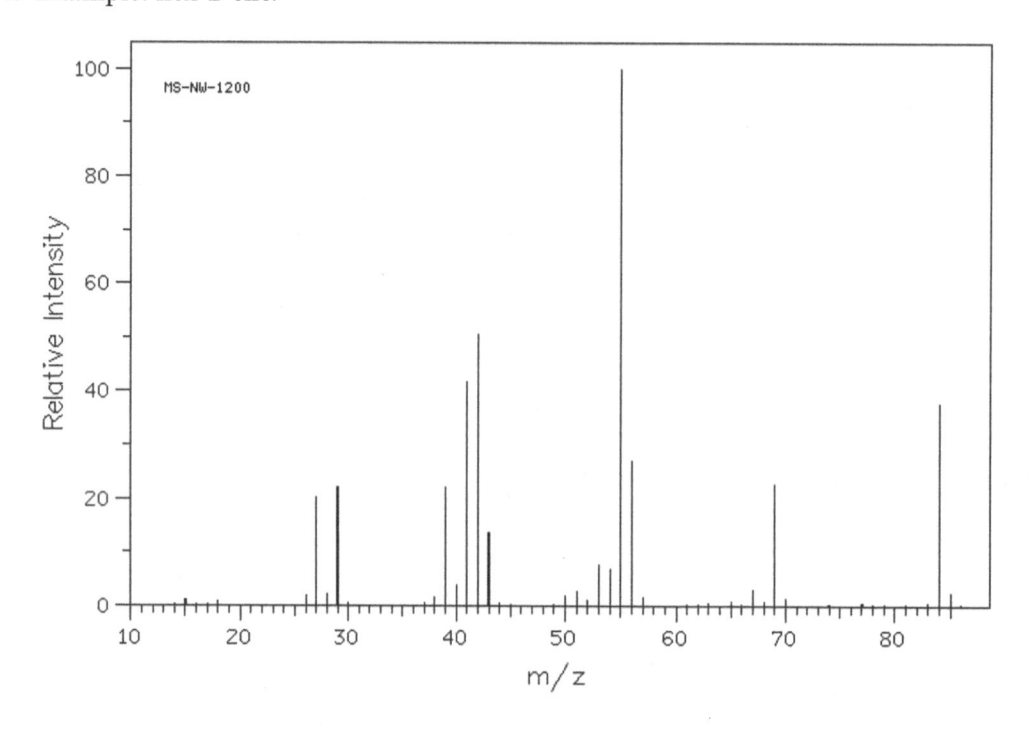

m/z	rel. I	m/z	rel. I	m/z	rel. I
15.0	1.1	41.0	41.9	55.0	100.
26.0	1.9	42.0	50.4	56.0	27.0
27.0	20.1	43.0	13.6	57.0	1.6
28.0	2.1	50.0	1.8	67.0	2.9
29.0	22.1	51.0	2.7	69.0	22.6
38.0	1.7	52.0	1.1	70.0	1.2
39.0	22.2	53.0	7.5	84.0	37.8
40.0	3.9	54.0	6.9	85.0	2.5

a. Resonance stabilized allylic carbocations form when possible.

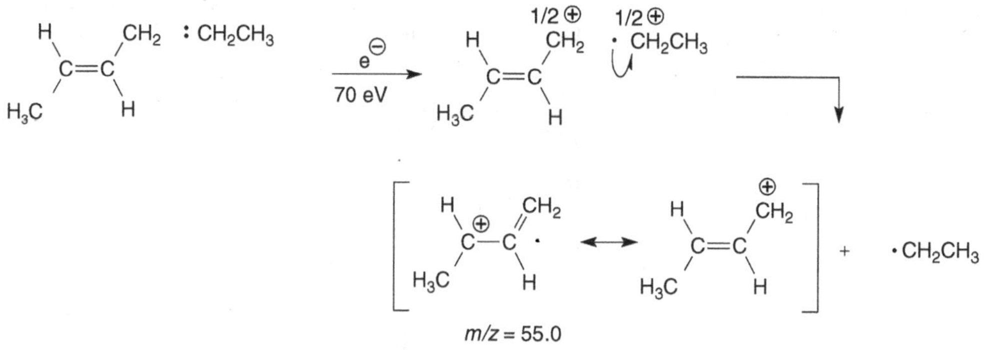

12.15.5. Example: 3-methylbutan-1-ol.

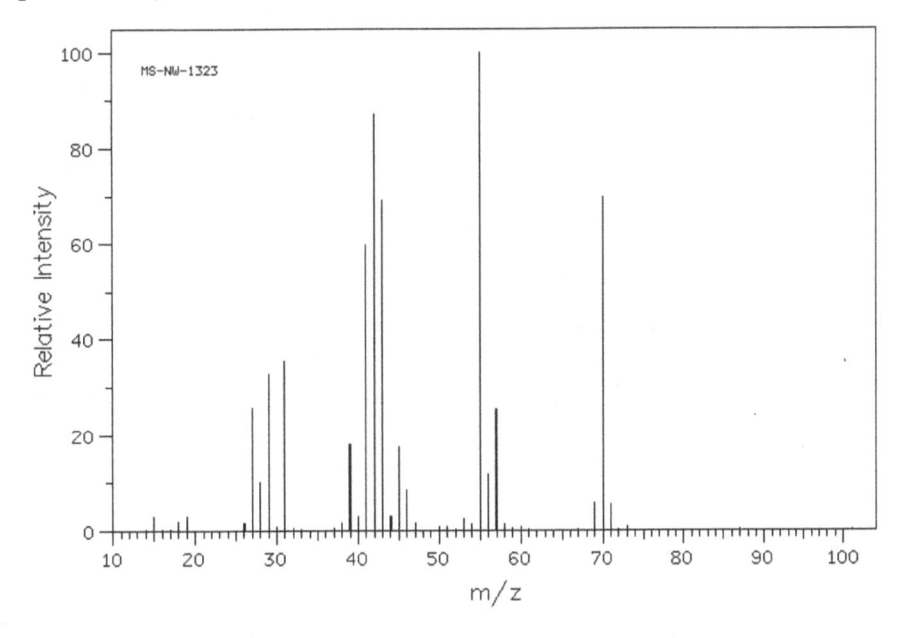

m/z	rel. I	m/z	rel. I	m/z	rel. I
15.0	2.9	39.0	18.1	53.0	2.4
18.0	2.0	40.0	3.1	54.0	1.3
19.0	3.0	41.0	59.9	55.0	100.0
26.0	1.7	42.0	87.1	56.0	11.8
27.0	25.7	43.0	69.0	57.0	25.4
28.0	10.0	44.0	3.0	58.0	1.2
29.0	32.8	45.0	17.5	69.0	5.6
31.0	35.5	46.0	8.3	70.0	69.7
38.0	1.5	47.0	1.6	71.0	5.7

a. Small, stable, even mass numbered molecules are often lost, such as H_2O, CO, CO_2, and ethene. H_2O loss is apparently so fast that $M^{+\bullet}$ cannot be detected. Some fragments might be accounted for by the following:

b. α-Cleavage is also a common pathway for alcohols.

$$H_3C-\underset{CH_3}{\underset{|}{C}}(\overset{\overset{\oplus\,..}{O}-H}{})-CH_3 \longrightarrow \left[H_3C-\underset{\underset{m/z = 59.0}{}}{\overset{\overset{\oplus\,..}{O}-H}{\|}{C}}-CH_3 \longleftrightarrow H_3C-\underset{\oplus}{\overset{\overset{:\ddot{O}-H}{|}}{C}}-CH_3 \right] + \cdot CH_3$$

12.15.6. Other functional group's mass spectra will be studied as they are encountered later in the course.

CHAPTER 13

13.1.1. Ether molecule – like water, an ether O atom is sp^3 hybridized, but both H atoms are replaced by R-groups.

  a. Ether properties – more polar than alkanes, but cannot hydrogen bond to themselves.

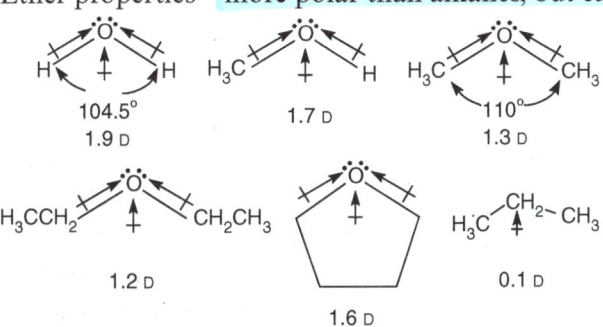

H–O–H 104.5° 1.9 D

H₃C–O–H 1.7 D

H₃C–O–CH₃ 110° 1.3 D

H₃CCH₂–O–CH₂CH₃ 1.2 D

(cyclic ether) 1.6 D

H₃C–CH₂–CH₃ 0.1 D

Compound	mol. mass	bp °C	Density g/mL	Solubility g / 100 g H₂O
H₃CCH₂OH	46.1	78.3	0.789	infinite
butan-1-ol	74.1	118	0.810	7.9
n-pentane	72.2	36	0.626	insoluble
H₃C-O-CH₃	46.1	−25	gas	infinite
H₃CCH₂CH₃	44.1	−42	gas	insoluble
(H₃CCH₂)₂-O	74.1	35	0.708	8.0
(H₃CCH₂CH₂)₂-O	102.2	91	0.736	
[(H₃C)₂CH]₂-O	102.2	68	0.725	
H₃COCH₂CH₂OCH₃	90.1	83	0.867	soluble
(tetrahydrofuran)	72.1	65	0.889	∞
(dioxane)	88.1	101	1.03	∞

399

13.1.2. Ether bp's ≈ alkane bp's. Ether (water) solubility ≈ alcohol (water) solubility due to crossed hydrogen bonding.

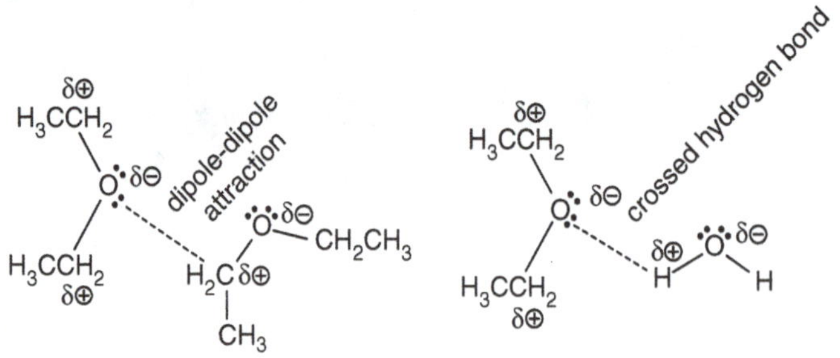

13.1.3. Ether was used as a general anesthetic (1842), but is very flammable, and causes nausea. N_2O, $CF_3CHBrCl$ and other new materials are better tolerated. Ether is easily administered, dissolves in fatty tissues where it quickly affects the central nervous system. It is less toxic than $CHCl_3$ since it metabolizes to H_3CCH_2OH.

13.2.1. Ethers are useful solvents:

  a. Unreactive towards oxidizing and reducing agents and strong bases.
  b. Dissolve many polar and nonpolar compounds.
   1. Nonpolar solutes are more soluble in ethers than in alcohols. Ethers have no H-bonding network (as alcohols do) to be broken up by a nonpolar solute.
   2. Polar H-bonding substances are about as soluble in ethers as in alcohols.
   3. Ionic compounds with small nonpolarizable (hard) anions (F^-) are only slightly soluble in ethers, but are more soluble in alcohols. Lithium iodides and carboxylates (soft anions) are moderately soluble in ethers, mainly due to lithium ion solvation.

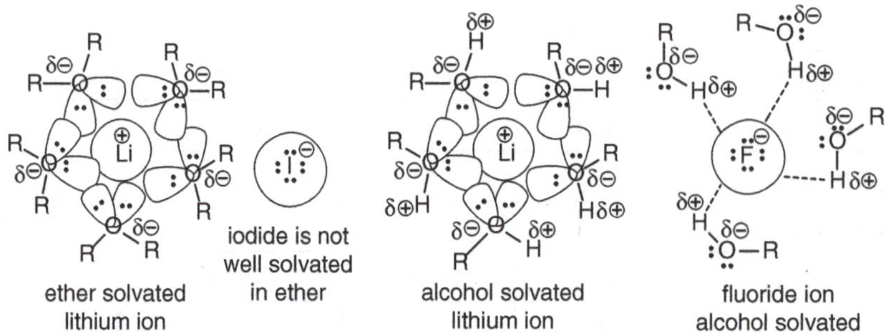

  c. Low boiling points – easy to distill out after reaction.
  d. Grignard reagent will not dissolve and reaction won't work without an ether solvent.

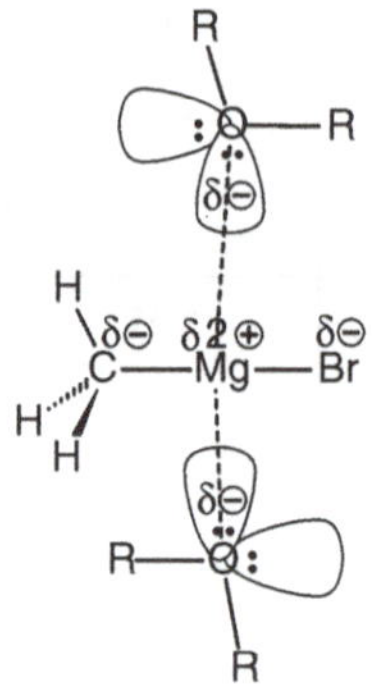

e. Explosive and toxic B₂H₆ (hydroboration/oxidation, section 8.7.1) and toxic BF₃ (a Lewis acid) are hard to handle gases. They form ether complexes that are easier and safer to use.

[Reaction showing B₂H₆ + 2 THF ⇌ 2 THF·BH₃]

[Reaction showing BF₃ + diethyl ether ⇌ boron trifluoride etherate]

13.2.2. The strongest base that can ever be dissolved in R-OH is R-O⁻.
  a. If a stronger base than R-O⁻ dissolves in an alcohol solvent, it just forms more R-O⁻:

  [B:⁻ + H—Ö—R ⇌ B—H + ⁻:Ö—R]
  stronger base + stronger acid ⇌ weaker acid + weaker base

  b. Alcohols therefore cannot be solvents for strongly basic reagents (such as Grignard).

13.2.3. C. J. Pedersen discovered crown ethers (1960, Nobel Prize, 1987) – examples: 12-crown-4, 15-crown-5, and 18-crown-6.

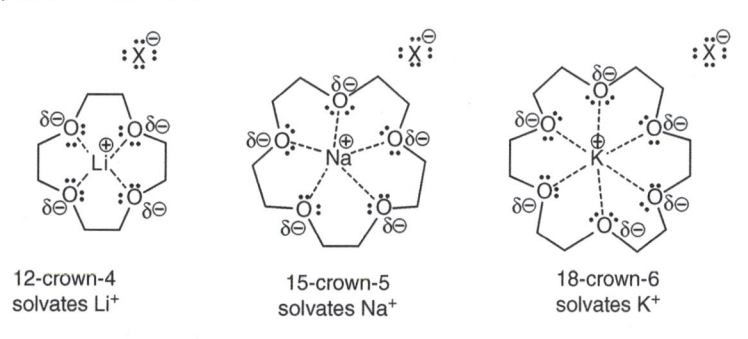

12-crown-4 solvates Li⁺   15-crown-5 solvates Na⁺   18-crown-6 solvates K⁺

Ionic complexes are soluble in nonpolar solvents.
X⁻ anions are poorly solvated (naked).

13.2.4. 18-Crown-6 has a hydrophilic (water loving) interior cavity 2.7 Å across, which strongly binds K⁺ ions (diameter 2.66 Å). Exterior CH₂ groups are hydrophobic (water hating). Other crowns are similar.
  a. Crown ethers can be phase transfer catalysts. A Li⁺, Na⁺, or K⁺ ion is hidden beneath the crown, allowing it to dissolve in less polar organic solvents (benzene, toluene, diethyl ether, or acetonitrile). A metal cation must take a "naked" anion (such as F⁻, CN⁻, R-COO⁻, MnO₄⁻) into solution with it to maintain charge balance. Naked anions are strong nucleophiles.
  b. Li⁺, Na⁺, or K⁺ ion is sometimes called the guest, and crown ether the host.
    1. Enzyme substrate binding is similar – complex crown ethers have been studied as enzyme models.
    2. Some antibiotics (nonactin) complex K⁺ or Na⁺ ions, allowing them to easily pass in and out of a lipophilic cell membrane. Ion balance is upset and eventually the cell dies.

402 SEQUENTIAL ORGANIC CHEMISTRY I

3. Oxygen and nitrogen containing macrocyclic rings often bind metal ions in biological processes, such as iron transport through bacterial membranes.
4. Modified 18-crown-6 structures have been used to extract radioactive Cs or Sr.

13.2.5. 15-Crown-5 preparation.

13.3.1. IUPAC and common ether names:
   a. Ethers have no IUPAC parent suffix – named as alkanes unless another functional group is present.
   b. Review: common functional group name priority table (section 2.17.4) – highest on top, lowest on bottom).
   c. Ether branch name is "alkoxy."
      1. Hydrocarbon substituent (branch) names (alkyl, aryl (phenyl), vinylic, allylic, and benzylic) review.

H₃C—   H₃CCH₂—   H₃CCH₂CH₂—   H₃C—CH—CH₃
                                        |
methyl   ethyl   n-propyl   2-propyl
                                        isopropyl

H₃CCH₂CH₂CH₂—   H₃CCH₂CHCH₃   CH₃   CH₃
                         | | |
n-butyl   sec-butyl   HC—CH₂—   H₃C—C—
                                | |
                              CH₃ isobutyl   CH₃
                                            tert-butyl

vinyl   allyl   phenyl   benzyl

d. Examples.

1-methoxymethane   1-methoxyethane   1-ethoxyethane
dimethyl ether   ethyl methyl ether diethyl ether

2-isopropoxypropane   3-methoxyhexane   2-methoxy-2-methylpropane
diisopropyl ether   methyl tert-butyl ether (MTBE)

higher priority { OH   lower priority
2-ethoxyethanol

3-ethoxy-1,1-dimethylcyclohexane

(1R,2R)-1-chloro-2-methoxycyclobutane
trans-1-chloro-2-methoxycyclobutane

e. Problem 1. Write condensed structural formulas for:

1. 2-isopropoxypropane

    $$H_3C-\underset{\underset{O-CH_2CH_2CH_3}{|}}{\overset{\overset{H}{|}}{C}}-CH_3$$

2. 1-tert-butoxy-2-methylpropane

    $$H_3C-\underset{\underset{CH_3}{|}}{\overset{\overset{CH_3}{|}}{C}}-O-CH_2CH\underset{CH_3}{\overset{CH_3}{<}}$$

3. 1-methoxypropan-2-ol

    $$H_3C-O\diagdown_{CH_2\underset{OH}{CH}CH_3}$$

f. Problem 2. Write IUPAC names for the following:

1. (H₃C)₂CHCH₂-O-CH₂CH(CH₃)₂

    $$H_3C-\underset{H}{\overset{CH_3}{|}}-CHCH_2-O-CH_2CH-CH_3$$
    $$CH_3$$

2. H₃CCH₂CH₂CH(OCH₃)CH₂CH₂CH₃

3. (H₃C)₃C-O-CH₂CH₃

13.3.2. Cyclic ether names. Tetrahydrofuran is a common solvent (more polar than ether) that is superior for some (Grignard) reactions.

oxirane / 1,2-epoxyethane / ethylene oxide

oxetane

furan

oxolane / tetrahydrofuran / THF

pyran

oxane / tetrahydropyran / THP

1,4-dioxane

a. Heteroatom is number 1 in these rings.

(2S,3R)-2-methoxy-3-methyloxirane
trans-2-methoxy-3-methyloxirane

2-ethyl-3,3-dimethyloxetane

3-methoxyfuran

4-methylpyran

b. TCDD (incorrectly known as dioxin) is an extremely toxic, carcinogenic substance made in trace amounts during agent orange (herbicide) synthesis, burning chlorinated plastics, or from forest fires.

dibenzodioxin (dioxin)

2,4,5-trichlorophenoxyacetic acid (2,4,5-T), or agent orange

2,3,7,8-tetrachorodibenzodioxin TCDD (incorrectly called dioxin)

13.4.1. Infrared spectra – C-O stretch is in the ambiguous fingerprint region. O-H and C=O stretch absences suggest ethers (if a formula has O).

Vibration	Functional Group	Wavenumber Range	Intensity / Shape
C-O stretch	ethers	1050	strong, 1° C-O bond
C-O stretch	ethers	1100	strong, 2° C-O bond
C-O stretch	ethers	1150	strong, 3° C-O bond
C-O stretch	aryl ethers	1200–1275 1020–1075	strong, broad weaker

13.4.2. Mass spectrum. Ethers often undergo α-cleavage to yield stable oxonium ion or carbocation fragments. Example: diethyl ether.

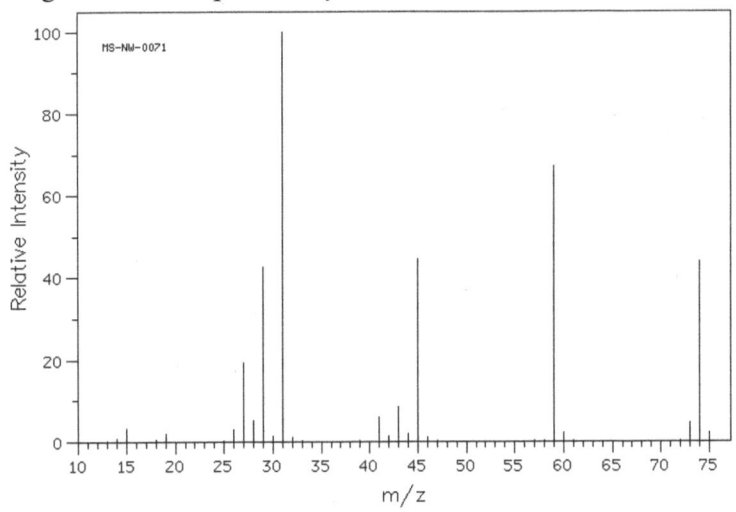

m/z	rel. I	m/z	rel. I	m/z	rel. I
15.0	3.2	31.0	100.0	46.0	1.1
19.0	1.8	32.0	1.1	59.0	67.1
26.0	3.1	41.0	6.1	60.0	2.2
27.0	19.4	42.0	1.2	73.0	4.7
28.0	5.1	43.0	8.3	74.0	44.0
29.0	42.7	44.0	2.0	75.0	2.1
30.0	1.3	45.0	44.6		

a. α cleavages account for peaks at 31.0, 45.0, and 59.0:

$H_3CCH_2-O-CH-CH_3 \rightarrow [\overset{\oplus}{O}=CH-CH_3 \leftrightarrow :\overset{\oplus}{O}-\overset{\oplus}{C}H-CH_3]$
m/z=45.0
+ ·CH$_2$CH$_3$

$H_3CCH_2-O-CH_2-CH_3 \rightarrow [H_2C-CH_2-\overset{\oplus}{O}=CH_2 \leftrightarrow H_3CCH_2-\overset{\oplus}{O}-CH_2]$
+ ·CH$_3$ m/z=59.0

$\downarrow$

$H_2C=CH_2 + [\overset{H}{\underset{\oplus}{O}}=CH_2 \leftrightarrow :\overset{H}{\underset{\oplus}{O}}-CH_2]$
m/z=31.0

13.4.3. ^{1}H NMR spectrum.

Functional Group	Proton Type	Chemical Shift (δ, ppm)
ether	H-C-O-R	3.5 – 4.0

13.4.4. ^{13}C NMR spectrum

Functional Group	^{13}C Nucleus Type	Chemical Shift (δ, ppm)
ether	H-^{13}C-OR	65 – 90

406 SEQUENTIAL ORGANIC CHEMISTRY I

a. Example:

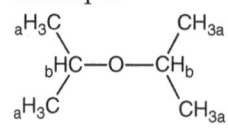

$_aH_3C$ \ CH_{3a}
   $_bHC-O-CH_b$
$_aH_3C$ / CH_{3a}

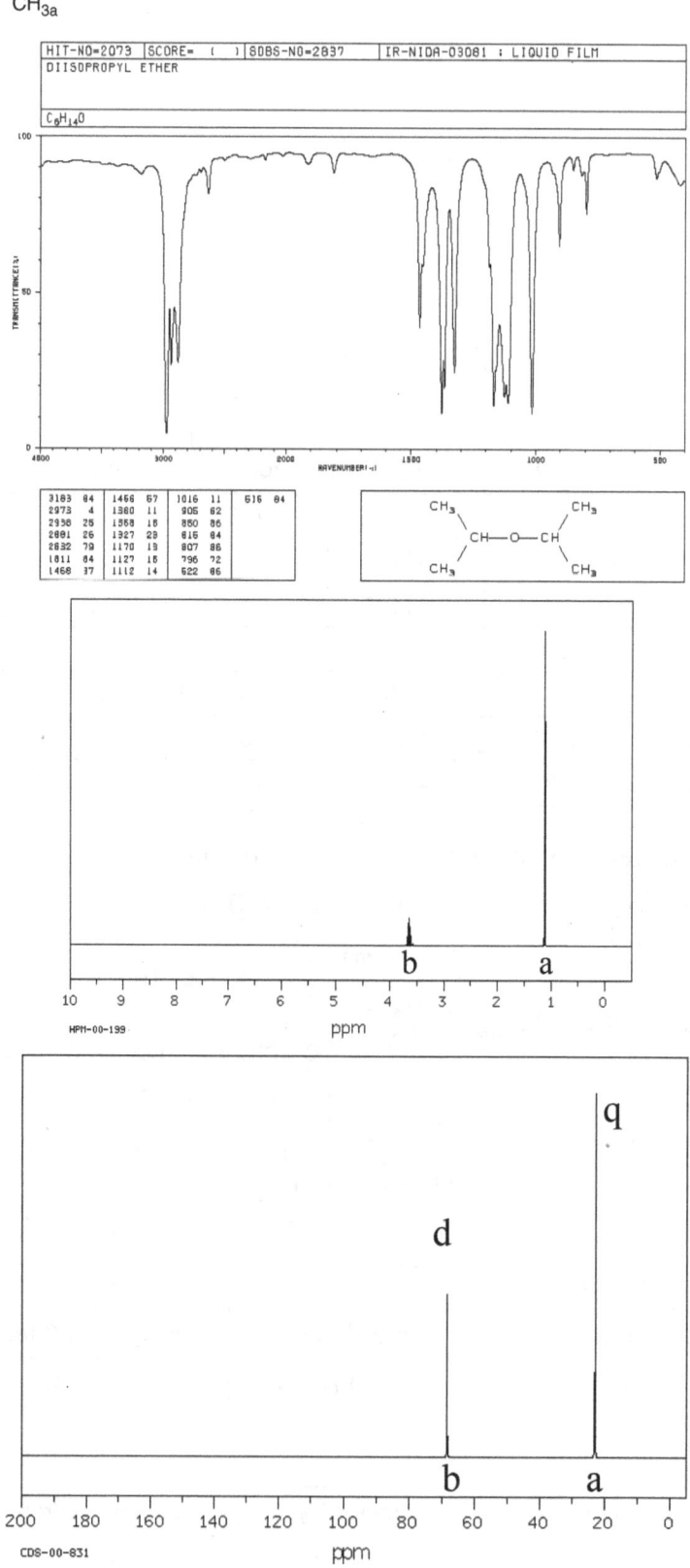

HIT-NO=2073 SCORE=( ) SDBS-NO=2837 IR-NIDA-03081 : LIQUID FILM
DIISOPROPYL ETHER
$C_6H_{14}O$

3183	84	1456	57	1016	11	616	84
2973	4	1380	11	905	82		
2958	25	1368	15	850	86		
2881	26	1327	29	615	84		
2832	79	1170	19	807	86		
1811	84	1127	15	796	72		
1468	37	1112	14	522	86		

13.5.1. Williamson ether synthesis – R-O⁻Na⁺ undergoes an S_N2 reaction with R'-X. ✽

    a. Example: choice (1) works, but (2) does not.

    b. Williamson ether synthesis yields are usually better than for bimolecular condensation (section 13.7), but R'-X cannot be 3°. S_N2 is slow for 2° and yields are often poor. E2 elimination competes and yields alkene byproducts.

    c. Problem 3. Use electron pushing arrows to write a mechanism for the reaction that occurs when 2-bromopropane is warmed with methanol. Include all elementary steps (collisions), intermediates, formal electrical charges, and involved lone electrons pairs.

    d. Problem 4. Optically pure R-(-)-octan-2-ol has a specific rotation [α] = −10.3° · dm⁻¹ · g⁻¹ · mL. Optically impure octan-2-ol whose [α] = −8.24° · dm⁻¹ · g⁻¹ · mL was reacted with Na metal to convert it into the alkoxide, which was subsequently warmed with bromoethane. The 2-ethoxyoctane product was optically active, [α] = −15.6° · dm⁻¹ · g⁻¹ · mL. What should be the reaction stereochemistry? What would be the specific rotation for optically pure 2-ethoxyoctane?

    e. Problem 5. Optically pure (R)-(-)-2-bromooctane has [α] = −39.6° · dm⁻¹ · g⁻¹ · mL. Optically impure 2-bromooctane whose [α] = −30.3° · dm⁻¹ · g⁻¹ · mL was warmed with sodium ethoxide in ethanol (solvent). The optically active 2-ethoxyoctane obtained had $[\alpha] = +15.3° \cdot dm^{-1} \cdot g^{-1} \cdot mL$. What mechanism is expected? Is the product configuration retained, inverted, or racemized? Why is $[\alpha]$ opposite that obtained in the previous problem?

f. Williamson aryl ether synthesis (section 17.15.14). The phenol must be the nucleophile. Aryl halides do not undergo S_N2.

13.6.1. Alkoxymercuration-demercuration (section 8.6.1):

a. Markovnikov orientation – nucleophile (ROH) attacks larger $\delta+$ (on 2° C atom).

13.7.1. Methanol or 1° alcohol bimolecular dehydration (section 11.10.3) is a cheap industrial method. Heating ethanol to 180°C with H_2SO_4(conc) yields ethene ($H_2C=CH_2$) as main product.

    a. Diethyl ether is the major product if ethanol is continuously added (in excess *vs.* H_2SO_4) at 140°C. H_2O byproduct removal drives reaction to completion.

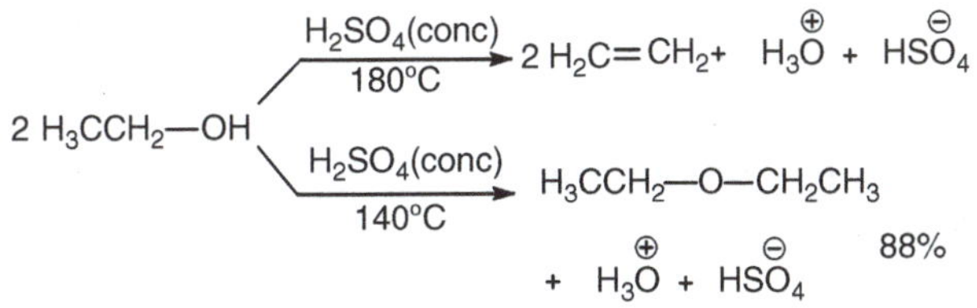

13.7.2. The mechanism is S_N2 for methanol and ethanol after fast reversible − OH protonation.

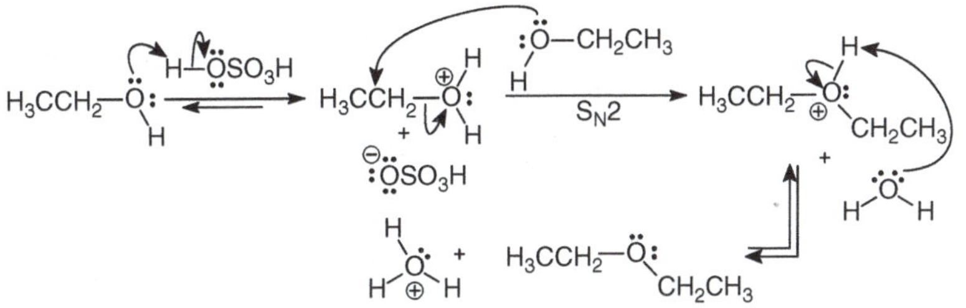

    a. Unimolecular (elimination) dehydration has a larger positive $\Delta S°_{rxn}$. One ethanol molecule ⇒ one ethene molecule, but two ethanol molecules ⇒ one ether molecule.) At 180°C, the $-T\Delta S°_{rxn}$ term for ethene formation dominates.

13.7.3. 1° and methyl alcohols react by acid catalyzed S_N2. 2° and 3° alcohols can form symmetrical ethers by acid catalyzed S_N1, but the major products are alkenes.

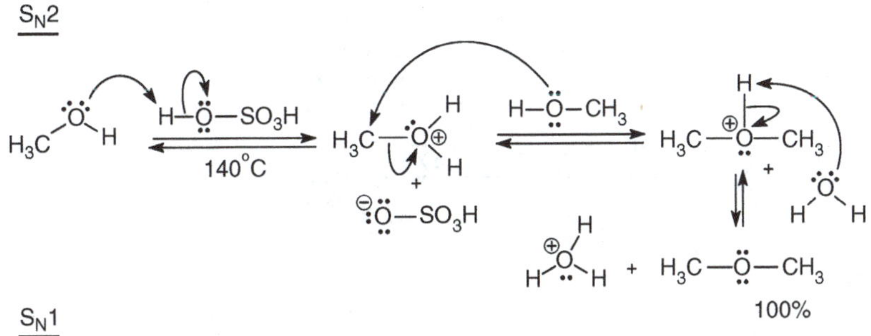

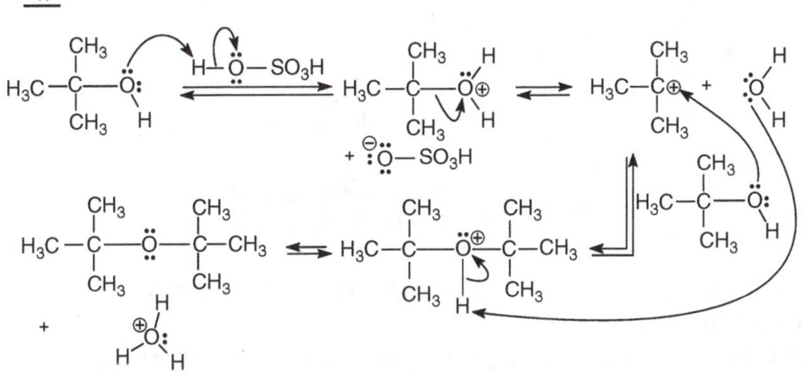

$\underline{S_N2}$

$\underline{S_N1}$

410 SEQUENTIAL ORGANIC CHEMISTRY I

13.7.4. 3° alkoxy groups (ethers) are used as alcohol protecting groups, since 3° C-O bonds cleave easily (after a synthesis step).

13.7.5. Problem 6.

  a. An equimolar ethanol / *n*-propanol mixture yields three different ethers when warmed with a trace of H_2SO_4. What are their structures?

  b. An equimolar 2-methylpropan-2-ol / ethanol mixture yields only one ether when warmed with a trace of H_2SO_4. What is this ether likely to be? Why is it formed in good yield?

13.8.1. Ether reactions – unreactive towards bases, but undergo acid catalyzed cleavage upon prolonged heating with concentrated acids (usually HI or HBr).

  a. First step – fast, reversible ether oxygen protonation (converts slow leaving group – OR into fast leaving group -$^+$OHR). Mechanism is then either S_N1 (if both R groups are 2° or 3°) or S_N2 (if one R group is methyl or 1°).

  b. HI reacts fastest since it is the strongest HX acid. I$^-$ is also the strongest X$^-$ nucleophile (aqueous solution). Reactivity:

  $$\text{faster } HI > HBr \gg HCl \gg HF \text{ slower}$$

  S_N1

  S_N2

  c. Problem 7. (*R*)-2-Methoxybutane was cleaved with H-Br(anhydrous) yielding chiefly bromomethane and (*R*)-butan-2-ol.

  1. Show chiral C atoms as three dimensional perspective drawings (wedges and dashes) and write the overall reaction.

2. Use electron pushing arrows to write a mechanism for this reaction that accounts for configuration retention. Show all elementary steps (collisions), intermediates, formal charges, and involved lone pair electrons.

13.9.1. Free radical peroxide formation with O_2 (air autoxidation):

a. A peroxide contaminant (explosion hazard) can be destroyed by Fe^{2+}(aq) reduction, or distillation from H_2SO_4(conc).
b. Absolute ether – diethyl ether that is free of peroxides, alcohols, and water (from H_2SO_4(conc) distillation).
c. Diethyl ether is extremely flammable – and peroxide explosions cause fires.

13.10.1. Thioethers (sulfides) and silyl ethers – sulfur ether analogues.

a. Thioethers are more easily oxidized and alkylated than ethers.

13.10.2. IUPAC naming: similar to ethers, but alkylthio replaces alkoxy.

methylthiomethane   methylthiobenzene   4-ethylthio-2-methylpent-2-ene

13.10.3. Thioether preparation – S_N2 reactions. Thiolate anions are strong nucleophiles (due to S's polarizability), but moderate bases.

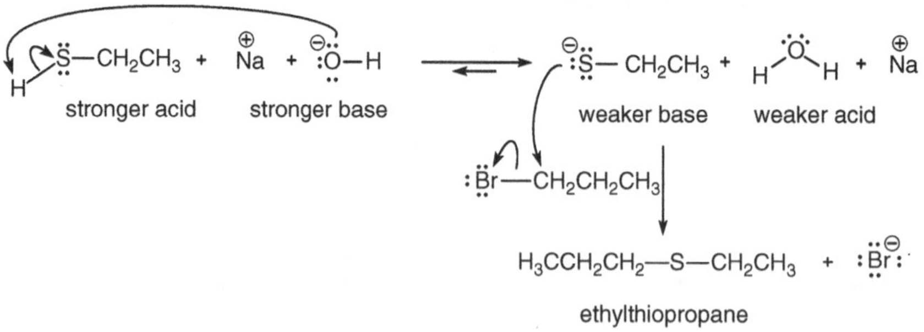

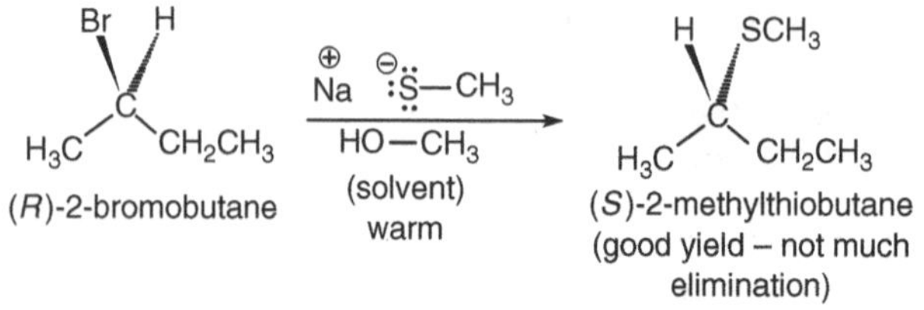

[Reaction scheme: H-S-CH₂CH₃ (stronger acid) + Na⁺ + ⁻O-H (stronger base) ⇌ ⁻S-CH₂CH₃ (weaker base) + H-O-H (weaker acid) + Na⁺; then thiolate attacks Br-CH₂CH₂CH₃ to give H₃CCH₂CH₂-S-CH₂CH₃ (ethylthiopropane) + Br⁻]

[Reaction: (R)-2-bromobutane + Na⁺ ⁻S-CH₃ in HO-CH₃ (solvent), warm → (S)-2-methylthiobutane (good yield – not much elimination)]

13.10.4. Reactions.

a. Thioethers are reducing agents (section 8.15.1a). Oxidation:

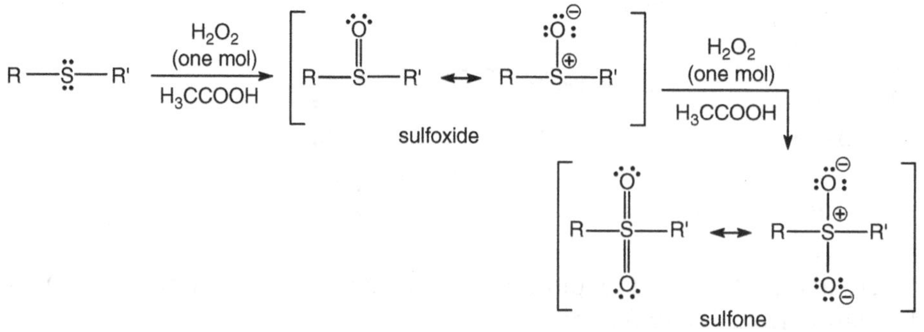

[Reaction: R-S-R' + H₂O₂ (one mol)/H₃CCOOH → sulfoxide (resonance structures shown) + H₂O₂ (one mol)/H₃CCOOH → sulfone (resonance structures shown)]

b. Sulfonium salt formation and alkylation (S_N2 reactions). Thioethers are much stronger nucleophiles than ethers.

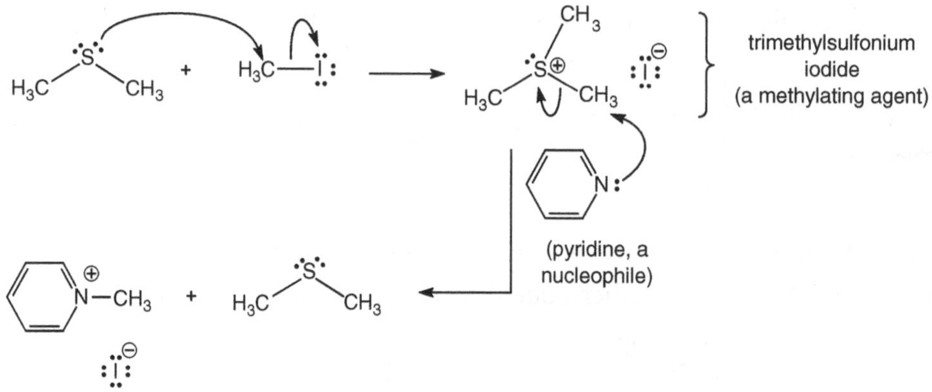

[Reaction: H₃C-S-CH₃ + H₃C-I → (CH₃)₃S⁺ I⁻ trimethylsulfonium iodide (a methylating agent); then pyridine (a nucleophile) attacks to give N-methylpyridinium + H₃C-S-CH₃ + I⁻]

1. Alkylations are important in biochemistry.

13.10.5. Protecting group use in Grignard synthesis was mentioned in section 10.10.4. Ethers can be alcohol protecting groups (see sections 13.7.3 and 18.12.20). Example:

a. Strong acid may cause side reactions during ether protecting group removal.

13.10.6. Triisopropylsilyl (TIPS) ether is a sterically hindered (toward nucleophiles) alcohol protecting group, unreactive towards acids, bases, oxidizing and reducing agents.

13.11.1. Oxiranes (epoxides) are three membered ring ethers (section 8.12). Angle strain (~105 kJ/mol) makes them unusually reactive, so they are treated separately from ethers. They are important in synthesis.

13.11.2. Oxiranes are made by alkene reaction with a peroxy acid (section 8.12).

H₂C=CH₂ + peroxybenzoic acid ⟶ H₂C—CH₂ (oxirane) + benzoic acid

cyclohexene + m-chloroperoxybenzoic acid →[H₂CCl₂ (solvent), dry] 1,2-epoxycyclohexane / cyclohexene oxide (100%) + m-chlorobenzoic acid

13.11.3. The mechanism is concerted (one step) and *syn* (stereospecific). Electron rich (substituted) alkenes react faster.

alkene + peroxyacid ⟶ oxirane + carboxylic acid

(E)-2-nitro-1-phenyl-propene + peroxybenzoic acid ⟶ (E)-2-methyl-2-nitro-3-phenyloxirane

a. Magnesium monoperoxyphthalate (MMPP/ethanol) is a relatively stable, neutral pH, "green" reagent, good for larger scale.

cyclohexene + [MMPP]₂ Mg²⁺ →[H₃CCH₂OH] cyclohexene oxide, 85%

b. From 1,2-chlorohydrins.
   1. Review: chlorohydrin formation.

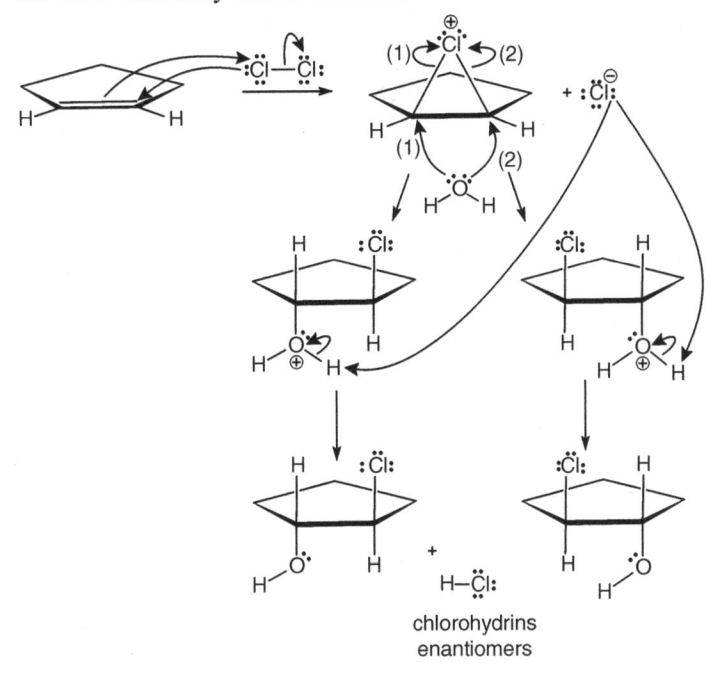

   2. Oxirane formation.

13.11.4. Four, five, six, and seven membered rings have been made similarly. A sterically hindered base (K^+ $^-O\text{-}C(CH_3)_3$ or 2,6-lutidine (2,6-dimethylpyridine)) helps prevent base from attacking the halide via S_N2.

13.12.1. Acid catalyzed oxirane ring opening (section 8.13).

a. Similar reactions review – halohydrin formation (X⁻ attacks the more hindered cyclic halonium ion C atom), and solvomercuration/demercuration (H₂O attacks the more hindered cyclic mercurinium ion C atom).

b. These reactions have some S_N1 character (δ+ charge involved). Transition states are similar to those in acid catalyzed oxirane ring opening.

c. Permanganate (or osmium tetroxide) alkene oxidation is a stereospecific *syn*-addition (section 8.14.5).

only product (no (R,R) or (S,S))

13.12.2. Acid promoted ring opening in alcohol solvent.

a. Example.

1. A 3° C is more sterically hindered, but it has a larger δ+ that is more attractive to a nucleophile (ethanol). A 3° C-O bond is also weaker than a 1° C-O bond.

[Diagram: σ-bond resonance (hyperconjugation) contributors to the oxonium ion intermediate — most important, less important (3°), least important (1°); hybrid with weaker bond shown with δ⊕ charges]

b. Another example.

[Mechanism diagram showing protonation of cyclic ether by HCl, nucleophilic attack by methanol, giving products 1 and 2]

13.12.3. Halohydrin formation.

[Mechanism: cyclic ether + HCl (excess) (dry gas?) → protonated oxonium + Cl⁻ → trans chlorohydrin + enantiomer]

13.12.4. A key step in cholesterol synthesis from squalene is an epoxidation. Ring opening promotes stereospecific cyclization under enzyme control.

13.13.1. Unlike ordinary ethers, strong bases cleave oxiranes via an S_N2 mechanism (section 8.13).

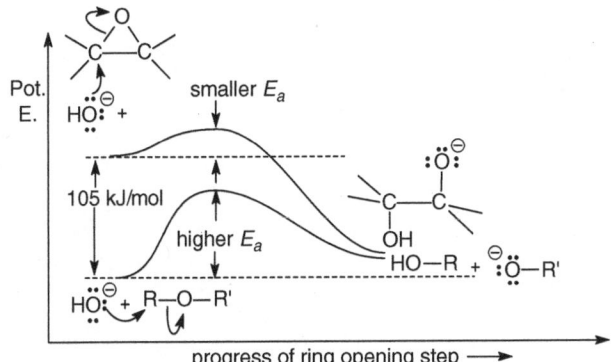

[Energy diagram: Pot. E. vs progress of ring opening step; HO⁻ + epoxide has smaller E_a pathway leading to HO–C–C–O⁻ (with OH); HO⁻ + R–O–R' has higher E_a leading to HO–R + ⁻O–R'; 105 kJ/mol difference indicated]

13.13.2. Steric factors control orientation (normal for S_N2). Nucleophile attacks from the *anti* face.

attack at 1° C preferred
(less steric hindrance)

a. Acid catalyzed ring opening occurs under milder conditions, and is preferred for synthesis unless an (another) acid sensitive group is present.

13.13.3. Oxirane can be anionically polymerized to polyethylene glycols and carbowaxes.

13.13.4. Ammonia or amines can open oxirane rings yielding aminoethanols.

a. Problem 8. What trick would you use to make 2-aminoethanol in good yield?

13.14.1. Oxirane ring opening in alcohol solvent.

    a. Base promoted.

13.15.1. Strongly basic and nucleophilic Grignard reagents react with oxiranes to yield 1° alcohols whose carbon chain is two C atoms longer.

    a. A Grignard reagent preferentially attacks the less substituted C atom, but mixtures form unless a C atom is very hindered. Organolithiums are more selective.

13.15.2. Ether chemical tests:

    a. dissolve in cold H_2SO_4(conc).

    b. cleave with HI, then identify alkyl iodides (ether derivatives).

13.16.1. Epoxy resins and glues involve oxirane ring chemistry.

13.16.2. Problem 9. Describe a simple laboratory (test tube) test that would distinguish an ether from the families below. Write the chemicals that you would use.
  a. an alkane

  b. an alkyl halide

  c. a 1° or 2° alcohol.

  d. a 3° alcohol

13.16.3. Problem 10. What simple chemical test could you use to distinguish:
  a. *n*-butoxybutane from *n*-pentanol

  b. ethoxyethane from *n*-octane

  c. 2-methylpropan-2-ol and 2,2-dimethylpropan-1-ol

13.16.4. Problem 11. Fill in the products for each reaction:

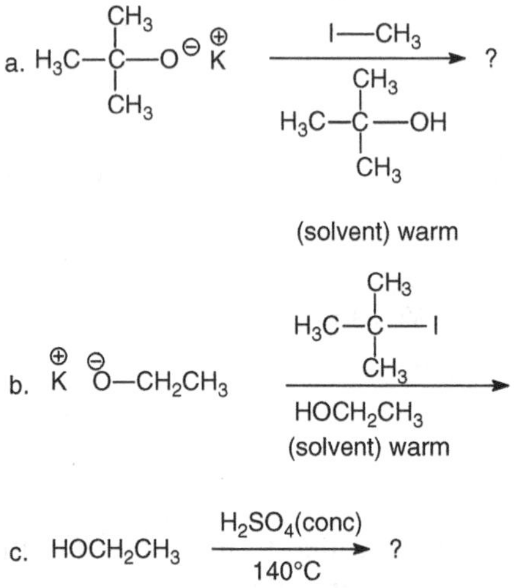

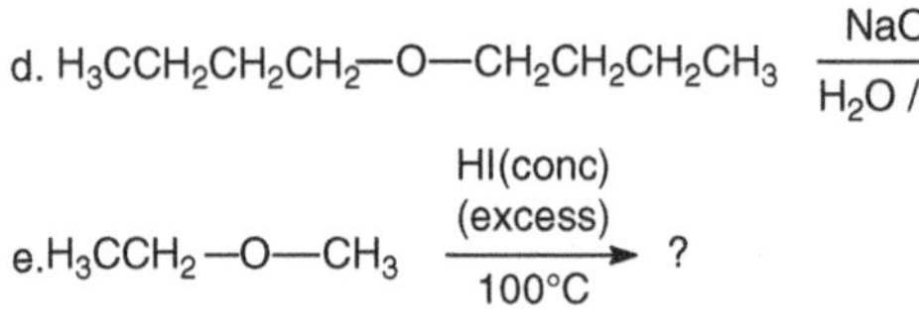

a. $H_3C-\underset{CH_3}{\underset{|}{\overset{CH_3}{\overset{|}{C}}}}-O^{\ominus} K^{\oplus}$ $\xrightarrow[\substack{H_3C-\underset{CH_3}{\underset{|}{\overset{CH_3}{\overset{|}{C}}}}-OH \\ \text{(solvent) warm}}]{I-CH_3}$?

b. $K^{\oplus} \; ^{\ominus}O-CH_2CH_3$ $\xrightarrow[\substack{H_3C-\underset{CH_3}{\underset{|}{\overset{CH_3}{\overset{|}{C}}}}-I \\ HOCH_2CH_3 \\ \text{(solvent) warm}}]{}$?

c. $HOCH_2CH_3$ $\xrightarrow[140°C]{H_2SO_4(conc)}$?

d. $H_3CCH_2CH_2CH_2-O-CH_2CH_2CH_2CH_3$ $\xrightarrow[H_2O / 100°C]{NaOH}$?

e. $H_3CCH_2-O-CH_3$ $\xrightarrow[100°C]{\substack{HI(conc) \\ (excess)}}$?

f. $H_3C-O-CH_3$ $\xrightarrow{Na}$?

g. $H_3CCH_2-O-CH_2CH_3$ $\xrightarrow[0°C]{H_2SO_4 (conc)}$?

13.16.5. Problem12. Show how you could synthesize a and b products from (*R*)-butan-2-ol and any other optically inactive reagents. Show reactants, reagents, and products for each synthetic step, but you do not need to show mechanisms.

  a. (*R*)-2-ethoxybutane

  b. (*S*)-2-ethoxybutane